9-9

ENVIRONMENTAL GEOLOGY

"There is one ocean, with coves having many names;
a single sea of atmosphere with no coves at all;
a thin miracle of soil, alive and giving life;
a last planet; and there is no spare."

David R. Brower

Rainbow Falls, Hawaii (Hawaii Visitors Bureau)

ENVIRONMENTAL GEOLOGY

Ronald W. Tank

New York · Oxford
OXFORD UNIVERSITY PRESS
1983

Copyright © 1983 by Oxford University Press, Inc.

Library of Congress Cataloging in Publication Data
Main entry under title:

Environmental geology.

  Includes index.
  1. Natural disasters—Addresses, essays,
lectures. 2. Geology—Addresses, essays,
lectures. 3. Mines and mineral resources—Addresses,
essays, lectures. 4. Conservation of natural
resources—Addresses, essays, lectures.
I. Tank, Ronald Warren.
GB5020.F63 1982   550   82-3448
ISBN 0-19-503288-8 (pbk.)   AACR2

This book is a revision of *Focus on Environmental Geology:
A Collection of Case Histories and Readings from Original Sources*,
edited by Ronald W. Tank.
Copyright © 1973, 1976 by Oxford University Press, Inc.

Printing (last digit): 9 8 7 6 5 4 3 2 1

Printed in the United States of America

To
Kris and Allequash
with love and thanks

Foreword

The term *environmental geology* is a relatively recent addition to our vocabulary. It is not a single discipline. Indeed, it is the application of all of the geological sciences to the study of the earth as it affects and is affected by human activities, aimed at providing knowledge and understanding that will help protect human health and safety, preserve the quality of the human environment, and facilitate wise use of the land and its resources. Use of geologic knowledge for such purposes is not new. Beginning near the end of the eighteenth century, William Smith, often called the Father of English Geology and of Stratigraphy, used his knowledge of areal geology to plan the routing of canals, stabilize landslides, and locate and evaluate coal deposits and construction materials. For many decades, geology has played an important part in the planning of engineering works, such as dams and tunnels, and in the evaluation of construction sites. All of the geological sciences, in fact, have a long history of application to human problems and endeavors.

What is new in the last twenty years or so is brought out in an increased focus of the geological sciences on problems related to the preservation of environmental quality and to the protection of human health and safety. Geologists have responded to the rapidly growing concern over the deteriorating quality of the environment in many areas, to the realization that many human activities are impacting the environment in ways that threaten human health, and to the recognition that natural hazards are becoming a threat to increasing numbers of people.

Fortunately, the power of the geological sciences to provide the knowledge and understanding necessary to cope with these problems is also growing rapidly. Within the last several years, for example, it has become possible to predict some earthquakes and some volcanic eruptions. Existing geologic knowledge is wholly adequate to guide most uses of the physical environment in a safe manner, provided it is gotten into the proper hands. Fortunately also, geologists are coming to recognize the need to translate their findings into terms that others—engineers, planners, policy makers, and the public at large—can understand. Geologists still have much to learn about how to communicate with the public, but it is fair to say also that the communications process is not a one way responsibility—the public must improve its understanding of earth science basics if it is to be able to participate in policy-making on issues related to the earth and its resources. This is a large order to be sure, and it will take years to achieve even if we work hard to bring it about. A good beginning would be to put earth science on the required list of courses for all college students.

Professor Tank's book is a significant contribution to public education in the geological sciences and their application to resource and environmental problems. Not only has he assembled easily readable writings by a wide range of specialists in each of the areas in which human activities affect or are affected by geologic processes, but in this edition he has added brief but illuminating descriptions of the basic geologic processes involved. The result is a text readily understandable by non-geologists. It could serve as an introductory text to earth science, and, with its focus on environmental geology and earth resources, it could help to enlarge public understanding of problems that are of growing importance.

V. E. McKelvey

Preface

The first edition of this book appeared in 1973, and the second edition in 1976, under the title *Focus on Environmental Geology*. The pace of research and publication in environmental geology continues to accelerate rapidly and, therefore, an updating and revision is in order.

But the present book goes beyond the customary practice of introducing new concepts, current case histories, contemporary ideas, and up-to-date information. It brings together, under one cover, the traditional textbook approach and an anthology of readings. I know of no other example of this type of approach.

Ordinarily, the task of the editor of an anthology is to blend into the landscape of the work he or she edits, to disappear among the co-accomplishments of the authors of selected articles. The decision to become a more visible part of the landscape by assuming the dual roles of author/editor is founded on the belief that at this time there is a need for a more versatile book than is generally available.

An anthology of readings has proved to be an asset in a variety of teaching situations. "Outside reading" has traditionally been a supplement to the textbook and has occasionally replaced it. At the introductory level, however, there is often a need for background information and for an explanation of basic principles and concepts. This type of textual material is now included in this revision. It is directed at the college undergraduate and the informed lay reader with little or no background in geology. It can be used as basic text material in many introductory courses in geology, physical geography, environmental studies, regional and urban planning, and engineering geology.

Many of the readings from the second edition have been retained. This introduces 15 new articles. The readings offer a stimulating and challenging point of departure often lacking in an introductory book.

They introduce the student to distinguished scientists through their ideas and accomplishments, present new perspectives, and analyze case histories. They can be used to illustrate basic principles and concepts and may be the basis for informed discussion or independent study.

The topical coverage has been expanded to include ore genesis and classification, classification of rocks, soils and surficial material, and engineering properties. The concepts of sediment yield and the Universal Soil Loss Equation are introduced, and there are new readings on expansive soils and the impact of Federal legislation. But the emphasis of the previous editions has been retained.

The prologue is an introduction to the subject of environmental geology; Part I illustrates the wide range of geologic hazards; Parts II and III deal with mineral and water resources; and Part IV considers waste disposal. Part V covers the relationship of geology to the urban environment and illustrates the overlapping nature of the problems presented in Parts I–IV. The lists of supplementary readings and films have been updated, and the scope of the review questions has been expanded.

I am indebted to those who offered comments on the articles included in the second edition and to those who offered suggestions for this revision. I also wish to thank the authors and publishers who have permitted the inclusion of their materials.

Appleton, Wisconsin   RWT

Contents

ENVIRONMENTAL GEOLOGY

PROLOGUE

"... and we can save 700 lira by not taking soil tests."

Courtesy of Western Technologies, Inc.,
formerly Engineers Testing Laboratories, Inc.

Ecology is the study of the intimate relationships between organisms and their environments. It involves the lithosphere, the hydrosphere, the atmosphere, and the biosphere. Environmental geology emphasizes one aspect of this complex subject—the interrelationship of humans and their geologic environment. It involves studies of hydrogeology, topography, engineering geology, and economic geology. It is therefore a multidisciplinary field which relies heavily upon traditional branches of geology. In the first article of this book, John Frye presents an overview of environmental geology. The role of geologic information in solving contemporary environmental problems is made clear, but, as Frye indicates, the collection of this information must be oriented toward environmental applications. Furthermore, "the results must be presented in a form that is readily understandable to, and usable by, planners and administrators who often are unfamiliar with science and, particularly, do not have a working knowledge of the geological sciences." This is a relatively new role for the geologist and one which serves to distinguish environmental geology from some of the more traditional branches of geology.

One of the more significant applications of environmental geology is in the area of urban planning. The importance of this application has not always been realized by planning agencies, and geologists have not always been effective in demonstrating the relevance of geologic data to the planning process. The unfortunate result of the lack of communication between planners and geologists is illustrated by Ernest Dobrovolny and Henry Schmoll in Reading 2. Evidence of a lack of communication can be found in many other situations. It is most obvious when people, either knowingly or unknowingly, become involved with geologic hazards. The loss of lives and property is staggering. It is less obvious, but nonetheless very costly, when valuable mineral, water, and agricultural resources are lost due to inadequately controlled urbanization or other land uses which have frequently taken place without considering the cost of foregone resources.

Dobrovolny and Schmoll also describe the development of special purpose derivative or interpretive maps which can aid urban planners in anticipating and providing predevelopment solutions to land-use problems. Experience has demonstrated that most geologic and hydrologic maps must be translated from technical to operational or practical terms and interpreted for the lay clientele. For some uses, information must be synthesized as well as translated and interpreted. Another type of special purpose map, the flood hazard map, is described in Reading 21, and references to the application of interpretive maps in other areas are included in the list of supplementary readings. These maps are products of some of the new developments in environmental geology and will be vital in the planning process. Also vital to the planning process is the support of government agencies in acquiring the needed data and incorporating this data into policy decisions.

1 A Geologist Views the Environment

John C. Frye

When the geologist considers the environment, and particularly when he is concerned with the diversified relations of man to his total physical environment, he takes an exceptionally broad and long-term view. It is broad because all of the physical features of the earth are the subject matter of the geologist. It is long term because the geologist views the environment of the moment as a mere point on a very long time-continuum that has witnessed a succession of physical and biological changes—and that at present is dynamically undergoing natural change.

Let us first consider the time perspective of the geologist, then consider the many physical factors that are important to man's activity on the face of the earth, and, third, turn our attention to specific uses of geologic data for the maintenance or development of an environment that is compatible with human needs.

The Long View

The earth is known to be several billion years old, and the geologic record of physical events and life-

Frye, J. C., 1971, "A Geologist Views the Environment," *Environmental Geology Notes Number 42*. (Illinois State Geological Survey). Reprinted by permission of the author and the Illinois State Geological Survey, Urbana, Illinois. It was prepared for "Voice of America," Earth Science Series, Dr. C. F. Park, Jr., Coordinator. Mr. Frye is Executive Secretary of the Geological Society of America and Former Chief of the Illinois State Geological Survey.

forms on the earth is reasonably good for more than the most recent 500 million years. Throughout this span of known time the environment has been constantly changing—sometimes very slowly, but at other times quite rapidly. Perhaps a few dramatic examples will serve as illustrations. Less than 20,000 years ago the area occupied by such North American cities as Chicago, Cleveland, Detroit, and Toronto were deeply buried under glacial ice. The land on which New York City is now built was many miles inland from the seashore. And part of the area now occupied by Salt Lake City lay beneath a fresh-water lake. Twelve thousand years ago glacial ice covered the northern shores and formed the northern wall of what was then the Great Lakes, and much of the outflow from those lakes was to the Gulf of Mexico rather than to the Atlantic Ocean through the St. Lawrence River, as it is now. Although firm scientific information is not available to permit equally positive statements about atmospheric changes during the past few tens of thousands of years, deductions from the known positions of glaciers and from the fossil record make it clear that the atmospheric circulation patterns were quite different from those of the present, and studies of radiocarbon show that the isotopic content of the atmosphere changed measurably through time.

The purpose of listing these examples is to emphasize that the environment is a dynamic system that must be understood and accommodated by man's activities, rather than a static, unchanging system that can be "preserved." The living, or biological systems of the earth are generally understood to be progressively changing, but much less well understood by the public is that the nonliving, physical aspects of the earth also undergo change at an equal or greater rate.

Clearly, a dynamic system is more difficult to understand fully, and it is more difficult to adapt man's activities to a constantly changing situation than to an unchanging or static system. On the other hand, the very fact of constant change opens many avenues of modification and accommodation that would not be available in a forever constant and unchanging system.

The Problem

Although it is important that we have in mind the long-term facts concerning earth history, modern man has become such an effective agent of physical and chemical change that he has been able to produce major modifications, some of which run counter to the normal evolution of our earthly environment, and to compress millennia of normal evolutionary changes into days. These rapid modifications are, almost without exception, made by man with the intention of producing improvements and advantages for people. Problems result from the fact that by-products and side effects do occur that are neither desirable nor pleasing, and at some times and places may be hazardous or even calamitous. In some cases the undesirable side effects are unknown or are unpredictable; in other cases they are tolerated as a supposed "necessary price" to

pay for the desirable end result. It is our intent to examine the role of the earth scientist in defining some of these problems and in devising ways of minimizing or eliminating them.

The ways in which man treats his physical surroundings, produces and uses the available nonliving resources, and plans for his future needs are, of course, social determinations. However, in order that social decisions can be made in such a way that we, and our children, will not find reason to regret them, they should be made in the light of all the factual information that it is possible to obtain. If we, collectively, decide to use the available supply of a nonrenewable resource— for example, petroleum—at a particular rate, we should know how long it will last and what substitute materials are available to replace it when the supply is exhausted; if we decide to dam a river, we should know what the side effects will be in all directions, how long the facility will last, and what the replacement facility might be; if we develop huge piles of discarded trash, we should know whether or not they will cause pollution of water supplies or the atmosphere, and whether or not the terrane is sufficiently stable to retain them; if we substitute one fuel for another with the desire to abate air pollution, we should know if it will make a net over-all improvement in the pollution problem, and if it will be available in the quantity required so that man's needs can be met; and if we plan expanding metropolitan areas we should have full information on the terrane conditions at depth and on the raw material resources that will be rendered unusable by urbanization.

Role of Earth Science in Solving Problems

When we consider the role of earth science in solving problems, we see that the earth sciences can and should develop answers to all of the questions we have asked. Many of the problem areas overlap one another, but it will be easier to discuss them if we class the contributions of the earth scientist to environmental problem solving in five general categories. The first of these is collecting data for planning the proper use of the terrane, or perhaps we should say the most efficient adjustment of man's use of the earth's surface to all of the physical features and characteristics at and below the surface—particularly in expanding urban areas. Second is determination of the factors that influence the safety and permanence of disposal of waste materials and trash of all kinds—both in the rocks near the surface and at great depth in mines and wells. Third is providing information for the planning and development of safe, adequate, and continuing water supplies in locations that will serve populated areas. Fourth is the identification of rock and mineral resources to provide for future availability of needed raw materials, or of appropriate substitute materials. And, fifth is the recognition of man as a major geologic agent by monitoring the changes he has caused in his environment and by providing remedies where these changes are, or may become, harmful.

Proper Use of Land

The first of these general categories—procuring data for physical planning of the proper use of the land surface and of the rocks below the surface—covers data provided by topographic and geologic maps, by engineering geology and soil mechanics investigations, by predictions of potential landslides and other geologic hazards, and by a complete inventory of available mineral resources and future potential water supplies. Much of the geologic data needed in this category can be produced by conventional methods of research, but, to be effective, the research program must be oriented toward environmental applications. Furthermore, the results must be presented in a form that is readily understandable to, and usable by, planners and administrators who often are unfamiliar with science and, particularly, do not have a working knowledge of the geological sciences. Perhaps the best way to explain what I mean by environmental orientation is to cite a few examples.

The first example of the use of geologic research oriented for planning is a laboratory study involving clay mineralogy, petrography, and chemistry. It was prompted by numerous incidences of structural failure of earth materials. In rapidly expanding suburban residential areas there has been a great increase in the use of septic tanks and, simultaneously, a rapid increase in the household use of detergents and water softeners. A laboratory research program was initiated to study the changes induced in the clay minerals by these chemical substances when they were introduced into the near-surface deposits by discharge from septic tanks. Preliminary results showed that the materials in septic tank effluent did, indeed, produce significant and undesirable changes in the properties of some earth materials. The data made it possible to predict changes that could occur in the structural characteristics of common surficial deposits and, thus, to prevent structural failure. Therefore, the conclusions were presented to planners, health officials, architects, and other groups that might have need of the information.

In strong contrast to such a sharply focused and specific research project is a second example provided by a study of a county at the northwest fringe of the Chicago metropolitan area, into which urbanization is spreading from that metropolis. Because of impending problems, the county government organized a regional planning commission, which called upon the State Geological Survey and other agencies to collect data on the physical environment that were essential to wise, long-range planning. Where some of the fields of activity of the agencies overlapped, they cooperated informally so that they could most effectively work with the planning commission. Essential to the project were modern topographic maps of the county, and, even though much of the county had been geologically mapped previously, several man-years of geologist time were devoted to the project.

This project to characterize the physical environment involved many types of geological study. These included (1) analysis of the physical character of the major land forms within the county; (2) interpretation of the relation between geologic units near the surface and the agricultural soil units; (3) establishment of the character of the many layers of rocks and glacial deposits penetrated at depth by drilling below the surface; (4) definition of the occurrence and character of water-bearing strata in the near-surface glacial deposits and the deeper bedrock layers; (5) determination of the geologic feasibility of water-resource management programs; (6) determination of the geologic feasibility of waste management programs; (7) delineation of the geographic occurrence and description of the characteristics of construction material resources; (8) location of commercial mineral resources and assessment of their economic value; (9) determination of the engineering characteristics of the geologic units near the surface; and (10) geologic evaluation of surface reservoir conditions and proposed reservoir sites.

The approach to such development of data includes field work by surficial geologists, engineering geologists, ground-water geologists, stratigraphers, and economic geologists. In the laboratory, chemists, mineralogists, and stratigraphers conduct studies of the subsurface by use of cores and cuttings of the deposits at all depths; make chemical, mineralogical, and textural analyses of all deposits and rocks; determine physical properties; compile statistics; and make economic analyses of the many mineral resource situations. The results of these studies are compiled on interpretative maps, which the planning commission combines with the results of studies by specialists of other agencies and by the commission itself for preparation of maps that show the recommendations for land use. These maps, together with explanatory, nontechnical text, serve as a basis for county zoning and long-range development planning.

Development of Waste Disposal Facilities
Our second category of environmental geologic information includes those geologic data needed for proper and safe development of waste disposal facilities. Man has the propensity to produce toxic and noxious waste materials in progressively increasing quantity and in an ever-increasing variety and degree. Waste products result from manufacturing, processing, and mining—but of even greater concern is the concentrated production of waste by the inhabitants of our large cities. Traditionally, man has used fresh water to dilute liquid waste and the atmosphere to dilute the gaseous waste products of combustion. He has often indiscriminately used the land or large bodies of water for disposal of solid waste. However, even the general public is now aware of the fact that we are exhausting the capacity of fresh waters and the atmosphere to absorb our waste products. Along the sea coasts there is still the ocean—although even the ocean is being restricted as a waste disposal medium—but in the vast region of the continental interior we have no ocean in which to dump our wastes. Instead, our choices are limited to (1) selective recycling, accompanied by essentially complete purification of the residue of waste materials; (2) selective recycling accompanied by land disposal of non-recyclable residues; or (3) the use of the rocks of the earth's crust for the total future expansion of waste disposal capacity. Geologists, who traditionally have been concerned with the discovery of valuable deposits of min-erals and their extraction from the crust, now also must concern themselves with the study of the rocks of the earth's crust as a possibly safe place for the disposal and containment of potentially harmful waste products.

Petroleum geologists were introduced to one aspect of the problem of large quantity underground disposal of waste material more than a quarter century ago when it became necessary to find methods for injecting into deep wells the increasing quantities of brines produced with petroleum in oil fields. However, it was not until population densities approached their present levels that we became genuinely concerned with the most critical problems of the future—that is, the safe disposal of industrial and human waste materials in large quantity, other than by dilution. As some of these undesirable materials are destined to increase at an exponential rate in the future, it is obvious that we must devote our best geologic effort to solution of the problems of disposal. The problem of disposal of high-level and intermediate-level radioactive wastes has already attracted a major research effort, probably because these radioactive materials are obviously so highly dangerous for such a long period of time, and a body of scientific data now exists concerning their safe management.

In my own state of Illinois, solid waste disposal is generally accomplished by sanitary landfill, and frequently the State Geological Survey is asked by state and local departments of health to make geologic evaluations of proposed sites. However, precise and universally accepted criteria for this type of evaluation are only now being developed. Several disposal sites of differing geologic character are being intensively studied by coring and instrumentation of test holes, and analyses are being made of the liquids leached from the wastes and the containing deposits. In addition to laboratory study of the obvious characteristics of the containing deposits—such as texture, permeability, strength, clay mineralogy, and thickness of the units that do not transmit water (and thus protect the aquifers)—studies must be made of the less obvious effects of the seepage of liquids on the structural character of the deposit, the removal of objectionable chemicals from water solutions by the clay minerals of the containing deposits, and the microflow patterns of water in earth materials surrounding the wastes. For some of these determinations it is necessary to know the chemical composition of the liquids that pass through the wastes, as well as the chemistry and mineralogy of the deposits that contain them.

Disposal near the surface by landfill or lagooning methods requires detailed studies of the earth deposits at and immediately below the surface, with only minor data required on the deeper bedrock. On the other hand, disposal of industrial wastes in deep wells requires studies of the character and continuity of all rock layers down to the crystalline basement. Geologic data needed for deep disposal involve a combination of the types of information needed for the exploration for both oil and water, plus a knowledge of the confining beds of shale or clay. It should also enable us to predict possible changes that might be produced by the injected wastes.

A different type of pollution problem is represented by sulfur compounds and other undesirable materials released by the burning of coal, oil, and gas and discharged into the atmosphere. The earth scientist contributes to the solution of this problem by studies of the mineral matter in the coal, studies of methods of processing the coal before it is burned, studies of means of removing the harmful materials from the effluent gases produced by combustion, and research on the conversion of coal into gaseous or liquid fuels from which much of the objectionable material can be extracted.

Planning Water Supplies
The third category of concern for the earth scientist is water, its occurrence, quality, continuing availability, and pollution; its use as a resource, as a diluent for waste materials, as a facility for recreation, and as an aesthetic attribute. Many of the problems and areas of data collection for water resources fall mainly in the province of the engineer, the chemist, the biologist, or the geographer. But, it would be unrealistic to exclude water from the subjects requiring significant and major data input by the geologist, because to a greater or lesser degree geologic data is needed for the proper development and management of each of the above aspects. The occurrence, quality, availability, development, and replenishment of ground water require major attention by the geological scientist because all of these aspects are directly controlled by the character of the rocks at and to considerable distances below the surface.

Of the categories of environmental data we are discussing, water resources and water pollution are the most widely discussed in the news media and most generally recognized by the public and by municipal planners and administrators. Furthermore, an extensive cadre of earth scientists specializing in water-related problems has developed within governmental agencies and industry. There are many examples of the application of geologic data to management of water resources, ranging from dam site evaluations to the mapping of aquifers and determination of areas suitable for artificial recharge. Before we leave this category, however, I should point out that, in contrast to the rock and mineral resources of the earth's crust, water is a dynamic and renewable resource and, therefore, is subject to management. Even the long-range correction of the effects of unwise practices in the past may be possible in some cases.

Future Availability of Rock and Mineral Materials

Our fourth category is, perhaps, the most complicated aspect of environmental geology. It is the problem of assuring adequate supplies of mineral and rock raw materials for the future, and especially of assuring their availability near densely populated areas where they are most needed. We have an increasing shortage of raw material resources to meet the needs of the increasing world population, and also a mounting conflict of interest for land use in and adjacent to our urbanizing regions. Conflict exists in populated regions because buildings and pavements commonly remove the possibility of extracting the rock or mineral raw materials underneath them. While the producer of rock or mineral products is exploring for the best deposit available from both a physical and economic standpoint, the urban developer may be planning surface installations without regard for the presence or absence of rock and mineral deposits, which he may be rendering unavailable. These unavailable resources may be urgently needed for community developments in the future, and it is important that the attitudes of the planner and mineral producer be brought into harmony, and that compatible working relations be evolved so that mineral resource development can move forward as an integral part of the urban plan. The geological sciences can supply an accurate and detailed description and maps of all of the rock and mineral resources, not only in but also surrounding urbanizing areas for a distance reasonable for transporting bulk commodities to the metropolitan centers. Grades of deposits that might have utility in the coming 25 to 50 years as well as grades currently being developed should be included in the study.

An equally important role of the earth scientists now, and more particularly in the future, lies in regions remote from the cities where exploration is needed for the raw materials and fuels required by modern society. . . . Furthermore, research by the earth scientist, directed toward the identification of substitute materials and toward meeting more exacting and different specifications for future needs, is called for if we are to keep pace with expanding human needs. Information about natural resources is just as essential a part of needed environmental data as are flood hazard maps, physical data maps for engineering projects, and aquifer maps indicating the occurrence of ground-water supplies.

Man as a Geologic Agent

Our fifth area is the recognition of man as a geologic agent, and here we have the culmination of the problems of Environmental Geology. Man's changes in his physical surroundings are made in order to obtain some real or imagined advantage for people. Some of these changes are designed to prevent natural events from happening and include levees and detention reservoirs to prevent flooding of land, revetments and terracing to prevent erosion, and irrigation to prevent the effects of droughts. But many other changes are intended to produce effects that are not in the natural sequence of things. In both cases, nature is liable to provide unplanned and undesirable side effects. It is the role of the earth scientist to determine the effects of man-made physical changes on all aspects of the physical environment before structural changes are made so that provision can be made to negate undesirable by-product effects—or, if the side effects are too severe, so that a decision against the environmental changes can be made.

A special facet of man-made changes is in the area of mineral and fuel resources. The public need and demand for energy, and for products based on mineral raw materials, is constantly increasing at the same time that increasing populations require that land resources be maintained at maximum utility. Here again, the earth scientist must add to his traditional role of finding and developing sources of energy and mineral resources the equal or more difficult role of devising methods of producing these resources in such a way that land resources have a maximum potential for other human uses. This has led to the concept of planning for multiple sequential use—that is, designing a mineral extraction operation in advance so that the land area will first have a beneficial use before the minerals need to be extracted, then be turned over to mineral extraction, and, finally, be returned to a beneficial use, perhaps quite different from the original use.

Conclusions

The earth scientist is concerned with the physical framework of the environment wherever it may be, with the supply of raw materials essential to modern civilization, and with the management of the earth's surface so that it will all have maximum utility for its living inhabitants. It is this last item that is a relatively new role for the earth scientist, and one in which he must work cooperatively with the engineer, the biologist, and the social scientist. The earth scientist must become the interpreter of the physical environment, and he must do it in the long-term context of a dynamically changing earth so that "architectural" designs will be in harmony with natural forces 50, 100, or 500 years from now, as well as with the conditions of the moment.

2 Geology as Applied to Urban Planning: An Example from the Greater Anchorage Area Borough, Alaska*

Ernest Dobrovolny and Henry R. Schmoll

Introduction

Geology commonly has been considered a study of rocks having as its main purpose the deciphering of past events, usually very remote ones. Although widely recognized as of great economic importance in the search for mineral and fossil fuel deposits, geology seemed to have little relevance to the general public and the everyday events of the present. Geologists, of course, have understood that geology deals as well with dynamic processes, and that these processes occur with as much vigor today as in the past. Except for application to engineering works of great magnitude, however, developers and the general public have often failed to recognize the relevance of these processes to many other facets of human endeavor involving use of the land.

It takes a dramatic and devastating event, like the Alaska earthquake of March 27, 1964, to forcefully draw the attention of the public to the dynamic

*Publication authorized by the Director, U.S. Geological Survey.

Dobrovolny, E., and Schmoll, H. R., 1968, Geology as Applied to Urban Planning: an Example from the Greater Anchorage Area Borough, Alaska, *Proceedings XXIII International Geological Congress*, vol. 12, pp. 39–56. Reprinted by permission of the authors and the Geological Survey, Prague, Czechoslovakia. The authors are on the staff of the U.S. Geological Survey.

processes of geology and the impact these processes can have on everyday life and on the economy and well being of a region. Although destruction from the earthquake was widespread throughout south-central Alaska, and some towns, notably Valdez and Seward, sustained much loss of life and property, the greatest volume of damage occurred in and around Anchorage, the largest city in the State and center of the greatest population concentration. Anchorage is located at considerable distance from the epicenter, but parts of the community were developed on areas particularly vulnerable to strong seismic shock of the duration encountered in the 1964 earthquake. Understandably, the first general public attention focused on geologic hazards associated with the seismic event, such as landslides, ground fracturing, generation of seismic sea waves, lurching, and shaking. This led in turn, however, to a wider if less immediate understanding of the role geology can and must play in the orderly development of our land and economic resources.

Existing Geologic Report

Anchorage was fortunate in having a geologic report covering the city and surrounding area (Miller and Dobrovolny 1959), which suddenly became a best-seller at least locally, and was widely cited as the chief source of background geologic information in the profusion of studies, geologic and otherwise, that rightfully accompanied the recovery from disaster. This new prominence for a geologic report, together with personal contacts with many geologists, served to make public officials aware of the importance of geology in fulfilling their responsibilities. The earthquake also caused a certain amount of soul searching on the part of the geologists, however, for it had become apparent that although the geologic report had correctly assessed potential dangers in the event of a strong seismic shock, little heed was paid to this warning. Obviously unless a different approach were adopted in the future there would be little need for preparing another geologic report only to have it go unused and unheeded. With this in mind it is instructive to review here the history of the old geologic report which emerged from a geologic mapping project started in 1949. A separate study of the occurrence, availability, and quality of ground water was started later the same year (Cederstrom, Trainer and Waller 1964; Waller 1964).

Anchorage was selected for an urban mapping project because the community was growing rapidly and because surficial geologic maps with engineering interpretations are of maximum usefulness in planning if available prior to a major period of development. Anchorage and vicinity was then the largest and by far the most rapidly growing community in Alaska. Much of the land surrounding the city limits was inaccessible to vehicular travel, but access roads were planned, and each year extended farther into areas suitable for homesteads and other development. And the more the surrounding land was opened up, the more Anchorage would grow as a center, with increasingly greater development of major support facilities. Thus data on sources of natural construction material, depth to bedrock, and foundation characteristics of unconsolidated surficial deposits were needed for the design of highways, railroads, bridges, airports, hydroelectric plants, water-distribution and sewage-disposal systems, and large buildings with special foundation requirements.

A map and brief description of the geology summarizing the fieldwork of 1949 was made available in 1950. The final report was published in 1959 as U.S. Geological Survey Bulletin 1093. In addition to the regular sales and exchange distribution, about 100 copies were distributed to individuals, post offices, and institutions in Alaska.

Bulletin 1093 contains a geological map at a scale of 1 inch to the mile (1:63,360), other illustrations, tables, and a text. The map shows 29 surficial and 2 bedrock units, the division made largely on the basis of age and origin. The text describes in some detail the fairly thick sequence of surficial materials upon which Anchorage is built and refers briefly to the bedrock exposed in the adjacent mountains. Of all the map units described, only one was given a formal name, the Bootlegger Cove Clay, a light-gray silty clay of Pleistocene age that is conspicuously exposed in the bluffs along Knik Arm. The physical properties and relationship of the Bootlegger Cove Clay to other units are presented in tables and diagrams. A text section on economic geology discusses known or potential mineral resources, construction materials, and engineering problems. Among the latter are: foundation conditions, excavation, slumps and flows, drainage, and frost heave. As an example of local conditions that should be considered in development, Figure 7 (p. 95) illustrates the relationship of groundwater movement to the upper surface of the Bootlegger Cove Clay where overlain by gravel, and the associated potential health hazard resulting from contamination of the water. Bluff recession caused by active shoreline erosion is documented in Figure 6. Conditions under which landslides can be activated in the Bootlegger Cove Clay are described under slumps and flows. These problems are closely related and in one way or another have a bearing on the geologic effects produced by the 1964 earthquake.

Awareness of the 1950 and 1959 reports was limited to parts of the engineering and geologic professions. The publications served as a guide for the Alaska Road Commission in selecting parcels of land underlain by sand and gravel to be set aside as a reserve for future highway construction needs. They were used as background information by foundation engineers in preliminary site studies for some of the larger buildings, and by geologists of the Alaska Department of Highways. The bulletin was quoted in a geological field guide with respect to geologic history, and has served as a teaching aid. For several reasons the bulletin was not used as background for planning. The planning department was relatively new and its early problems concerned more pressing matters. The report is a general treatment and did not zone or classify the ground except by geologic map units. The map is not a document the planner can use directly without interpretation by a geologist. There were no geologists on the planning staff.

After the earthquake of March 27, 1964, Bulletin 1093 was read by many and frequently cited for its warning about the susceptibility to landsliding in the Bootlegger Cove Clay. A report by the Engineering Geology Evaluation Group (1964, p. 8–10) established that all of the major landslides in the city of Anchorage were related to the physical properties of the Bootlegger Cove Clay and its distribution in relation to local topography. In a similar way Bulletin 1093 was used by the Scientific and Engineering Task Force of the Federal Reconstruction and Development Planning Commission for Alaska as a basis for preparing a preliminary map classifying "high risk" areas (Federal Reconstruction and Development Planning Commission for Alaska, 1964, maps, p. 59, 61). The principal criteria for the classification were the presence of the Bootlegger Cove Clay along steep slopes and fractures associated with landslides induced by the earthquake. The preliminary map was later modified as more definitive information derived from an emergency program of detailed geologic investigation became available.

It is evident that prior to the earthquake there was little or no communication between the authors of the geologic report and the planners and other public officials who could have used the information in it. The geologists produced their report, which was published in customary manner, and they went on to other assignments. The planners zoned the area under their jurisdiction for various and seemingly appropriate uses. The two groups went on their separate ways, essentially in ignorance of each other, largely because there was no common meeting ground for them. The geologic map was a classification of the various deposits largely in terms of origin and age; engineering characteristics that were given were largely buried in information that seemed nonrelevant to the unitiated (nongeological) reader. Furthermore, the report not only was not read by the nongeologist, but was probably not even known to him. Thus, the geologists were writing only for other geologists, as is their custom. The planners meanwhile were proceeding, as is their custom, without much consideration of the naturally occurring materials that underlie the land, without sufficient awareness of the importance of such a consideration, and without the understanding that they did need to know about the geologic map and its accompanying descriptions.

The events of the earthquake brought the two groups face to face, working in their own ways to repair the damage and look to the future. In this process the planners and other public officials recognized the need for geologic information, and the geologists saw that the conventional format of the standard geologic report was not adequate to meet this need. Consequently in any new geologic work to be undertaken, new products would have to be designed to present geologic information relevant to planning in a form that would be more readily intelligible to the nongeological professional worker.

It was in this context that the Chairman of the Greater Anchorage Area Borough, a new political unit much more extensive than the city, asked the U.S. Geological Survey to begin geologic investigation that would encompass the entire Borough. Interpretive maps were especially requested, though the specific kinds of interpretations could not be defined at the time the request was made.

New Geologic Projects

The Greater Anchorage Area Borough is located in south-central Alaska (Fig. 1) and occupies most of a roughly triangular piece of land between the two estuaries at the head of Cook Inlet, Knik Arm to the northwest, and Turnagain Arm to the southwest. The western apex of this triangle and a narrow strip along Knik Arm lie within the Cook Inlet-Susitna Lowland physiographic province (Wahrhaftig 1965) and contain the city of Anchorage and its environs, including most of the land suitable for urban expansion. The remainder of the Borough lies within the Kenai-Chugach Mountains province, and is dominated by rugged, partly glacier-clad mountains with relief of as much as 6,000 feet; narrow valleys and some glacially planed shoulders provide only scattered and discontinuous areas suitable for extensive development. Thus, the greatest interest lies in a relatively small part of the Borough near the city of Anchorage, but as the Borough develops further, more and more of the marginally suitable land within the mountainous area will be brought into use, and it is vital to have basic information on this area at hand before unplanned

FIG. 1. Perspective diagram of upper Cook Inlet region showing physiographic setting of Anchorage. Boundaries of Greater Anchorage Area Borough and City of Anchorage shown by *solid line*; area of Figures 2, 3, 4, and 5 shown by *dashed line*.

opportunistic development is allowed to spread into unsuitable territory. The total area of the Borough is approximately 1,730 square miles (4,480 km^2), of which about 240 square miles (620 km^2) lies in the lowland and the remainder in the mountains; 330 square miles (855 km^2) of this part is underlain by glacier ice.

Of the two coordinated and in part interdependent U.S. Geological Survey projects now in progress in the Borough, one is directed toward the general geologic mapping of the area, with emphasis on the engineering aspects thereof, and the other toward the hydrologic aspects of the geology, especially the subsurface geology of the unconsolidated deposits in the lowland. The objective of the engineering geology project is the preparation of detailed and reconnaissance general-purpose geologic maps to provide data basic to understanding the physical features of the ground and to establish a framework within which rational land development plans can be formulated by borough officials. The hydrologic project aims to assemble and integrate the facts about the hydrologic system so that appropriate city and borough officials and their technical associates can design procedures for orderly development of the water resources.

The use of geologic maps in the search for mineral and fossil fuel deposits is well known. Geologic maps and interpretations drawn from them are also widely used in the engineering profession, generally in connection with site selection and location of construction materials for major engineering projects. Geologists are usually employed on such projects to do the geological work. The use of geologic maps in regional planning, however, has not yet become standard practice, nor, with a few notable exceptions, are geologists normally included on planning staffs. This is perhaps because the field is rather new and still developing, and because the need for this service has not had wide enough demonstration. Geologic maps are recognized by the planning profession, but not significantly discussed in their publications (Chapin 1965, p. 254). There is further reason for neglect of geologic maps by planners. To the untrained user geologic maps are complicated; geologic information is usually plotted on a topographic base, requiring that the user see through a mass of three-dimensional data presented on a flat surface. Geologic units are shown by different colors and patterns to increase legibility, but on many such maps where contacts are intricate, the maps have the appearance of inscrutable clutter.

Nonetheless it is apparent that use of geologic maps would be of considerable benefit to the planner. Among the many considerations of regional land utilization planning are (1) development of the land for agricultural, residential, industrial, recreational, and other uses; (2) site selection for planning engineering works such as dams, bridges, highways, and other major structures; and (3) assessment and development of the hydrologic system for efficient use of water resources. Intelligent planning for land utilization therefore requires knowledge of features of the ground which may be categorized as topography, geology, and hydrology. Geologic maps on the topographic base, together with accompanying hydrologic information, provide just such knowledge, and are basic to any rational understanding of the ground conditions that the planner must have.

By itself, however, the geologic map does not convey to a planner information that is directly useful in his work. It is therefore necessary to interpret the geologic map for the planner. In parallel with the other uses of geologic maps cited above, it would be logical for the planning staff to include a geologist to interpret the geologic map at any time and for any purpose, and to participate fully in the planning process. This practice would indeed be desirable, and in the case of some large planning staffs it is followed; in the usual case of a smaller staff, however, this may not be practical. A useful alternative is the interpretive map, or a series of such maps, designed for the special needs of the planner, and based on the facts contained in the geologic map. It is this new and not at all standard approach that is being employed by the current U.S. Geological Survey projects in the Anchorage area, and interpretive maps are being produced to meet the needs of planners and other professional workers, and to some extent, the general public. Maps of this type have not had wide use in civilian application in the United States, although similar maps have been used by the military for many years (Hunt 1950).

The primary product of the engineering geology project will be the general geologic map of the entire Borough at a scale of 1 inch to the mile (1:63,360). The hydrologic project ultimately plans to produce an electric analog or mathematical model of the hydrologic system. The model can be used to examine changes in the system that would result from various stresses applied, so that reasonable forecasts can be made of the effects of various methods of operating the system for years in advance.

Each interpretive map is restricted in coverage to single or closely related topics. All maps will have a similar format, and are designed to overlay each other for easy reference. The series will be designed for publication at a scale of about 1 inch to 2 miles (1:125,000) so that the entire Borough can appear on a single sheet of manageable size. Maps of selected topics of particular importance for the lowland area may have the larger scale of 1:24,000. Some of the topics to be covered in this series are:

Slopes
Construction materials
Foundation and excavation conditions
Stability of natural slopes
Recreation areas
Areas of potential development of ground and surface water
Availability of ground-water
Depth to unconfined aquifers
Depth to confined artesian aquifers
Water-table contours
Piezometric surface
Saturated thickness
Principal recharge areas
Chemical quality of water

Each project will also produce a final report in one of the Geological Survey's standard series of papers, in which the general geology and hydrology will be discussed, and the applications thereof summarized. This report will discuss only the topographic, geologic, slope, slope stability, and construction materials maps.

Topographic Maps

The topographic map is an absolute necessity as a base on which to plot data about the land. Planimetric maps used in this report were constructed from a topographic map. Figure 1 illustrates the configuration of land surface in the area. The Greater Anchorage Area Borough is covered by topographic quadrangle maps at a scale of 1:63,360 with a 100-foot contour interval in the mountains and 50-foot contour interval in most of the lowland. A special map at a scale of 1:24,000 with 20-foot contour interval covers the Anchorage metropolitan area in the lowland. Neither map shows changes caused by the 1964 earthquake. Separate topographic maps will not be included among the other maps in the series to be produced, as these maps are already widely distributed and readily available.

Geologic Map

For most of the Greater Anchorage Area Borough a conventional geologic map at a scale of 1:63,360 is adequate to show the geology. Emphasis is placed on the surficial deposits, as these are quite widespread even in the mountains in this glaciated terrane, and comprise most of the materials most critical to development. The bedrock units, on the other hand, are relatively uniform over wide areas, and for the purpose of the current project are being mapped in reconnaissance fashion. The preliminary version of the geologic map currently under

revision includes six bedrock units of Mesozoic and Tertiary age, chiefly eugeosynclinal sedimentary and volcanic rocks; 23 units of Pleistocene age, chiefly of glacial origin; and 28 units of Recent age, including glacial, alluvial, and colluvial deposits. Figure 2 is a simplified version of a part of this map, at reduced scale.

Slope Map

The generalized slope map (Fig. 3) is a special kind of topographic map that summarizes the continuously variable slope information shown on the standard topographic map. Categories of slope can be chosen to meet various needs, and the map divided into units with the same slope characteristics. Such a map is necessarily generalized, because except at very large scales there usually are local slopes that do not fall within the assigned category. Inasmuch as the distribution of slopes is not random but systematic, however, it is practical to make a meaningful map. Though the slope map can be made directly from and entirely dependent upon the topographic map, its accuracy can be considerably enhanced by field observation and use of aerial photographs.

For planning purposes three primary categories of slope have been chosen: (1) slopes less than 15 percent; (2) slopes 15 to 45 percent; (3) slopes greater than 45 percent. In addition, the upper and lower categories have been subdivided, so that large areas with slopes less than 5 percent, and significant areas with many slopes greater than 100 percent are also shown. The most important division

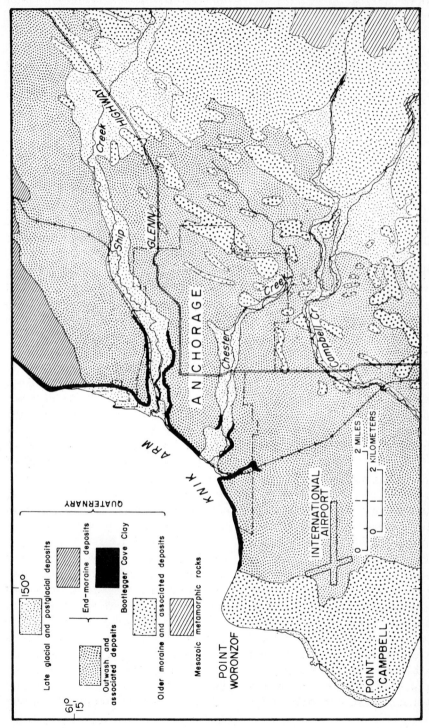

FIG. 2. Geologic sketch map of City of Anchorage and vicinity, Alaska.

here is at 45 percent, which is the reasonable upper limit for maneuvering tracked vehicles. Figure 3 shows a part of the slope map that has been made for the entire borough.

The principal reason for making a slope map is to identify areas having the same range of slope percent, because the angle of slope is a major factor in estimating slope stability. Landslides occur mainly on steep slopes and along sharp topographic discontinuities such as bluffs. They do not occur in areas of low relief that are far from breaks in topography. Identification of steep slopes is the first step in the process of isolating areas where landslides are more likely to occur. Used in conjunction with the geologic map, the slope map is a tool that leads to appraising the landslide potential of natural slopes.

Slope Stability Map

An interpretation of the stability of natural slopes in the Anchorage area is shown on Figure 4 and the map units are summarized in Table 1. The potential for landslides, especially in reponse to seismic shock, is of principal concern here, but other types of ground movement, such as solifluction and rockfall, are also considered. The map is derived essentially from the slope map and the geologic map. The primary criterion for determining instability is the degree of slope; areas of steep and very steep slope will be the chief sites of instability. However, the degree of instability depends considerably on the geologic material underlying the slope. Thus by combining elements of the two maps the slope stability map can be prepared, showing varying degrees of stability. Generally the areas of low and moderate slopes are lumped together as having high stability. Parts of these areas adjacent to zones of low stability, however, may have potentially low stability because, for example, of the possibility that a landslide can work back from an area of steep slope and "eat" into an adjacent area of low slope; this actually happened during the 1964 earthquake. Also, the low slopes extending outward from the base of a bluff may be overridden by landslide debris or disturbed by pressure ridges. Such features cannot be projected on the scale of a map used here, but the potential problem area is outlined as to general location.

Figure 4 clearly shows (a) those areas which are unsuited for most development because of slope stability problems that cannot feasibly be surmounted except in unusual circumstances (units 5 and 6); (b) those areas in which slope stability presents problems that must be considered in planning and design but for which a solution can be worked out (units 3 and 4); and (c) those areas which are relatively free of slope stability problems. It must be emphasized, however, that the map is generalized, that locally within each unit the described situation may not exist, and that for particular development detailed site investigations must be made. The map is designed primarily to delineate those areas in which particular problems can be expected.

Construction Materials Map

The construction materials map is one of the most direct interpretations that

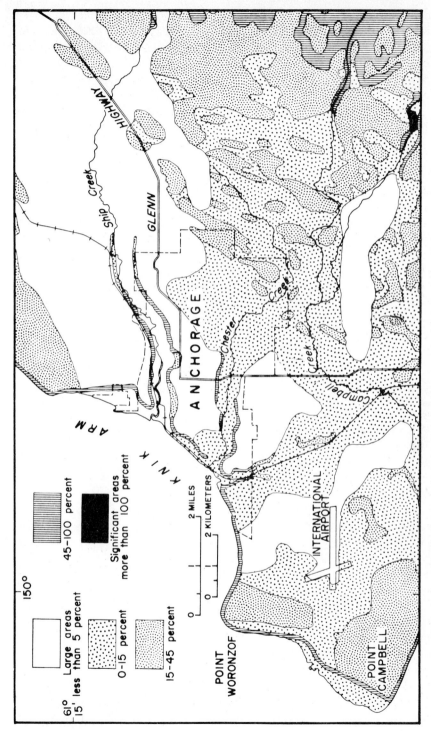

FIG. 3. Generalized slope map of City of Anchorage and vicinity, Alaska.

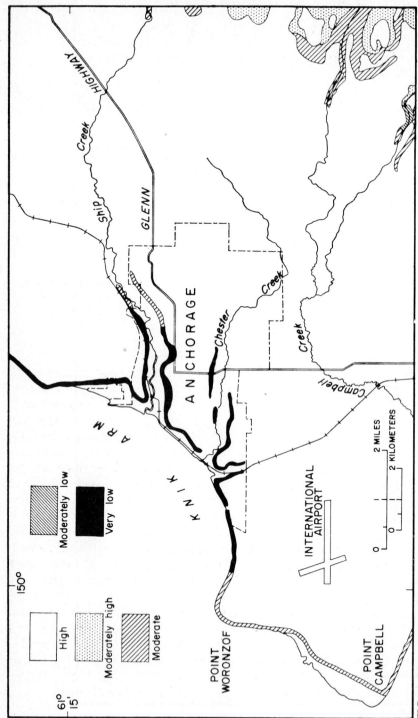

FIG. 4. Map showing stability of naturally occurring slopes, City of Anchorage and vicinity, Alaska. Map units are described in Table 1.

Table 1 Description of map units for stability map of naturally occuring slopes

Map unit	Slope stability	Landslide potential	Solifluction potential	Slopes (percent)	Geology
1	High	Very low	Very low	<45	Undifferentiated
2	Moderately high	Generally low	Generally low	45–100	Generally stable bedrock
3	Moderate	Moderately low to moderately high	Moderately low	45–100	Medium- to coarse-grained unconsolidated deposits
4	Moderately low	Moderately low to moderately high	High	45–100	Soft, frost-susceptible bedrock, chiefly argillite
5	Low (This unit is not present in the area covered by Fig. 4)	Moderately high	High (chiefly rockfalls and snow avalanches)	Commonly >100	Bedrock, fractured and in part faulted
6	Very low	High	High (chiefly slumps and earthflows)	15–>100	Unconsolidated deposits underlain by sensitive clay and silt, or incorporating ice

can be made from the geologic map. It is essentially a lithologic map, in which geologic units of various ages and origins that have similar lithologies, and therefore similar utility as construction materials, are grouped together in a single map unit. In some cases single units on the geologic map have been split into more than one lithologic unit, reflecting usually a gradational change of material within the geologic unit. Sometimes such a breakdown within a geologic unit requires additional fieldwork, and even so may have to be made arbitrarily. Thus, the transformation is not purely mechanical but requires refined geologic interpretation.

A construction materials map has been made for Anchorage based on the geologic map published in Bulletin 1093, and is confined largely to the lowland area; it has been produced at a scale of 1:24,000 and required additional field checking and laboratory analysis beyond what was done in the first survey. Six of the seven map units are for unconsolidated materials; the seventh includes all bedrock, which covers only a small part of the area and has no particular utility as a construction material. The six unconsolidated units range from gravel and commonly interbedded gravel and sand, which are of great economic importance in the area, to sand, silt, till, and swamp deposits which have relatively little utility. During the current project the entire Borough will be so mapped at smaller scale, and the large-scale map may be further refined. Figure 5 shows a part of the small-scale map.

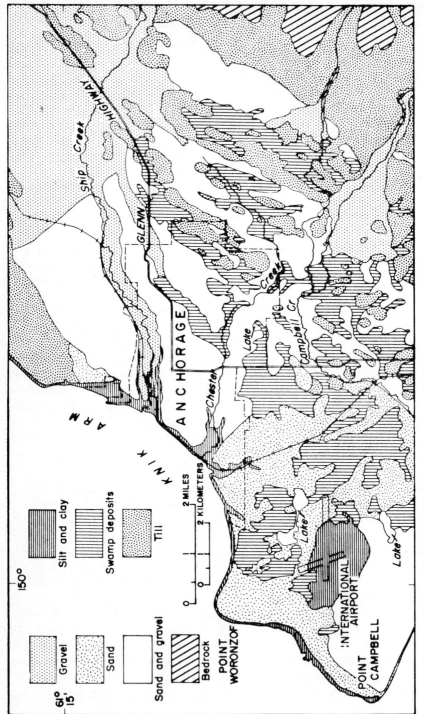

FIG. 5. Construction materials map, City of Anchorage and vicinity, Alaska.

On such a map, areas that are of interest for potential use as sources of granular construction materials can be readily seen. Though much of the land has already been preempted for other purposes, this map can still be used to select from the unoccupied areas those deposits which appear to warrant more detailed investigations. The more desirable deposits can be reserved for future use and, as part of a systematic plan, exploited prior to development of the land for other uses. It is not the intent of the map to serve as a resource map in the sense of estimating gravel reserves, or detailing the precise uses for which a particular deposit is most suited. More detailed site investigations are required for such determinations. The map does serve primarily as a guide so that the planner can know which areas are to be selected for further investigation, and which have no construction materials and can be considered for other types of development.

Recreation Areas

It has been considered appropriate to present a map showing areas particularly suited for recreational purposes, particularly those areas that are of interest for their scenic beauty or natural history, but including other areas that are less spectacular but well suited for such activities as hiking and camping, and able to accommodate large numbers of people. Chiefly esthetic considerations are involved here, and these are necessarily subjective in nature. They can less readily be derived precisely from measurements of slope or behavior of geologic materials. Nonetheless, elements of topography, geology, and hydrology play a major role in determining which areas have high esthetic appeal, and we regard that in any study of these fields such considerations should not be overlooked.

The Greater Anchorage Area Borough is fortunate in having many areas of unusual scenic interest, some of which are widely known and frequently visited; others are presently inaccessible to most people and totally undeveloped. Of special interest is Lake George, a multistage glacial lake that forms when water from a tributary valley is impounded behind Knik Glacier in the spring. In early summer the water overtops the glacier, and the river cuts a narrow gorge between steep rock and ice walls, almost emptying the lake.

The recreation potential map is intended as a guide to the planning staff, to aid in assigning priorities to development and determining degree of development that should be permitted, so that the recreation potential of the area can be realized in an orderly manner.

Conclusions

There are many elements of an economic, social, and political nature that must be considered in the planning process; these have been deciding factors in the past, and no doubt will continue to be so. Nonetheless, the physical environment deserves increased consideration, especially in areas where that environment is not as hospitable as it is in other areas where more planning experience has accumulated. This is true of large areas in the western United States that are in a general way characterized by dynamic geological processes of greater magnitude and intensity than those in the more highly developed East. It is also true, however, that in the older established cities the less problematical areas have been developed first, and those areas into which new development will extend are more likely to present physical problems. Thus there is need for wide application of the principles of planning in this era of rapid urbanization, and need for these principles to include consideration of the natural physical environment. This is true for those areas that are already well developed such as the United States, and also for the less well developed nations that will grow at ever increasing rates in the near future. And the greater the intensity of development, the greater the need for this type of planning.

In order that considerations of the physical environment gain greater prominence, geologists and others who study the physical environment should promote understanding of their work. They should meet with planners and other responsible public officials on a personal working-level basis, they should learn the jargon and needs of those individuals, and they should design their own products to be of optimum use. This statement does not imply that there should be any lessening in their own scientific pursuits, without which their work would have less and less value, but the geologist does have to bear the added burden of carrying his work to the user so that its potential utility can be fully realized. It is with this goal in mind that we believe the work of the geologic and hydrologic projects in the Greater Anchorage Area Borough can make a contribution to the wider application of the geological sciences in the expanding area of orderly urban development.

References

Cederstrom, D. J., Trainer, F. W. and Waller, R. M. (1964): Geology and ground-water resources of the Anchorage area, Alaska. U.S. Geol. Survey Water-Supply Paper 1773, 108 p.

Chapin, F. S., Jr. (1965): Urban land use planning (2d ed.). Urbana, Univ. Illinois Press, 498 p.

Hunt, C. B. (1950): Military geology, in Paige, Sidney, chm., Application of geology to engineering practice. Geol. Soc. America, Berkey Volume, p. 295–327.

Engineering Geology Evaluation Group, 1964, Geologic report—27 March 1964 earthquake in Greater Anchorage area. Prepared for and published by Alaska State Housing Authority and the City of Anchorage, Anchorage, Alaska, 34 p., 12 figs., 17 pls.

Federal Reconstruction and Development Planning Commission for Alaska, 1964, Response to disaster, Alaskan earthquake, March 27, 1964. Washington, U.S. Govt. Printing Office, 84 p.

Miller, R. D. and Dobrovolny, E. (1959): Surficial geology of Anchorage and vicinity, Alaska. U.S. Geol. Survey Bull. 1093, 128 p.

Wahrhaftig, C. (1965): Physiographic divisions of Alaska: U.S. Geol. Survey Prof. Paper 482, 52 p. [1966].

Waller, R. M. (1964): Hydrology and the effects of increased ground-water pumping in the Anchorage area, Alaska. U.S. Geol. Survey Water-Supply Paper 1779-D, 36 pp.

Editor's note: The Arctic Environmental Information and Data Center of the University of Alaska has published an *Environmental Atlas of the Greater Anchorage Area Borough, Alaska* (edited and coordinated by Lidia L. Selkreff, 1972) which contains a variety of special purpose interpretive maps useable by the informed layman, planning personnel, and administrators who may not have a formal background in the geological sciences.

Supplementary Readings

Cargo, D. N., and Mallory, B. F., 1974, *Man and His Geologic Environment*, Addison-Wesley Publishing Co., Inc., Reading, Mass., 548 pp.

Coates, D., 1981, *Environmental Geology*, John Wiley & Sons, New York. Hutchinson & Ross, Inc., Stroudsburg, Pa., 483 pp.

Costa, J., and Baker, V., 1981, *Surficial Geology: Building with the Earth*, John Wiley & Sons, New York.

Flawn, P. T., 1970, *Environmental Geology—Conservation, Land-Use Planning, and Resource Management*, Harper and Row, New York, 313 pp.

Geological Survey of Alabama, 1971, *Environmental Geology and Hydrology, Madison Co., Alabama: Meridianville Quadrangle: Geol. Survey Alabama, Atlas Series 1*, 72 pp.

Griggs, G., and Gilchrist, J., 1977, *The Earth and Land Use Planning*, Duxbury Press, Boston.

Howard, A., and Remson, I., 1978, *Geology in Environmental Planning*, McGraw-Hill, New York.

Keller, E., 1982, *Environmental Geology*, 3d ed., C. E. Merrill Publishing Co., Columbus, Ohio.

Laporte, L., 1975, *Encounter with the Earth*, Canfield Press, San Francisco.

Larson, E., and Birkeland, P., 1982, *Putnam's Geology*, 4th ed., Oxford Univ. Press, New York.

Leveson, D., 1980, *Geology and the Urban Environment*, Oxford Univ. Press, New York.

McKenzie, G. D., and Utgard, R. O., eds., 1975, *Man and His Physical Environment*, 2d, ed., Burgess Publishing Co., Minn. 388 pp.

Menard, H. W., 1974, *Geology, Resources and Society*, W. H. Freeman and Co., San Francisco, 621 pp.

National Academy of Sciences—Committee on Geological Sciences, 1972, *The Earth and Human Affairs*, Canfield Press, San Francisco, 142 pp.

Pessl, F., Jr., Langer, W. H., and Ryder, R. B., 1972, Geologic and Hydrologic Maps for Land-Use Planning in the Connecticut Valley with Examples from the Folio of the Hartford North Quadrangle, Connecticut, *U.S. Geological Survey Circular 674*, 12 pp.

Robinson, G., and Spieker, A., eds., 1978, Nature To Be Commanded, *U.S. Geological Survey Professional Paper 950*, U.S. Govt. Printing Office.

Selkregg, L. L., ed., 1972, *Environmental Atlas of the Greater Anchorage Area Borough, Alaska*, Univ. of Alaska, 105 pp.

Strahler, A. N., and Strahler, A. H., 1973, *Environmental Geoscience: Interaction Between Natural Systems and Man*, Hamilton Publishing Co., Calif., 511 pp.

Turk, L. J., ed., 1975, *Environmental Geology*, Springer-Verlag, New York.

U.S. Dept. of Interior, 1973, Resource and Land Information, South Dade County Florida, *Geological Survey Investigation I-850*, 66 pp.

Wermund, E. G., ed., 1974, *Approaches to Environmental Geology, Bureau of Economic Geology Report of Investigations, No. 81*, Univ. of Texas, Austin, 286 pp.

Young, Kieth, 1975, *Geology, The Paradox of Earth and Man*, Houghton Mifflin Co., Boston, 526 pp.

I
GEOLOGIC HAZARDS
AND HOSTILE
ENVIRONMENTS

Earthquake Damage, Anchorage, Alaska. (U.S. Army.)

It seems appropriate to continue our analysis of the geologic aspects of environmental problems by reviewing the impact of geologic processes on the environment and on the activities of humans. These processes can be highly beneficial. They may, for example, produce valuable ore deposits or provide rich soil and abundant water resources.

On the other hand, these same geologic processes represent natural hazards or hostile environments. When people become involved with these hazards, they may suffer—and the price can be high (Fig. 1).

Given the highly sophisticated nature of modern technology, it would appear that we could simply exploit the benefits and control the hazards, thereby exercising a high degree of control over our own destiny. Such control is far from complete, however, because we have not understood or have frequently failed to make allowances for a variety of geologic hazards.

Part I deals with the hazards associated with six geologic agents. There are other potential geologic hazards, but the examples cited serve to illustrate the dynamic aspects of geologic agents and their impact on humans. Most of the problems associated with these hostile forces can be avoided or minimized by utilizing adequate geologic data. If we understand the processes and have information about their occurrence in both time and place, we should be able to assess the risk potential and develop techniques for reducing damage, predicting their occurrence, and in some cases, controlling them.

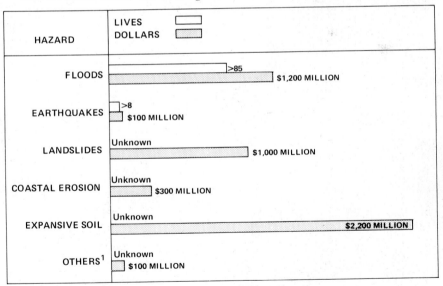

[1]Others: subsidence, creep, fault displacement, liquefaction of sand and clay, dust, waves caused by earthquakes, volcanoes.

FIG. 1. Mean annual losses in the United States from geological hazards. (From *U.S. Geological Survey Professional Paper 950.*)

Volcanism

> *"All mountains, islands, and level lands have been raised up out of the bosom of the earth into the position they now occupy by the action of subterranean fires."*
>
> Lazzaro Moro

On August 24, A.D. 79, the apparently extinct volcano Vesuvius suddenly exploded and, in two days, completely buried the cities of Pompeii, Herculaneum, and Stabiae. Sixteen thousand people were killed. On May 18, 1980, Mount St. Helens exploded with a blast of about the same magnitude and spread death and destruction in the Cascade Range in the northwestern United States. Both volcanoes had been quiet for decades, their surfaces covered with mature vegetation and their flanks populated by resort communities. There are more than 500 other modern volcanoes which together have been responsible for more than 200,000 deaths and countless millions of dollars in property damage within the past 500 years.

Causes of Volcanic Activity

The word "volcano" comes from the small island of Vulcano in the Mediterranean Sea north of Sicily. The people living in the area believed that the island was the chimney of the forge of Vulcan, the blacksmith of the Roman gods, and that the smoke and hot ash erupting from Vulcano came from Vulcan's forge as he hammered out weapons for the gods. Contemporary theory suggests otherwise.

Although volcanoes are prominent topographic features, they owe their origin to subsurface (*endogenic*) processes. A volcano is a mountainous feature that builds up around a vent, or series of cylindrical conduits, which connects with reservoirs of molten rock called *magma*. The origin of the molten rock and the manner in which it makes its way to the surface are somewhat subject to speculation.

Eruption of El Chichón volcano, Mexico, April 2, 1982. A 3658-m- (12,000-ft-) column of steam and ash can be seen at the slope of the volcano. The initial explosion occurred on March 28, 1982, and was the first to occur in historic time. Tephra falls were heavy near the volcano, causing tens of thousands to flee the area. As many as 187 persons were killed and hundreds were injured. A cloud of dust, ash, and sulfur dioxide reached an altitude of 16.9 km (10.5 mi), and within a few months it covered more than one-quarter of the earth's surface. Between the equator and 30° N, the debris blocked out 10 percent of the sun's total radiation. (United Press International)

Table 1 History's major volcanic disasters

Year	Locality	Deaths
A.D. 79	Vesuvius, Italy	16,000
1669	Mt. Etna, Sicily	20,000
1783	Skaptar Jokull, Iceland	10,000
1815	Tambora, Indonesia	12,000
1883	Krakatoa, Indonesia	36,000
1902	Mt. Pelée, Martinique	30,000

Most modern volcanoes (79 percent) are concentrated in the "ring of fire" that circumscribes the Pacific Ocean Basin. A second belt (13 percent) is associated with the rift zone of the Mid-Atlantic Ridge, and a third belt, accounting for 4 percent, extends from the Mediterranean area eastward to Indonesia.

The major volcanic belts coincide with the boundaries between rigid plates that make up the earth's crust (Fig. 1). The plates may move relative to each other in several different ways. They may slide past each other, move apart (diverge), or converge, with one plate moving under the adjacent plate. The movement of the plates is facilitated by thermal *convection currents.* Volcanic ridges form where plates diverge (*zones of divergence* or *rift zones*); thus these ridges are sites where new crust is being formed. Where plates converge—*zones of convergence* or subduction zones—deep trenches are formed and old crustal material is consumed (Fig. 2). In both zones, magma may be formed in various

FIG. 1. The earth's lithospheric plates and world pattern of zones of convergence and divergence. Arrows indicate relative movement of plates.

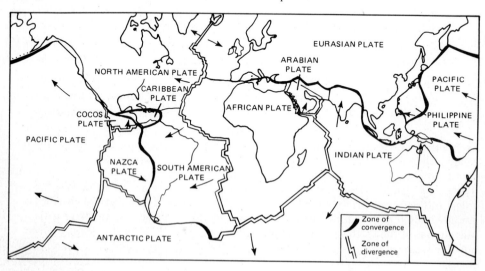

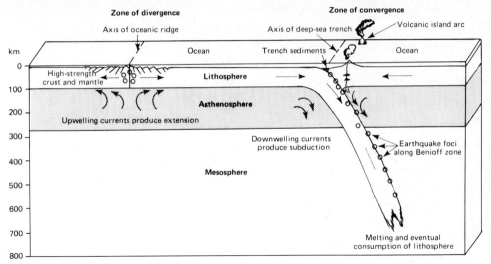

FIG. 2. Hypothetical model of relationships between zones of divergence and convergence, and upwelling and downwelling convection currents. New crustal material is formed in the zone of divergence (rift zone); and old crust is consumed in the zone of convergence (subduction zone). Plates of the lithosphere sink into the asthenosphere and mesosphere, causing earthquakes along a steeply inclined plane called the *Benioff zone*. (Adapted from *Putnam's Geology,* 4th ed., 1982. © Oxford University Press.)

ways. During the slow convergent sliding of one plate against another, heat is produced, which contributes to the melting of crustal material. As the plate is subducted, or thrust deeper and deeper into the earth, it encounters temperatures high enough to complete the melting, thereby producing magma. In zones of divergence, on the other hand, the heat flow associated with upwelling convection currents is high. In these areas the fracturing of the earth's crust aids in the generation of molten material by reducing pressure, which in turn reduces the melting point of crustal material.

The idea that the earth's crust is made up of rigid plates which float on an underlying layer and move relative to each other in response to convection currents is known as the *theory of plate tectonics.* The theory not only explains the origin of magma but also accounts for earthquake activity (see p. 78).

Not all volcanoes occur over plate boundaries, however. Mid-plate volcanoes, like the Hawaiian Islands, may lie over radioactive hot spots or thermal plumes. It is thought that heating within the mantle leads to a localized reduction of the density of the mantle material, which then begins to rise toward the surface in a feather-like plume.

The rising plume may produce magma which in turn may lead to a volcanic eruption. As the plate moves across the plume, a chain of volcanoes grows.

Once magma forms, it behaves as a relatively mobile body. It can rise both vertically and laterally and can travel great distances. The liquid magma is less dense than the surrounding solid country (preexisting) rock and, therefore, a buoyant force is created. Movement is also facilitated by a process called *stoping*. Country rock simply falls into the magma and is assimilated, thereby providing room for upward movement. The magma may solidify before it reaches the surface, creating rock bodies of varying sizes and shapes. Or the magma may continue to move upward and reach the surface, creating either a volcano or a lava flow, depending on the composition of the molten material. Final movement to the surface is frequently aided by fracture zones, sudden decreases in pressure, and rapid expansion of gases dissolved in the magma. Rock bodies formed with the subsurfaces are known as *intrusive* structures whereas those formed on the surface are known as *extrusive* structures.

Intrusive Structures

When magma solidifies in the magma chamber or makes its way into the surrounding country rock and solidifies, a structure known as a *pluton* is formed. If a pluton conforms to the layering of the surrounding rock, it is classified as *concordant;* if it cuts across the layering, as *discordant.* The most common types are

DISCORDANT PLUTONS
    Batholith
    Stock
    Volcanic neck
    Dike

CONCORDANT PLUTONS
    Sill
    Laccolith
    Lopolith

Shape also plays a role in pluton classification. *Batholiths* and *stocks* are large, irregularly shaped, discordant plutons associated with the core of folded mountain ranges. They increase in size as they extend downward, and only the upper portions of some batholiths have been exposed by erosion. The surface exposure of a batholith is greater than 100 km^2 (38.6 mi^2) whereas a stock is smaller. The solidified material that fills the cylindrical conduit connecting reservoirs of molten rock (magma chambers) with the surface is called a *volcanic neck*. Tabular-shaped plutons called *dikes* may radiate from a volcanic neck and cut across the country rock or layers within a volcanic cone. Other tabular-shaped plutons may branch off from the dike and follow the bedding planes in country rock or a volcanic cone. These structures are known as *sills*. If the lateral flow of magma cannot keep pace with the supply, a potential tabular-shaped structure will be transformed into either a structure which has a flat base and a dome-shaped upper surface (*laccolith*), or a structure with a flat upper surface and a dome-shaped base (*lopolith*). Figure 3 illustrates the relationships between the most common structures.

Although intrusive structures form at great depths, they may eventually be exposed as a variety of landforms through uplift and erosion. Plutons frequently serve as the host rock for many metallic mineral deposits.

Extrusive Volcanism

The essential elements of a volcano are the magma chamber and the conduit which brings the magma to the surface. If magma flows or explodes from a cylindrical vent at the top of the conduit, it starts to build a volcanic edifice. If it flows from a fissure it simply floods the surrounding area. One such flood covered an area of 52,000 km^2 (20,077 mi^2) in the Columbia Plateau of eastern Washington and Oregon in prehistoric times. Such magma is commonly of basaltic composition (high in ferromagnesian minerals) and the deposits are called *flood basalts*.

When magma is extruded onto the surface it is called *lava*. When lava of low viscosity solidifies, it often produces a relatively smooth, ropy surface called *pahoehoe*. Flows of higher viscosity produce a rough, clinkery surface known as *aa*. The violent separation of gas from lava may produce a rock froth called *pumice*, which is so light it floats on water.

The main factors that determine whether a volcano will erupt in a smooth outpouring of lava or as a violent explosion are the chemical composition of the magma and the characteristics of the gases dissolved in it. Basaltic magmas, for example, usually erupt in a highly fluid state, and their dissolved gases escape easily, so that the eruption is relatively quiet. More viscous, acidic magmas with a high silica

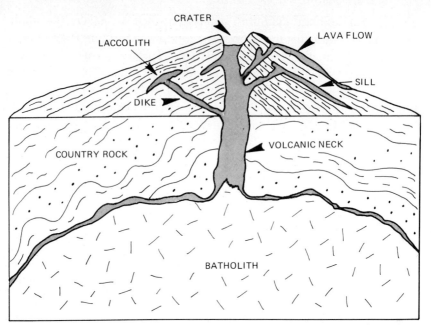

FIG. 3. Hypothetical cross-section of volcanic cone and associated discordant structures (batholith, volcanic neck, and dike) and concordant structures (sill and laccolith).

content usually erupt explosively. The solid material which accompanies an explosive eruption is referred to as *pyroclastics, tephra,* or *ejecta* and ranges in size from very fine-grained dust and ash (less than 4 mm), to medium-sized material known as *lapilli* (4 to 64 mm), to very large objects (>64 mm) called *volcanic bombs.* Occasionally, large chunks of country rock (*volcanic blocks*) are torn loose and become part of the pyroclastic debris.

A number of schemes have been used to classify volcanoes. It is, for example, tempting to classify them according to period of activity. This scheme recognizes *active* (erupted within historic time or the past 10,000 years), *dormant* (with a potential for eruption), and *extinct* volcanoes. The problem with this approach is that many dormant volcanoes have extremely long periods of dormancy, so are often mistakenly presumed to be extinct.

Another classification scheme is based on topographic form. This scheme recognizes four types of volcanoes: (1) cinder cones, (2) shield volcanoes, (3) composite cones, (4) lava domes.

Cinder cones are steep-sided, conical features dominated by pyroclastic debris that has collected about a central vent. Wizard Island in Crater Lake National Park is a good example. Cinder cones are usually indicative of young volcanoes or volcanoes with a short life span.

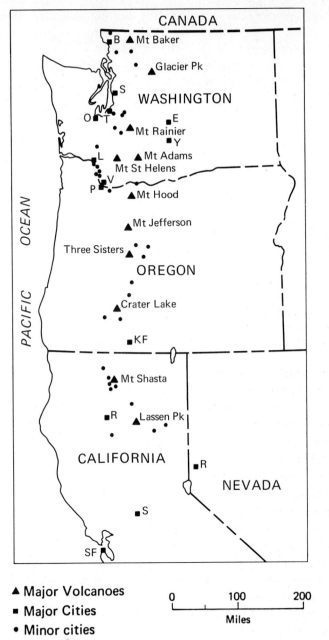

▲ Major Volcanoes
■ Major Cities
• Minor cities

FIG. 4. Index map of the Cascade Range, showing location of major volcanoes. Lassen Peak is a lava dome while most of the remaining volcanoes are classified as composite cones. (From Crandell and Waldron, *Geologic Hazards and Public Problems Conf. Proceedings*, 1969.)

Shield volcanoes are broad, gently sloping structures dominated by lava flows. The Hawaiian Islands are composed of clusters of active shield volcanoes.

A *composite cone,* or *stratovolcano,* is built up of alternate layers of lava and pyroclastics. This is the characteristic type of volcano in the Cascade Range (Fig. 4).

Lava domes, the fourth type of volcano, are built of highly viscous lavas that tend to pile up in a rough knob rather than to flow outward as they emerge from the vent. Mount Pelée in Martinique, and Lassen Peak in California are examples of lava domes.

When volcanoes erupt, they usually eject material in all three states: gaseous, liquid, and solid. The dominant gas is water vapor. Other gases commonly present include carbon dioxide, carbon monoxide, hydrogen sulfide, nitrogen, and various halogens such as chlorine and fluorine.

Types of Hazards

A variety of hazards are associated with volcanism but, since each volcano is somewhat unique, the specific hazards vary from volcano to volcano. There are even major differences between eruptions of the same volcano. For purposes of discussion it is convenient to recognize two major classes of hazards: (1) primary hazards, and (2) secondary hazards. The most dangerous primary hazards are pyroclastics, nuées ardentes, lava flows, and poisonous gases. Secondary hazards include mudflows (lahars), jökulhaups, and tsunamis. Many eruptions present a combination of both primary and secondary hazards. For example, historical eruptions in the Cascade Range have produced extensive pyroclastics (Fig. 5) and mud flows (Fig. 6). The eruption of Mount St. Helens on May 18, 1980, repeated this behavior. Table 2 is a summary description of volcanic hazards; Table 3 summarizes the relative degree of danger associated with different geologic and geographic factors related to volcanoes. Jökulhaups and tsunamis are unusual hazards. *Jökulhaups* are glacial outburst floods that result from the partial melting of a glacier by volcanic heat. Meltwater builds up within the glacier until the water pressure exceeds the strength of ice, resulting in an enormous burst of water. *Tsunamis,* or seismic sea waves, may be produced as portions of the ocean floor are displaced during an eruption. The explosion of Krakotoa in 1883 generated a sea wave with a height of 40m (131.2 ft) which destroyed several hundred villages and drowned 36,000 people.

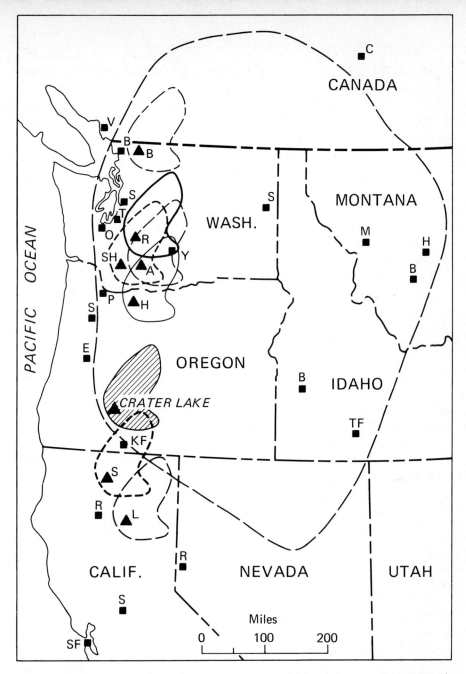

FIG. 5. Map of Cascade Range, showing area covered by pumice eruption at site of Crater Lake about 7000 years ago. The outer line shows maximum limits of ash fall; inner line (pattern) shows area covered by 6 inches or more of pumice. The same 6-inch thickness line is shown superimposed on the other major volcanoes of the range. Data from reports by Howel Williams and H. A. Powers and R. E. Wilcox. (From Crandell and Waldron, *Geologic Hazards and Public Problems Conf. Proceedings,* 1969.)

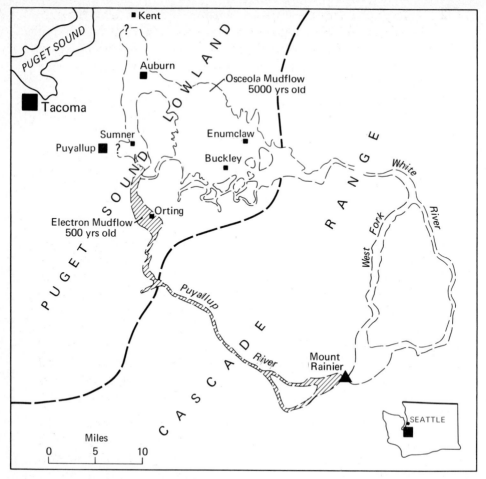

FIG. 6. Map of Mount Rainier and vicinity, showing extent of the Osceola Mudflow (dashed line) and the Electron Mudflow (pattern). The Osceola Mudflow moved 80 km (130 mi) from the volcano while the Electron Mudflow moved 56 km (90 mi). More than 33,000 people now live in the area of these two flows. (From Crandell and Waldron, *Geologic Hazards and Public Problems Conf. Proceedings*, 1969.)

Prediction Techniques

A long-range goal of some volcanologists is to predict the time, place, and magnitude of eruptions. These predictions will be based partly on the rock record of historic eruptions and partly on careful monitoring of precursory events.

Geophysical Precursors

1. Earthquakes. Nearly every eruption which has been monitored has been preceded by minor earthquakes or "microseismic events."

2. Topographic changes. As the magma increases in volume or moves toward the surface there will be a bulging of the crust or an increase in the steepness of the slope of the volcano. These changes can be detected by sensitive leveling instruments called *tiltmeters*.
3. Changes in gravity field. Magmas with densities different than the surrounding country rock will produce changes in the local gravity field as they move toward the surface.
4. Changes in magnetic and electrical fields. Apparently, high temperatures influence both the magnetic susceptibility and the electrical resistivity of the rocks.

Chemical Precursors
1. Changes in gaseous emanations. Changes in both the volume and composition of gases frequently occur prior to an eruption.
2. Temperature changes. Temperatures of crustal rocks increase as the magma makes its way toward the surface. There may also be changes in the behavior of fumaroles and hot springs.

Behavioral Precursors
1. Animal behavior. Changes in the behavior patterns of birds and other animals commonly occur prior to an eruption. Animals may leave an area or they may appear to be nervous or confused.

Detecting the precursory phenomena is within the range of the state of the art of modern scientific instrumentation. The deployment and monitoring of the instruments on a worldwide scale is, however, prohibitively expensive. It is, therefore, necessary to concentrate efforts on those volcanoes which pose the greatest threat to lives and property. For example, the U.S. Geological Survey has conducted an extensive monitoring program at the volcano Kilauea in the Hawaiian Islands, and successfully predicted the time and place of the eruptions of 1955, 1959–60, and 1971. The U.S. Geological Survey has also monitored potentially dangerous volcanoes in the Cascade Range. Seismometers have been installed on Mt. Baker, Mt. Rainier, Mount St. Helens, and Lassen Peak. Tiltmeters and geodimeter networks have also been employed, and experiments involving the monitoring of fumarole activity have been conducted. The major blast of Mount St. Helens on May 18, 1980, was preceded by minor eruptions, seismic activity, and the development of an anomalous bulge on the north flank of the volcano. These precursors enabled volcanologists to issue warnings, and local officials proceeded with evacuation plans

Table 2 Description of volcanic hazards

	Lava flows	Hot avalanches, mud flows, and floods	Volcanic ash (tephra) and gases
Origin and characteristics.	Result from non-explosive eruptions of molten lava. Flows are erupted slowly and move relatively slowly; usually no faster than a person can walk.	Hot avalanches can be caused directly by eruption of fragments of molten or hot solid rock; mudflows and floods commonly result from eruption of hot material onto snow and ice and eruptive displacement of crater lakes. Mud flows also commonly caused by avalanches of unstable rock from volcano. Hot avalanches and mud flows commonly occur suddenly and move rapidly, at tens of km per hour.	Produced by explosion or high-speed expulsion of vertical to low-angle columns or lateral blasts of fragments and gas into the air; materials can then be carried great distances by wind. Gases alone may issue non explosively from vents. Commonly produced suddenly and move away from vents at speeds of tens of km per hour.
Location	Flows are restricted to areas downslope from vents; most reach distances of less than 10 km. (6 mi.) Distribution is controlled by topography. Flows occur repeatedly at central-vent volcanoes, but successive eruptions may affect different flanks. Elsewhere, flows occur at widely scattered sites, mostly within volcanic "fields."	Distribution nearly completely controlled by topography. Beyond volcano flanks, effects of these events are confined mostly to floors of valleys and basins that head on volcanoes. Large snow-covered volcanoes and those that erupt explosively are principal sources of these hazards.	Distribution controlled by wind directions and speeds, and all areas toward which wind blows from potentially active volcanoes are susceptible. Zones around volcanoes are defined in terms of whether they have been repeatedly and explosively active in the last 10,000 years.
Size of area affected by single event.	Most lava flows cover no more than a few square miles. Relatively large and rare flows probably would cover only hundreds of square kilometers.	Deposits generally cover a few square km to a few hundreds of square km. Mud flows and floods may extend downvalley from volcanoes many tens of km.	An eruption of "very large" volume could affect tens of thousands of square km, spread over several States. Even an eruption of "moderate" volume could significantly affect thousands of square miles.
Effects	Land and objects in affected areas subject to burial, and generally they cause total destruction of areas they cover. Those that extend into areas of snow, may melt it and cause potentially dangerous and destructive floods and mud flows.	Land and objects subject to burning, burial, dislodgement, impact damage, and inundation by water.	Land and objects near an erupting vent subject to blast effects, burial, and infiltration by abrasive rock particles, accompanied by corrosive gases, into structures and equipment. Blanketing and infiltration effects can reach hundreds of miles downwind. Odor, "haze," and acid

of areas endangered by future eruptions.	volcanoes. Elsewhere, only general locations predictable.	central-vent volcanoes and are restricted to flanks of volcanoes and valleys leading from them.	nates mostly at central-vent volcanoes; its distribution depends mainly on winds. Can be carried in any direction; probability of dispersal in various directions can be judged from wind records.
Frequency, in conterminous United States as a whole.	Probably one to several small flows per century that individually cover less than 16 km^2 (10 mi^2). Flows that cover tens to hundreds of square km probably occur at an average rate of about one every 1000 years. (In Hawaii, eruption of many flows per decade would be expected.)	Probably one to several events per century caused directly by eruptions. Probably only about one event per 1000 years caused directly by eruption at "relatively inactive" volcanoes.	Probably one to a few eruptions of "small" volume every 100 years. Eruption of "large" volume may occur about once every 1000 to 5000 years. Eruption of "very large" volume, probably no more than once every 10,000 years.
Degree of risk in affected area.	To people, low. To property, high.	Moderate to high for both people and property, near erupting volcano. Risk relatively high to people because of possible sudden origin and high speeds. Risk decreases gradually down-valley and more abruptly with increasing height above valley floor.	Moderate risk to both people and property near erupting volcano; decreases gradually downwind to very low.

From *U.S. Geological Survey Prof. Paper 1240-B*, 1981

Table 3 Summary of relative degree of danger associated with different geologic and geographic factors related to volcanoes [Modified from Decker, 1978]

Degree of Danger	
Low ◄─────────────────────────────────► **High**	
Rift zones	Subduction zones
Shield volcanoes	Stratovolcanoes
Lavas	Pyroclastics
Magmas low in silica	Magmas high in silica
Magmas low in water	Magmas high in water
Short period of dormancy	Long period of dormancy
Upwind location	Downwind location
Ridge setting	Valley setting

that may have saved as many as 10,000 lives. Very few accurate predictions of when other volcanoes will erupt have been made. No simple precursor appears to be the master key to prediction and each individual volcano is almost unique.

Risk Assessment

Volcanologists have been more successful at mapping areas where volcanic hazards exist than predicting when an eruption will occur. These maps portray the relative degree of risk (e.g., high, moderate, or low) for each of the anticipated hazards (such as ash falls, lava flows, or lahars), and are based on geologic studies of the rock record of relatively recent activity (i.e., within the past 10,000 years) of modern volcanoes. The record may reveal "evolutionary stages," a periodicity associated with historical eruptions, the products of these eruptions, and the extent of the area affected. It is believed that many volcanoes tend to repeat, in a general way, their past performances, and that this information would be useful in risk assessment (Fig. 7). It is important to note that risk assessment, or evaluating the hazards, is significantly different from predicting a specific eruption. It is somewhat analogous to the difference between mapping the aerial extent of a modern floodplain and predicting the time and extent of the next flood. However, when one considers the difficulty of predicting volcanic eruptions, the mapping of volcanic risk zones can probably save more lives than can predictions.

Reducing Damage

Volcanic eruption cannot be prevented, but the damage associated with an eruption can be reduced in several ways:

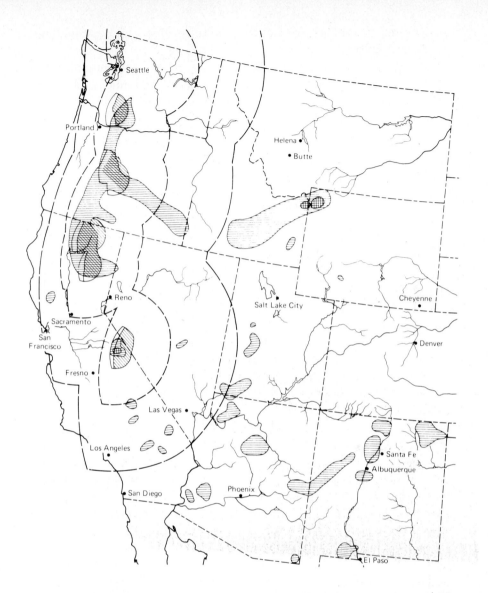

FIG. 7. Volcanic hazard zones in the western states. Shaded zones with vertical lines
are volcanic vent areas in which one or more extremely explosive and voluminous
eruptions have taken place within the last 2 million years. Zones with horizontal lines
are subject to lava flows and small volumes of ash from groups of volcanic vents called
volcanic fields. Zones with diagonal lines would receive most of the ashfall from nearby
relatively active and explosive volcanoes. The inner dashed line encloses areas subject
to 5 cm or more of ash from a large eruption and the outer dashed line encloses areas
subject to 5 cm or more of ash from a very large eruption. (U.S. Geological Survey.)

1. By passing zoning laws that control the development permitted in hazardous areas.
2. By planning for evacuation and disaster relief in communities in high-risk areas. Evacuation was highly successful during the 1973 eruption on the island of Heimaey, Iceland and during the 1980 eruption of Mount St. Helens.
3. By construction of protective structures such as diversion walls, levees, and ditches. This approach is being implemented to protect Hilo in the Hawaiian Islands.
4. By controlling lava flows by spraying water on the advancing front or bombing the flow in order to control the direction of the movement.

Benefits of Volcanism

The physical impact of volcanism can be beneficial as well as destructive. Volcanic deposits have played a major role in building the earth's lithosphere, and volcanic terrains are often highly scenic. During the early history of our planet, gaseous volcanic emissions were responsible for the evolution of the hydrosphere and the atmosphere. Both extrusive and intrusive volcanism provide major metalliferous mineral deposits (see p. 282) and soils derived from the weathering of volcanic materials are among the richest in the world. These benefits are, however, a mixed blessing because they encourage development in hazardous settings.

Young mountain ranges and areas of active volcanism contain vast resources of geothermal energy in the form of steam or hot water. The "natural" steam vents at Lardello, Italy, have been generating electricity since 1904, and most of the buildings in Reykjavik, Iceland, have been heated by steam and hot water for decades. The Geysers geothermal field in northern California is one of the world's largest. (Reading 5 reviews the technological and environmental problems associated with the development of various types of geothermal resources.)

Readings and Case Histories

The Eruption of Mount St. Helens—
Entering the Era of Real-Time Geology
Before 1980 the only major volcanic explosion within the 48 conterminous states in historic time occurred at Mt. Lassen in 1914–15. There were a few small-scale eruptions at Mt. Rainier between 1820 and

1894, and in March 1975 anomalously high levels of steam were emitted at Mt. Baker. Mount St. Helens had erupted in 1857 but had long been silent. Volcanologists, however, considered it the most hazardous volcano in the Cascades. The rock record demonstrated that it had been more active and more explosive during the past 4500 years than the others.

The Disaster Relief Act of 1974 gave the U.S. Geological Survey the responsibility for warning state and local officials of impending geologic hazards. Many years of field investigations in the Cascades combined with the monitoring of selected dormant volcanoes enabled the Survey to warn people of the imminent danger of Mount St. Helens. As a result, as many as 10,000 lives may have been saved during the May 18, 1980, eruption.

In Reading 3, Robert L. Wesson vividly describes the catastrophic eruption of Mount St. Helens and reviews the role of the earth scientist in the era of "real-time geology." Increasingly, geologists are being asked to predict when a hazardous event will take place and to forecast the impact of the event. These are severe challenges to the scientific community, for they involve a thorough understanding of basic geologic processes and research of the highest quality. But for the practitioner of real-time geology, this is only part of the challenge. The eruption of Mount St. Helens has taught us that geologists must learn to deal with government officials, the media, and the public. They will need to communicate and offer guidance if society is to benefit from predictive information about geologic hazards.

Iceland Chills a Lava Flow

The people in the vicinity of Heimaey, Iceland, received almost no warning when Kirkjufell erupted at 2 A.M. on January 23, 1973. A volcanologist had in fact pronounced the parent volcano for the fissure eruption "dead" because it had been inactive for more than 600 years. There was, however, a plan for evacuation of the island in the event of an eruption, and within six hours 5300 inhabitants were evacuated. In Reading 4 Richard Williams, Jr., and James Moore describe the single most ambitious program to date to control a lava flow and lessen the damages.

Geothermal Energy and
Our Environment

In Reading 5, the Department of Energy reviews the technological and environmental problems associated with the development of three broad types of geothermal resources: hydrothermal reservoirs, geopressured zones, and hot dry rock. Although geothermal energy may provide as much as 5 percent of our energy needs by the year 2000, there is a wide range of environmental problems to be identified and addressed if we are to satisfy both environmental and energy goals.

3   The Eruption of Mount St. Helens—Entering the Era of Real-Time Geology

Robert L. Wesson

Time generally means something different to geologists than it does to others. The geologists' common view of time is like looking through the wrong end of a pair of binoculars. All the myriad events that mark the passing of our daily lives—the passage of the seasons, rains and thaws, floods and droughts, growth and death—are tightly compressed into a murky record of mountain building and erosion, of the deposition of sediments and the weathering of exposed rocks, of evolution and extinction. The geologic history which has been pieced together by the investigations of geologists extends back 4.5 billion years to the origin of the Earth. This history is characterized by spatial changes on a grand scale—the movement of continents, the opening and filling of ocean basins. Generally, however, these movements take place so slowly as to be virtually imperceptible to man—without the aid of instruments developed only recently. The general uniformity and continuity of geologic pro-

Reprinted from U.S. Geological Survey Yearbook, Fiscal Year 1980, U.S. Government Printing Office.

cesses through this 4.5 billion years are central to geologists' approach to understanding the Earth, its resources, and its hazards. However, not all geologic processes occur at continuous imperceptibly slow rates. Indeed major geologic changes can be accomplished in days or even minutes. No more powerful recent reminder of the Earth's capacity for sudden and catastrophic change can be cited than the explosive and disasterous eruption of Mount St. Helens on May 18, 1980.

U.S. Geological Survey scientists had their collective eye on Mount St. Helens over the years and, as a consequence, were able to advise local, State, and Federal officials about the threats from the volcano such that many hundreds, perhaps thousands, of lives were saved. This experience of warning vividly demonstrates the geologist's increasing capability to switch his geologic vision from the long view, thousands or millions of years, to the short view, days and weeks; from a history of Mount St. Helens over the last 40,000 years and especially the last 4,500 years to an estimate of what the mountain will do today—and tomorrow.

In the past, geologists—and the Geological Survey—have used their understanding of geologic history, and the processes which shape it, principally to answer "where" and "what" questions—questions about the location and physical description of geologic deposits or features, such as, where are mineral deposits or petroleum resources located? Increasingly, geologists are being asked "when" questions: When will the next major earthquake strike California? When will we run out of ground water in the high plains of the west? When will the next Cascade Range volcano erupt explosively? The new tenor of these questions presents serious challenges to the earth sciences, and to the Geological Survey in particular. The scientific challenges are severe, requiring answers to questions, many of which, so far, at least, are beyond our grasp: but no less difficult are the challenges to the Geological Survey, and to society, of how to use predictive information about geologic hazards.

What are these challenges? and how is the Geological Survey meeting them? The scientific challenges are the challenges of understanding; the institutional and societal challenges are of communication, education, and the understanding of risk.

Research and the Long View: Basis for Prediction

There can be no scientifically credible warning without a thorough understanding of the geologic history and of the geologic processes at work. The path leading to this understanding is not a straight one nor one which can be cut short. It is, in all likelihood, a circuitous and torturous path which must be followed step by step, flashes of insight followed by deadends and backtracking, each step, whether forward or backward, adding some small bit to the total understanding. The investment of research does not always pay off immediately, nor is the final payoff always evident while the work is being done. With the eruption of Mount St. Helens, three long-term research efforts paid off handsomely. Only one of these was aimed directly at the volcanic hazards posed by Mount St. Helens. Perhaps even that was a long shot. What were these areas of research? and how did they come to pay off?

Hazard Studies of Mount St. Helens

Geologic studies of the volcanoes of the Cascade Range began in 1841 with Charles Wilkes' U.S. Exploring Expedition. However, the documentation of the hazards posed by them, and particularly by Mount St. Helens, was done by the career-long efforts of Geological Survey geologists Dwight R. Crandell, Donal R. Mullineaux, and their collaborators. The investigations of Crandell and Mullineaux included the construction of a detailed history of the activity of Mount St. Helens over the last 4,500 years. During this time, the volcano experienced several apparently dormant intervals of about 100 to 150 years and some dormant intervals as long as 400 to 500 years. Throughout this 4,500-year period, however, the dormant intervals were repeatedly interrupted with periods of activity which included numerous explosions, eruptions of ash, hot avalanches of rock debris, the building of domes, emission of lava and the flows of volcanic debris, and mud and ice off the glacier-mantled slopes of the volcano. This history led Crandell and Mullineaux to conclude in their hazards report, published in 1978, that, "Mount St. Helens has been more active and more explosive during the last 4,500 years than any other volcano in the conterminous United States . . . Mount St. Helens has had a long history of spasmodic explosive activity, and we believe it to be an especially dangerous volcano because of its past behavior and the relatively high frequency of its eruptions during the last 4,500 years. In the future, Mount St. Helens will probably erupt violently and intermittently just as it has in the recent geologic past, and these future eruptions will affect human life and health, property, agriculture, and general economic welfare over a broad area."

The record of Mount St. Helens history was deciphered by a combination of classical geologic mapping and stratigraphic studies with modern techniques. These studies included geochemical analysis to trace deposits from prehistoric falls of ash by their "mineralogical fingerprints" and to determine the age of deposits by radiocarbon analyses of pieces of wood found within them. Soils beneath the ferns and Douglas fir of the forests contain deposits of volcanic ash and hold the clues to determine the pattern of past eruptions of the volcano. Deposits of rock debris from prehistoric eruptions were examined in the area around the volcano to determine their origin, extent, and age. By mapping the distribution of these deposits and analyzing their mineral content and grain size, it is possible to trail them

back to their source: the volcano from which they were spewn.

But how, from the record of past eruptions, can future hazards be estimated? A critical assumption was made that eruptions in the future will be of the same general type and have the same general effects as those in the past. Using this kind of approach, Crandell and Mullineaux described a set of "hazard zones," first, on the flanks of the volcano and along valleys leading from it, which would be endangered by lava flows, mudflows, pyroclastic flows (avalanches of hot rock debris ash), and floods and second, in the nearby and downwind areas which would be endangered by the ash fall. These zones were further differentiated by the frequency with which parts of the zone were affected by past eruptions. The potential for flooding caused by flows of mud into the reservoirs behind power dams on the Lewis River, which drains the south and east sides of the mountain, was viewed as a particularly serious problem. Finally, Crandell and Mullineaux concluded their analysis of potential hazards by indicating what hints might be given by the volcano if it were preparing for an eruption, such as the occurrence of numerous small earthquakes or swelling of the volcano, both caused by the underground movement of molten rock, and by listing actions that could reduce possible effects on people and property should warning signs appear or an eruption begin.

The published work of Crandell and Mullineaux was circulated in December 1978 to Federal, State, and local officials and was supplemented by a meeting with key officials and with personal discussions. But, understandably, concern about an event in a sparsely settled region—an event which might have localized impact and be of passing interest only every 100 years or so and have impacts distant from the volcano itself only every 500 to 3,000 years—was accorded relatively little attention. The probability of a catastrophic event was low compared to frequent floods, not to mention the day to day crises that affect us all.

Studies at the Hawaiian Volcano Observatory

Fountains, cascades, and rivers of glowing lava are images of the active volcanoes of Kilauea and Mauna Loa on the Island of Hawaii. Since 1912, Geological Survey scientists have studied, and been trained, in this unique natural laboratory of volcanic processes. The very fluid lava of Hawaii, as contrasted with the thick viscous lava of most Cascade Range volcanoes, is revealed by the smooth gently sloping forms of the Hawaiian volcanoes as compared with the steep, sometimes rugged character of the Cascade Range volcanoes. The difference in the fluidity of lava also accounts for the difference in the behavior of the volcanoes. The investigations at the Hawaiian Volcano Observatory have demonstrated that eruptions in Hawaii are preceded by systematic swelling of the volcano, swarms of small earthquakes, and a periodic trembling detected on seismographs called "harmonic tremor." Eruptions of Hawaiian-type volcanoes are rarely explosive but, nevertheless, have potentially serious impacts for residents on the island. Repeated successful warnings of impending eruptions based on swelling of the volcano and swarms of earthquakes gave a generation of Geological Survey scientists experience, not only with the active science of volcanology, albeit in quite a different geologic setting from Mount St. Helens, but also experience in dealing with officials, the press, and the public before, during, and after an eruption.

Earthquake and Geothermal Studies in the Cascade Range

The active geologic process in the Cascades—of which Mount St. Helens is a part—has other manifestations: earthquakes and geothermal energy. Since 1973, the scientists at the University of Washington, supported by and in cooperation with the Geological Survey, have studied the occurrence of earthquakes in the Puget Sound-Cascade region of Washington State. The core of these studies has been the establishment of a tool for basic research, a seismographic network. Studies of the potential for geothermal power associated with the active volcanic regions of the Cascade Range led to an expansion of the network, with the motive of using the occurrence of earthquakes—and studies of the propagation of seismic waves from them—for locating "hot" regions of the Earth's crust and of developing an understanding of the origin of these hot regions. As it turned out, this seismic network and the joint University of Washington-Geological Survey project were to give the first clues that Mount St. Helens was about to awaken from her 123-year sleep.

Earth Scientists Refocus from the Long View to the Short View

On March 20, 1980, the vicinity of Mount St. Helens was shaken by a magnitude-4 earthquake. The location of the earthquake, as determined by the seismograph network which included one station near the volcano, was at shallow depth immediately northwest of the summit. However, this earthquake was only the first of an intense sequence or swarm of earthquakes. Both the magnitude and number of these earthquakes were unusual for the Pacific Northwest, and their location directly beneath the volcano was immediately recognized as the possible symptoms of an impending eruption. What was the probability of an eruption? and, if the mountain did become active, what precisely would happen? No one knew. The possible scenarios for an eruption, or noneruption, were many, but no one knew how to assess the relative probabilities of the possible scenarios. Geological Survey and University of Washington scientists, working together, began to install portable seismograph equipment and to expand the permanent seismograph network, an exercise which has become commonplace following almost every significant earthquake or earthquake sequence in the United States. One of the principal aims of this effort was to try to determine whether this earthquake sequence was indeed symptomatic of molten rock beneath the volcano. As the days passed, the number of earthquakes increased, and the comparison with preeruptive sequences at other volcanoes became more compelling. Snow avalanches, triggered by the earthquakes, led to the closing of the upper slopes of the volcano by the U.S. Forest Service. By March 25, the frequency of magnitude-4 earthquakes reached a level of as many as eight per hour. Sightseers flocked to the area hoping for a glimpse of some activity. On March 26, the immediate vicinity of the volcano was closed on the basis of the potential hazards from an eruption.

On March 27, the Geological Survey issued a cautiously worded but formal "Hazards Watch," indicating that Survey scientists did not "have adequate hard information to determine whether an immediate volcanic eruption will or will not occur. Furthermore, it is not possible at this time to indicate which of the possible geologic effects of the eruption might take place or whether these effects might be experienced only very locally on the volcano or over a wider area." However, the Hazards Watch summarized potential dangers as indicated in the work of Crandell and Mullineaux.

That day the volcano began to erupt steam and ash. Following the actions recommended in the Survey publication, U.S. Forest Service and State and local and emergency officials evacuated people from the immediate area.

It was a chaotic time. The event captured the imagination of the Pacific Northwest and the Nation. No volcano had been active in the contiguous United States since 1921. The pressure from the news media was intense. Demands for interviews were so great that scientists could not comply with all and still perform their work. Sightseers jammed the roads leading from the volcano and blocked possible evacuation routes. In addition, many sightseers evaded road blocks hoping to get a closer look.

Analyses of ash indicated that so far only old rock fragments were involved; no new molten rock had been ejected. But, on April 1 and again on April 2, University of Washington and Geological Survey scientists recognized the first sign of harmonic tremor, which normally is associated with the movement of molten rock. On April 3, the Geological Survey updated its Hazards Watch, indicating that "the harmonic tremors . . . are the best indications so far that magma (molten rock) is involved and moving underground. We still cannot predict, however, whether this apparent subsurface movement of magma will break through the surface to produce an eruption of molten material." The Governor of Washington called out the National Guard to help maintain the closure of an area around the volcano to all but property owners and scientists. These moves were unpopular and contested, particularly by loggers excluded from the forest.

Following the first phase of the steam and ash eruption of March 27, a major system of cracks across the summit crater was observed, suggesting the spreading of the north flank of the mountain. Throughout the first 3 weeks of April, evidence of the bulging of the northern slope of the mountain accumulated; visual observations of the cracking and slumping of the glaciers and bedrock were substantiated by photogrammetric surveys and geodetic measurements. By April 25, instruments were in place to monitor the expansion of the bulge on a daily basis. The measurements carried on through early morning on May 18 indicated steady northward expansion of the bulge at rates of 0.5 to 2 meters per day. As the bulge grew, so did concern about the potential hazards it posed. On April 30, the Geological Survey again updated its Hazards Watch, indicating that the apparently unstable mass, if triggered by an eruption or earthquake, could lead to a massive avalanche, which in turn could lead to mudflows and floods in the valleys draining the north side of the mountain. "USGS scientists cannot pinpoint the exact cause of the bulge with the available data," the press release said, "but suspect that the bulging may reflect a combination of swelling from the upward movement of viscous (sticky) magma and gravitationally induced downward creep. . . ."

Did the U.S. Geological Survey tell—or indeed recognize—the whole story? In hindsight, it seems possible to say that the appearance and expansion of the bulge added substantial weight to the hypothesis that significant molten material was accumulating in the volcano and, consequently, that the odds of a significant eruption were substantially increased. However, the scientists could not reach a consensus. Although many individual geologists and geophysicists made this judgment, even aloud at the volcano, the Survey, through its institutional procedures of hazard warning, was not moved to reach this judgment. Why? Scientific uncertainty, institutional fear of being wrong, substitution of the more visible threat of a massive avalanche for the less visible threat of an eruption? Herein quantifiable uncertainty? Where does an established scientific interpretation stop and hunch begin?

Reacting to the advice from the

Geological Survey, Governor Ray established two "hazard zones" around the volcano. A "red zone" included the area within a radius of approximately five miles from the mountain. Admission to this zone, which included the vacation community of Spirit Lake, was prohibited to all except law enforcement officials and scientists. A "blue zone," which extended 5 miles beyond the red zone in some places, was under the same prohibitions as the red zone except that logging operations and visits by property owners were permitted, but no overnight stays were permitted. The zones were unpopular. They were violated by journalists and scientists in helicopters and by curiosity seekers. Loggers and property owners, excluded from their homes, were particularly critical. The bulge, ominous as it was to geologists, did not seem to bother old timers of the region, who seemed unable to grasp the danger from a potentially explosive volcano, or even some of the scientists who worked in its shadow.

As scientists monitored the bulge and considered its implications, and as picnickers gathered at road blocks waiting for a view of a plume of steam or ash, owners of private property began to demand access to their cabins and houses to gather their possessions. Finally on May 17, property owners were allowed into the red zone by signing releases and agreements that they would be out before nightfall.

The Picnic Becomes a Nightmare

On the morning of May 18, the picnic became a nightmare. Mount St. Helens erupted catastrophically. The north side of the mountain collapsed in an avalanche, almost certainly triggered by a magnitude-5 earthquake, releasing the pent-up power of the magma with its gas under high pressure. The suddenly unconfined molten rock and gas exploded like a bottle of warm champagne suddenly uncorked. The resulting blast devastated everything in a sector extending as far as 16 miles north of the volcano and nearly 20 miles wide. A massive debris avalanche filled the valley of the North Fork Toutle River for a distance of about 17 miles downstream. Mudflows continued on down the Toutle carrying extremely large loads of sediment, logs, and debris on to the Cowlitz and Columbia Rivers. The deposited sediments blocked the shipping channel in the Columbia and dangerously reduced the ability of the Cowlitz River to carry water within its banks without flooding. Volcanic ash streamed higher than 12 miles into the atmosphere by the force of the explosion and was carried eastward by the wind to deposit ash in a plume across eastern Washington and into Idaho and beyond. What had formerly been a symmetrical cone—the "Mount Fuji of the United States"—was now a truncated cone, marred by a gaping north-facing amphitheater. The questions of "when" and "how big" had finally been answered. Geological Survey scientists, prior to the eruption, in describing the range of possibilities, had sometimes, in passing, mentioned the explosion of Mount Mazama, which formed the famous Crater Lake in Oregon, as an example of an eruption well beyond that which could reasonably be expected at Mount St. Helens. Indeed, the 6-mile-

Mount St. Helens, May 18, 1980. A massive explosive eruption sends a column of ash to an altitude of more than 19 km (63,360 ft.). (U.S. Geological Survey.)

wide Crater Lake is a vivid example of raw destructive power of the explosive Cascade volcanoes. The eruption of Mount St. Helens was small by comparison, but closer than the public had seemed willing to believe was likely.

What was the most "expectable" size of the eruption and how did the May 18 eruption compare? So far, our understanding of the history of Cascade volcanoes is insufficient to answer the question of the *most* expectable size, either on the basis of statistics of past eruptions or on the basis of the kinds of premonitory phenomena observed; that is, the earthquakes, the bulge, the steam eruptions. But Crandell and Mullineaux did make some estimates about the range of possible eruptions and their relative probabilities based on their 4,500-year history of Mount St. Helens. They estimated that eruptions depositing an inch or less of ash within 50 miles or so of the volcano would have a frequency of 1 per 100 years. They estimated that eruptions depositing a few inches of ash at distances of 100 miles and more, as did the May 18 eruption, would have a frequency of 1 per 2,000 to 3,000 years or less. So the May 18 eruption was not a "common" event, but closer to the "worst case." Previous eruptions, however, have erupted much larger volumes of ash.

The course of potentially catastrophic geologic processes such as eruptions and earthquakes may depend on many random factors. Indeed, the most surprising aspect of the eruption was the magnitude of the disastrous lateral blast, which devastated the landscape north of the mountain. Although some evidence for much smaller prehistoric lateral blasts was available, geologists had expected the main force of an eruption to be directed vertically upward. Consequently, it may be very difficult, if not impossible, to know in advance from the symptoms of an impending event whether the event will be a "common one" or a "worst case." We would like to think we can solve this problem, perhaps by determining the volume of molten material available in a volcano or the size of the area over which anomalous phenomena are observed before an earthquake, but there is certainly no guarantee that these ideas will be right. Indeed, prior to May 18, some geologists began to feel that the molten rock, indicated inside the volcano by the bulge and the earthquakes, might solidify inside the volcano in a system of what geologists call "dikes" and, aside from the threat of avalanches and mudflows, be of no particular threat at all. Will the observed symptoms of a geologic process lead to a "worst case," a "common" event or a "nonevent?" This is another key dilemma facing the real-time geologist.

On May 18 and the days after, impacts of the eruption that had not been anticipated by the public were keenly felt by residents of the Pacific Northwest. Although Governor Ray, at the end of April, had declared the entire State an emergency area because of the possibility of widespread ash fall, the volume of the ash which settled like an ominous dark snow over eastern Washington and Idaho created problems of unexpected proportion and nature. Who had prepared for the problems of operating motor vehicles and machinery in an atmosphere of constant dust stirred up by wind or the movement of vehicles, an atmosphere in which the particles of ash are far more abrasive than common dust? And who in Portland, Oregon, where the spectacular plume of ash on May 18 had provided a thrilling display, expected that the mud and rocky debris carried down the Toutle and Cowlitz Rivers would block the shipping channel in the Columbia River, temporarily closing the important port facilities in Portland? Or who anticipated the added impact of the thousands of logs stacked at loading yards along the Toutle Valley which were picked up by the mudflows and swept away the bridges along the river as if they were made of paper? Because many of the phenomena dealt with by the real-time geologist occur so infrequently—from the perspective of man—the impacts are commonly unexpected. Preparations require preparing for the unexpected.

As an example, the Cowlitz River channel was found to be clogged with sediment deposits following the floods of May 18—average flows would cause flooding. Meanwhile, snowmelt runoff from the higher elevations of the Cascades was rapidly filling upstream reservoirs. Any significant runoff either from accelerated snowmelt or rainfall would exceed reservoir capacity and result in flooding along the clogged reach of the lower river. Geological Survey hydrologists created a computer model of the river system using new channel surveys, parts of which were provided by the Corps of Engineers and private engineering firms. Output of the model allowed immediate assessment of inundation limits for any Cowlitz River flow rates and provided a realistic basis for warning and evacuating flood-threatened residents. Hundreds, perhaps thousands, of lives were saved by the establishment of the hazard zones. Some of those who died were in the closed areas in defiance of the closure. However, people are skeptical about statements contrary to their common experience, particularly when their livelihood is at stake. Fortunately, the eruption occurred on Sunday when logging operations permitted in the "blue zone," much of which was devastated, were shut down. Real-time geologists must learn that they, like Dr. Stockmann of Henrik Ibsen's *An Enemy of the People*, will not always be either popular or even believed by segments of society.

Questions in the Aftermath

What would the volcano do next? Was the debris flow stable? Would the waters dammed behind it be released in further floods? Would the now clogged Cowlitz River be able to contain, without flooding, more water should the reservoir upstream become full? These were the questions asked immediately after May 18. The prospect of the volcano continuing to be active for 10 to 20 years, as it had in the nineteenth century, forms a background against which a whole new set of questions began to take shape. When could people go back in to salvage and rebuild? What should be rebuilt and what should be relocated? Would the volcano give a short-term warning of further eruptions as it seemed not to have done on May 18? Could the geologists predict subsequent eruptions in time for people to be evacuated safely from the hazard zones? What would be the effects of the ash on crops and on the health of animals and humans in the affected regions? Would the easily erodible mass of ash and debris in the tributaries of the Cowlitz be mobilized and eroded by winter rains, carrying yet more debris into the Cowlitz and Columbia?

As the months passed, some questions were answered, others deferred. Measurements were made in the clogged channels of the Cowlitz River, and computer models calculated to determine the amount of water which could be released safely from upstream reservoirs and to determine the inundation limits for potential floods. How long will the activity last? Will the volcano rebuild itself? The volcano, as expected from experience elsewhere, continued its activity, with eruptions on May 25, June 12, July 22 and 28, August 7 and 15, and October 16–17. Volcanic domes were erupted into the crater, then exploded during eruptions of ash. Varying patterns of seismic activity have emerged which seem, so far, diagnostic of these continuing eruptions, enabling warning and evacuation of workers in endangered zones. Other possible precursors, such as gas emissions, seem promising. The debate about the efficacy of attempts to contain sediment in the Toutle Valley and to preserve the gains made by dredging in the Cowlitz and Columbia Rivers continues. Flash floods from new mudflows or the breach of dammed lakes were recognized as a threat. An elaborate system of stream and lake gages equipped with real-time data transmission systems has been installed to monitor the movement of water and sediment downstream. It is hoped that sufficient warning of floods can be given to communities along the Cowlitz.

Prior to the eruption, Geological Survey scientists worked closely with the U.S. Forest Service and, as the situation developed, with State and local officials. After the eruption, the number of individuals and entities needing contact with the Geological Survey scientists, elected officials, and their staffs grew rapidly—the Federal Emergency Management Agency, the Corps of Engineers, the Environmental Protection Agency, and many more. The real-time geologist works in a fish bowl. Important decisions must be made by a multitude of persons based upon individual interpretation. Commonly, the decisions must be made immediately, and, even more commonly, questioners must be given help to rephrase their questions so that meaningful answers are possible. The real-time geologist requires a capability for translation and public relations so that the results of his scientific investigations and interpretations can be used by people needing the answers.

Sharpening the Near Vision: Lessons and Challenges for the Real-Time Geologist

The eruption of Mount St. Helens bears many lessons for the Geological Survey—and for practitioners of real-time geology. Certainly some lessons are yet to be made clear. Implicit in each of these lessons is a challenge for the future. The lessons and challenges are not limited to the prediction of volcanic eruptions but have wide applicability to anyone who would try to forecast the future behavior of the Earth and the processes which shape it, the prediction of earthquakes, glacier surges or retreats, ground water exhaustion or contamination, or future climatic changes.

The capability for looking into the future, even in an uncertain probabilistic way, will come from basic research into the nature of the geologic processes which may present future hazards; research to determine the history of past events, their nature, their frequency, and their effects; and research into the mechanics, the fundamental physics, and chemistry of the processes. Only through improved understanding can we improve our ability to take the Earth's pulse and estimate its future activity, benign or deadly. It will not be easy to know exactly what research to do, and the value of the research may not be obvious as it is being done. But the experience of Mount St. Helens emphasizes the value of combining the classical methods of investigation with the modern, of combining historical investigations with the process oriented, and of combining the geological, geophysical, and geochemical approaches. The real-time geologist must utilize the best of the classical methods along with the best of the most modern.

For a long time to come, problems of real-time geology will be at the leading edge of understanding in earth science. From a scientific point of view, this will make them fascinating. From the point of view of society, this will make these problems very difficult because uncertain scientific judgments will require flexible public policies, policies which may not always be as cost effective as they might be if their scientific basis were certain. The possibility must be faced squarely that a predicted event may fail to occur. The

science of real-time geology is new. Little, if any, public policy is in place to utilize its results. Policy is exceedingly difficult to develop in advance, in the abstract. The challenge to the real-time geologist is that the public policy will be developed as we go along. Further, the real-time geologist will be asked to help make the policy, an area where geologists have little experience or expertise.

Many problems in real-time geology involve phenomena which pose substantial hazards to life and property. Consequently, the public pays much closer attention to studies in real-time geology than to classical investigations. This poses many problems for the earth scientist unaccustomed to doing his research under the bright lights of the television cameras. The pressures of the media can easily distort the perspectives of the scientists. In other cultures where less value is placed an openness, public attention need not be a problem, particularly if the impending phenomenon has few if any symptoms visible to the naked eye. But in the United States, the public is perceived as having "a right to know," even to know many things which at that moment may be unknowable or, at least, very uncertain. Emergency preparedness requires leadership. Preparations against an unseen hazard, such as earthquakes, rarely spring from enlightened self interest. So public attention to the scientific findings, uncertain as they may be, is required to motivate public action. At the same time, the public attention and discussion can magnify sometimes subtle differences in judgment among scientists, reducing credibility in the public eye. Intense public attention makes the job of the real-time geologist no easier.

The natural instinct of a scientist is to follow his intuition, wherever the scent of discovery may lead him. Commonly, this instinct creates a tension with the need to plan. Contingency plans, however, must be in place to respond quickly and effectively to the threat or occurrence of a potentially hazardous geologic event.

The list of scientific problems ahead for real-time geologists in the Geological Survey is immensely challenging. What does the future hold for the other potentially explosive volcanoes of the Cascade Range and Alaska? When and where will the next major earthquake strike California? When will the Ogallala aquifer supplying water to the High Plains be exhausted? How can systems be designed and monitored to assure that hazardous wastes do not reach ground water supplies? What is the future of our climate?

The eruption of Mount St. Helens is only the beginning. We have entered the era of real-time geology. The challenges before the U.S. Geological Survey will require the highest quality of geologic research. However, these challenges will also require substantial growth in our ability to communicate our results and to assist in their interpretation and implementation.

4   Iceland Chills a Lava Flow

Richard S. Williams, Jr., and James G. Moore

One of the most destructive volcanic eruptions in the history of Iceland seems to be waning, but only after heavily damaging the country's main fishing port, Vestmannaeyjar, on the island of Heimaey. The eruption is

Williams, R. S., Jr., and Moore, J. G., 1973, Iceland Chills a Lava Flow, *Geotimes*, v. 18, pp. 14–17. Reprinted by permission of the authors and the American Geological Institute. The authors are on the staff of the U. S. Geological Survey. Williams spent Feb. 7, 1973, on Heimaey, and Moore was there from May 3 to 5. This report is based on their observations, information from the Icelandic Ministry for Foreign Affairs, Icelandic scientists' reports through the Smithsonian Institution's Center for Short-Lived Phenomena, and published scientific reports. Icelandic scientists and officials provided logistical support and discussions of scientific and engineering aspects.

also notable as leading to a major attempt to control a lava flow by chilling it with sea water.

The effusive eruption is the fourth in Iceland in little more than a decade. (Earlier events: Askja, Oct. 26 to about Dec. 6, 1961; Surtsey, Nov. 14, 1963, to May 5, 1967; and Hekla, May 5 to July 5, 1970.) It is the second major eruption—the other being Surtsey—definitely known to have occurred in the Vestmann Islands archipelago since settlement of Iceland (about A.D. 874). Thorarinsson has documented at least 13 offshore eruptions (14 including Heimaey) and about 100 onshore since settlement. 10 of the 13 previous offshore eruptions occurred off the Reykjanes Peninsula, along Reykjanes

Ridge, the submarine extension of the peninsula.

This ridge lies along a parallel fracture system about 175 km west of the northeast-trending Vestmann Islands archipelago. The Vestmann Islands group, on the southwestern continuation of the eastern volcanic zone, follow the same structural trend as the fissures, grabens, and crater rows on the mainland in the eastern neovolcanic zone, a zone of very productive and historically active volcanoes (Fig. 1.)

The Heimaey eruption began about 2 A.M. Jan. 23, 1973, on the eastern side of the island and about 1 km from the center of town. A fissure about 1.5 km long lengthened rapidly to nearly 2 km, traversing the island from one shore to another. A continuous curtain of fire predominated all along the fissure in the early phases, but fountaining soon retracted to a small area about 0.8 km northeast of Helgafell. Within 2 days a cinder-spatter cone grew to more than 100 m above sea level and was informally named "Kirkjufell" after a farmstead, Kirkjubaer, which it had obliterated. During the early phase, the outpouring of tephra and lava averaged about 100 cubic meters a second. A few days after the eruption began, the combination of a high rate of tephra production and strong easterly winds resulted in a major fall of tephra in Vestmannaey-

FIG. 1. Lower left, the Vestmann Islands (after Thorarinsson and Icelandic Surveying Department). Far right, the Island of Heimaey, with the original shoreline shown as a solid line (Geodetic Institute, Copenhagen), the eruptive fissure of Jan. 23, a dashed line, and limits of the lava flows of Jan. 28 and April 30 (after Thorarinsson and others, 1973, and from maps by the Science Institute, University of Iceland).

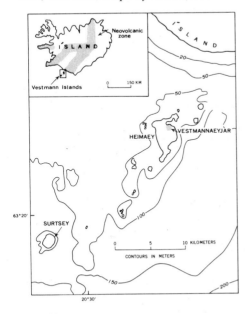
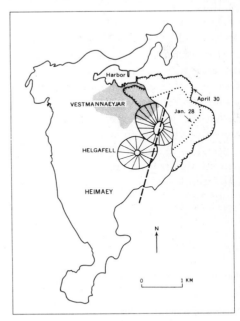

jar, burying houses in the eastern part of the town. By early February the tephra fall had slackened markedly, but lava flows continued unabated, making inroads into the edge of town and threatening to fill in the harbor of Iceland's biggest fishing port.

The volcano was continuously monitored from the air and on the ground by Icelandic geologists and geophysicists; they acquired satellite thermography and imagery of the eruption from the NOAA-2 satellite and the ERTS-1 satellite under an experiment by the U.S. Geological Survey (EROS Program) and the University of Iceland (Science Institute).

By the end of February, the spatter-cinder cone reached more than 200 m above sea level. The central crater also fed a massive blocky lava flow that moved north, northeast, and east. By early May, the flow was 10–20 m high at its front, and averaged more than 40 m thick—in places 100 m. Its upper surface is littered with scoria and volcanic bombs, as well as large masses of the main cone, which broke off and were carried along with the flow. Some of these masses of welded scoria are 200 m across and stand 20 m above the general lava surface. They were rafted as much as 1 km from their original sites. Photogrammetric measurements (from aerial photographs) and geodetic measurements, made from the end of March to the end of April, indicate that it moved as a unit about 1 km long and 1 km wide at 3–8 m per day.

In addition to the advance of the flow to the north and east, large slump blocks broke loose from the cone Feb. 19 and 20 and moved west toward the southeast part of the town. In late March, a second large lava flow moved northwest on the west side of the main flow and covered many houses, a large fish-processing plant, and the power plant in the northeast part of town (Fig. 2).

By Feb. 8, the outpouring of lava had dropped from an initial estimate of 100 cubic meters per second to 60; by mid-March, to 10; and by mid-April, to about 5. By early June, activity had virtually ceased.

Thermocouple measurements of the lava indicated a temperature of 1,030°–1,055°C during the first week of the eruption, increasing later (Feb. 17) to as much as 1,080°C.

Samples of volcanic gas of widely varying composition were collected at several places, indicating that gas fractionation operates effectively over short distances. Gas collected at sea along the submerged part of the active eruptive fissure is dominantly CO_2; gas collected at sea, bubbling up from cooling submerged lava flows, is about 70% H_2.

Poisonous gas accumulated in low areas in eastern Vestmannaeyjar and concentrated in houses partly buried by ash and scoria. This gas is 98% CO_2 but also contains CO and methane. One person died as a result of breathing the gas in a building, and several have been overcome in houses or on the street.

Within 6 hours after the eruption began, nearly all the 5,300 residents of Heimaey had been safely evacuated to the mainland, partly as a result of the fishing fleet's being in port, but also, and more important, as a result of a foresighted evacuation plan by Almannavarnir Ríkísins, Iceland's State Civil Defense Organization.

FIG. 2. Panoramic view across the fishing port of Vestmannaeyjar. The outskirts (foreground) were burned and buried under some of the estimated 300 million cubic yards of volcanic debris. (Photo courtesy U.S. Geological Survey.)

Except for the damage to houses and farmsteads close to the eruptive rift, the first severe property loss was caused by the heavy tephra fall a few days after the onset of the eruption. Many houses, public buildings and businesses were buried by tephra, set afire by lava bombs, or overridden by the advancing front of lava flows. Many structures collapsed from the weight of the tephra, but the shoveling of accumulated tephra from roofs saved dozens of others.

By early February, the lava had begun filling the harbor, a situation that threatened the future use of Vestmannaeyjar as Iceland's prime fishing port. The harbor on Heimaey is the best along the south coast of Iceland and is in the midst of some of the richest fishing grounds in the North Atlantic. Approximately 20% of Iceland's fish catch is landed and processed on Heimaey, which has only 2.5% of Iceland's population. As fish products represent nearly 80% of Iceland's export, the loss of the harbor would have a severe impact on the economy. Also, the flow into the harbor severed a 30,000-volt submarine power cable and broke one of the two fresh-water pipelines from the mainland.

In late March, a new surge of lava into the eastern edge of Vestmannaeyjar destroyed a large fish-freezing plant (and damaged 2 others), the local power-generating facility, and 60 more houses. By early May, some 300 buildings had been engulfed by lava flows or gutted by fire. Sixty to 70 more houses have been buried by tephra. It is estimated that half the structures on the island can be salvaged.

Short- and long-term costs will total tens of millions of dollars, a stupendous amount compared with Iceland's gross national product in 1971 of $500 million. The location of housing and other services for 5,300 persons would be equivalent in impact to finding homes for 5.3 million Americans.

Icelanders have always had to contend with volcanic eruptions, jökulhlaups ("glacier bursts"), and periods of deteriorating climate (usually caused by sea ice off the north and east coasts).

Of great interest to Icelanders and earth scientists, then, was the decision by government officials, after advice from Icelandic geologists and geophysicists, to fight the lava flows. Drawing on small experiments initiated by Thorbjörn Sigurgeirsson on Surtsey in 1967, where lava-wetting was used in an attempt to stop the flow of lava from the north slope of Surtur I toward the scientific observatory, Sigurgeirsson, with the support of other scientists, recommended that

such a technique, on a much grander scale, be used on Heimaey.

As lava flows threatened to close the harbor and encroach on the heart of Vestmannaeyjar, a determined effort was begun to restrict the flow of lava to the west and north by building barriers and by pumping large amounts of sea water to cool the flows. This effort became the most ambitious program ever attempted by man to control volcanic activity and to minimize the damage caused by a volcanic eruption. Consequently the experiment is of great importance to other communities exposed to volcanic hazards.

An experiment using city water in late February indicated that spraying water on the flow did indeed slow its advance, causing the flow front to thicken and solidify. Hence, in early March, a pump ship which could deliver much more water was taken into the harbor. Finally, in early April, a group of fuel pumps were purchased largely from the United States to deliver water not only to the flow front but to many sites on the surface of the flow as well.

In late April, 47 pumps mounted on barges in the harbor were delivering a total of 1 cubic meter per second of sea water to various parts of the flow at an average height of 40m. This could lower by 200°C a volume of lava several times that of the water (Fig. 3).

The water was pumped directly on the flow front at sea level and was also pumped through 3 main 12-inch plastic pipes as much as 1 km south to the surface of the flow. Each 12-inch pipe fed several 5-inch plastic pipes, each of which could deliver as many as 200 liters per second.

The most difficult aspect of the

FIG. 3. Rescue workers pump nearly a million gallons of water per hour over tongues of advancing molten lava in a successful effort to prevent the lava from blocking Iceland's major fishing port. (Photo courtesy U.S. Geological Survey.)

cooling program was to deliver large volumes of sea water to the surface of the flow some distance back from the flow front. The water effectively increases viscosity, producing internal dikes or ribs within the flow, causing the flow to thicken and ride up over itself.

First, the margin and surface of the flow is cooled with a battery of fire hoses fed from a 5-inch pipe. Then a bulldozer track is made up onto the surface of the slowly moving (up to 1 m per hour) flow. The water produces large volumes of steam, which reduces visibility and makes roadbuilding difficult. Then the larger plastic pipes are snaked up on the flow; they will not melt as long as water is flowing in them. Small holes in the wall of the pipes help cool particularly hot spots.

In each area 50–200 liters per second of water is pumped onto the lava flow, where it generally has little effect for about a day. Then the flow begins to slow in that area. Water is poured onto each point for about 2 weeks until steaming stops near the point of discharge, because by then much of the flow in a small area is cooled below 100°C. Water is delivered about 50 m back from the edge of the flow margins and front, so that a thick wall of cool rubbly lava is created at the margin, allowing the flow to thicken behind it.

The cooling of the flow margin is used in conjunction with bulldozed diversion barriers of scoria adjacent to the flow margin. The cooled flow tends to pile up against the barriers rather than rush under them as it would if the flow were more fluid.

This water-cooling program distinctly changes the surface of the main lava flow. Before watering, the surface is blocky and covered with partly welded scoria and volcanic bombs, the whole having a distinct reddish oxidized color. The general surface has a local relief of 1 m or less—the size of many of the blocks. However, large masses of welded scoria broken from the main cone stand 10–20 m above the flow surface.

After watering, the general flow surface is much more jagged, having a local relief of about 2 m. Cooling has apparently formed internal ramp structure as the more plastic interior of the flow breaks upward and rides over itself. The flow surface is black to grey and difficult to traverse. In places, closely spaced joints, perpendicular to larger joints and shears, resemble the joints in pillow basalts. In other places, salt coats fractures that were formerly deeper in the flows where cooling sea water was heated and evaporated. The change in surface texture can be readily identified on aerial photographs.

The Heimaey eruption is especially suitable to the methods that are being employed to control the lava flows. First, the initial eruptive fissure was only 1 km from the center of a large town with an important harbor; consequently, it was in the national interest to minimize damage as much as possible by supporting a control program. Second, the main lava flow was viscous and slow moving; this allowed time to plan and carry out the control program. Third, sea water was readily available. Fourth, transport by sea as well as by a local road system is good, and it was easy to move in with pumps, pipe, and heavy equipment. As a result, an ambitious program of control was attempted. This program has had an important effect on the lava flows and undoubtedly restricted their movement and reduced property loss. A final analysis of the program must await completion of many studies under way by Icelandic scientists, but other communities faced with volcanic hazards are looking to the lessons gained from the Heimaey experience.

Author's note: There was no warning of the volcanic eruption on Heimaey. Earthquakes of small magnitude in the Vestmannaeyjar archipelago were recorded on seismographs of the University of Iceland's Science Institute, but it was not until posteruption research was carried out by Sveinbjörn Björnsson that they were specifically related to Heimaey.

Reconstruction efforts were implemented within a few weeks after the eruption began and were carried out in earnest after the cessation of flow of lava into the town of Vestmannaeyjar in late March 1973. All during the eruption the Vestmannaeyjar fishing fleet was active and continued to use the port. By the summer of 1974, about 2600 residents (about one-half the total permanent population) had returned, and plans were underway to complete excavation of all homes which were salvageable and to construct, within the next two to three years, 450 homes to replace those inevocably lost under tons of lava and tephra. Heimaey has now become a vigorous fishing community, a laboratory for geologists, a major tourist attraction, but most importantly a monument and a testimony to the perseverance and courage of the Vestmann Islanders to turn, with the help of other Icelanders and foreign friends, a seemingly hopeless situation into a very bright future. (R. S. Williams, Jr., January 1975.)

5   Geothermal Energy and Our Environment

U.S. Department of Energy

Mankind has always been fascinated by natural phenomena such as volcanoes, geysers and hot springs—all manifestations of geothermal energy. Today, methods for using energy derived from the earth's heat are being actively explored. Geothermal energy—abundant and relatively clean—is one of many alternative domestic energy sources currently being developed to replace our dwindling supplies of oil and gas.

By tapping into and harnessing pockets of natural heat thousands of feet below the earth's surface, experts believe that the United States could develop up to 20 million kilowatts of electrical generating capacity over the next two decades. This would supply energy equivalent to about four percent of this country's present crude oil production.

Often at shallower depths of as few as hundreds of feet or less, geothermal resources can be used for direct heat applications such as heating homes and greenhouses, drying crops and wood, distilling ethanol for fuel,

Reprinted from *Geothermal Energy and our Environment*, DOE/EV-0088, Spring 1980, U.S. Department of Energy.

and processing food. Direct heat uses tap the large and abundant low-to-moderate temperature (less than 300°F) reservoirs present in many areas of the United States. Scientists expect these hydrothermal resources to supply 1 quad of direct heat energy by the year 2000. This is enough to displace 480,000 barrels of crude oil per day.

The production of geothermal energy, however, is not totally free from adverse environmental impacts. Measures for protecting the environment must go hand-in-hand with the development of geothermal technology.

Three broad types of geothermal resources are being considered for near-term development in the United States: hydrothermal reservoirs, geopressured zones, and hot dry rock. Each type poses its own particular technological and environmental problems, and each specific site has its own special environmental impacts.

Hydrothermal

Hydrothermal systems consist of steam or hot water reservoirs trapped in fractured rocks or sediments below impermeable surface layers of the earth. They are presently being used to produce electricity on a commercial basis.

"Dry steam" is the preferred hydrothermal resource, but it is also the rarest. To harness energy from a dry steam reservoir, a hole is drilled into the reservoir, the released steam is filtered to eliminate solid material, and the steam is then piped directly to a turbine to generate electricity. The only commercial geothermal power stations driven by dry steam are at The Geysers in northern California.

Hot-water reservoirs, which are considerably more common, produce a mixture of steam, scalding water, and dissolving solids. A major hot-water reservoir lies beneath Southern California's Imperial Valley, one of the nation's most important agricultural regions. Other important locations of hot-water geothermal developments are Valles Caldera, New Mexico; Snake River Plain, Idaho; Carson Desert, Nevada; and Roosevelt Hot Springs, Utah.

Low-to-moderate temperature hydrothermal reservoirs are now being used in over 180 locations in the United States for direct heat applications. In Nevada, hydrothermal energy serves to clean and dehydrate onions; in Idaho, these resources heat homes and fish ponds; in South Dakota, they dry crops and heat farm buildings; and in Oregon they are used to grow mushrooms and trees, and to heat school buildings and hospitals. Soon cities in Idaho, Colorado, Oregon and California will have district heating systems powered by hydrothermal energy.

FIG. 1. Known and potential hydrothermal resources.

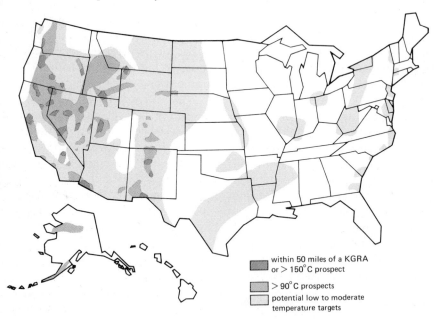

within 50 miles of a KGRA or > 150°C prospect

> 90°C prospects

potential low to moderate temperature targets

Geopressured Zones

Geopressured zones are high-temperature, high-pressure reservoirs of water (often saturated with natural gas) trapped beneath impermeable beds of shale or clay. Such zones are known to exist along the Texas and Louisiana gulf coasts. This resource is especially promising because three types of energy can be obtained; electrical (from high-temperature fluids), mechanical or hydraulic (from the high pressure), and chemical (from the natural gas). Like hydrothermal systems, geopressured resources are tapped by drilling, but the extremely high pressures create additional technical problems. Development of geopressured zones is currently at the exploratory, resource-assessment stage.

Hot Dry Rock

The potential for hot dry rock geothermal developments occur where the magma—the molten rock within the earth—has intruded into the earth's crust and is heating subsurface rock to high temperatures. Hot dry rock is potentially the largest and most widely distributed geothermal resource in the country. Exploration of this resource is underway at Fenton Hill near Valles Caldera, New Mexico, and Coso Hot Springs near China Lake, California. Other sites are being evaluated on the eastern coastal plain, in the Midwest, and on the Pacific Coast. Development of hot dry rock resources will require drilling into the rock, inducing fracturing, injecting fluid, and removing the heated fluid through a separate well. Commercial development must await the solution of technical problems associated with the fracturing process.

Environmental Issues

Geothermal resources represent a relatively clean energy source, but environmental changes have accompanied the development of hydrothermal resources in various parts of the world. With adequate attention (proper monitoring and controlling) to environmental concerns exploitation of geothermal resources need not adversely impact air and water quality, land stability, or the social and economic environments.

Air Quality

During the process of extracting heat from hydrothermal reservoirs, a variety of gases may be released into the atmosphere. Of these, the greatest nuisance and a potential hazard is hydrogen sulfide, which is toxic to humans at high concentrations and has a very disagreeable "rotten egg" odor at the very low concentrations associated with geothermal areas. There are no federal standards regulating such emissions; however, hydrogen sulfide concentrations at The Geysers have at times exceeded California air quality standards, resulting in complaints from residents of downwind communities. To deal with this problem, new hydrogen sulfide removal units have been designed at The Geysers. Improved "scrubber" systems are being developed by the Department of Energy (DOE).

Other gases are emitted at The Geysers—including carbon dioxide, methane, ammonia, arsenic, mercury and radon. Emissions of boron have caused leaf burn on sensitive trees in the immediate vicinity of cooling towers, but the small amounts of other gases do not appear to present hazards. However, emission levels can vary substantially from site to site, depending on the chemical composition of the geothermal fluid. Better methods are being developed to evaluate and, if necessary, control air pollution emissions from hydrothermal projects.

Geopressured systems present fewer air pollution problems than hydrothermal systems. Although fluids in geopressured reservoirs contain large amounts of methane, this gas can be recovered and used beneficially. Hot dry rock experiments at Fenton Hill, New Mexico, indicate that air emissions will not be a significant problem with this resource.

Water Quality

Hydrothermal systems must also overcome the problem of surface and ground water pollution which can result from the disposal of excess steam condensate and water. Fluids at the Imperial Valley have a high salt content, ruling out their discharge into surface waters. At The Geysers, the excess steam condensate had been released into lakes and streams until the resulting temperature changes in the water were found to be harmful to aquatic life.

The most environmentally attractive disposal scheme is injection of these used fluids back into the geothermal reservoir. Injection of excess steam condensate has been carried out for several years at The Geysers, with excellent results, and long-term injection tests underway in the Imperial Valley have been successful so far.

Many hydrothermal reservoirs which are now being used have waters clean enough to be drinkable. At a Nevada vegetable dehydration plant, for instance, plant engineers use the hot water to sterilize equipment and wash onions. In Idaho, hydrothermal waters are used without treatment to raise channel catfish. Hydrothermal resources of this quality cause few, if any, environmental problems. Most problems are associated with the higher temperature resources, which are much less abundant than those with low or moderate temperatures.

At geopressured sites, the great depths and pressures involved may make it economically unfeasible to inject fluids back into the reservoirs, and the expected high solids content of the fluids would render them useless for most industrial purposes. Possible means of wastewater disposal include discharge into existing saline bodies of water, injection into subsurface saline aquifers, or desalinization. Any system for disposal in areas such as the Gulf of Mexico would have to be accomplished without endangering the wetlands ecosystem.

Land Stability

The withdrawal of geothermal fluids from the earth may cause sinking of the land (subsidence), an environmental problem long associated with ground-water wells, oil field development and mining operations.

Land subsidence has not occurred during the development of the two existing vapor-dominated geothermal fields at The Geysers and Lardarello, Italy, due to the sturdy geologic structures under which such systems form. Hot-water systems appear more likely to cause subsidence. For example, at a power plant in Wairakei, New Zealand, a rate of subsidence of up to 1–3 feet per year has been measured. In the Imperial Valley, subsidence could disrupt the gravity-flow irrigation and drainage systems, damaging an area of highly productive farmland. However, it is expected that fluid-reinjection systems will prevent subsidence.

Much of the land overlying geopressured reservoirs along the Gulf Coast is only a few feet above sea level, and significant subsidence could not only result in the direct inundation of the land, but could make the area susceptible to seasonal flooding.

At hot dry rock sites, the great depths (about 3,500 meters), small fractured areas, and the use of externally introduced fluids make subsidence unlikely.

It is possible that injection of spent geothermal fluids could itself induce seismic activity (earthquakes). However, field trials indicate that the pressures required to maintain injection of spent fluids are well below those demonstrated to cause minor earthquakes.

Earthquakes caused by tapping geopressured-reservoirs are not likely because the geological structure of the Gulf Coast is earthquake-resistant. In contrast, the massive fluid injection and withdrawal that will mark hot dry rock development could increase seismic activity, but none has been detected to date by sensitive measurements at the Fenton Hill, New Mexico, site.

Noise

Noise from the venting of geothermal wells was a significant problem at The Geysers, causing complaints from nearby residents until it was effec-

tively reduced by using appropriate muffling systems. Studies of the possible impact of noise on wildlife are currently underway at The Geysers.

At hot water hydrothermal sites, the noise problem is not expected to be important. Steam pressures are lower, and noise levels associated with well tests are much lower than at The Geysers. Monitoring programs will be necessary, however.

The high pressures of geopressured reservoirs may cause noise to be a problem with this geothermal resource. Significant noise during development of hot dry rock sites is not anticipated.

Land/Water and Economy

Competition for water may result where geothermal development must be accommodated with residential, commercial and recreational uses of area. The coastal wetlands, areas of great value in terms of recreation, commercial productivity, and cultural and scientific interest, are sensitive to disruptions from technological activities. The semi-arid regions of the West may also experience land use conflicts.

In many areas where geothermal energy has been developed on a commercial scale, it has proved to be cost-competitive with other energy sources.

Large scale geothermal developments may have a significant impact upon regional economics and community structure as well. A "boomtown" atmosphere could develop in a community with a formerly quiet, rural lifestyle. Because geothermal development is expensive in these large-scale projects, tax bases would be broadened. Geothermal reservoirs can be used for many different purposes, however, and since such multiple uses of the same energy supply reduce costs dramatically, the economic base of a community could be diversified. Early planning with affected communities could increase local employment and provide for social services.

Most geothermal developments, however, will be in the area of direct heat applications. This type of development is normally small in scale, and should have only a minor impact on local economies or water and land use patterns.

DOE Geothermal Environmental Program

The goal of the Department of Energy's geothermal/environmental program is to ensure that the adverse environmental impacts described above are averted or corrected during the demonstration, development, and commercialization of geothermal technologies. A basic feature of the program is that environmental research and development proceeds in parallel with technology development projects. In this way, technology development can be guided down environmentally acceptable paths.

The environmental research and development program consists of a wide variety of projects, grouped into several categories:

A complex program of measuring sites to determine the normal state of the environment; the amount and behavior of potential pollutants; and the effects of pollutants on the environment.

Laboratory studies of the ways in which pollutants and environmental entities interact.

Analyses of laboratory and field data in order to predict short- and long-term effects.

Assessment of potential environmental, health, and safety consequences.

To provide a basis for managing environmental aspect of the energy programs under DOE's jurisdiction, Environmental Development Plans (EDPs) are prepared. These plans identify the environmental issues and constraints that must be addressed and are used to develop strategies for the requisite environmental research. They play an important role in guiding the technology development as it progresses so that the technology will be assured of compliance with the National Environmental Policy Act and with present or anticipated standards and regulations of appropriate governmental agencies.

Conclusion

Although geothermal energy may provide only three to five percent of our energy needs by the year 2000, it is one of the cleaner energy sources available, offering significant environmental advantages over coal, oil, and nuclear fuels. In addition, unlike fossil fuels and nuclear power production, geothermal energy does not require a massive network of facilities and equipment and large amounts of input energy. Land disruption is far less extensive than from the surface mining of coal or uranium. Furthermore, the health and safety issues related to operating relatively small geothermal power plants are minor compared with those associated with coal or nuclear power production.

Given a national policy that seeks to balance our energy requirements with the need to preserve and enhance our environment, geothermal energy can make a valuable contribution toward satisfying both environmental and energy goals.

Earthquakes and Tectonic Movements

Earthquake damage, Santa Barbara, California, June 29, 1925. Part of the Arlington Hotel collapsed from severe shaking. (Putnam Studios, courtesy Bruce Bolt.)

Earthquake Location

Earthquakes are natural vibrations caused by the distortion and rupturing of earth materials. The rupturing usually occurs along weak
zones called *faults,* and the resulting movement along the faults releases energy in the form of an earthquake. Seismology, the study of
earthquakes, is one of the most active areas of scientific research
today.

Earthquakes can occur anywhere on earth, but they are concentrated in two major belts—the Circum-Pacific Belt (80 percent) and the
Mediterranean-Trans-Asiatic Zone (15 percent) (Fig. 1). These belts
coincide generally with zones of active volcanism and plate boundaries. Significant earthquakes have, however, occurred within the interior of continents.

The point of initiation of rupture is called the *focus* or *hypocenter.* The
focus may be as deep as 700 km (435 mi), but most earthquakes originate at depths less than 70 km (43.5 mi) (*shallow focus* earthquakes).
Those originating at depths between 70 and 300 km (43.5 and 186.4 mi)
have an *intermediate focus,* while those originating at greater depths are
known as *deep focus* quakes. Once rupture begins, it will propagate
along the fault and may continue for many miles. The point on the
earth's surface vertically above the focus is designated the *epicenter* and
defines the geographic position of the earthquake.

Earthquake Waves

In the focal region of an earthquake, strain energy is converted and
released as heat and seismic wave motion. The seismic waves radiate
in all directions from the focus. The initial waves travel through the
interior of the earth and are called *body waves.* When the energy
reaches the surface, *surface waves* are generated. There are two types
of body waves (primary and secondary waves) and two types of surface waves.

Body Waves

A *primary (P) wave* is longitudinal, propagates through both liquids
and solids, and is usually the first signal that an earthquake has

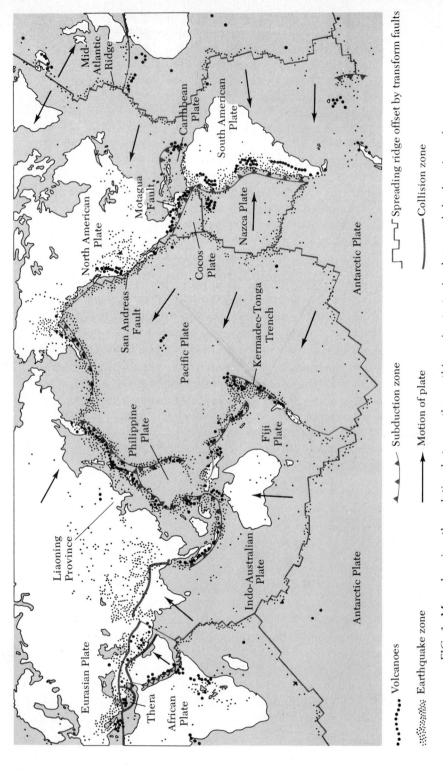

FIG. 1. Map showing the relation between the earth's major tectonic plates and the location of recent earthquakes and volcanoes. Earthquake epicenters are shown by small dots. (From *Earthquakes: A Primer* by Bruce A. Bolt. Reproduced by permission of W.H. Freeman and Company, © 1978.)

•••••••• Volcanoes

∴∴∴∴∴ Earthquake zone

◄◄◄ Subduction zone

➤ Motion of plate

⊐⊏ Spreading ridge offset by transform faults

── Collision zone

occurred. The velocity of a P wave is usually less than 7 km/sec (nearly 15,000 mph) in crustal material but increases to 14 km/sec (more than 30,000 mph) in the earth's core. The compressional phase of the P-wave motion pushes particles together and displaces them away from the focal region. Next, the rarefaction phase dilates the particles and displaces them toward the earthquake source. These movements result in a series of pushes and pulls parallel to the wave path.

A *secondary (S) wave* is transverse, which means that particles in the wave's path move from side to side or up and down at right angles to the wave's advance. The S wave travels much slower than the P wave and is the second signal to arrive at a recording station. It is easy to recognize the arrival of an S wave because it has twice the period and amplitude of the associated P wave. S waves advance by shearing displacements (e.g., displacements that change shape without changing volume) which require a rigid medium. This is significant because their failure to travel through the outer core of the earth (Fig. 2) gives credence to the theory that this part of the core behaves as a fluid.

Surface Waves

Surface waves, which are also called L waves (long waves), are of much greater length and period and transmit a large proportion of earthquake energy. They are largely responsible for damage to surface structures.

The arrival times of the P and S waves, the ratio of their speed, and their amplitudes can be recorded by instruments called *seismographs*. This information is used to determine location of the epicenter, focal depth, and earthquake severity. L waves are also recorded by the seismograph.

Earthquake Severity

Intensity

The severity of an earthquake may be described in terms of its intensity or its magnitude. Intensity is an indication of an earthquake's apparent severity at a specified location, as determined by experienced observers through interviews, damage surveys, and studies of earth movements. The *Modified Mercalli Intensity Scale* (1931) measures earthquake intensity based on effects on people, objects, and buildings, and grades observed effects into twelve classes (Table 1). The scale is easily understood by most people, but is a relative gauge and is only indirectly related to the amount of energy released by an

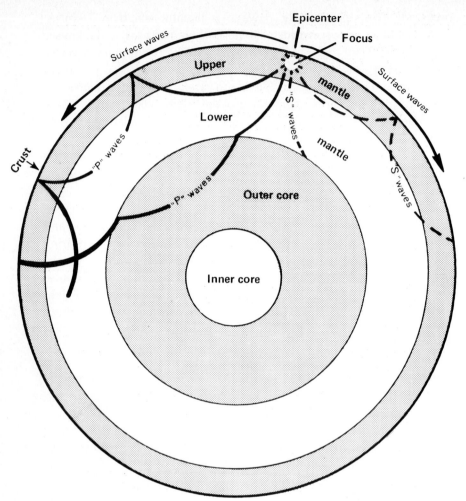

FIG. 2. Hypothetical cross section of the earth showing the path of movement of the main types of earthquake waves. (U.S. Geological Survey.)

earthquake. Variables such as distance from epicenter, focal depth, local geology, and building design also influence intensity values.

Magnitude

Magnitude expresses the amount of energy released by an earthquake as determined by measuring the amplitudes produced on seismographs. The *Richter scale*, developed by Charles Richter in 1935, is the most commonly used scale of earthquake magnitude. The Richter scale is logarithmic. An earthquake of magnitude 8 (M = 8), for example, represents seismograph amplitudes ten times larger than

Table 1 Modified Mercalli Intensity Scale

I. Not felt except by a very few under especially favorable circumstances.

II. Felt only by a few persons at rest, especially on upper floors of buildings. Delicately suspended objects may swing.

III. Felt quite noticeably indoors, especially on upper floors of buildings, but many people do not recognize it as an earthquake. Standing automobiles may rock slightly. Vibration like passing of truck. Duration may be estimated.

IV. During the day, felt indoors by many, outdoors by few. At night some awakened. Dishes, windows, doors disturbed; walls make cracking sound. Sensation like heavy truck striking building. Standing automobiles rock noticeably.

V. Felt by nearly everyone, many awakened. Some dishes, windows, etc., broken; a few instances of cracked plaster; unstable objects overturned. Disturbances of trees, poles, and other tall objects sometimes noticed. Pendulum clocks may stop.

VI. Felt by all, many frightened and run outdoors. Some heavy furniture moved; a few instances of fallen plaster or damaged chimneys. Damage slight.

VII. Everybody runs outdoors. Damage negligible in buildings of good design and construction; slight to moderate in well-built ordinary structures; considerable in poorly built or badly designed structures; some chimneys broken. Noticed by persons driving automobiles.

VIII. Damage slight in structures designed to withstand earthquakes; considerable in ordinary substantial buildings, with partial collapse; great in poorly built structures. Panel walls thrown out of frame structures. Fall of chimneys, factory stacks, columns, monuments, walls. Heavy furniture overturned. Changes in well water levels. Drivers of automobiles disturbed.

IX. Damage considerable in specially designed structures; great in substantial buildings, with partial collapse. Well-designed frame structures thrown out of plumb. Buildings shifted off foundations. Ground cracked conspicuously. Underground pipes broken.

X. Some well-built wooden structures destroyed; most masonry and frame structures destroyed with foundations; ground badly cracked. Rails bent. Landslides considerable from river banks and steep slopes. Water splashed (slopped) over banks.

XI. Few, if any structures remain standing. Bridges destroyed. Broad fissures in ground. Underground pipelines completely out of service. Earth slumps and land slips in soft ground. Rails bent greatly.

XII. Damage total. Practically all structures damaged greatly or destroyed. Wave forms seen on ground surface. Lines of sight and level are distorted during the event. Objects thrown upward into the air.

those of a magnitude 7 earthquake, 100 times larger than those of a magnitude 6 earthquake, and so on. There is no highest or lowest value. The largest earthquakes of record have been rated at M = 8.9; the smallest, about M = minus 3. Approximate relationships between magnitude, energy release, incidence, area and distance felt, and intensity are shown in Table 2.

Acceleration of Ground Shaking

Another measure of earthquake severity is the degree of acceleration of ground shaking. This is measured by instruments called *strong motion accelerometers*. Accelerations in excess of 1.0 g would, for example, destroy most structures.

Effects of Earthquakes

The impact of an earthquake is determined by the following factors: (1) amount of energy released; (2) frequency, orientation, and duration of shaking; (3) distance from epicenter; (4) physical properties of bedrock and surficial materials; and (5) building design.

The nature of the bedrock and surficial materials is extremely important. For example, earthquake wave amplitudes are greater and vibrations last longer in unconsolidated surficial material. Water-saturated unconsolidated material is especially vulnerable to strong shaking whereas earthquake waves are dampened and travel quickly

Table 2 Approximate relationships between energy released, magnitude, and intensity [NOAA]

Magnitude	Energy released[3] (Ergs)	Expected annual incidence[1]	Felt area (mi^2)	Distance felt (statute mi)[2]	Intensity (maximum expected, Modified Mercalli)[3]
3.0–3.9	9.5×10^{15}–4×10^{17}	49,000	750	15	II–III
4.0–4.9	6×10^{17}–8.8×10^{18}	6,200	3,000	30	IV–V
5.0–5.9	9.5×10^{18}–4×10^{20}	800	15,000	70	VI–VII
6.0–6.9	6×10^{20}–8.8×10^{21}	120	50,000	125	VII–VIII
7.0–7.9	9.5×10^{22}–4×10^{23}	18	200,000	250	IX–X
8.0–8.9	6×10^{23}–8.8×10^{24}	1	800,000	450	XI–XII

[1]B. Gutenberg and C. F. Richter, *Seismicity of the Earth and Associated Phenomena*, Princeton University Press, Princeton, N.J., 1954, p. 18.
[2]H. Benioff and B. Gutenberg, "General Introduction to Seismology," *Earthquakes in Kern County During 1952*, California Division of Mines, Bull. 171, San Francisco, 1955, p. 133.
[3]C. F. Richter, *Elementary Seismology*, W. H. Freeman and Co., San Francisco, 1958, pp. 353, 366.

FIG. 3. Buildings on the unconsolidated "old channel" material collapsed while those on the bedrock "bench" materials remained standing (Varto, Turkey). (Photo R. Wallace, courtesy U.S. Geological Survey.)

through granite bedrock. Figure 3 shows the influence of surficial materials.

Topographic Effects

The most significant topographic effects of earthquakes are upwarping or subsidence, offset along faults, and ground cracking. Usually these changes are minor and quickly "healed" by gradational agents. Some changes, however, can be quite profound and long lasting. For example, the upwarping and subsidence associated with the New Madrid, Missouri, earthquakes of 1811–12 altered the course of the Mississippi River, created new lakes, and drained numerous swamps. In California, almost continuous offset along segments of the San Andreas Fault has left a linear scar in the landscape extending southward from San Francisco Bay to Los Angeles.

Ground shaking reorients grains in unconsolidated sediments, which may lead to landslides or soil liquefaction. *Liquefaction* is a

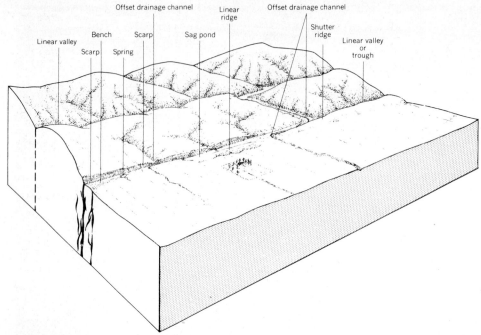

FIG. 4. Block diagram showing landforms developed along recently active strike-slip faults. (From Wesson et al., 1975.)

process which transforms material which behaves as a solid into material which behaves as a liquid. Apparently, the reorientation of grains temporarily transfers the load from grain-to-grain contact to the pore water. When pore fluid pressure is increased, movement can occur down almost imperceptible slopes.

Topographic features associated with recent earthquakes include fault scarps, sag ponds, offset drainage channels, linear troughs, and hummocky topography if landslides have occurred (Figs. 4 and 5).

Tsunamis

Tsunamis are large sea waves which may be generated by submarine earthslides, volcanic eruptions, or earthquakes. They are also called seismic sea waves and tidal waves (incorrectly, because they have nothing to do with tides). These waves are capable of traveling at speeds of 960 kph (596.5 mph) and as they approach the coast they may be 15 m (49.2 ft) high or higher. They not only bring death and destruction to coastal communities, but they also cause extensive shoreline erosion. The tsunami generated by the Lisbon earthquake of 1755 killed more than 60,000 people.

FIG. 5. Topographic features along a segment of the San Andreas Fault in southern California. Note offset drainage and linear ridges, valleys, and scarps. (Photo R. Wallace, courtesy U.S. Geological Survey.)

Impact on Humans

In the last thousand years, earthquakes have been responsible for the loss of more than 3 million people and more property damage than any other geologic hazard. An average of 10,000 people are killed each year, but in 1976 between 750,000 and 1 million lost their lives in earthquakes. Much of the loss is related to primary forces such as ground shaking, fault movement, and crustal warping, which are responsible for the collapse of buildings and other structures. Losses

Table 3 Some of history's most severe earthquakes

Year	Locality	Magnitude (Richter scale)	Intensity (Modified Mercalli)	Lives lost
1556	Shensi, China	?	XII	830,000
1737	Bengal, India	?	XII	300,000
1755	Lisbon, Portugal	8.7	XII	70,000
1897	Assam, India	8.7	XII	1,500
1906	San Francisco, California	8.3	XI	800
1923	Tokyo, Japan	8.3	XII	143,000
1927	Tsinghai, China	8.3	XII	100,000
1933	Sanriku Coast, Japan	8.9	XI	3,000
1950	Assam, India	8.7	XII	1,600
1964	Prince William Sound, Alaska	8.6	XI	131
1976	Hopei, China (Tangshan)	7.8	?	650,000

are commonly magnified by such secondary aspects as fires, land-slides, and tsunamis. Table 3 lists some of history's most severe earthquakes, and Figure 6 shows the location of notable destructive earthquakes in the United States.

Building design and materials are critical factors influencing loss of lives and property damage. Wooden buildings are generally more stable than rigid brick or concrete structures. Adobe huts of mud and

FIG. 6. Location of notable destructive earthquakes in the United States. The losses associated with these earthquakes can be related to ground shaking, surface faulting, landslides, tsunamis, and fire. (From Hays, 1981.)

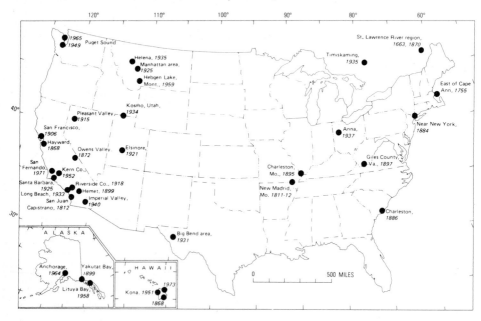

straw commonly fail. Multistory buildings must be designed to with-stand substantial horizontal stress. When neighboring buildings of different heights oscillate, they may knock against each other, and the damage from this hammering effect is often extensive. Many multi-story buildings in earthquake-prone areas are being outfitted with strong motion accelerometers in order to record building response to earthquake vibrations accurately. Improvements in earthquake-resis-tant design may have to wait until these buildings experience a strong earthquake.

Causes of Earthquakes

Elastic Rebound Theory

In 1908 Harry F. Reid, an American seismologist who investigated the 1906 earthquake in San Francisco, formulated the elastic rebound the-ory to describe the release of energy responsible for earthquakes. According to this theory, energy is stored in rocks that are being elastically deformed. When the elastic limit and frictional resistance of the rock are exceeded, strain energy is released by slippage along planes of weakness (faults) and by a tendency for the deformed rocks to rebound to their original shape (Figs. 7 and 8). The energy is

FIG. 7. Four types of fault movement. The characteristic movement along the San Andreas Fault is right-lateral strike slip. (From *U.S. Geological Survey Professional Paper 941-A.*)

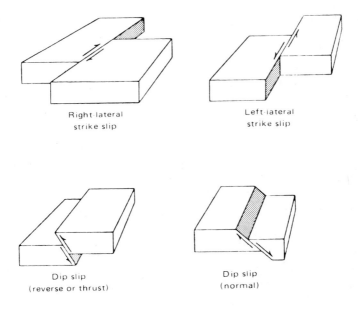

Right lateral
strike slip

Left-lateral
strike slip

Dip slip
(reverse or thrust)

Dip slip
(normal)

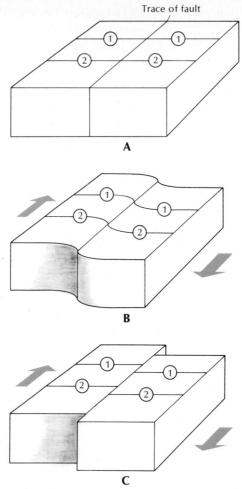

FIG. 8. Stages in the movement of a right-lateral strike-slip fault, as postulated in Reid's elastic rebound theory. (A) Undeformed fault blocks. (B) Elastic deformation due to movement along fault plane. (C) After rupture, offset, and rebound to original shape.

transferred from the area of failure (the focus) through the earth by body waves. Reid based his theory on observations along the San Andreas Fault, which is characterized by shallow focus earthquakes and strike-slip faulting. His theory is also supported by laboratory experiments on crustal rocks subjected to pressures equivalent to those in the earth's crust.

The elastic rebound theory has two major limitations: (1) it does not explain the origin of the forces that deformed the rocks, and (2) it does not apply to intermediate and deep focus earthquakes, which occur at depths and pressures characteristic of plastic deformation rather than elastic deformation.

Deforming Forces

The geographic distribution of earthquakes suggests that the deforming forces may be related to convection currents and plate movements. The greatest number of earthquakes occur along convergent (subducting) and strike-slip plate boundaries (see Fig. 1). Some volcanologists have suggested that temperature or phase changes within deep-seated magma chambers may also be a source of energy.

Intraplate earthquakes may be related to other forces, such as isostatic adjustments to melting of continental ice caps or erosion of large volumes of sediments.

It has also been suggested that earthquakes occur in swarms, and that the timing of these events may be related to a triggering mechanism such as the gravitational pull associated with the alignment of the earth, sun, and moon, or the alignment of the earth and selected planets. This pull is said to be the force which precipitates final failure of rocks that have already been severely strained.

Dilatancy Model

Observations of a large number of earthquakes since Reid first proposed his theory show that marked changes take place in the properties of rocks in the vicinity of the epicenter prior to an earthquake. These observations have contributed to a more complete understanding of the earthquake mechanism and to the formulation of two new models, one proposed by Russian seismologists and the other by Americans. Both models, which describe the response of crustal rocks to earthquake forces, suggest that an increase in elastic strain in rocks leads to the formation of a large number of open fractures, causing an increase in volume (dilatancy), before failure. The next stage, according to the Russians, is marked by an avalanche of fractures followed by movement along the fractures, which is recorded as an earthquake.

The American version of the model relies on diffusion of pore water (water in the openings or pore spaces in the rock) and is called the *dilatancy-diffusion* model. According to this model, initial fracturing is accompanied by a decrease in pore pressure and a temporary strengthening of the rocks, followed by a movement of pore water to the open fractures (diffusion). An increase in pore water pressure weakens the rocks and facilitates movement along the fractures which is recorded as an earthquake.

Earthquake Prediction

In 1966, geoscientists Frank Press and William F. Brace wrote in *Science:* "A few years ago the subject of earthquake prediction fell under the purview of astrologers, misguided amateurs, publicity seekers, and religious sects with doomsday philosophies. No wonder the occasional scientist who ventured an opinion on the subject did so with trepidation and then with conservatism lest he be disowned by his colleagues." Since then, the situation has changed dramatically. Earthquakes have been successfully predicted in China, Japan, the U.S.S.R., and the United States, and seismologists are urging support for earthquake monitoring and prediction programs throughout the world.

There are two general approaches to earthquake prediction: geophysical monitoring and risk assessment. The first approach involves close monitoring of a given site, which requires a high density network of sophisticated geophysical instruments designed to detect earthquake precursors. The most diagnostic precursors are given below.

Geophysical Precursors
1. Changes in the velocity ratio of P and S waves.
2. Changes in the number and frequency of foreshocks (micro-seisms).
3. Changes in electrical resistivity.
4. Changes in magnetic susceptibility.
5. Changes in acoustics.
6. Emission of radon gas.

Topographic Precursors
1. Ground level deformation (Fig. 9).
2. Changes in water levels.
3. Change in rate or direction of fault creep.

Anomalous Animal Behavior
1. Examples include restlessness, loss of appetite, howling, and barking.

Not all of these precursors are associated with every earthquake, and some of them may be related to other processes, such as volcanism.

It is interesting to note that the dilatancy-diffusion model can account for many of the precursors. For example, the rock fracturing associated with dilatancy could lead to ground level deformation, changes in water levels, emission of trapped gases, and changes in

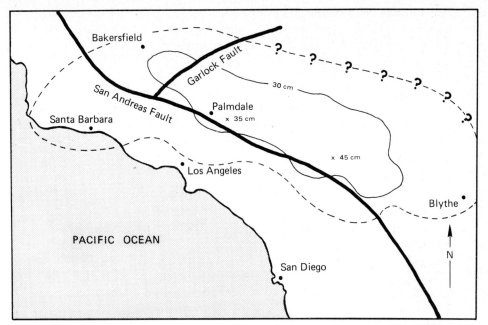

FIG. 9. Map of the "Palmdale Bulge" in southern California, discovered by U. S. Geological Survey seismologists in 1976. The area totals more than 80,000 km^2 (~32,000 mi^2). The dashed line shows where there was zero elevation change between 1959 and 1974. The solid contour encloses an area uplifted more than 30 cm. (U.S. Geological Survey map.)

the velocity of seismic waves. Diffusion of pore water would also change the velocity of seismic waves and electrical resistivity. Microfracturing and slippage along mineral grains before total failure may account for the observed foreshocks.

Risk Assessment

Only about 8 percent of the people in the United States are safe from earthquake hazards. Seventy million people live in areas with a significant earthquake risk, while another 115 million are exposed to a less significant, but not negligible, danger. Risk assessment, which is based on seismological history, local geology, records of previous damage, and design of current structures, rates the degree of damage an area might experience in an earthquake. The importance of local geology and building design has been noted in previous sections.

Seismological history focuses on identifying seismic gaps and recurrence rates of earthquakes in a given region. A *seismic gap* is a zone where few earthquakes have occurred in a known active belt. The faults in this belt are said to be "locked," and strain energy may be accumulating to the point where it will be released as a high-magnitude earthquake. Statistics on earthquake recurrence give valuable information on level of activity and return intervals for specific areas.

Seismic Risk Maps

Figure 10, a map of four zones of relative seismic risk in the United States, is based on historic earthquakes, evidence of strain release, and geologic structures. Figure 11 indicates earthquake frequency and horizontal acceleration, a measure of ground shaking as a percent of acceleration of gravity (g)—often the most destructive aspect of an earthquake.

Some seismologists argue that it is necessary to consider a longer time span than that available from the historical record of earthquakes. They suggest that the record of late Quaternary faulting (500,000 years before present) may be more helpful in identifying hazard zones in areas where earthquake recurrence intervals are extremely long. Quaternary age faulting, for example, may be identified by the features shown in Fig. 4, which are more likely to persist in arid and semiarid regions than in humid areas.

Earthquake Control

A proposed strategy for controlling earthquakes calls for releasing strain in stressed areas in small controlled increments either by fluid injection or by detonation of nuclear blasts.

A most startling demonstration of strain release by fluid injection was produced inadvertently when liquid waste was injected into a well at the Rocky Mountain Arsenal near Denver in 1962. The correlation between fluid injection and earthquakes is documented in Reading 7. Fluid injection is also practiced in some oil recovery programs. The U.S. Geological Survey has demonstrated that earthquakes can be turned off and on in the Rangely oil field of northwestern Colorado by altering reservoir pressures. Apparently, low pore pressures strengthen rock, thereby reducing the potential for earthquakes, while an increase in fluid pressure can weaken rock, thereby producing an earthquake. These observation are consistent with the dilatancy-diffusion model discussed earlier. The effects of pore pressure may also explain why earthquakes are sometimes induced by the damming of rivers or impouding of large reservoirs (water leaks into the pore spaces, increasing pressure).

After nuclear tests at the Nevada site activated old faults and pro-

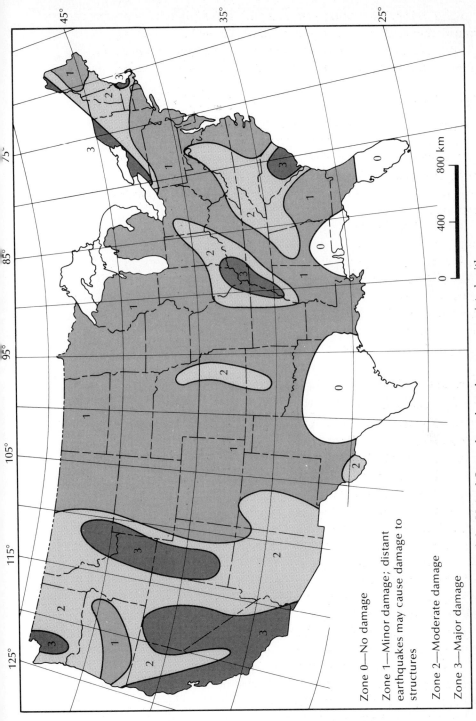

Zone 0—No damage

Zone 1—Minor damage; distant earthquakes may cause damage to structures

Zone 2—Moderate damage

Zone 3—Major damage

FIG. 10. Seismic risk zones of the United States based on damages associated with historic earthquakes. (From Algermissen, 1969.)

800 km

400

0

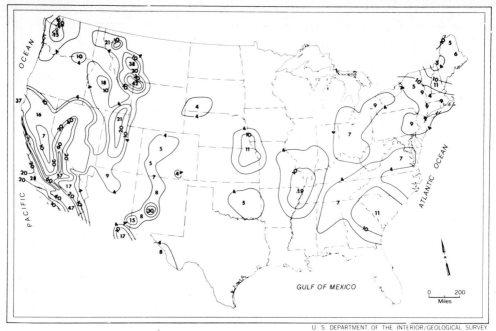

FIG. 11. Expected levels of earthquake shaking hazards. Contour lines show ground-shaking levels in percentages of the force of gravity (g) for the maximum likely to occur at least once in 50 years. All percentages are at the 90 percent probability level. For example, a contour of 20 percent of gravity means that the odds are 1 in 10 that the region will experience ground shaking more than 20 percent of the force of gravity during any 50-year period. (*U.S. Geological Survey Open File Rep. 76-46, July 1976.*)

duced new ones, some scientists were led to believe that underground nuclear explosions might be employed to release strain in rocks, thereby reducing the severity of earthquakes. Application of either technique—injecting fluids or detonating explosions—is bound to be highly controversial, however.

Reducing Earthquake Damage

Earthquake control is not yet practicable, and implementation of an extensive prediction program is years away. Several techniques, however, might be implemented now:

1. Passage of zoning laws that would control the types of development in areas of high seismic risk.
2. Planning for evacuation and disaster relief in communities in high risk areas.
3. Informing the public of the risks to which they are exposed and the recommended response should an earthquake occur (Table 4).

Table 4 Earthquake safety rules [NOAA]

An earthquake strikes your area, and for a minute or two the "solid" earth moves like the deck of a ship. What you do during and immediately after the tremor may save your life and those of others. These rules will help you survive.

During the shaking:

1. Don't panic. The motion is frightening but, unless it shakes something down on top of you, it is harmless. The earth does not yawn open, gulp down houses, and slam shut. Keep calm and ride it out.
2. If you are indoors, stay there. Take cover under a desk, table, bench; or in doorways, halls; or against inside walls. Stay away from glass.
3. Don't use candles, matches, or other open flames, either during or after the tremor. Douse all fires.
4. If the earthquake catches you outside, move away from buildings and utility wires. Once in the open, stay there until the shaking stops.
5. Don't run through or near buildings. The greatest danger from falling debris is just outside doorways and close to outer walls.
6. If you are in a moving car, stop as quickly as safety permits, but stay in the vehicle. A car is an excellent seismometer, and will jiggle fearsomely on its springs during the earthquake; but it is a good place to stay until the shaking stops.

After the shaking:

1. Check your utilities, but do not turn them on. Earth movement may have cracked water, gas, and electrical conduits.
2. If you smell gas, open windows, shut off the main valve, and then leave the building. Report the leakage to authorities. Don't reenter the house until a utility official says it is safe.
3. If water mains are damaged, shut off the supply at the main valve.
4. If electrical wiring is shorting out, close the switch at the main meter box.
5. Turn on your radio or television (if conditions permit) to get the latest emergency bulletins.
6. Stay off the telephone except to report an emergency.
7. Don't go sight-seeing.
8. Stay out of severely damaged buildings; aftershocks can shake them down.

4. Improving the design of earthquake-resistant structures and reinforcing more adequately structures that are not earthquake resistant.

5. Combining insurance programs with hazard-reduction methods to spread the risk and the economic impact of a disaster and to discourage unsafe building practices. Governments would subsidize premiums for those who reduce the hazards by improved design, zoning, and so on.

Tectonic Movements

Tectonic movements include the faulting, folding, or regional warping of the earth's crust. Earthquakes are often associated with rather sudden movement along localized fault planes. The San Francisco earthquake of 1906, for example, produced horizontal displacements of more than 7 m (23 ft) near Tomales Bay, and the Prince William Sound (Alaska) earthquake of 1964 caused vertical displacements of over 13 m (41.6 ft) in some areas. In New Madrid, Missouri, earthquakes in 1811–12 were accompanied by crustal warping over an area of 13,000 km^2 (5019 mi^2). Reelfoot Lake in Tennessee occupies part of the downwarped area.

Not all tectonic activity, however, is accompanied by earthquakes. It is also characterized by slow, almost imperceptible movements known as *tectonic creep*. They may be primarily horizontal and related to movement along faults, as along some segments of the San Andreas Fault, or vertical and related to adjustment of the land surface as a great overburden of glacial ice melts. The weight of continental glaciers during the Ice Age was great enough to cause a downwarping of the land they covered. When the glaciers started to melt only a few thousand years ago, the land slowly started to rebound to its original elevation. Examples of vertical movement may be found in Scandinavia and in the Great Lakes area. In any case, tectonic creep presents problems in the design and maintenance of engineering structures. Figure 12, a map showing that there are few stable areas in the United States, is based on measurements made over the past 100 years and does not discriminate between movements such as subsidence (frequently caused by human activity) and movements caused by natural internal forces.

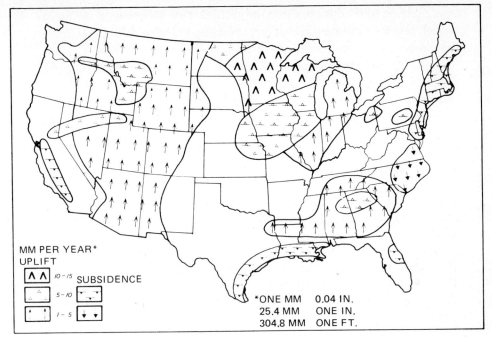

FIG. 12. Map of crustal movement showing probable vertical movements of the earth's surface. (National Geodetic Survey, Vertical Network Division.)

Case Histories and Readings

The San Andreas Fault: Geologic and Earthquake History

In Reading 6, Gordon Oakeshott describes the San Andreas Fault system and its attendant earthquakes. He reviews some of the unsolved problems associated with the fault and offers the intriguing suggestion that although the solution of these problems might enable us to predict earthquake activity, this ability might in itself create a completely new set of problems. This is because success in developing a reliable prediction system will not be without serious social, political, and economic repercussions.

Man-made Earthquakes in Denver

In 1962 a deep well at the Rocky Mountain Arsenal near Denver was used for disposing of liquid wastes. Soon afterward, a series of earthquakes began—the first in the area since minor "natural" quakes in 1882. In November 1965 David Evans, a consulting geologist, presented data which strongly indicated that the quakes were being triggered by the injection of fluid wastes into the ground. His disclosure (Reading 7) which reveals a strong sense of responsibility to Denver's understandably anxious citizens, hastened official action and led to a more detailed monitoring of waste disposal and tremors. These studies supported Evans's contention that the army's method of waste disposal was a major cause of the recent earthquakes. With subsequent suspension of the disposal program, the region appears to have regained its former stability. The events discussed in the reading demonstrate the urgent need for full use of geologic knowledge before we interfere with an environment that is in equilibrium.

A Federal Plan for the Issuance
of Earthquake Predictions and
Warnings

As seismologists expand earthquake monitoring programs, it will be their responsibility to evaluate the significance of new precursors and to inform governmental authorities whose responsibility it will be to issue predictions and warnings. In Reading 8, Vincent E. McKelvey, former director of the U.S. Geological Survey, outlines a tentative plan for evaluating scientific data and disseminating predictions and warnings.

Possible Loss-Reduction Actions
Following an Earthquake Prediction

The primary goal of a preduction program is to reduce the loss of lives and property. In Reading 9, Charles Thiel of the National Science Foundation reviews what can be done to reduce the losses from an anticipated earthquake. His analysis, which is concerned with four basic time-spans (3, 30, 300, and 3000 days), considers the types of action that could be taken to protect buildings and their contents, and such special structures as bridges, dams, nuclear reactors, and pipelines.

Earthquake Prediction:
The Scientist's Responsibilities
and Public Response

Gordon Oakeshott points out in Reading 6 that earthquake prediction is bound to have significant economic, legal, political, social, and psychological implications. In Reading 10, Eugene Veek considers the scientist's responsibilities and public response to prediction.

Creep on the San Andreas Fault:
Fault Creep and Property Damage

Reading 11 documents the occurrence of tectonic creep in the San Andreas Fault system. This paper shows that buildings and other structures may be used to document and quantify tectonic creep. It should be noted, however, that topographic expressions of prehistoric tectonic activity, which are numerous, can alert developers or planners to hazardous settings (see Fig. 4).

Seismologists have also been able to predict episodes of tectonic creep along some portions of the San Andreas fault. It is obvious, however, that some developers and planners either ignore the topographic indicators or do not recognize them because much construction is going on in areas of active creep. It should also be noted that the occurrence of fault creep does not preclude sudden rupture and its attendant earthquakes.

6 San Andreas Fault: Geologic and Earthquake History

Gordon B. Oakeshott

The San Andreas fault is California's most spectacular and widely known structural feature. Few specific geologic features on earth have received more public attention. Sound reasons for this are found in the series of historic earthquakes which have originated in movements in the San Andreas fault zone, and in continuing surface displacements, both accompanied and unaccompanied by earthquakes. This active fault is of tremendous engineering significance, for no engineering structure can cross it without jeopardy and all major structures within its potential area of seismicity must incorporate earthquake-resistant design features. Recently a proposal for a great nuclear power plant installation on Bodega Head, north of San Francisco, was abandoned because of public controversy over the dangers of renewed movements and earthquakes on the nearby fault. Expensive design features are being incorporated into the State's plan to transport some of

Adapted largely from Oakeshott, Gordon B., 1966, San Andreas fault in the California Coast Ranges province: California Division of Mines and Geology Bulletin 190, pp. 357–373.

Oakeshott, G. B., 1966, "San Andreas Fault: Geologic and Earthquake History," *Mineral Information Service*, vol. 19, no. 10, pp. 159–66. Abridged by permission of the author and reprinted by permission of California Division of Mines and Geology, Sacramento, California.

Dr. Oakeshott is Deputy Director of the California Division of Mines and Geology.

northern California's excess of water to water-deficient southern California in order to ensure uninterrupted service across the fault in the event of fault movements and earthquakes in the Tehachapi area.

Geologists and seismologists the world over have directed their attention to the San Andreas fault because of: (1) the great (Richter magnitude 8.25)[1] San Francisco earthquake of 1906 and many lesser shocks which have originated in the fault zone; (2) development of the "elastic rebound" theory of earthquakes by H. F. Reid; (3) striking geologic effects of former movements and continuing surface movements in the fault zone; and (4) postulated horizontal displacements of hundreds of miles—east block moving south.

The San Andreas has been frequently and widely cited in the scientific and popular literature as a classic example of a strike-slip fault with cumulative horizontal displacement of several hundred miles; however, the geologic evidence that can be documented is highly controversial.

Location and Extent

The San Andreas fault strikes (bears) approximately N. 35° W. in a nearly straight line in the Coast Ranges province and extends southward for a total length of about 650 miles from Shelter Cove on the coast of Humboldt

County to the Salton Sea. This takes it completely across geologic structures and lineation of the Coast Ranges at a low angle, then south across the Transverse Ranges and into the Salton Trough. Latest movement in the fault zone, as noted by the late Professor A. C. Lawson who named and traced it, has thus been clearly later than all major structural features of those provinces. This recent movement may, however, be an expression of renewed activity along an older fault zone that antedated differentiation of the geologic provinces now in existence. If so, we need to distinguish between such an older, or "ancestral," San Andreas fault zone and latest movements on the modern, or Quaternary, San Andreas fault proper.

The long northwesterly trend of the fault zone is interrupted in three places (see Fig. 1): (1) At Cape Mendocino, where it turns abruptly westward to enter the Mendocino fault zone, as reflected in the Mendocino Escarpment; (2) at the south end where the Coast Ranges adjoin the Transverse Ranges and the fault turns to strike east into the complex knot of major faults in the Frazier Mountain area and on emerging splits into the 50-mile-wide system of related faults, including the San Andreas fault proper, in southern California; and (3) in the San Gorgonio Pass area where the San Andreas fault proper appears to change direction again and butt into the Mission Creek-Banning fault zone which continues into the Salton Trough.

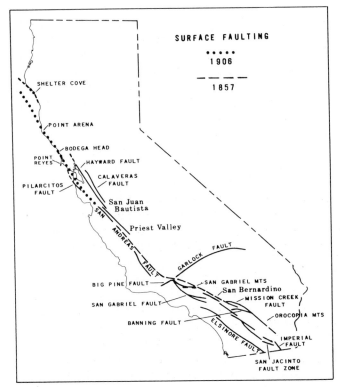

FIG. 1. San Andreas fault zone.

Earthquake History

Earthquake history of California is extremely short. The earliest earthquake in written records was felt by explorer Gaspar de Portola and his party in 1769 while camped on the Santa Ana River about 30 miles southeast of Los Angeles. The earliest seismographs in use in California, and also the earliest in the United States, were installed by the University of California at Lick Observatory on Mount Hamilton, and at the University at Berkeley in 1887. The earliest seismograms of a major California earthquake are those of the San Francisco earthquake of 1906, which was recorded at seven California stations as well as elsewhere throughout the world.

One of the Bay area's largest earthquakes centered on the Hayward fault (within the San Andreas fault *zone*) in the East Bay on June 10, 1836. Surface faulting (ground ruptures) took place at the base of the Berkeley Hills from Mission San Jose to San Pablo. On October 21, 1868 another large earthquake centered on the Hayward fault with surface faulting for about 20 miles from Warm Springs to San Leandro. Maximum right-lateral (east block moving south) offset was about 3 feet.

In June of 1838 a strong earthquake originating on the San Andreas fault was accompanied by surface rupturing from Santa Clara almost to San Francisco. This damaged the Presidio at San Francisco and the missions at San Jose, Santa Clara, and San Francisco. Another strong earthquake centered on the San Andreas fault in the Santa Cruz Mountains on October 8, 1865. This was accompanied by ground cracks, land-slides, and dust clouds; buildings were damaged in San Francisco and at the New Almaden mercury mine, which was only a few miles east of the active part of the fault.

On April 24, 1890, a strong earthquake damaged Watsonville, Hollister, and Gilroy. Joe Anzar, who was a young boy living in the San Andreas rift valley in the nearby Chittenden Pass area at the time of that earthquake, was interviewed in 1963 by Olaf P. Jenkins and the writer. Mr. Anzar clearly remembered ground breakage, which caused Anzar Lake to drain, and landslides, which closed the railroad and highway where the fault trace crosses Chittenden Pass. He judged the motion to be stronger (at his home) than during the San Francisco earthquake of 1906.

The famous San Francisco earthquake, 5:12 A.M. local time, April 18, 1906, was probably California's greatest. Visible surface faulting occurred from San Juan Bautista to Point Arena, where the San Andreas fault enters the ocean. At the same time surface faulting also occurred 75 miles north of Point Arena at Shelter Cove in Humboldt County, probably along an extension of the San Andreas fault. The 1906 scarp viewed at Shelter Cove in 1963 clearly shows upthrow of 6 to 8 feet on the east side; there was no evidence of a horizontal component of displacement. However, offset of a line of old trees and an old fence viewed east of Point Arena in 1963 gave clear evidence of right-lateral displacement on the order of about 14 feet. The epicenter of the earthquake was near Olema, at the south end of Tomales Bay, near where a road was offset 20 feet in a right-lateral sense.

Richter magnitude is generally computed at about 8.25. Damage has been estimated at from $350 million to $1 billion. An estimated 700 people were killed. A large part of the loss was due to the tremendous fires in San Francisco, which resulted from broken gas mains and lack of water owing to numerous ruptures in the lines. Most extensive ground breaking in the city was near the waterfront in areas of natural Bay mud and artificial fill.

Another of California's great earthquakes, comparable in magnitude to the San Francisco 1906 earthquake, was caused by displacement on a segment of the San Andreas fault extending through the southern part of the Coast Ranges province and on beyond across the Transverse Ranges. This Fort Tejon earthquake of January 9, 1857, probably centered in the region between Fort Tejon in the Tehachapi Mountains and the Carrizo Plain in the southern Coast Ranges. Surface faulting extended for 200 to 275 miles from Cholame Valley along the northeast side of the Carrizo Plain through Tejon Pass, Elizabeth Lake, Cajon Pass, and along the south side of the San Bernardino Mountains. Accounts of this earthquake are unsatisfactory and inconclusive, but horizontal surface displacement almost certainly amounted to several feet in a right-lateral sense.

Perhaps among California's three greatest earthquakes was that in Owens Valley on March 26, 1872. At Lone Pine, 23 out of 250 people were killed and 52 out of 59 adobe houses were destroyed. The shock was felt from Shasta to San Diego. Surface faulting at the eastern foot of the Sierra Nevada produced scarps with a maximum net

vertical displacement of about 13 feet and horizontal right-lateral offset of about 16 feet. Surface faulting extended for perhaps 100 miles. This fault, of course, has no direct relation to the San Andreas.

There have been in historic times two great earthquakes (Fort Tejon and San Francisco) originating on the San Andreas fault, each accompanied by more than 200 miles of surface ruptures: one at the southern end of the Coast Ranges and one in the north. Between is left a segment, roughly 90 miles long, in the southern Coast Ranges, which has not been disrupted by surface faulting in historic time. It is interesting to note that the two ends of this segment—the Hollister area and the Parkfield area—are now the most seismically active in the southern Coast Ranges. The extreme southern segment—south of the Tehachapi Mountains—is quiet on the San Andreas fault proper, but very active on the closely related San Jacinto, Elsinore, Inglewood, and Imperial faults. In the segment marked by surface rupture in 1906, many earthquakes have originated in the central and southern part on the San Andreas fault and its auxiliary faults in the East Bay—the Hayward and Calaveras faults. However, since 1906 there have been no earthquakes on the most northerly segment from Marin to Humboldt Counties. The strongest earthquake in the Bay area since 1906 was the San Francisco earthquake of March 22, 1957, of magnitude 5.3. It originated at shallow depth near Mussel Rock, off the coast a few miles south of San Francisco; there was no surface faulting. No lives were lost, but minor damage to many homes in the Westlake-Daly City district totalled about a million dollars.

Land Forms in the Fault Zone

Extensive activity along the San Andreas fault zone in Quaternary time (last one million years of geologic time) has developed a linear depression, marked by all the features of a classic rift valley, extending the entire length of the fault and encompassing a width from a few hundred feet to over a mile and a half. Rift-valley features are particularly well expressed in the San Francisco Bay area, in the arid Carrizo Plain, and along the north side of the San Gabriel Mountains in southern California. Within the rift zone fault gouge and breccia always occur as well as a disorganized jumble of fault-brecciated rocks from both the eastern and western blocks, the result of hundreds of repeated ruptures on different fault planes in late Pleistocene and Recent time. Features of the rift valleys have resulted from: (1) Repeated, discontinuous fault ruptures on the surface, often with the development of minor graben, horsts, and pressure ridges; (2) land-sliding, triggered by earthquake waves and surface faulting; and (3) erosion of brecciated, readily weathered rock. Within the rift-valley troughs, it is common to find late Pliocene to Recent sediments.

Many of the observations made after the earthquake of 1906 are of great significance in understanding the origin and development of rift valleys and the nature of movement on the San Andreas fault: (1) open ruptures were mapped along the fault trace from San Juan Bautista to Point Arena, and at Telegraph Hill north of Shelter Cove in Humboldt County; (2) individual fault ruptures were not continuous, but extended for a few feet to a mile or a little more, with the continuations of the displacements being picked up along *en echelon* breaks; (3) the ruptures were often complex, with small grabens (downdropped blocks) and horsts (uplifted blocks) developed between breaks; (4) apparent movements were dominantly right lateral, with lesser vertical displacements; and (5) the amount of displacement varied irregularly along the fault trace, but in a gross way decreased in both directions from the maximum at the south end of Tomales Bay.

North of San Francisco, across Marin County, the fault follows a remarkably straight course approximately N. 35° W. The most prominent features are Bolinas Bay and the long, linear Tomales Bay which lie in portions of the rift valley drowned by rising sea waters following the Pleistocene glacial epoch. Between these bays, the rift zone is a steep-sided trough, in places as deep as 1,500 feet, with its lower levels characterized by a remarkable succession of minor, alternating ridges and gullies parallel to the general trend of the fault zone. Surfaces of the ridges and gullies are spotted by irregular hummocks and hollows; many of the hollows are undrained and have developed sag ponds, which are common along the San Andreas rift. Geologically Recent adjustment of the drainage in the rift zone leaves little positive evidence of the amount and direction of Recent displacement, except for that which took place in 1906. Offset lines of trees in this area still show the 13- to 14-foot horizontal right slip of 1906, and just south of Point

Arena trees also serve to show the 1906 offset. In the long stretch northward from Fort Ross to a point a few miles south of Point Arena, the broad expression of the rift zone is clear, but minor features within the zone have been obscured by erosion of the Gualala and Garcia Rivers and by the dense forest cover of the area.

South of San Francisco across San Mateo County the San Andreas fault zone follows the same trend as to the north but is less straight and is complicated by several subparallel faults. Near Mussel Rock, where the fault enters the land south of San Francisco, are great landslides which obscure the trace, and for a few miles to the southeast is a succession of sag ponds, notched ridges, and rift-valley lakes within a deeply trenched valley. The long, narrow San Andreas Lake and Crystal Springs Lakes are natural lakes which were enlarged many years ago by the artificial dams built to impound San Francisco's water supply. Similar rift-valley features mark the fault southward to the Tehachapi Mountains; because of the local aridity, they are particularly clear and striking in the Temblor Range area, in the Cholame Valley, and in the Carrizo Plain. As the fault enters its eastward bend in the San Emigdio Mountains area, the rift-valley features become less striking, perhaps because the contrast between the basement rocks in the east and west blocks disappears where the fault lies wholly within granitic rocks and older schists.

South of the Tehachapi the striking rift-valley land forms continue, through sag ponds like Elizabeth Lake and Palmdale Reservoir, along the northern margin of the San Gabriel Mountains through the Cajon Pass, and along the south side of the San Bernardino Mountains. None of the offsets and other rift-valley features in this southern segment of the great fault appears to be younger than 1857.

Displacement on the Fault

In 1953, geologists Mason L. Hill and T. W. Dibblee, Jr., advanced the possibility of cumulative horizontal right-lateral displacement of possibly 350 miles since Jurassic time (135,000,000 years ago) on the San Andreas fault. This hypothesis has received very wide acceptance among earth scientists, has intrigued geologists, and has been an important factor in stimulating work on the fault. Hill and Dibblee compared rock types, fossils, and gradational changes in rock characteristics in attempting to match units across the fault. By these methods they developed suggestions for horizontal displacement (east block moving south) of 10 miles since the Pleistocene (few thousand years), 65 miles since upper Miocene (about 12 million years), 225 miles since late Eocene (about 40 million years), and 350 miles since the Jurassic. At the opposite extreme, the late Professor N. L. Taliaferro of the University of California felt less confident about this "matching" of rock units and stated unequivocally that horizontal movement on the northern segment of the San Andreas fault has been less than 1 mile! Taliaferro believed that the principal movements on this great fault have been vertical.

Thus, geologic evidence is so varied that geologists have drawn conflicting interpretations of the geologic history and characteristics of the fault; at one extreme are those who believe that there has been several hundred miles of right slip since Late Jurassic time, and at the other are those who consider that there has been large vertical displacement on an ancient San Andreas fault and relatively small horizontal displacements in late Pliocene and Quaternary time.

Some of the latest work by geologists and engineers on the San Andreas fault in the San Francisco Bay area, and the closely related Hayward fault in the East Bay, shows that these faults are still active. In several places surface "creep," or slippage, is taking place. At the Almaden Winery a few miles south of Hollister, for example, creep occurs in spasms of movement of small fractions of an inch, separated by intervals of weeks or months. Average displacement, east block moving relatively south, is a half-inch a year. Several cases of well-substantiated right-lateral creep on the order of an eighth to a quarter of an inch per year have now been recognized along the 1868 trace of the Hayward fault from Irvington (Fremont) to the University of California stadium. Frequent earthquakes, with epicenters on these faults, also show that present-day movements are taking place.

Unsolved Problems

In spite of the interests of geologists, and the very considerable amount of time and attention given by geologists and seismologists to study of the San Andreas fault, it remains very incompletely known and understood. There is no agreement on answers to such interesting and fundamental questions

as: When did the fault originate? Should the late Quaternary and "ancestral" San Andreas be regarded as different faults, developed by different stresses, and with entirely different characteristics and displacements? Have the sense and direction of displacement (presently, right slip—east block moving south) always been the same, or has great vertical movement taken place? If the latter, which is the upthrown block? (Or has this changed from one side to the other in some segments during geologic time?) If dominantly right-lateral strike slip, has the present rate of displacement or strain been about the same for the last 100 million years? Is the cumulative displacement on the fault a few thousand feet or several hundred miles? To what depth does the faulting extend— 5 or 6 miles, as suggested by the depth of earthquake foci, or several times this? Is the San Andreas fault becoming more, or less, active? Are earthquakes, which center in the San Andreas fault zone, relieving stresses and thus lessening the chances of future earthquakes, or do the continuing earthquakes merely indicate a high level of seismic activity portending many future earthquakes? When may the next earthquake be expected?

Answers to these problems, so vital to our generation and generations of Californians to come, await the intensive work of geologists, seismologists, and other scientists of many disciplines. At present, fault movements and earthquakes are unpredictable; perhaps, however, our problems will become even more acute when we reach the state of knowledge which will allow prediction of earthquakes in time and place!

Notes

[1] Dr. Charles F. Richter, in 1935, devised a means of comparing the total energy of earthquakes expressed in terms of a figure now called the "Richter magnitude." The logarithm of the maximum trace amplitude in thousandths of a millimeter is taken from the measurement of earthquake waves on the seismogram of a certain standard seismometer at a standard distance from the epicenter. Constants have been worked out to make the figures comparable for other seismometers at other distances. On this scale magnitude $M = 2$ is the smallest earthquake felt. Earthquakes of $M = 4\frac{1}{2}$ to 5 cause small local damage, and $5\frac{1}{2}$–6 may cause an acceleration of one-tenth gravity and cause considerable damage. Earthquakes of 7 or more are called "major" earthquakes, and those of $7\frac{3}{4}$ and over are "great" earthquakes. Long Beach, 1933, with $M = 6.3$, was a "moderate" earthquake (but a very damaging one), Arvin-Tehachapi, 1952, at $M = 7.7$, was a major earthquake, and San Francisco, at 8.25 in 1906, was a great earthquake. Local size or strength has long been measured by an *intensity* scale based on how the earthquake is felt and its apparent damage. The commonest intensity scale in use is the Modified Mercalli.

References

Allen, C. R., St. Amand, P., Richter, C. F., and Nordquist, J. M., 1965, Relationship between seismicity and geologic structure in the southern California region: Seismological Society of American Bull., v. 55, no. 4, p. 753–797.

Bateman, Paul C., 1961, Willard D. Johnson and the strike-slip component of fault movement in the Owens Valley, California, earthquake of 1872: Seismological Society of America Bull., v. 51, p. 483–493.

Blanchard, F. B., and Laverty, C. L., 1966, Displacements in the Claremont Water Tunnel at the intersection with the Hayward fault: Seismological Society of America Bull., v. 56, no. 2, p. 291–294.

Bolt, Bruce A., and Marion, Walter C., 1966, Instrumental measurement of slippage on the Hayward fault: Seismological Society of America Bull., v. 56, no. 2, p. 305–316.

Bonilla, M. G., 1966, Deformation of railroad tracks by slippage on the Hayward fault in the Niles district of Fremont, California: Seismological Society of America Bull., v. 56, no. 2, p. 281–289.

Byerly, Perry, 1951, History of earthquakes in the San Francisco Bay area: California Division of Mines Bull. 154, p. 151–160.

California Division of Mines and Geology, 1958–1965, Geologic map of California, Olaf P. Jenkins edition, published sheets: California Division of Mines and Geology, scale 1:250,000.

California Resources Agency, 1964, Earthquake and geologic hazards conference, December 7 and 8, 1964, San Francisco, California, 154 p.

California Resources Agency, 1966, Landslide and subsidence conference, Los Angeles, California.

Cluff, Lloyd, 1965, Evidence of creep along the Hayward fault: Association of Engineering Geologists, First Ann. Joint Meeting San Francisco-Sacramento Section, paper delivered at Berkeley, Sept. 25, 1965.

Cluff, Lloyd S., and Steinbrugge, Karl V., 1966, Hayward fault slippage in the Irvington-Niles districts of Fremont, California: Seismological Society of America Bull., v. 56, no. 2, p. 257–279.

Hill, M. L., and Dibblee, T. W., Jr., 1953, San Andreas, Garlock, and Big Pine faults, California—a study of the character, history, and tectonic significance of their displacements: Geological Society of America Bull., v. 64, no. 4, p. 443–458.

Lawson, A. C., and others, 1908, The California earthquake of April 18, 1906, Report of the State Earthquake Investigation Committee: Carnegie Inst. Washington Pub. 87, v. 1, pts. 1–2, 451 p.

7 Man-made Earthquakes in Denver

David M. Evans

From April 1962 to November 1965, Denver experienced more than 700 earthquakes. They were not damaging—the greatest magnitude was 4.3 on the Richter scale—but the community became increasingly concerned. More and more people took out earthquake insurance. There was talk in the press that Denver might be removed from the list of possible sites for a $375 million accelerator to be built by the Atomic Energy Commission because it was becoming known as an earthquake area.

In November 1962, I publicly suggested that there was a direct relationship between the earthquakes and contaminated waste-water being injected into a 12,045-foot disposal well at the Rocky Mountain Arsenal, northeast of Denver.[1] Representative Roy McVicker of Colorado immediately called for a full scientific investigation of the tremors, and on March 19, 1966, the U. S. Geological Survey released the results of its studies in coöperation with the Colorado School of Mines, Regis College in Denver, and the University of Colorado.[2] The USGS concluded that "The pumping of waste fluids into a deep disposal well at the Rocky Mountain Arsenal near Denver appears to be a significant cause of a series of minor earthquakes that have occurred just north of Denver since the spring of 1962."

Since 1942, the Rocky Mountain Arsenal has manufactured products on a large scale for chemical warfare and industrial use, under direction of the Chemical Corps of the U. S. Army. One by-product of this operation is contaminated waste-water and, until 1961, the water was disposed of by evaporation from earthen reservoirs.

When it was found that the waste-water was contaminating the ground water and endangering crops, the Chemical Corps tried evaporating the water from water-tight reservoirs. That failed, so the Corps decided to drill an injection disposal well.

It commissioned E. A. Polumbus Jr. & Associates Inc. to design the well, supervise drilling and completion, provide the necessary engineering-geology services, and manage the project. Louis J. Scopel, as an associate, was the project geologist. Another associate, George R. Downs, contributed to the initial design and acted as an adviser.

The well was drilled in NW¼ NE¼ sec. 26, T2S, R67W, Adams County, Colorado. It was completed in September 1961 at a total depth of 12,045 feet.[3]

Regional Geology

The Rocky Mountain Arsenal disposal well is on the gently dipping east flank of the Denver-Julesburg Basin, just a few miles east of the basin axis. As

Evans, D. M., 1966, "Man-made Earthquakes in Denver," *Geotimes*, vol. 10, no. 9, pp. 11–18. Lightly edited by permission of the author and reprinted by permission of the American Geological Institute. Copyright 1966 by A. G. I.

Mr. Evans is a consulting geologist in Denver, Colorado.

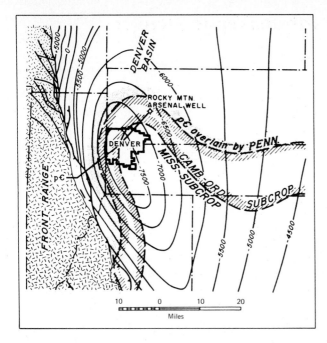

FIG. 1. Structural map of a part of the Denver-Julesburg Basin shows the location of the Rocky Mountain Arsenal Well. (After Anderman and Ackman[4])

indicated in Fig. 1, it is in a region of the . subcrop of Cambro-Ordovician rocks near the area where those rocks are truncated and overlain by Pennsylvanian sediments.[4] Figure 2 is a cross-section that shows the subsurface geology from the Arsenal well to the outcrop of Precambrian granite gneiss west of Denver.[5]

About 13,000 feet of structural relief exists between the top of the Precambrian in the Arsenal well and the Precambrian outcrops.

Injection in Precambrian Rocks

According to Scopel,[3] Precambrian rocks were penetrated in the Arsenal well from 11,950 feet to the total depth of 12,045. He described the rocks as bright green weathered schist from 11,950 to 11,970 and as highly fractured hornblende granite gneiss containing pegmatite intrusions from 11,970 to the bottom of the hole.

As a part of the USGS study, Sheridan, Wrucke, and Wilcox[2] analyzed the core and cuttings from the lower part of the well, and concluded that the top of the Precambrian is at 11,935. They describe the section from 11,970 to 12,045 as "migmatitic gneiss: rock containing fine-to medium-grainedhornblendic biotite-quartz-feldspar rock, containing steeply dipping open fractures, and thin calcite and ankerite-filled veinlets and microbreccias." They point out the striking similarity between the fractured Precambrian gneiss of the Arsenal well and the breccia-reef faults and fracture zones in the Precambrian outcrop of the Front Range west of Denver.

In the Arsenal well, a 5½-inch liner was cemented 5 feet into the Precambrian gneiss at 11,975 feet, and 5½-inch tubing was run to a depth of 9,011 feet,

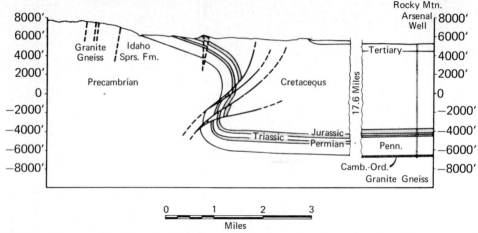

FIG. 2. Cross-section shows subsurface geology from the Arsenal well to the outcrop of Precambrian granite gneiss west of Denver. (After M. F. and C. M. Boos and H. H. Odiorne) The line of cross-section is shown in Fig. 1.

to complete the well for injection into the almost vertically fractured gneiss from 11,975 feet to 12,045.

Pumping and pressure-injection tests were made from November 1961 to February 1962 to obtain reservoir fluid samples and to determine rates and injection pressures at which the reservoir would take fluid.

A conventional oil-field pump was run in the well, and pumping tests were made. After pumping out 1,100 barrels of salt water more than the fluid lost in the hole during drilling, the well pumped down and recovery became negligible. It was concluded at the time of testing that fluid recovery was from fractures. It was believed further that as fluid was withdrawn from these fractures they were squeezed shut by compressive forces, which restricted fluid entry into the well bore.

When fluid injection tests were made, it was noticed that as fluid was injected the calculated drainage radius and formation capacity increased.

That was interpreted as an indication that the reservoir consisted of fractures that expanded as additional fluid was injected.

In March 1962 the Arsenal disposal program began, and 4.2 million gallons of waste was injected. The Denver earthquakes started the next month.

The monthly volume of waste injected is shown in the lower half of Fig. 3. From March 1962 until September 1963, the maximum injection pressure is reported to have been about 550 pounds per square inch, with an injection rate of 200 gallons a minute.

At the end of September 1963 the injection well was shut down, and no fluid was injected until September 17, 1964. During the shut-down, surface evaporation from the settling basin was sufficient to handle the plant output.

From September 17, 1964, until the end of March 1965, injection was resumed by gravity discharge into the

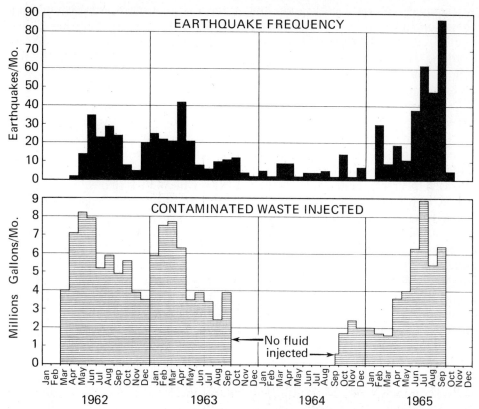

FIG. 3. Upper half: number of earthquakes per month recorded in the Denver area. Lower half: monthly volume of contaminated waste water injected into the Arsenal well.

well. No well-head pressure was needed to inject the maximum of 2.4 million gallons of waste per month into the well. Beginning in April 1965 larger quantities of fluid were injected. During April and May a maximum pressure of 1,050 pounds was required to inject 300 gallons a minute.

The Denver Earthquakes

The U. S. Coast & Geodetic Survey reports that on November 7, 1882, an earthquake was felt in Denver and nearby Louisville and Georgetown, and in southeast Wyoming. According

to Joseph V. Downey, director of the Regis College Seismological Observatory, no earthquake epicenters were recorded in the Denver area by either the C&GS or Regis between 1882 and the first earthquake in April 1962. (The Regis Observatory has been operating since 1909.)

From 1954 to 1959 a seismic station was operated at the University of Colorado in Boulder, directed by Warren Longley. As a part of the recent USGS investigation, Harold L. Krivoy and M. P. Lane analyzed the records from that station.[2] They found a few small events that might have been earthquakes in the Derby area, but they concluded that, since all those events occurred during weekday working hours, they were probably due to construction blasting or explosives disposal at the Arsenal.

From April 1962 to the end of September 1965, 710 earthquakes with epicenters in the vicinity of the Arsenal were recorded at the Cecil H. Green Observatory, Bergen Park, Colo., which is operated by Colorado School of Mines.[6]

The total number of earthquakes reported in the Denver area is plotted in the upper half of Fig. 3. The magnitude of the earthquakes reported range from 0.7 to 4.3 on the Richter scale. About 75 were intense enough to be felt. Yung-liang Wang[6] calculated the epicenters and hypocenters of the 1963–65 Denver earthquakes, and Fig. 4 shows the results of his calculations.

Most of the epicenters are within 5 miles of the well. All epicenters calculated from four or more recording stations are within 7 miles of the well.

Wang[6] calculated the best-fitting plane passing through the zone of hypocenters determined from four or more recording stations. He concluded that this plane might be a fault along which movement was taking place. The plane dips east and passes beneath the Arsenal well about 6.5 miles below the surface. (See Fig. 4)

In the USGS study,[2] J. J. Healy, W. H. Jackson and J. R. Van Schaack report that Wang's data were compiled from records of available seismographs and that most of the earthquakes plotted were with fewer than four stations and that only a few were located with four stations. Also, the four stations available to Wang in his study were not optimally placed to locate earthquakes in the vicinity of the Arsenal. Therefore the USGS set up a seismic network around the Arsenal well that would greatly improve the accuracy of earthquake location. Up to eight seismic-refraction units were in operation at the same time, and during the study from one to 20 micro-earthquakes were recorded every day.

Healy, Jackson and Van Schaack concluded that the precise USGS work showed the epicenters clustered even more closely around the Arsenal well than Wang had reported. The epicenters located by the USGS outline a roughly ellipsoidal area (which includes the well) about six miles long and three miles wide, suggesting the presence of a fault or fracture zone trending about N 60° W. The epicenters of the events studied in detail were between 4.5 and 5.5 km deep.

Pressure Injection and Earthquake Frequency

Pressure injection began in March 1962. The first two earthquakes with epicenters in the Arsenal area were recorded in April 1962.

The lower half of Fig. 3 is a graph of the monthly volume of waste injected

FIG. 4. Earthquake hypocenters are shown here for 1963–64 as computed by seismological stations in the Denver area. All epicenters calculated from four or more recording stations are within 7 miles of the Arsenal well. All hypocenters calculated from four or more recording stations are within the area indicated on section A-A. (After Wang[6])

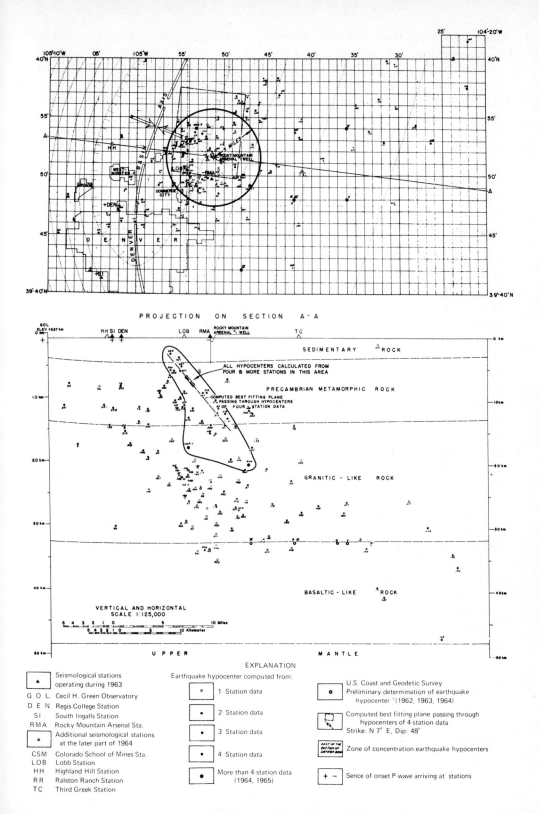

PROJECTION ON SECTION A-A

SEDIMENTARY ROCK

ALL HYPOCENTERS CALCULATED FROM
FOUR & MORE STATIONS IN THIS AREA

PRECAMBRIAN METAMORPHIC ROCK

COMPUTED BEST FITTING PLANE
PASSING THROUGH HYPOCENTERS
OF FOUR STATION DATA

GRANITIC - LIKE ROCK

BASALTIC - LIKE ROCK

VERTICAL AND HORIZONTAL
SCALE 1:125,000

UPPER   MANTLE

EXPLANATION

Earthquake hypocenter computed from:

- Seismological stations operating during 1963
G O L   Cecil H. Green Observatory
D E N   Regis College Station
SI   South Ingalls Station
RMA   Rocky Mountain Arsenal Sta.
- Additional seismological stations at the later part of 1964
CSM   Colorado School of Mines Sta.
LOB   Lobb Station
HH   Highland Hill Station
RR   Ralston Ranch Station
TC   Third Greek Station

o   1- Station data
•   2- Station data
•   3- Station data
•   4- Station data
●   More-than 4-station data (1964, 1965)

U.S. Coast and Geodetic Survey Preliminary determination of earthquake hypocenter '(1962, 1963, 1964)

Computed best fitting plane passing through hypocenters of 4-station data
Strike: N 7° E, Dip: 48°

Zone of concentration earthquake hypocenters

+ −   Sence of onset P-wave arriving at stations

into the Arsenal well. The total number of earthquakes recorded in the Arsenal area is plotted for each month in the upper half of the graph.

During the initial injection period, from March 1962 to the end of September 1963, the injection program was often suspended for repairs to the filter plant. In this period there does not appear to be a direct month-by-month correlation. However, the high injection months of April, May and June 1962 seem to correlate with the high earthquake frequency months of June, July and August. The high injection months of February and March 1963 may correlate with the high earthquake month of April.

The period of no injection from September 1963 to September 1964 coincides with a period of minimum earthquake frequency. The period of low-volume injection by gravity flow, from September 1964 to April 1965, is characterized by two months (October and February) of greater earthquake frequency than experienced during the preceding year.

The most direct correlation of fluid injection with earthquake frequency is during the months of June through September 1965. That period was characterized by the pumping of 300 gallons a minute, 16 to 24 hours a day, at a pressure of 800 to 1,050 pounds.

There have been five characteristic periods of injection (see Fig. 5):

April 1962–April 1963; high injection at medium pressure.

May 1963–September 1963; medium injection at medium pressure.

October 1963–September 1964; no injection.

September 1964–March 1965; low injection at no pressure (gravity feed).

April 1965–September 1965; high injection at high pressure.

The average numbers of earthquakes per month are shown in Fig. 5 above the average volumes of fluid injected per month for each of those five periods. The injection for March 1962 is not used in the averages because the exact day injection was started is unknown.

Figure 5 indicates that there is a direct correlation between average monthly injection and earthquake frequency when an injection program has been carried out for several months.

The period of October, November and December 1965 provided the first check period of the correlation between earthquakes and fluid injected. From October 1 to December 20 an average of 3.8 million gallons a month were injected at an average pressure of 1,000 pounds. On December 20 the pressure was reduced to 500 pounds. From Fig. 5, it can be seen that during May–September 1963 approximately the same amount of fluid was injected at roughly half the pressure and an average of 12 earthquakes a month were recorded. With an injection pressure of 1,000 pounds, about twice as many tremors (as recorded during April–September 1965) would be expected. Allowing for the 10 days in December when the pressure was reduced, an average of about 25 earthquakes a month would be predicted. Actually, 68 shocks were recorded, for an average of slightly less than 23 a month.

During January 1966, about 2.4 million gallons were injected, and 19 shocks were recorded. On January 20, 1966, pumping was stopped; during

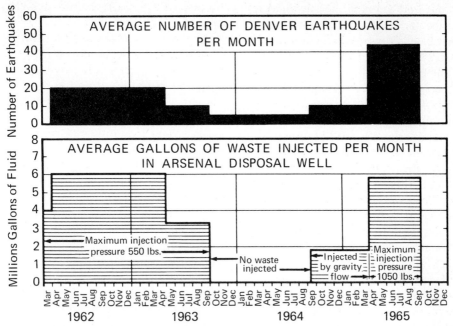

FIG. 5. Relationships of earthquake frequency and waste injection are shown here for five characteristic periods.

February, 200,000 gallons were injected by gravity flow. On February 20 the well was shut in. Ten earthquakes were recorded during February at the Cecil H. Green Observatory (whose earthquake count has been used in this report).

George Bardwell[7] has made a statistical analysis of the relationship between fluid injection at the Arsenal well and earthquake frequency in the area. Even though his study did not include the effect of injection pressure, he concluded that the probability of the injection-earthquake relationship being due to random fluctuation was about 1 in 1,000.

Fluid Pressure and the Arsenal Earthquakes

Evidence gained from drilling and testing the Arsenal disposal well indicates that the Precambrian reservoir is composed of a highly fractured gneiss that is substantially impermeable. It indicates that as fluid was pumped out of the reservoir the fractures closed, and as fluid was injected the fractures opened. In other words, the pumping and injection tests indicated that rock movement occurred as fluid was withdrawn or injected at relatively low pressures.

The pressure-depth relations of the Precambrian reservoir, showing hydrostatic and lithostatic pressure variations with depth, are shown in Fig. 6. Those data were determined from a drill-stem test. As shown on the chart, the observed pressure of the Precambrian reservoir is almost 900 pounds less than the hydrostatic pressure.

Hubbert and Rubey[8] have devised a simple and adequate way to reduce by the required amount the frictional resistance to the sliding of large overthrust blocks down very gentle slopes. It arises from the circumstance that the weight of such a block is jointly supported by solid stress and the pressure of interstitial fluids. As the fluid pressure approaches the lithostatic pressure, corresponding to flotation of the overburden, the shear stress required to move the block approaches zero.

If high fluid pressures reduce frictional resistance and permit rocks to slide down very gentle slopes, it follows that as fluid pressure is decreased frictional resistance between blocks of rock is increased, permitting them to come to rest on increasingly steep slopes. The steeper the slope on which a block of rock is at rest, the lower the required increase in fluid pressure necessary to produce movement.

In the case of the Precambrian reservoir beneath the Arsenal well, the rocks were at equilibrium on high-angle fracture planes with a fluid pressure of 900 pounds less than the hydrostatic pressure before injection began.

As fluid was injected into the Precambrian reservoir, the fluid pressure adjacent to the well bore rose, and the frictional resistance along the fracture planes was thereby reduced. When, finally, enough fluid pressure was exerted over enough area, movement occurred. The elastic energy released was recorded as an earthquake.

Since the formation fluid pressure is 900 pounds subhydrostatic, merely filling the hole with contaminated waste (mostly salt water) raises the formation pressure 900 pounds, or to the equivalent of hydrostatic pressure. Any applied injection pressure above that of gravity flow increases pressure to a total higher than hydrostatic pressure. For example, an injection pressure of 1,000 pounds would raise the reservoir pressure adjacent to the well bore to 1,900 pounds, or by the amount to bring the pressure to hydrostatic (by filling the hole) plus 1,000 pounds.

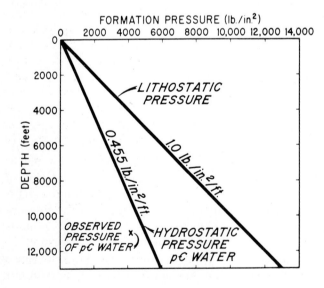

FIG. 6. Pressure-depth relations are shown here for the Precambrian reservoir at the Arsenal well.

Apparently a rise in fluid pressure within the Precambrian reservoir of 900 to 1,900 pounds is enough to allow movement to occur.

Open Fractures

The hypocenters in the Arsenal area plotted from data derived from four or more recording stations indicate that movement takes place 1.5 to 12 miles below the well. If the Precambrian fracture system extends to a depth of 12 miles, then fluid pressure could be transmitted to that depth by moderate surface injection pressure as long as the fracture system is open for transmission of that pressure.

Secor[9] concluded that open fractures can occur to great depths even with only moderately high fluid pressure-overburden weight ratios. It appears possible that high-angle open fractures may be present beneath the Arsenal well at great depths with much lower fluid pressure-overburden weight ratios than has formerly been considered possible.

Time Lag Between Fluid Injections and Earthquakes

The correlation of fluid injected with earthquake frequency (Fig. 3) suggests that the two are separated by a time lag. Bardwell[7] notes that the frequency of Denver earthquakes appears to lag waste injection by one to four months. This phenomenon is probably the same as that described by Serafim and Del Campo.[10] They describe the observed time lag between water levels in reservoirs and the pressures measured in the foundations of dams, and ascribe it to an unsteady rate of percolation through open joints in the rock mass, due to the opening and closing of the passages resulting from internal and externally applied pressures.

The time lag between waste injection in the Arsenal well and earthquake frequency is probably due to an unsteady rate of percolation through fractures in the Precambrian reservoir due to the opening and closing of these fractures resulting from the applied fluid pressure of the injected waste. The delayed application of this pressure at a distance from the well bore is believed to trigger the movement recorded as an earthquake.

Earthquakes During Shut-Down Period

In considering the earthquake frequency during the year the injection well was shut down, unfortunately neither periodic bottom-hole pressure tests nor checks of the fluid level in the hole were made. If these measurements had been made, we would know how long it took bottom-hole pressure to decline.

By the end of September 1963, about 102.3 million gallons of fluid had been injected into the well. It is believed that this injection had raised the fluid level pressure in the reservoir for some distance from the well bore. During shut-down, this elevated pressure was equalizing throughout the reservoir and at increasing distance from the well bore. As this fluid pressure reduced the frictional resistance in fractures farther from the well, movement occurred, and small earthquakes resulted.

Conclusion

The Precambrian reservoir receiving the Arsenal waste is highly fractured gneiss of very low permeability. The fractures are nearly vertical. The fracture porosity of the reservoir is filled with salt water. Reservoir pressure is 900 pounds subhydrostatic.

It appears that movement is taking place in this fractured reservoir as a result of the injection of water at pressures from 900 to 1,950 pounds greater than reservoir pressure.

Hubbert and Rubey[8] point out that rock masses in fluid-filled reservoirs are supported by solid stress and the pressure of interstitial fluids. As fluid pressure approaches lithostatic pressure, the shear stress required to move rock masses down very gently dipping slopes approaches zero.

These principles appear to explain the rock movement in the Arsenal reservoir. The highly fractured rocks of the reservoir are at rest on steep slopes under a condition of subhydrostatic fluid pressure. As the fluid pressure is raised within the reservoir, frictional resistance along fracture planes is reduced and, eventually, movement takes place. The elastic wave energy released is recorded as an earthquake.

In the present case, I believe that a stable situation in this Precambrian reservoir was made unstable by fluid pressure. It is interesting to speculate that the principle of increasing fluid pressure to release elastic wave energy could be applied to earthquake modification. That is, it might be possible to relieve the stresses along some fault zones in urban areas by increasing the fluid pressures along the zone, using a series of injection wells. The accumulated stress might thus be released at will in a series of non-damaging earthquakes instead of eventually resulting in one large event that might cause a major disaster.

Notes

[1] David M. Evans, 1966, The Denver area earthquakes and the Rocky Mountain Arsenal disposal well: *The mountain geologist*, v. 3, no. 1, p. 23–36.

[2] J. H. Healy and others, 1966, Geophysical and geological investigations relating to earthquakes in the Denver area, Colorado: U. S. Geological Survey open-file report.

[3] L. J. Scopel, 1964, Pressure injection disposal well, Rocky Mountain Arsenal, Denver, Colo.: *The mountain geologist*, v. 1, no. 1, p. 35–42.

[4] G. G. Anderman and E. J. Ackman, 1963, Structure of the Denver-Julesburg Basin and surrounding areas: in Rocky Mountain Association of Geologists' *Guidebook to the geology of the northern Denver Basin and adjacent uplifts*.

[5] C. M. Boos and M. F. Boos, 1957, Tectonics of the eastern flank and foothills of the Front Range, Colorado: American Assn. of Petroleum Geologists *Bulletin*, v. 41, p. 2,603–2,676.

[6] Yung-liang Wang, 1965, *Local hypocenter determination in linearly varying layers applied to earthquakes in the Denver area:* unpublished DSc dissertation, Colorado School of Mines.

[7] G. E. Bardwell, 1966, Some statistical features of the relationship between Rocky Mountain Arsenal waste disposal and frequency of earthquakes: *The mountain geologist*, v. 3, no. 1, p. 37–42.

[8] M. King Hubbert and W. W. Rubey, 1959, Role of fluid pressure in mechanics of overthrust faulting; part 1, Mechanics of fluid-filled porous solids and its application to overthrust faulting: Geological Society of America *Bulletin*, v. 70, p. 115–166.

[9] D. T. Secor Jr, 1965, Role of fluid pressure in jointing, *American journal of science*, v. 263, p. 633–646.

[10] J. L. Serafim and A. del Campo, 1965, Interstitial pressures of rock foundations of dams: *Journal* of the Soil Mechanics & Foundations Division, Proceedings of the American Society of Civil Engineers v. 91, no. SM5.

8   A Federal Plan for the Issuance of Earthquake Predictions and Warnings

V. E. McKelvey

We are now entering an age when scientific instruments are detecting signals that can be interpreted to forecast earthquake occurrence. We have not advanced to the stage of full-scale deployment of earthquake-prediction systems. In fact, we now have in operation only an experimental system covering a small area of central California. However, advance indications of a coming earthquake can be detected on fairly inexpensive instruments that would give a scientific basis for making a prediction.

With the increased deployment of geophysical sensors the number of scientifically based predictions is increasing. Scientists recognize that the data are difficult to interpret at this stage, but nevertheless it is clear that the observations must be reported to the public, and the best interpretation possible must be attempted. Developing a plan to issue predictions may appear to be premature when the capability is not really operational, but the impact that a prediction can have requires that even the most fragmentary data be processed in a careful and responsible manner.

Our plan is tentative and intended as a basis for discussion. We expect it to evolve as progress is made in

The author was the director of the U.S. Geological Survey when this article was written. Reprinted with light editing from *U.S. Geological Survey Circular 729* (1976), pp. 10–12.

prediction research and as experience is gained in issuing predictions.

There is a difference between a *prediction* and a *warning*. A *prediction*, as we are using the term here, is a forecast that an earthquake will occur at a certain time and place and will have a certain magnitude capable of causing certain kinds of effects. A *warning* is a recommendation or an order to take some defensive action, such as to reduce the water level in a reservoir or to evacuate a building. As you will see in the plan, the U.S. Geological Survey (USGS) has the responsibility for issuing a prediction, but State and local officials have the responsibility for issuing a warning. The plan deals with formulating the prediction and transmitting it to the local officials for issuance of a warning to the public.

The steps in this plan are shown in Fig. 1. The starting point is with the scientists who are receiving data from field instruments and interpreting them. This is also the starting point of contact with the public, for it is our policy that the raw data be made available to the public. The scientific interpretation of data will be reported through talks at scientific meetings, publications, and information releases. Care will be taken, however, to distinguish between an individual scientist's interpretation of data and the interpretation of his data and other relevant data by his peers that might result in a USGS prediction.

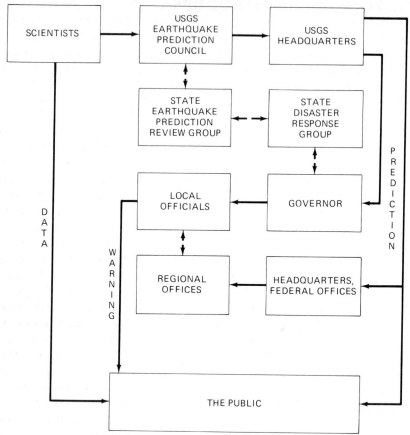

FIG. 1. Proposed Federal plan for the issuance of earthquake predictions and warnings. It provides for continual public release of scientific data but ensures that there are firm bases for an official prediction. In the plan, the U.S. Geological Survey has the responsibility for issuing a prediction (statement that an earthquake will occur), whereas state and local officials have the responsibility for issuing a warning (recommendation or order to take defensive action).

Peer review will be provided by the USGS Earthquake Prediction Council composed of five to ten USGS scientists and scientists from outside the USGS whose experience covers all aspects of earthquake-prediction technology. The purpose of this review is to assure the public that, in the judgment of scientific experts, the basis for the prediction is sound. The role of the Council is to review all relevant data and to report its conclusions. The report need not be a consensus report, and it might not agree with the conclusion of the scientist presenting the data. He, of course, could continue to argue his case, but he must make clear that his is not a USGS position.

The report of the Earthquake Prediction Council would go to USGS head-quarters, where a decision would be made whether and how to issue a prediction. If the case is not sufficiently strong, a decision could be made to issue an advisory notice, stating, for example, that possible precursors have been detected in a certain area and that that area is under intensive study. The nature of the headquarters action would be tailored to the particular situation.

The statement issued by the USGS headquarters would be communicated to the governor of the state potentially affected, to Federal agencies with responsibilities for disaster preparedness and response, for example, the Federal Disaster Assistance Administration and the Defense Civil Preparedness Agency, and to the public. This does not necessarily mean that the public would be notified simultaneously with the others, but any delay should be short. It may be judged that the negative impact of a prediction could be lessened if responsible state and Federal officials received prior notice. A strong case can be made that a warning should be issued with a prediction, so that the public is not left without any recommendation for appropriate action.

Upon receipt of a prediction, we anticipate that the governor's office would refer the prediction to the state office concerned with disaster response, and the governor may choose to call together his own group of experts to evaluate the evidence for the prediction. In California, for example, the prediction would probably be referred to the Office of Emergency Services (OES) and then to the gover-

nor's/OES Earthquake Prediction Review Group. USGS scientists would certainly be available for discussion with state personnel, as indicated on the chart by the dashed line.

The governor's decision about the prediction would be transmitted to local officials and a warning issued. USGS personnel would be available for consultation at every stage of this process. The procedures adopted will surely vary from state to state.

The prediction going to the headquarters of other Federal agencies would be transmitted to their regional offices, where coordination would be effected with state personnel.

Scientists not funded by the USGS who find evidence of an earthquake precursor are not specifically considered in this plan. We believe, however, that given the mechanisms I have described, other scientists would be willing to enter the system, by discussing their data with either the USGS Earthquake Prediction Council or the state earthquake prediction review group. In either way, the findings would be evaluated as a basis for issuing a prediction.

Perhaps this is sufficient discussion of the plan on my part. As I mentioned earlier, this is only a proposal, and we are here to receive your suggestions for improvements. We want a procedure that assures responsible, open, and credible treatment of earthquake-prediction information. We realize that the transition to the age of earthquake prediction will be difficult, but a carefully developed prediction and warning plan can ease the difficulties and yield great savings in both life and property.

9 Possible Loss-Reduction Actions Following an Earthquake Prediction

Charles C. Thiel

I propose to discuss what can be done—what mitigating public and private actions can be taken in anticipation of the event—when a prediction is of such scientific validity that it constitutes an official warning. Whether one agrees with the proposition that earthquake prediction is good (as I do) or not (and there is some dispute) is really a moot point. There are going to be predictions, some scientifically based and others from seers and fortunetellers. It will be the responsibility of the public official, the engineer, the applied scientist, and the entrepreneur to take the information that an earthquake will occur and turn this into public and private policies and actions that allow the individual and the community to reduce the net impact of the earthquake in terms of loss of life, injury, economic cost, and social disruption.

A natural tendency is to think in terms of earthquakes posing a major risk only to the Western states. Without delving into the specifics, 70 million people in 39 states live in regions of major and moderate earthquake risk. This conference is directed at the representatives of the nine Western states at principal risk, each having a major vulnerability to loss of life and property destruction.

The author was the research applications director of the National Science Foundation when this article was written. Reprinted from *U.S. Geological Survey Circular* 729 (1976), pp. 13–16.

I have summarized the extent of the vulnerability of these states in the West by listing in Table 1 the population residing in major- and moderate-damage-potential zones and the projected state property values. Although California clearly has the largest number of people at risk, each of the other states have large parts of their population exposed. There is some likelihood that earthquakes will be predicted in the future in each of these states.

It would be nice if, in dealing with earthquake prediction, we could merely decide to leave the area, wait for the earthquake to take place, and then go back and pick up the pieces. Unfortunately, this is impractical. As an example, if the 1906 San Francisco earthquake were to recur in the year 2000, it could cause about $20 billion in damage, 9,000 deaths, and 400,000 injuries. Even so, this does not represent the destruction of the Bay area—it represents the loss of less than 25 percent of the value of structures in the region. The damage would be widely distributed, and those persons in San Jose could not seek refuge in Oakland, San Francisco, or even Marin County. I could just as well show you similar losses for Seattle, Salt Lake City, or Los Angeles. Obviously we cannot count on mass evacuation as a means of protecting the public when the population of the area affected is large; we must seek other means to protect life and property. We cannot run away from the problem.

Table 1 Population and property at high earthquake risk in the Western United States

State	Population (thousands, 1970) Major risk	Moderate risk	State's property value (billion dollars, 1980 est.)
Alaska	270	25	2.9
California	17,317	2,636	292.1
Hawaii	63	39	19.2
Idaho	200	513	8.1
Montana	142	313	7.0
Nevada	189	300	8.2
Utah	972	48	9.0
Washington	2,169	1,240	46.4
Wyoming	5	19	7.4

An earthquake can cause damage in several ways. First, I want you to understand that it is not the fault rupture but the shaking of the ground that causes most of the damage. Extensive ground shaking can cause the disruption of a building's contents and its collapse, the collapse of a dam, the rupture of a pipeline, and many other types of structural damage. Ground shaking can cause soils to lose their capacity to support buildings, causing them to fail. The damage can be from secondary sources—the ruptured dam can cause downstream inundation, toxic chemicals can be released from an industrial facility, or a falling parapet can strike a pedestrian. Old unreinforced buildings are generally the most vulnerable to collapse, and there are lots of them, even in Los Angeles and San Francisco where a long-term attempt has been made to build earthquake-resistant structures. Finally there is fire. Conflagration can be a companion to earthquakes, and the resulting devastation can be nearly total.

Given that an earthquake is going to take place, a variety of actions and procedures can be undertaken to reduce its direct impacts and hasten the restoration of the community.

We can evacuate hazardous buildings or sites. We can reinforce or replace structures that will not perform adequately. Note this last word, "adequately," because we want a hospital to remain operational but may not care if a shed collapses. We can remove the contents of a structure so that they will not be damaged. We can change the pattern of use of a facility or area—for instance, not use a theater, or move a clinic to a "better" building. We can activate emergency-preparedness plans and distribute emergency materials. We can review insurance options. We can adopt tax policies that benefit owners who upgrade their facilities. Finally, we can provide for the relief of the victims and rehabilitation of the community. This list is far from complete. It is meant to show that many methods are available to decrease earthquake impacts.

The actual nature of the specific response to an earthquake prediction will be determined by the size of the predicted earthquake, how long until the event is to take place, the time of year, and whether the earthquake will occur in an urban, suburban, or rural region. It will also depend on the technical and managerial skills of the resident population and its economic and material capabilities. Generally, the longer the period of forewarning, the more the community will be able to do to decrease the impact on the community. As a caution, please remember that, in most cases, it is unrealistic to expend more than can be expected in losses. We must always be careful to balance social and economic costs and resultant benefits in reduced human suffering and property damage.

From an engineering point of view, there seem to be four basic time spans that should be discussed when considering the types of specific actions to be taken. These are 3, 30, 300, and 3,000 days. Table 2 lists a group of actions that could be initiated according to the various stated lead times. I have assumed that we are dealing with an urban event of major magnitude, and have distinguished the actions that could be taken in the time allowed to protect buildings, their contents, lifeline facilities (for example, bridges, communications, water, and hospitals), and special structures (for example, dams, nuclear reactors, and pipelines). If the prediction is a few days in advance, say three, the options are limited to the somewhat obvious.

We can evacuate previously identified hazardous buildings and selectively remove contents. Special facilities such as reactors can be closed down. Petroleum pipelines can be emptied and shut down. We can deploy emergency materials and identify staging areas. Use of mass-assembly buildings such as theaters and schools can be restricted.

When there is a 30-day warning period, building inspections can be performed to identify hazardous buildings and conditions. In most cases there will be neither the professional expertise, skilled labor, or materials available to reinforce hazardous buildings in this short time. Reservoirs can be emptied within this time but not much faster. Hospital patients and prisoners can be moved to facilities beyond the area to be affected. Clinics, emergency communications, and emergency response and relief personnel and materials can be moved to less vulnerable sites. Some toxic, incendiary, and explosive materials can be removed from industrial facilities to places where they pose no major hazards.

Only when the period of warning is of the order of 300 days is it realistic to expect that substantial numbers of structures can be upgraded to reduce their vulnerability. During this period an earthquake-prediction response plan can be formulated. Unfortunately it is unlikely that such a plan, which responds to more than obvious opportunities, can be formulated and put into operation in a much shorter time.

Beyond these time periods, in the 3,000-day or about 10-year range, the potential to reduce the vulnerability of the community is great. Building

Table 2 Engineering responses to an earthquake prediction

Lead time	Buildings	Contents	Lifelines	Special Structures
3 Days	Evacuate previously identified hazards	Remove selected contents	Deploy emergency materials	Shut down reactors, petroleum products pipelines
30 days	Inspect and identify potential hazards	Selectively harden (brace and strengthen) contents	Shift hospital patients; alter use of facilities	Draw down reservoirs, remove toxic materials
300 days	Selectively reinforce		Develop response capability	Replace hazardous storage
3,000 days	Revise building codes and land-use regulations: enforce condemnation and reinforcement			Remove hazardous dams from service

codes can be adopted and enforced that protect the occupants from unacceptable risks of injury. Land-use policies can be adopted that decrease the density of occupations in hazardous landslide areas or filled areas that are likely to be subject to soil failure. These procedures, although useful and obvious, are not the ones that will yield the greatest benefit. *For the next decades the earthquake vulnerability of virtually all this nation's cities will be dominated by those older structures now standing that cannot be expected to withstand a major shake. The biggest challenge to the public official is to develop economically and politically realistic procedures and policies for the condemnation of substandard structures, and for their reinforcement, replacement, or abandonment.* I wish to call to your attention that these are the same major policies and procedures that we in the engineering profession urge you to take when a specific earthquake has not been predicted but can be expected to occur in the not-too-distant future.

It is logical to ask what the potential impact of all these adjustments during the warning period might be. When the period is but a few days, the major benefit will be in the saving of lives. Property damage will be essentially unchanged. When 30 days warning is available, the damage might be reduced by 20 percent. In 300 days, we might be capable of reducing impacts by 40 percent, and in 10 years, the reduction could be 60 percent or more. These reductions do not consider economic costs incurred in improving physical performance and presume that all available technology is applied.

I would be remiss in my professional responsibilities if I did not point out to all of you that many of the procedures that an engineer or architect might use to design or reinforce a structure economically are yet to be formulated. The same lack of clear-cut procedure is even more true for public-policy matters. Just as a substantial effort must be expended to achieve earthquake prediction, there must be a significant companion research, development, and educational program in the engineering, economic, and social sciences. They must all be pursued together.

In summary, I wish to highlight a few of the public-policy problems, or should I say challenges, that will present themselves:

1. Within the built environment, a particular building or piece of land may be within the jurisdiction of any of a large number of groups, often having overlapping responsibilities. The point is that the control of the physical structures in a community is vested in a vast maze of generally unknown groups.
2. You must start to plan now how you will respond to a prediction, and formulate the basic plans and procedures that you intend to pursue while there is time to rationally contemplate objectives, rather than be forced to respond to the need for immediate action.
3. During the warning period you will have to mobilize trained personnel and materials and protect the public from unscrupulous practitioners pretending to possess capability in the professions dealing with earthquakes.
4. The public will need information on how they can protect themselves and their property. It is logical to assume that they will look to government officials for the information.
5. Finally, there will be the problems of equity. We must formulate and carry out public programs that do not overly benefit some at the expense of others.

10 Earthquake Prediction: The Scientist's Responsibilities and Public Response

Eugene B. Veek

The headline of a newspaper article about a recent National Security Council study of earthquake preparedness in California announced "We're not ready for the big one!" How true. But why should we be prepared? After all, the odds are only 50:50 and the time frame given is 30 years for a Richter magnitude 8 event somewhere in southern California. Property damage, the study concluded, could amount to $70 billion (Associated Press, 1981).

The author is associated with the Institute for Marine and Coastal Studies, University of Southern California, Los Angeles. This selection is reprinted from *California Geology*, v. 37, n. 7, pp. 154–155, (1981) by permission of the author and California Division of Mines and Geology.

But what if there were more certainty (perhaps 90:10) associated with the prediction? Or the time frame was given as a month and year (perhaps June, 198x)? Or the predictions were for a specific area and the events were to occur soon? There would be sociological (including economic) and psychological ramifications as well as physical ones.

Sociological Aspects

Society is affected to the extent that risk management results in an appropriate balance of benefits and costs. Reliable is the operative word. Public and private perceptions are that reliable predictions are *not* possible at this time.

Risk management is only as good as the knowledge base. Man has learned gross lessons with regard to modifying and protecting his environment (don't build across earthquake fault zones; do avoid constructing underengineered dams, bridges, canals), but basically he still tries to shape his environment to suit his need. Man relies on technological fixes to avoid or ameliorate hazards which exist and can adversely affect his activities. When he chooses to construct buildings, dams, and land-fills, he has learned that there are economic and societal penalties to be weighed against the benefits to be gained. The extra financial costs, environmental consequences, and hazards and potential hazards to life and property can be more adequately quantified if he has access to a reliable earthquake prediction system. This he can deal with in a rational manner through legislative, technological, and educational means.

Psychological Aspects

On either an individual or group basis, there are choices:

1. Deny that there is risk (or disbelieve its validity).
2. Acknowledge the risk and take actions (cost/benefit weighed) to mitigate or eliminate the risk. This is the proper course of action to take.
3. Acknowledge the risk and avoid the area (thus carrying out option 2 to the ultimate degree). Irrational and exceptionally fearful persons or institutions (or particularly vulnerable ones such as nuclear power plants) can be expected to choose this option.
4. Acknowledge the risk and take no (or minimal) action. This is a form of a fatalistic view—"what will be, will be." It is a form of denial of reality. Many individuals and some institutions and businesses now hold this view.

If a reliable prediction capability were to exist (and were perceived as such by the public), the consequences would be manifested in the following ways:

1. Severe economic disruption before the event depending on the predicted magnitude, certainty of its predicted consequences, the confidence level of the prediction, and the time interval between prediction and event. This would occur due to falling real estate values and/or expenses required to mitigate against loss or damage to life and property.
2. Mass exodus of persons (and institutions/businesses), who are unable, unwilling, or unsure of being protected, will occur.
3. Mass hysteria and other psychological symptoms will probably increase during the period immediately prior to the predicted event. Crime can be expected to increase.
4. Planning and protection to avoid overload of critical facilities such as hospitals and fire fighting, police, and utility services would be expensive but required. It could, given too short a lead time, tax even Federal ability to cope.

Scientists' Responsibilities

The scientific community has a particular responsibility to critique the illusions of certainty with regard to earthquake prediction. The phenomenon of social denial of reality and the clinging to illusions of stability and status quo will continue to confront us. The purpose of foreknowledge of impending risks is to permit the mitigation and prevention of physical or economic hardships resulting from existing or predicted conditions. Any scientist (or other person or institution) who believes that such foreknowledge is available has a responsibility to share that information with the scientific community. If it is validated, the information should be passed to the political bodies for dissemination to the community at large. The alternative to notification is to accept the consequences of withholdings a valid prediction and suffer legal (if discovered), or mental, penalties for the results of failing to warn.

The alternative to making an invalid prediction also has consequences and political decision makers would have to share that blame.

References

Associated Press (Washington, D.C.). 1981, *in* The Daily Breeze, Torrance, California, January 19, p. 1.

Kerr, Richard A., 1981, Prediction of huge Peruvian quakes quashed: *Science*, v. 211, no. 808.

Sullivan, Walter, 1980, *in* The Daily Breeze, Torrance, California: New York Times News Service, November 23, p. E5.

11   Creep on the San Andreas Fault: Fault Creep and Property Damage

Karl V. Steinbrugge and Edwin G. Zacher

Introduction

In the course of a building inspection of the W. A. Taylor Winery in April, 1956, Mr. Edwin G. Zacher noticed fractures in reinforced concrete walls

Steinbrugge, K. V. and Zacher, E. G., 1960, "Creep on the San Andreas Fault," Article 1: "Fault Creep and Property Damage," *Bull. Seismol. Soc. Amer.*, vol. 50, no. 3, pp. 389–96. Reprinted with light editing by permission of the senior author and the publisher. Copyright 1960, S. S. A.

Mr. Steinbrugge is professor of structural design at the University of California, Berkeley. Mr. Zacher is a consulting structural engineer in San Francisco, California.

and displacement of concrete slabs which could not be explained by landslide or attributed to other conventional causes. The winery is on the Cienega Road about 7 miles south of Hollister, California.

An examination of geologic maps placed it in the San Andreas fault zone. On a relatively detailed geologic map Taliaferro (1949) shows the fault as a line going through the winery buildings. Since conventional explanations failed, a study was made of a possible connection between the damage and earthquakes and between the damage and possible fault movements.

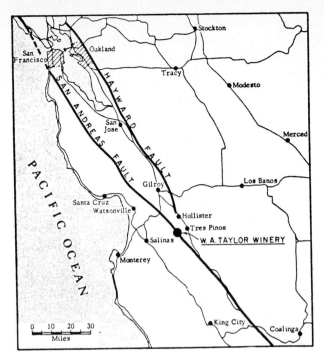

FIG. 1. Location map of W. A. Taylor Winery in California.

Observed Damage

The observed building damage is found more or less along a straight line as may be seen in Fig. 2, and this line is oriented in the direction of the San Andreas fault. Evidence at all damage locations indicates that the westerly portion of the main building is moving northward with respect to the easterly portion; in geologic terms, this is right-lateral movement. The total movement of one portion of the building with respect to the other between 1948, when the building was constructed, and December, 1959, is almost 6 inches.

Winery employees with many years' service have known of the damage and were aware of its growth. However, its growth was slow and gave no alarm.

This shearing movement has caused reinforced concrete walls to break at three places. Figure 3 shows an offset between wall and floor slab and is one indication of the total movement. Some columns along the line of creep were so badly affected that major reconstruction was required in 1954. Concrete floor slabs have moved along construction joints or have broken, again in a right-lateral sense.

Damage outside the main building is consistent with the linearity and type of motion taking place within the building. A concrete-lined drainage ditch south of the winery, constructed about 1943, has been ruptured. In the fall of 1957 two waterlines between the winery and the drainage ditch were broken. When subsequently they were under repair, it was observed that several ruptures had occurred here in previous years. Another pipeline

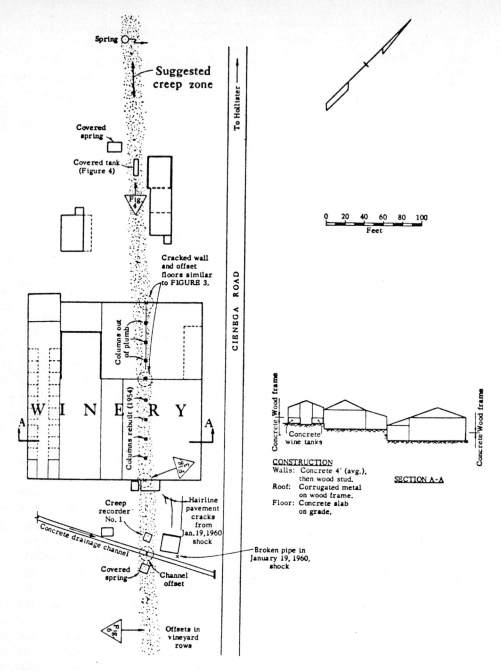

FIG. 2. W. A. Taylor Winery, showing locations of creep damage.

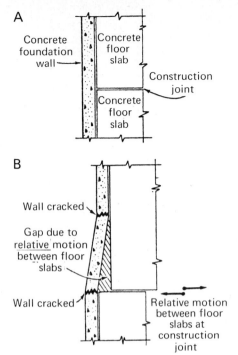

A

Concrete foundation wall

Concrete floor slab

Construction joint

Concrete floor slab

B

Wall cracked

Gap due to relative motion between floor slabs

Wall cracked

Relative motion between floor slabs at construction joint

FIG. 3. Diagrammatic plans show the effects of creep, and may be considered as typical of all three broken walls. A: as originally built. B: effect of creep.

break occurred in the January 19, 1960, earthquake (Fig. 2). A covered tank in the ground and adjacent to the office building north of the winery also shows distortion due to right-lateral movement. Vineyard rows just south of the main building have offsets.

Measurements within the building have been made periodically by the authors since 1956, using as references the face of the walls, the edges of the floor slabs, and marks chiseled in the surface of the floor slabs. The data are plotted in Fig. 4, and indicate about one-half inch right-lateral movement per year.

Damage from Earthquakes

This area has had a number of felt shocks, probably more than the average for the Pacific coast region; however, the historic record makes no mention of surface faulting at this location. Four felt shocks are of special interest.

The buildings at this winery (then known as Palmtag's Winery) were damaged in the April 18, 1906, San Francisco shock (Lawson *et al.*, 1908): "At Palmtag's winery, in the hills southwest of Tres Pinos, the shock seems to have been more severe than elsewhere in the vicinity of that village. Furniture was moved, water was thrown from troughs, and an adobe building was badly cracked. One low brick winery was unharmed." A local resident who was in or near the buildings at the time, has stated to one author that "loose adobe brick" then fell from a wall. The 1906 surface faulting ended about 11 miles to the northwest.

On June 24, 1939, a local earthquake occurred which had a Modified Mercalli Intensity of VII, and its field epicenter was in the vicinity of this winery. Records of the U. S. Coast and Geodetic Survey (1939*a*, *b*) relate that an adobe wall pulled away from a side wall; girders pulled away from brickwork and the brickwork was badly cracked; new cracks were formed and old ones opened 1 or 2 inches in width; many fresh ground cracks appeared in the neighborhood of the building, and their general trend appeared to be northwest-southeast and at right angles thereto. The observer did not comment further on the ground cracks.

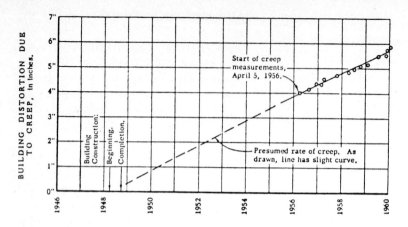

FIG. 4. San Andreas fault creep as measured by building distortions. The average creep is one-half inch per year.

The U. S. Coast and Geodetic Survey report (1939b) with its accompanying photographs has been reviewed by Mr. John Ohrwall, the present W. A. Taylor Company superintendent, who was present at the time of the 1939 shock. He says that the ground cracks were in the same location as the present movement. The parts of the winery which now lie across the creep zone have been built since 1939.

Another local earthquake, occurring on August 10, 1947, was also studied by the U. S. Coast and Geodetic Survey (1948). Their report regarding the winery states: "There was a small differential settlement at the junction of the old and new portions of one of the buildings, the older section having settled with respect to the newer. Furthermore, there was a transverse crack extending all the way across a concrete platform about 4 feet wide." Examination of the concrete-platform crack indicated local damage of no particular significance.

On January 19, 1960 (January 20,

1960, GCT), a strong local earthquake caused minor damage at the winery and also resulted in one-eighth inch of "instantaneous creep" in the building, a movement that was recorded at the time of the shock by University of California instruments at the winery. This movement caused cracks in the floors and walls to widen, and some concrete spalling occurred. Hairline ground cracks were noted in the paving to the south of the winery (Fig. 2), and an underground pipe was broken. The ground cracks and the pipe breakage were undoubtedly associated with the one-eighth inch movement. Typical other damage: some objects were thrown from shelves, a hollow concrete block chimney cracked, catwalks over large wooden wine tanks pulled apart, and some wine barrels worked loose from their chocks.

The published historic record, plus interviews with winery employees, ruled out the creep damage as the result of an earthquake or some other obvious event, except for the January 19, 1960, earthquake.

Fault Creep

Since the winery was in the San Andreas fault zone and since the damage had been progressive, the possibility of fault creep was considered almost immediately and measurements were taken accordingly (Steinbrugge, 1957).

The winery is on a sloping site, and there are springs near and under the main building. There has been some conjecture that what is taking place may be somehow related to landslide. Springs beneath the building could facilitate the movement. However, the direction of the motion is essentially at right angles to the slope, and no noticeable downhill component of motion can be found. It would seem that landslide and other gravity effects cannot be present.

Practically all damage locations fall in a narrow straight zone as shown in Fig. 2. The bearing of this zone is essentially that of the strike of the San Andreas fault. Near-by springs, such as are often noted on a fault trace, also fall in the zone. Geologic mapping, which was done independently of the building damage, places the fault through the buildings (Taliaferro, 1949). The right-lateral movement is consistent with the right-lateral faulting known to have occurred on the San Andreas fault and is also consistent with possible creep resulting from the known regional strain pattern. The creep through the building appeared to be gradual as far as visual observation and the aforementioned crude measurements could determine, except for the earthquake of January 19, 1960.

All the foregoing is consistent with the theory of fault creep.

There is local opinion to the effect that the previous building at this site had been damaged in a similar manner. The description of the damage (skewed roof trusses), and wall offsets of about two feet in perhaps half a century, suggest that the rate of creep of one-half inch per year may have been fairly constant for many years.

Fences to the north and south of the winery show no damage or offsets which could be the result of fault creep. It is not known if the creep is markedly local, or if the fences are so situated as not to be indicative of creep.

References

Lawson, A. C., et al., 1908. *The California Earthquake of April 18, 1906; Report of the State Earthquake Investigation Commission* (Washington, D. C.: Carnegie Institution of Washington, 2 vols. plus atlas).

Steinbrugge, Karl V., 1957. *Building Damage on the San Andreas Fault*, report dated February 18, 1957, published by the Pacific Fire Rating Bureau for private circulation.

Taliaferro, N. L., 1949. "Geologic Map of the Hollister Quadrangle, California," Plate 1 of California Division of Mines *Bulletin 143* (text not published).

U. S. Coast and Geodetic Survey, 1939a. *Abstracts of Earthquake Reports from the Pacific Coast and the Western Mountain Region, MSA-22, April 1, 1939, to June 30, 1939.* 1939b. "Central California Earthquake of June 24, 1939" (unpublished manuscript prepared by Dean S. Carder). 1948. *Abstracts of Earthquake Reports from the Pacific Coast and the Western Mountain Region, MSA-55, July, August, September, 1947.*

Mass Movement

Multiple rotational slip resulted in downslope movement along the Genessee River near Avon, New York, after several days of rain. The cliff is 21 m (69 ft) high and consists of surficial material rich in clay. Behind the cliff is a plowed field. (*The Rochester Democrat & Chronicle*, April 8, 1973. Burr Lewis, photographer)

"Nature to be commanded must be obeyed."
Francis Bacon

Nature of the Problem

Mass movement may be defined as the downslope movement of surface material in response to gravitational forces. The most visible type of mass movement and the one that poses the greatest threat to humans is the landslide. Inasmuch as the earth's surface is dominated by slopes, the movement of material en masse is potentially the most dangerous and widespread geologic hazard. Figure 1, a map of landslide severity in the United States, shows that only a few areas are without problems stemming from these hazards. In many areas, in fact, landslides represent a significant risk to life and property, with annual damages in the United States amounting to more than one billion dollars and several times that throughout the world. During the present century, there has been an upward trend in property damage that can be related to urban sprawl in hilly areas, and to a trend toward building larger and more expensive structures. Property losses include residences, business and industrial structures, highways, railroads, utility lines, dams and reservoirs, and pipelines.

Loss of lives from mass movement, however, has decreased markedly during the present century despite a growing world population and a strong trend toward urbanization. This may be related to improvements in risk assessment, grading codes, and to more informed and rapid response warnings. According to statistics compiled by several Federal agencies, less than 100 persons lost their lives in mass movement accidents in the United States between 1925 and 1975.

Some of history's major landslides are listed in Table 1.

Classification and Morphology

Landslide is a general term covering a wide variety of landforms and processes involved in mass movement. Landslides usually occur over a relatively confined surface of shear but may vary widely in scale, pattern and rate of movement, material, and morphology. Terminology and classification may reflect any one of the many variables.

Landslides are commonly distinguished on the basis of their dominant type of movement: falling, flowing, or sliding (Fig. 2). A further

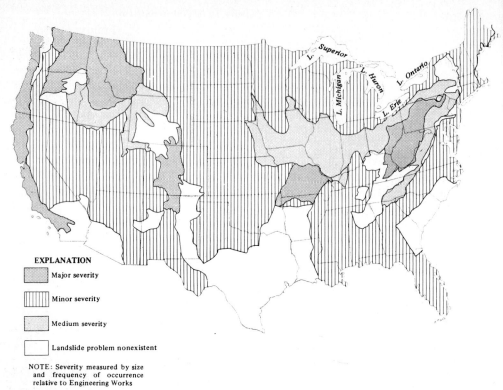

EXPLANATION

▨ Major severity

▥ Minor severity

▧ Medium severity

□ Landslide problem nonexistent

NOTE: Severity measured by size
and frequency of occurrence
relative to Engineering Works

FIG. 1. Landslide severity in the United States. (Printed with permission of R. F. Baker and R. Chieruzzi, Engineering Experiment Station, Ohio State University, 1957.)

distinguishing characteristic is the type of material being moved— bedrock, soils, or sediments. Some classification schemes emphasize rate of movement—slow or rapid. Almost any combination of falls, flows, and slides may be involved in a single failure, and that failure may include a variety of materials moving at changing velocities (Table 2).

Falls involve a very rapid downward movement of earth materials

Table 1 History's major landslides

Year	Location	Deaths
1618	Mount Conto, Switzerland	2,430
1893	Vaerdalen, Norway	111
1920	Kansu Province, China	200,000
1945	Kure, Japan	1,154
1963	Vaiont, Italy	2,600
1970	West-central Peru	21,000

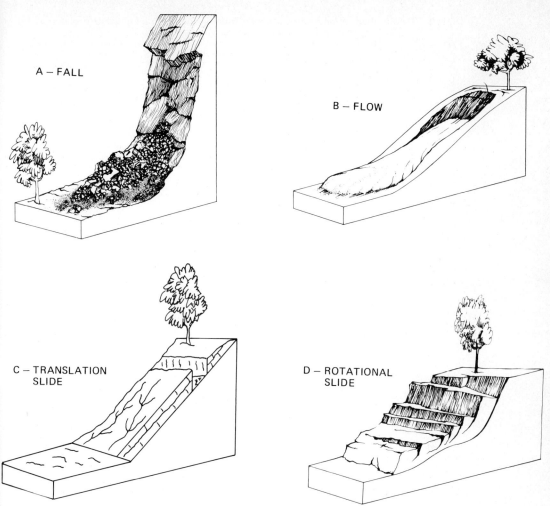

FIG. 2. Common types of landslides. (A) *Fall*. Rock that has moved primarily by free fall or bouncing. (B) *Flow*. Unconsolidated material that has moved in a manner similar to a viscous fluid. (C) *Translation slide*. Downslope movement by sliding along a bedding plane. (D) *Rotational slide*. Downslope movement by rotational slip. (A,B, and D from *U.S. Geological Survey Professional Paper 941-A*.)

bouncing downslope or traveling mostly through the air by free fall. The debris that accumulates at the base of the slope or cliff is called *talus*.

Flows move at variable rates, consist of unconsolidated material, and resemble a viscous fluid. Often they have a distinctive hourglass shape, with a bowl-shaped source area leading to a relatively narrow neck followed by a bulge at the base. Most flows occur during or after

Table 2 Classification of landslides. [Adapted from Committee on Landslide Investigations, Highway Research Board Special Report 29, 1958]

Type of movement		Type of material			
		Bedrock		Soils	
Falls		Rockfall		Soilfall	
Slides	Few units	Rotational slump	Translation block glide	Translation block glide	Rotational slump block
			Rock slide	Debris slide	Failure by lateral spreading
Flows	Slow	Rock creep		Soil creep	
	Rapid	Saturated unconsolidated materials Earthflow (sand or silt) Debris avalanche (mixed) Mud flow (mostly plastic)			
	Complex	Combinations of materials or type of movement			

periods of heavy rainfall. Movement can occur on fairly gentle slopes and cover distances of several kilometers. *Lahars* are flows composed of pyroclastics, and materials involved in liquefaction move as flows.

Slides involve very slow to very rapid movement along surfaces of shear failure. Two basic patterns of movement are associated with slides—rotational and translation. *Translation slides* occur along planes of weakness, such as bedding planes, fractures, and clay lamina. The slip surface follows the plane of weakness. In a *rotational slide*, the slip surface, or surface of rupture, is concave upward or spoonshaped, leading to a backward rotation in the displaced mass. A steep scarp and flanking walls are commonly formed as are subsidiary scarps and numerous cracks and ridges that give rise to a hummocky surface (Fig. 3). The term *slump block* usually refers to a rotational slide.

Landslides, in the strict sense, are distinguished by the presence of a surface of rupture. *Creep*, which involves the slow downslope movement of soil mantle and bedrock at imperceptible rates, occurs without surface rupture. It is a type of mass movement but technically it is not a landslide. Creep is commonly caused by expansion and contraction of earth materials in response to diurnal temperature changes. Movement is at right angles to the slope and may range from only 1 to 10 mm a year.

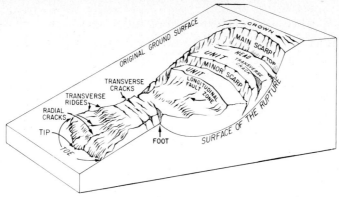

FIG. 3. Landslide nomenclature. The surface of rupture is typically concave upward, and numerous scarps are commonly developed. Cracks and ridges near the toe of the slide give rise to a hummocky surface. (From Eckel, 1958.)

Causes of Landslides

A landslide occurs when the driving forces exceed the sum of the resisting forces that tend to keep material in place. This can be expressed by the equation

$$SF = \frac{\text{resisting forces (shear strength)}}{\text{driving forces (shear stress)}}$$

"SF" is the safety factor. The driving forces include the weight of the material, the gradient of the slope, and ground vibrations such as earthquakes or blasting. Resisting forces include strength, cohesion of materials, and friction. The safety factor is reduced when (1) there is a reduction of the factors that resist failure (shear strength), or (2) there is an increase in factors that promote failure (shear stresses). When the safety factor is less than or equal to 1.0, movement is likely to take place.

The factors leading to landslides may be viewed as either external or internal (Terzaghi, 1950). The most common *external* factors are (1) steep topographic slope, (2) removal of support at the toe, (3) overloading the crown of the slope, (4) ground vibrations, and (5) changes in vegetative cover. Some of these factors may be related to natural processes. Storm waves beating against a sea cliff may produce vibrations and erode the base of the cliff. Lightning may cause fires which remove vegetative cover, and tectonic movements may increase the gradient of slope. In construction projects, cut-and-fill operations may remove vegetation and lead to overloading and removal of support while at the same time producing ground vibrations.

The most common *internal* factors are (1) an increase in moisture content as it relates to weight and pore-water pressure; (2) structures such as bedding planes, joints, foliation, cleavage, faults, cavities, and crushed zones (a *dip-slope* relationship is extremely hazardous—the dip of the bedding plane approximates the topographic slope, thereby facilitating a translation slide); (3) composition, such as the presence of water soluble minerals and expansible clays; and (4) weathering by solution, hydration, freeze-thaw, and so on.

Triggering Mechanisms

Both external and internal factors may act together over a long period of time to greatly reduce slope stability. However, abrupt failure may be triggered by a single event such as a heavy rainstorm, earthquake, volcanic eruption, rapid drawdown of a reservoir, or a large-scale excavation. The distinction between fundamental causes and a triggering mechanism may be critical in insurance settlements or legal disputes.

Risk Assessment

Risk assessment involves identifying where landslides are most likely to occur and describing the nature and behavior of potential landslides. Risk assessment is, therefore, a valuable guide for planners and those concerned with reducing the hazards of landslide terrain. It should be noted that it is significantly different from predicting *when* a landslide will occur.

In undeveloped areas, a *slope stability map* can be a valuable aid in risk assessment. These maps are based on the physiographic expression of ancient and modern landslides, bedrock geology, surficial deposits, and vegetation (Fig. 4, Reading 2). We have already seen that landslides may produce such features as hummocky or steplike topography, ground cracks, and frontal bulges. These features are indicative of slope instability and are easily identifiable on aerial photographs or reconnaissance ground surveys. Analysis of bedrock geology and surficial deposits is more site specific, focusing on factors that represent the internal causes of slope failure (e.g., moisture conditions, structure, composition, and weathering). Abrupt changes in vegetation are often associated with landslide deposits, and bent tree trunks may be indicative of active sliding.

In areas already developed, a large variety of cultural features reflect slope stability. Leaning fence posts and utility poles, cracked walls and foundations, and windows and doors that stick and jam are associated with structures built on active landslides.

In making risk assessments it is important to remember that changing physical or climatic conditions may have decreased or increased slope stability. For example, ancient slides produced under dramatically different climatic conditions may now be stabilized. Or, the construction of buildings and roads on stable slopes may significantly reduce the safety factor through oversteepening, overloading, saturation from septic systems, or undercutting for roads.

Prediction

The prediction of landslides is site specific and the goal is to predict when a landslide will take place. The techniques are costly and are only employed where there is a high risk of loss of life.

Installation of monitoring equipment is preceded by calculating the slope safety factor. This requires studies of subsurface borings and ground-water conditions and tests of mechanical properties of the earth materials at the site. A safety factor of less than or equal to 1.0 indicates that a slide is imminent; 1.1 to 1.25 indicates that the slope is conditionally stable. Higher safety factors would not ordinarily require constant monitoring.

Monitoring techniques include (1) periodic surveying of reference markers in the area of the potential slide, (2) observing changes in growth of cracks or fissures, (3) installation of electric circuit breakers which react when excessive movement occurs, (4) installing strain meters to record changes in strain.

Another approach is to determine thresholds for mass movements through measuring rainfall intensity and duration, total seasonal rainfall, amount of snowpack, or level of ground vibrations. When the thresholds are reached, mass movement is imminent. Russell H. Campbell, an expert on landslides, has proposed a warning system (1975) consisting of (1) rain gages recording total rainfall hourly, (2) a weather-mapping system to identify centers of high intensity rainfall, and (3) an administrative and communications network to collate and interpret the data and warn residents. He suggests that such a system is within the capability of existing technology and could be applied at present to the Santa Monica Mountains of southern California, an area prone to soil slips and debris flows.

Landslide Prevention

There are a number of practical and cost-effective ways to prevent some landslides:

1. Passage of zoning laws that would control the development of unstable areas.
2. Establishment of grading codes that would maintain slope integrity during construction projects.
3. Control of drainage through diversion ditches, shallow wells, or horizontal tiles.
4. Reducing the load on the head of a slope.
5. Grading a slope to reduce the gradient.
6. Buttressing the toe of a slope with retaining structures.
7. Planting slopes with vegetation with extensive root systems and high rates of evapotranspiration.
8. Treatment of surficial materials with chemical solutions which promote stabilizing ion exchanges.
9. Grouting rock formations with cement to reduce pore water and increase shear strength.
10. Insurance could prevent damage to the extent that premiums based on level of risk might discourage development in high risk areas. Insurance coverage for mudslides is included in the federal government's Flood Disaster Protection Act of 1973 and is also available from private companies.

F. B. Leighton (1976) has suggested that potential dollars to be lost from landsliding will exceed the dollars spent on preventive planning by average factors ranging from 50 to 110 and that a 95 to 99 percent reduction in damage caused by landslides is attainable today through risk assessment and implementation of techniques for reducing the damage. This suggestion is supported by a dramatic decrease in number of homes damaged in the Los Angeles area since grading codes were introduced.

Case Histories and Readings
Quick Clays and California's Clays:
No Quick Solutions

In Reading 12, Quintin Aune describes an unusual clay deposit that led to earthquake-triggered landslides 75 miles from the epicenter of the Good Friday Alaska earthquake in 1964. Geologists had warned that these clay deposits were hazardous, but the area was nevertheless developed as a fashionable suburb of Anchorage (Reading 2). Similar deposits have plagued Scandinavia and other areas, such as the St. Lawrence River valley, where marine clays composed essentially of glacial rock flour have been elevated above sea level as a result of glacial rebound. The electrolytes have been naturally leached from these clays, leaving them highly unstable. Mr. Aune asks if the activities of human could also lead to the leaching of other clay deposits and thereby set the stage for the generation of sensitive clays and major landslides. His suggestions should be carefully considered by developers, civil engineers, and others who evaluate sites for new developments.

The Vaiont Reservoir Disaster

One of the worst dam disasters in history resulted from a huge rock slide that displaced water in the Vaiont reservoir in northern Italy in 1963. More than 2600 people lost their lives in less than six minutes. The hazardous setting was well documented before construction. Development was a calculated risk. Although there were warnings, and the site was monitored up to the time of failure, the data was misinterpreted and tentative plans for evacuation were not implemented. In this reading, note the specific external and internal factors accounting for instability in the area, and also how the construction of the reservoir might have accounted for final slope failure.

Permafrost

Reading 14 reviews the nature and distribution of permafrost. Permafrost is an unusual deposit which becomes unstable when disturbed. A variety of construction activities can cause permafrost to thaw, which in turn may lead to creep, landslides, slumping, and subsidence. As human activities expand in areas underlain by permafrost, it is important that we recognize the hazards and develop techniques for avoiding them.

Expansive Soils— The Hidden Disaster

Expansive soils are materials that increase and decrease significantly in volume as moisture content changes. Volume changes are controlled primarily by the type of clay minerals present in the soils. For example, the swelling pressures associated with the clay mineral montmorillonite easily exceed the loads imposed by single family homes. Other factors that influence volume changes are soil density, structure, and moisture content. The damage potential of an expansive soil may be reduced by controlling surface water and soil moisture and by reinforcing concrete slabs and building foundations.

In Reading 15, D. Earl Jones, Jr., and Wesley Holtz review the nature of the damage associated with expansive soils in the United States. They point out that although these soils are found in every state and inflict more than twice the damage from floods, hurricanes, tornadoes, and earthquakes, virtually nothing is spent on damage prevention or on developing methods to deal with the problem of expansive soils.

12 Quick Clays and California's Clays: No Quick Solutions

Quintin A. Aune

The March 27 Good Friday earthquake of 1964 caused extensive damage at Anchorage, Alaska, 75 miles from the earthquake hinge belt and epicenter. Much of the damage can be related to earthquake-triggered landslides. What is there about the geologic setting of Anchorage which led to localized landslide activity? U. S. Geological Survey Water-Supply Paper 1773, Geology and Ground-water Resources of the Anchorage Area, Alaska (Cederstrom et al., 1964), although written before the Good Friday earthquake, contains several clues.

The accompanying map, (Fig. 1) adapted from the ground-water study, shows distribution of earthquake-triggered landslides relative to subsurface distribution of the Bootlegger Cove Clay. These landslides contributed to a substantial part of the damage at Anchorage; they are known to have been caused by local failure of the Bootlegger Cove Clay. Major landslide failure was not uniform over the entire area underlain by the clay, but was restricted to low escarpments paralleling Knik Arm and Ship Creek in the western part of the Anchorage area.

Aune, Q. A., 1966, "Quick Clays and California's Clays: No Quick Solutions," *Mineral Information Service*, vol. 19, no. 8, pp. 119–23. Lightly edited by permission of the author and reprinted by permission of California Division of Mines and Geology, Sacramento, California.

Quintin Aune is a geologist on the staff of the California Division of Mines and Geology.

What caused failure of the Bootlegger Cove Clay? Cederstrom et al. (1964, p. 32, and Table 4) indicate that sand in lenses in the Bootlegger Cove Clay "commonly becomes quicksand when penetrated by the (well) drill." Quicksand has the peculiar property of losing all its cohesive strength and acting as a liquid when disturbed, as by the sudden jar of a man's footsteps on the surface, or the sudden vibration of a drill—or an earthquake—beneath the surface. Some clays develop similar properties, and are called quick clays because of their analogous behavior. As we shall see below, there is reason to suspect that part of the clay fraction of the Bootlegger Cove Clay assumed the properties of a quick clay.

Quick clays are generally confined to the far north areas. They are known to have caused excessive damage in eastern Canada and Scandinavia in the past (Kerr, 1963; Liebling and Kerr, 1965). They result when a marine or brackish water clay composed essentially of glacial rock flour is elevated above sea level. Such post-Pleistocene elevation is common in the far north due to isostatic adjustment of the land surface resulting from the melting of the great overburden of glacial ice only a few thousands of years ago.

Clay derived from rock flour consists of the minutely ground up particles of many minerals, generally including several clay minerals. It is not excessively responsive to disturbance as

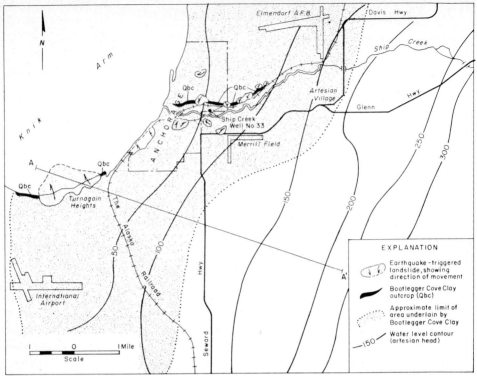

FIG 1. Geologic map of the Anchorage area, Alaska, showing critical features of the geology pertaining to slope failures precipitated by the Good Friday Earthquake. (From *U. S. Geological Survey Water-Supply Paper 1773*, 1964.)

long as the intermolecular water—the formation water—contains dissolved salt (sodium chloride). The salt acts as an electrolytic "glue" which adheres to the clay particles and provides cohesiveness and structure to the clay (Kerr, 1963, p. 134). Once the clay is elevated above sea level the salts—this natural "glue"—may be progressively flushed out by fresh groundwater. Cederstrom et al. (1964, p. 72) cite evidence to show current loss of sodium electrolyte to well water (Well #33) that penetrates the Bootlegger Cove Clay in the Anchorage area. The sodium salt may be flushed out, reducing or destroying the clay's cohesive

properties, or its sodium may be replaced by calcium, greatly reducing the clay's cohesiveness. Because this process has gone on only in the past few thousand years, only portions of the clay have recently become extremely responsive to disturbance. There was, therefore, little obvious geological evidence of the clay's present instability—such as landslide scarps—until the major earthquake disturbance disrupted it at Anchorage.

As shown in the cross-section the Bootlegger Cove Clay, (Fig. 2) as defined in well-logs (Cederstrom et al., 1964), lies athwart the path of westward-flowing ground-water. It was

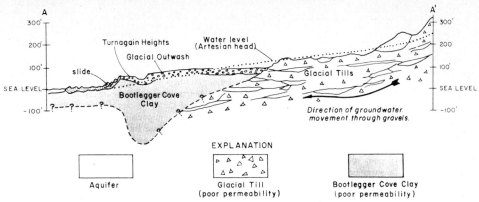

FIG. 2. Geologic cross section, A-A′, from Fig. 1.

EXPLANATION

| Aquifer | Glacial Till (poor permeability) | Bootlegger Cove Clay (poor permeability) |

indeed fortunate for Anchorage that the salt leaching or exchange process, sensitizing the Bootlegger Cove Clay, is a slow, incomplete one. Impermeability of the clay and position of much of the formation below sea level have significantly retarded the invasion of fresh water and "leaching" or exchange of the sodium ions.

Shown also on the cross-section and on the map is the artesian head, the level to which water from a well will rise when confined in a standing pipe. It may be seen from the cross-section that this level is well above the ground surface in the area along the eastern margin of the Bootlegger Cove Clay but drops off rapidly as one travels west across the area underlain by the clay.

In the Anchorage area, artesian water travels in a crudely interconnected system of gravels, confined from above by impermeable till (east half of the cross-section). The westward passage of the water is impeded by the Bootlegger Cove Clay. If the clay were absent, water in the underground channels or "pipe system" would pass unobstructed to the sea and no high confining pressure could be built up, but such is not the case.

Theoretically, this relationship can be very dangerous. Confined groundwater in a unified hydraulic system has much the effect of a hydraulic ram. Here, it exerts a constant pressure against the clay barrier—the Bootlegger Cove Clay—to the west. As long as that formation is coherent, and has shearing strength, it will hold together, and its molecules will not "collapse" into an unoriented liquid substance at the onset of a major shock, as an earthquake. If rendered incoherent, however, as it was in the disastrous quick-clay slides in eastern Canada and Scandinavia (Kerr, 1963), entire sections of the Bootlegger Cove Clay could break and glide laterally in translation movement, settling ultimately into a disheveled landslide "mush." Such a mass might flow seaward, carrying part or even all of Anchorage with it, because nearly all of Anchorage is underlain by the Bootlegger Cove Clay.

The slides at Anchorage were minor, compared to the above-inferred catastrophe, because the main earthquake shock was far distant, and because much of the Bootlegger Cove Clay is not yet a quick clay. As shown by Well

#33 (Fig. 1), it is still "in the making." Perhaps the limited disaster was fortunate for Alaska. It will stimulate the residents, scientists, and engineers to recognize the problems they face, and to learn through industry and science how they can control or work around their problems at Anchorage and elsewhere in Alaska (*Time*, 1965).

Could a similar "Good Friday" earthquake-landslide disaster happen to California? Because of the Ice-Age genesis of known quick clay deposits, California probably has no quick clays as such. It has no geological environment in which natural forces are actively flushing out the electrolytic "glue" to create structural hazards. Then does the Alaskan example apply to California? Many poorly consolidated marine and evaporite clay-bearing formations in California contain electrolytes—in the form of saline formation waters—which add to their

stability. It has often been demonstrated in California that this stability may readily be disrupted during the substitution of fresh water for saline waters by a drilling mud during the drilling of oil wells (Morris et al., 1959).

California's recent geologic history is one of frequent volcanism. Volcanoes produce ash which settles into the sea or becomes concentrated in saline evaporite lakes, and later becomes part of the soil or of underlying marine lake bed formations or terrace deposits. Eventually the ash alters into illite and montmorillonite clay minerals. Such clays, while not reaching the extremes of quick clays, are nevertheless potentially sensitive and may even approach quick clays in this respect. As with quick clays, the sensitivity of these clays to shock or disturbance may be greatly increased by the flushing out of saline formation water with its electro-

FIG. 3. Cross-section of a hypothetical California landscape analogous to Anchorage. Schematic buildings represent what could be a major subdivision; irrigation losses may represent runoff from thousands of acres. Leaching of sodium ions from clay may reduce its coherence, set stage for major landslide.

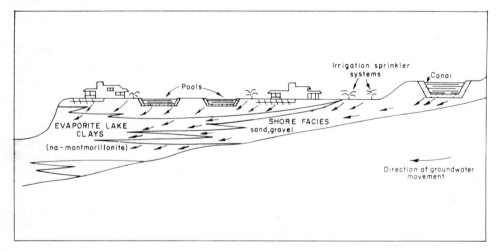

lytic "glue," and substitution of fresh, "unglued" solutions in its place.

Geological processes strive toward stability. Stability of a clay deposit is a relative state. It is relative to the nature of the formation water, the type and condition of the clay, and the nature of the slope. This stability may be affected by: (1) a change in the clay through the addition of abnormal amounts of water. Certain montmorillonites will adsorb water until the "rock"—it actually becomes a gel—contains as much as 20 parts of water to one part of clay; (2) a change in salinity of the formation water, as at Anchorage; (3) a change in the slope or load conditions, by the driving of a pile or the excavation for a building foundation.

All three of these possible means of clay stability change are active to some extent in California today, in the form of man-made building excavations; canal construction and subsequent leakage of fresh canal water into saltbearing clays; and in related activities of man. A concrete structure may give off free calcium in solution, to substitute for sodium in a clay formation electrolyte, changing the sensitivity of the clay. The number of stability changes in potentially unstable clays may be expected to accelerate greatly in California in the immediate future, as a direct result of the transportation of vast quantities of fresh northern California water to the south to satisfy the needs of that rapidly expanding area.

This water will be delivered to areas of need, many of which are the loci of poorly consolidated clay-bearing formations, such as in the southern San Joaquin Valley, southern Coast Ranges, vast areas of the Mojave Desert,

and in Los Angeles Basin region. Many of these areas contain saline clay-bearing sedimentary basins which may become vulnerable, through wastage, leakage, and plan, to fresh water invasion. This can be somewhat analogous to the Anchorage area. While disasters of the magnitude of Anchorage are unlikely, nevertheless substantial losses in life and property may result unless care is exercised.

What is our lesson from the Good Friday Alaskan earthquake? It is that change—even innocuous change, such as substitution of a fresh formation water for saline water—may breed instability in clay-bearing beds. A new residential property, and especially a "view" property which abuts against an escarpment or "free face" susceptible to landslide failure, invites disaster even without a "Good Friday" earthquake if the foundation conditions underlying the property are subjected to uncompensated stability changes. With the occurrence of a major earthquake—a potentiality in most populated areas of California—the stakes are indeed high.

Clay technology and clay mineralogy are new and dynamic branches of scientific study. Only in the past dozen years have up-to-date texts become available on these subjects. Few geologists are trained in clay problem studies and fewer laboratories are equipped to cope with them. Yet, there is a "need to know" these subjects. Great cities are growing. New ones will grow. The press is outward, to marginal development sites, to unproved development areas. Drill hole subsurface data, qualitative information on clay and pore water properties, and geologic know-how are lacking in many of these areas.

If expansion and development are to be judicious and safe, they must be accompanied by careful planning. Such planning must be based on geologic maps, which show locations of potential hazard due to past slides or presence of potentially sensitive clay formations. It must be based on complementary maps showing clay and groundwater properties of various geologic map units, so that appropriate care may be taken where potentially sensitive clays are encountered. Following such planning, competent development of an area will anticipate and prevent possible man-made "natural" disasters by utilization or preparation of foundation conditions that can resist them, and by avoiding construction of high population density developments in areas where foundation conditions are or may become too poor or unreliable to be remedied.

References

Cederstrom, D. J., Trainer, F. W., and Waller, R. M., 1964, Geology and ground-water resources of the Anchorage area, Alaska: U. S. Geological Survey Water-Supply Paper 1773.

Coulter, H. W., and Migliaccio, R. R., 1966, Effects of the earthquake of March 27, 1964 at Valdez, Alaska: U. S. Geological Survey Professional Paper 542-C, p. C1–C36.

Grim, Ralph E., 1962, Applied clay mineralogy: McGraw-Hill Book Co., Inc., New York, 422 p. Chapter 5, "Clay Mineralogy in relation to the engineering properties of clay materials," gives an excellent technical discussion of clays and foundation problems resulting from instability within clay-water systems.

Hansen, W. R., 1965. Effects of the earthquake of March 27, 1964, at Anchorage, Alaska: U. S. Geological Survey Professional Paper 542-A, p. A1–A64.

Kachadoorian, Reuben, 1965, Effects of the earthquake of March 27, at Whittier, Alaska: U. S. Geological Survey Professional Paper 542-B, p. B1–B21.

Kerr, Paul F., and Drew, Isabella M., 1965, Quick-clay movements, Anchorage, Alaska (Abstract): Geological Society of America Program, 1965 annual meetings, p. 86–87.

Kerr, Paul F., 1963, Quick clay; Scientific American, Vol. 209, no. 5 (November, 1963), p. 132–142.

Liebling, Richard S., and Kerr, Paul F., 1965, Observations on quick clay: Geological Society of America Bulletin, vol. 76, no. 8, p. 853–877.

Mielenz, Richard C., and King, Myrle E., 1955, Physical-chemical properties and engineering performance of clays: in Clays and clay technology.

Mitchell, James K., 1963, Engineering properties and problems of the San Francisco Bay mud: in California Division of Mines Special Report 82, p. 25–32.

Morris, F. C., Aune, Q. A., and Gates, G. L., 1959, Clays in petroleum reservoir rocks: U. S. Bureau of Mines Report of Investigations 5425, 65 p.

Oakeshott, Gordon B., 1964, The Alaskan earthquake: Mineral Information Service, July issue, p. 119–121, 124–125.

Schlocker, Julius, Bonilla, M. G. and Radbruch, Dorothy H., 1958, Geology of the San Francisco North quadrangle, California: U. S. Geological Survey Miscellaneous Geologic Investigations Map I-272.

Time Magazine, 1965. Anchorage's feet of clay: Time, vol. 86, no. 25, Dec. 17, 1965, p. 62.

13 The Vaiont Reservoir Disaster

George A. Kiersch

The worst dam disaster in history occurred on October 9, 1963, at the Vaiont Dam, in Italy, when some 2600 lives were lost. The greatest loss of life in any similar disaster was 2,209 in the Johnstown Flood in Pennsylvania in 1899. The Vaiont tragedy is unique in many respects because:

1. It involved the world's second highest dam, of 265.5 meters (875 ft).
2. The dam, the world's highest thin arch, sustained no damage to the main shell or abutments, even though it was subjected to a force estimated at 4 million tons from the combined slide and overtopping wave, far in excess of design pressures.
3. The catastrophe was caused by subsurface forces, set up wholly within the area of the slide, 1.8 kilometers long and 1.6 km. wide.
4. The slide volume exceeded 240 million cu. m. (312 million cu. yd.), mostly rock.
5. The reservoir was completely filled with slide material for 1.8 km. and

Kiersch, G. A., 1965, "The Vaiont Reservoir Disaster," *Mineral Information Service*, vol. 18, no. 7, pp. 129–38. Abridged by permission of the author and reprinted by permission of California Division of Mines and Geology. The report was first published in *Civil Engineering* (vol. 34, no. 3, 1964) and the American Society of Civil Engineers has permitted its reprinting in a revised form.

Professor Kiersch is Professor Emeritus of Engineering Geology at Cornell University, Ithaca, New York.

up to heights of 150 m. (488 ft.) above reservoir level, all within a period of 30 to 60 sec. (A point in the mass moved at a speed of 25 to 30 m. per sec.)
6. The slide created strong earth tremors, recorded as far away as Vienna and Brussels.

The quick sliding of the tremendous rock mass created an updraft of air accompanied by rocks and water that climbed up the right canyon wall a distance of 240 m. (780 ft.) above reservoir level. (References to right and left assume that the observer is looking downstream.) Subsequent waves of water swept over both abutments to a height of some 100 m. (328 ft.) above the crest of the dam. It was over 70 m. (230 ft.) high at the confluence with the Piave Valley, one mile away. Everything in the path of the flood for miles downstream was destroyed.

A terrific, compressive air blast preceded the main volume of water. The overtopping jet of water penetrated all the galleries and interior works of the dam and abutments. Air currents then acted in decompression; this tensional phase opened the chamber-locked safety doors of all the galleries and works and completed destruction of the dam installations, from crest to canyon floor.

This catastrophe, from the slide to complete destruction downstream, occurred within the brief span of some 7 min. It was caused by a combination of: (1) adverse geologic features in the reservoir area; (2) man-made condi-

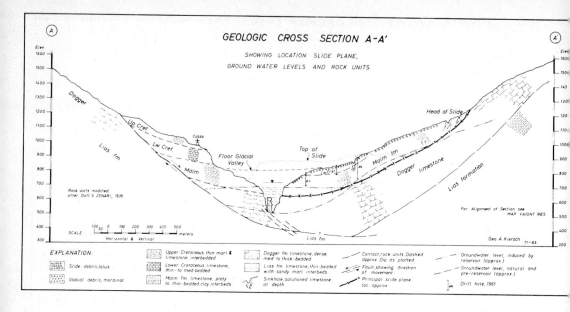

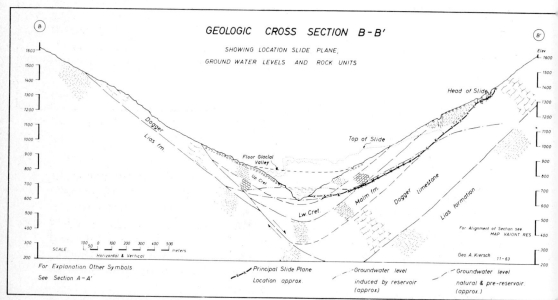

FIG. 1. On geologic cross-sections of slide and reservoir canyon, running from north to south, principal features of the slide plane, rock units and water levels are shown. For location of Sections A-A' and B-B', see Fig. 2.

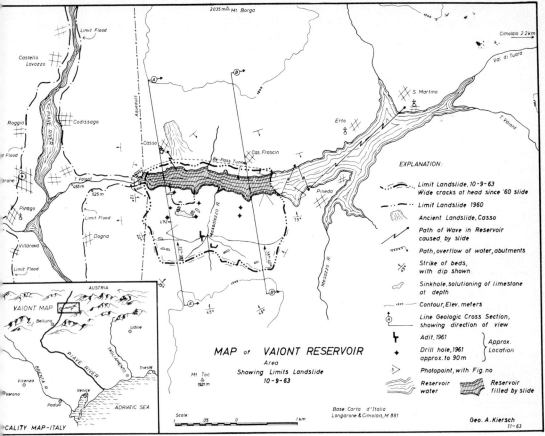

FIG. 2. Map of Vaiont Dam area and Piave River valley shows geographic features, limits of slide and of destructive flood waves.

tions imposed by impounded water with bank storage, affecting the otherwise delicately balanced stability of a steep rock slope; and (3) the progressive weakening of the rock mass with time, accelerated by excessive groundwater recharge.

Design and Construction

Vaiont Dam is a double-curved, thin-arch, concrete structure completed in the fall of 1960. The dam is 3.4 m. (11.2 ft.) wide at the top and 22.7 m. (74.5 ft.) wide at the plug in the bottom of the canyon. It has an overflow spillway, carried a two-lane highway on a deck over the crest, and had an underground powerhouse in the left abutment. Reservoir capacity was 150 million cu. m. (196 million cu. yd., or 316,000 acre-ft.).

The way in which the dam resisted the unexpected forces created by the slide is indeed a tribute to designer Carlo Semanza and the thoroughness of construction engineer Mario Pancini.

Design and construction had to overcome some disadvantages both of the site and of the proposed structure. The foundation was wholly within limestone beds, and a number of unusual geologic conditions were noted during the abutment excavation and construction. A strong set of rebound (relief) joints parallel to the canyon walls facilitated extensive scaling within the destressed, external rock "layer." Excessive stress relief within the disturbed outer zone caused rock bursts and slabbing in excavations and tunnels of the lower canyon. Strain energy released within the external, unstable "skin" of the abutment walls was recorded by seismograph as vibrations of the medium. This active strain phenomenon in the abutments was stabilized with a grout curtain to 150 m. (500 ft.) outward at the base—and the effects were verified by a seismograph record. Grouting was controlled through variations of the elastic modulus.

The potential for landslides was considered a major objection to the site by some early investigators; others believed that "the slide potential can be treated with modern technical methods."

The Geologic Setting

The Vaiont area is characterized by a thick section of sedimentary rocks, dominantly limestone with frequent clayey interbeds and a series of alternating limey and marl layers. The general subsurface distribution is shown in the geologic cross sections.

Retained Stress

The young folded mountains of the Vaiont region retain a part of the active tectonic stresses that deformed the rock sequence. Faulting and local folding accompanied the regional tilting along with abundant tectonic fracturing. This deformation, further aided by bedding planes and relief joints, created blocky rock masses.

The development of rebound joints beneath the floor and walls of the outer valley is shown in an accompanying figure. This destressing effect creates a weak zone of highly fractured and "layered" rock, accentuated by the natural dip of the rock units. This weak zone is normally 100 to 150 m. (330 to 500 ft.) thick. Below this a stress balance is reached and the undisturbed rock has the natural stresses of mass.

Rapid carving of the inner valley resulted in the formation of a second set of rebound joints—in this case parallel to the walls of the present Vaiont canyon. The active, unstable "skin" of the inner canyon was fully confirmed during the construction of the dam.

The two sets of rebound joints, younger and older, intersect and coalesce within the upper part of the inner valley. This sector of the canyon walls, weakened by overlapping rebound joints, along with abundant tectonic fractures and inclined bedding planes, is a very unstable rock mass and prone to creep until it attains the proper slope.

Causes of Slide

Several adverse geologic features of the reservoir area contributed to the landslide on October 9:

Rock units that occur in a semicircular outcrop on the north slopes of

Mt. Toc are steeply tilted. When deformed, some slipping and fault movement between the beds weakened frictional bond.

Steep dip of beds changes northward to Vaiont canyon, where rock units flatten along the synclinal axis; in three dimensions the area is bowl-shaped. The down-dip toe of the steep slopes is an escarpment offering no resistance to gravity sliding.

Rock units involved are inherently weak and possess low shearing resistance; they are of limestone with seams and clay partings alternating with thin beds of limestone and marl, and frequent interbeds of claystone.

Steep profile of the inner canyon walls offer a strong gravity force to produce visco-elastic, gravitational creep and sliding.

Semicircular dip pattern confined the tendency for gravitational deformation to the bowl-shaped area.

Active dissolving of limestone by ground-water circulation has occurred at intervals since early Tertiary time. The result has been subsurface development of extensive tubes, openings, cavities and widening of joints and bedding planes. Sinkholes formed in the floor of the outer valley, particularly along the strike of the Malm formation on the upper slopes; these served as catchment basins for runoff for recharge of the ground-water reservoir. The interconnected ground-water system weakened the physical bonding of the rocks and also increased the hydrostatic uplift. The buoyant flow reduced gravitational friction, thereby facilitating sliding in the rocks.

Two sets of strong rebound joints, combined with inclined bedding planes and tectonic and natural fracture planes, created a very unstable rock mass throughout the upper part of the inner canyon.

Heavy rains in August and September produced an excessive inflow of ground-water from the drainage area on the north slopes of Mt. Toc. This recharge raised the natural ground-water level through a critical section of the slide plane (headward part) and subsequently raised the level of the induced water table in the vicinity of their junction (critical area of tensional action). The approximate position of both water levels at the time of the slide is shown in the accompanying figure (Fig. 1).

Excessive ground-water inflow in early October increased the bulk density of the rocks occurring above the initial water table; this added weight contributed to a reduction in the gross shear strength. Swelling of some clay minerals in the seams, partings and beds created additional uplift and contributed to sliding. The upstream sector is composed largely of marl and thin beds of limestone with clay partings—a rock sequence that is inherently less stable than the downstream sector.

The bowl-shaped configuration of the beds in the slide area increased the confinement of ground water within the mass; steeply inclined clay partings aided the containment on the east, south, and west.

Two exploratory adits driven in 1961 reportedly exposed clay seams and small-scale slide planes. Drill holes bored near the head of the 1960 slide were slowly closed and sheared off. This confirmed the view that a slow gravitational creep was in progress following the 1960 slide and probably even before that—caused by a combination of geologic causes. Creep and the accompanying vibrations due to stress relief were later described by Muller.

Effects of Man's Activities

Construction of Vaiont Reservoir created an induced ground-water level which increased the hydrostatic uplift pressure throughout a triangular subsurface mass aided by fractures and the interconnected system of solution openings in the limestone.

Before April 1963, the reservoir was maintained at El. 680 m. or lower, except for two months in 1962 when it was maintained at 690–700 m. In September, five months after the induced water table was raised 20 m. or higher (700–710 m.), the slide area increased its rate of creep. This action has three possible explanations: (1) a very delicate balance existed between the strength of the rock mass and the

FIG. 3. On sketch of inner Vaiont canyon and remnants of the outer glacial valley, are shown rebound joints—old and young set—from stress relief within the walls of the valley to depths of 100 to 150 m. (330 to 500 ft.).

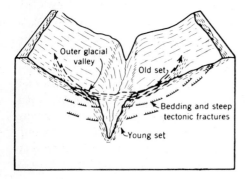

internal stresses (shear and tensile), which was destroyed by the 20-m. rise of bank storage and accompanying increase in hydrostatic pressure; (2) the same reaction resulted from the large subsurface inflow in early October due to rains; or (3) the induced ground-water level from the reservoir at El. 680 m. during 1961–1962 did not attain maximum lateral infiltration until September 1963, when creep accelerated. In any case, the rate of ground-water migration into bank storage is believed to have been critical.

Evidence indicates that the immediate cause of the slide was an increase in the internal stresses and a gross reduction in the strength of the rock mass, particularly the upstream sector where this mass consists largely of marl and alternating thin beds of limestone and marl. Actual collapse was triggered by an excess of ground water, which created a change in the mass density and increased the hydrostatic uplift and swelling pressures along planes of inherent weakness, combined with the numerous geologic features that enhanced and facilitated gravitional sliding.

The final movement was sudden—no causes from "outside" the affected area are thought to have been responsible.

Sequence of Slide Events

Large-scale landslides are common on the slopes of Vaiont Valley; witness the ancient slide at Casso and the prehistoric blocking of the valley at Pineda. Movement at new localities is to be expected periodically because of the adverse geologic setting of the valley. The principal events preceding the movement on October 9 were:

In 1960, a slide of some 700,000 cu. m. (one million cu. yd.) occurred on the left bank of the reservoir near the dam. This movement was accompanied by creep over a much larger area; a pattern of cracks developed upslope from the slide and continued eastward. These fractures ultimately marked the approximate limits of the October 9 slide. The slopes of Mt. Toc were observed to be creeping and the area showed many indications of instability.

In 1960–1961, a bypass tunnel 5 m. (16.4 ft.) in diameter was driven along the right wall of the reservoir for a distance of 2 km. (6,560 ft.), to assure that water could reach the outlet works of the dam in case of future slides.

As a precaution, after the 1960 slide the reservoir elevation was generally held at a maximum of 680 m. and a grid of geodetic stations on concrete pillars was installed throughout the potential slide area extending 4 km. (2.5 miles) upstream, to measure any movement.

The potential slide area was explored in 1961 both by drill holes and by man-sized adits (see map for depth). Reportedly, no confirmation of a major slide plane could be detected in either drill holes or adit. An analysis now indicates that the drill holes were too shallow to intercept the major slide plane of October 9, and what was in all probability the deepest plane of gravitational creep started by the 1960 slide and active thereafter.

Gravitational creep of the left reservoir slope was observed during the 1960–1961 period, and Muller reports "movement of 25 to 30 cm. (10 to 12 in.) per week (on occasion) which was followed in close succession by small, local earth tremors due to stress relief within the slope centered at depths of 50 to 500 m. (164 to 1,640 ft.). The total rock mass that was creeping was about 200 million cu. m. (260 million cu. yd.)."

During the spring and summer of 1963, the eventual slide area moved very slowly; scattered observations showed a creep distance of 1 cm. (⅜ in.) per week, an average rate since the 1960 slide.

Beginning about September 18, numerous geodetic stations were observed to be moving 1 cm. a day. However it was generally believed that only individual blocks were creeping; it was not suspected that the entire area was moving as a mass.

Heavy rains began about September 28 and continued steadily until after October 9. Excessive run-off increased ground-water recharge and surface inflow; the reservoir was at El. 700m. or higher, about 100 ft. below the crest.

About October 1, animals grazing on the north slopes of Mt. Toc and the reservoir bank sensed danger and moved away. The mayor of Casso ordered townspeople to evacuate the slopes, and posted notice of an expected 20-m. (65-ft.) wave in the reservoir from the anticipated landslide. (The 20 m. was also the estimate of engineers for the height of the wave that would follow such a slide, based on experience of the slide at nearby Pontesi Dam in 1959.)

Movements of geodetic stations throughout the slide area reported for about three weeks before the collapse showed a steady increase from about 1 centimeter per day in mid-September to between 20 and 30 centimeters to as much as 80 centimeters on the day of failure.

About October 8, engineers realized that all the observation stations were moving together as a "uniform" unstable mass; and furthermore the actual slide involved some five times the area thought to be moving and expected to collapse about mid-November.

On October 8, engineers began to lower the reservoir level from El. 700 m. in anticipation of a slide. Two outlet tunnels on the left abutment were discharging a total of 5,000 cfs. but heavy inflow from runoff reduced the actual lowering of the water. The reservoir contained about 120 million cu. m. of water at the time of the disaster.

On October 9, the accelerated rate of movement was reported by the engineer in charge. A five-member board of advisers were evaluating conditions, and authorities were assessing the situation on an around-the-clock basis. Although the bypass outlet gates were open, oral reports describe a rise in the reservoir level on October 9. This is logical if lateral movement of the left bank had progressed to a point where it was reducing the reservoir capacity. These reports also mention difficulty with the intake gates in the left abutment (El. 591 m.) a few hours before the fatal slide.

Movement, Flood and Destruction

Those who witnessed the collapse included 20 technical personnel stationed in the control building on the left abutment and some 40 people in the office and hotel building on the right abutment. But no one who witnessed the collapse survived the destructive flood wave that accompanied the sudden slide at 22 hours 41 min. 40 sec. (Central European Time). However, a resident of Casso living over 260 m. (850 ft.) above the reservoir, and on the opposite side from the slide, reported the following sequence of events:

About 10:15 p.m. he was awakened by a very loud and continuous sound of rolling rocks. He suspected nothing unusual as talus slides are very common.

The rolling of rocks continued and steadily grew louder. It was raining hard.

About 10:40 p.m. a very strong wind struck the house, breaking the window panes. Then the house shook violently; there was a very loud rumbling noise. Soon afterward the roof of the house was lifted up so that rain and rocks came hurtling into the room (on the second floor) for what seemed like half a minute.

He had jumped out of bed to open the door and leave when the roof collapsed onto the bed. The wind suddenly died down and everything in the valley was quiet.

Observers in Longarone reported that a wall of water came down the canyon about 10:43 p.m. and at the same time a strong wind broke windows, and houses shook from strong earth tremors. The flood wave was over 70 m. (230 ft.) high at the mouth of Vaiont canyon and hit Longarone head on. Everything in its path was destroyed. The flood moved upstream in the Piave Valley beyond Castello Lavazzo, where a 5-m. (16-ft.) wave wrecked the lower part of Codissago. The main volume swept downstream from Longarone, hitting Pirago and

Villanova. By 10:55 p.m. the flood waters had receded and all was quiet in the valley.

The character and effect of the air blast that accompanied the main flood wave at the dam have been described in the introduction. The destruction wrought by the blast, the jet of water, and the decompression phase are difficult to imagine. For example, the steel I-beams in the underground power-house were twisted like a corkscrew and sheared; the steel doors of the safety chamber were torn from their hinges, bent, and carried 12 m (43 ft.) away.

Seismic tremors caused by the rock slide were recorded over a wide area of Europe—at Rome, Trieste, Vienna, Basel, Stuttgart, and Brussels. The kinetic energy of the falling earth mass was the sole cause of the seismic tremors recorded from Vaiont according to Toperczer. No deep-seated earthquake occurred to trigger the slide. The seismic record clearly demonstrates that surface waves ($L_1 = 3.26$ km. per sec., or about 730 mph.) were first to arrive at the regional seismic stations, followed by secondary surface waves ($L_2 = 2.55$ km. per sec. or 570 mph). There was no forewarning in the form of small shocks and no follow-up shocks—which are typical of earthquakes from subsurface sources. No P or S waves were recorded.

Pattern of Sliding

The actual release and unrestricted movement of the slide was extremely rapid. Seismological records show that the major sliding took place within less than 30 sec. (under 14 sec. for the full record of the L_1 wave) and thereafter

sliding ceased. The speed of the mass movement (25 to 30 m. per sec.) and the depth of the principal slide are strikingly demonstrated by the preservation intact of the Masselezza River canyon and the grassy surface soil with distinctive "fracture" pattern.

Wave Action Due to Slide

Sketchy reports from observers at Erto described the first wave by stating that "the entire reservoir for 1.8 km. (1.1 miles) piled up as one vast curving wave" for a period of 10 sec. The strong updraft of air created by the rapid slide was confined in movement by the deep Vaiont Valley encircled by high peaks. The updraft within the confined outer valley sucked the water, accompanied by rocks, up to El. 960 m. (885 ft. or more above the original reservoir level) and accounted for part of the force possessed by the initial wave.

At the dam, the initial wave split on hitting the right canyon wall, after demolishing the hotel building at El. 780 m. (300 ft. above the reservoir surface). Some of the water followed the canyon wall downstream and moved above and around the dam. The major volume, however, seems to have bounced off the right wall, swept back across the canyon to the left abutment and moved upslope and around the dam to at least El. 820 m. (460 ft. above the reservoir level).

The overflow waves from the right and left abutments were joined in the canyon by the main surge, which overtopped the dam, and together these constituted the flood wave that hit Longarone. Water overtopped the dam crest on the left side for some hours after the slide, strongly the next morning, and during this time also displaced water drained from pools scattered over the slide surface.

Upstream the wave generated by the slide moved first into the area opposite Pineda, where it demolished homes, bounced off the canyon wall and moved southward, hitting the Pineda peninsula. On receding from there, the wave moved northeastward across the full length of the lake and struck San Martino with full force, bypassing Erto, which went unharmed.

Conditions Since the Slide

The water level just behind the dam dropped at the rate of 50 to 80 cm. (20 to 32 in.) per day during the first two weeks after the slide. This loss is believed due in part to leakage through the intake gates for the bypass aqueduct and powerhouse conduits. Geologically, there was a substantial loss due to the new conditions of bank storage, subsurface circulation and saturation of material filling the canyon.

A pond that formed at the Massalezza River canyon, along the foot of the slide plane, dropped in level rapidly and was dry on October 24, confirming the idea of ground-water recharge to slide material and the establishment of a water table within the newly formed mass. Smaller ponds initially formed upstream from the dam along the zone of contact between the slide and the right reservoir bank. These likewise dried up by October 24 as a result of groundwater recharge and readjustment in the water table within the slide mass.

The lake level behind the slide dam rose steadily from the inflow of tributary streams. For example, two weeks after the slide, the reservoir was 13 m. (43 ft.) higher than the water level at the dam—a major problem in the future operation of Vaiont Dam.

Strong funneling craters developed during the first days after the slide in the soil and glacial debris concentrated near the toe of the slide. This cratering was of concern to some as indicating large-scale movement to come, but other conditions are the probable causes of the surface subsidence. Large blocks of rock, with some bridging action, fill the canyon and create much void space in the lower mass. Some of these spaces are filled by normal gravity shifting of fines, and ground-water circulation also distributes fines into these void spaces. Formation of craters is restricted to the section of the slide that fills the former canyon, and craters appear at intervals along its entire length. They are most extensive in the slope behind the dam.

Numerous small, step-like slide blocks occur at different levels on the main slip plane. These blocks were loosened by the movement on October 9 and have since moved slowly down the slip plane, some to the bottom of the escarpment. Talus runs are common from small V-notched canyons along the edge of the steep eastern sector of the slide.

Future of the Reservoir

The steeply dipping beds along the head of the slide will undoubtedly fail from time to time as a result of gravitational creep. Ultimately the upper most part of the slip plane will be flattened and thereby will attain a stable natural slope.

The Italian Ministry of Public Works has announced that Vaiont Dam will no longer be used as a power source. The cost of clearing the reservoir would be prohibitive because of the volume involved, the distance of 4 or 5 km. (3 miles) that waste material would have to be hauled to the Piave Valley, and the 300-m. (1,000-ft.) lift required to transport the waste over the divide west of the dam.

The bypass tunnel in the right wall of the reservoir could be ultimately used to pond water behind the dam for release through the existing outlet works. Another alternative would be to divert the reservoir water southeastward to the Cellina River drainage by a tunnel driven from the upper end of the lake. Such diversion would develop the upper catchment area of Cellina and utilize Vaiont storage, behind the slide dam, as a multi-seasonal storage astride the Piave and Cellina catchments.

Vaiont in Retrospect

Vaiont has tragically demonstrated the critical importance of geologic features within a reservoir and in its vicinity— even though the site may be otherwise satisfactory for a dam of outstanding design.

In future, preconstruction studies must give thorough consideration to the properties of a rock mass as such, in contrast to a substance, and particularly to its potential for deformation with the passage of time. An assessment that is theoretical only is inadequate. The soundest approach is a systematic appraisal that includes:

An investigation of the geologic setting and its critical features

An assessment of past events that have modified features and properties of the site rocks

A forecast of the effects of the engineering works on geologic features in the area and on the strength of the site rocks

The geologic reaction to changed conditions in the process of time.

Project plans should set forth a system for acquiring data on the interaction between geologic conditions and changes induced by project operation.

Time, in terms of the life of the project, is a key to safety and doubtless was a controlling factor at Vaiont. Since 1959, eight major dams around the world have failed in some manner. It seems imperative that the following factors be recognized:

1. Rock masses, under changed environmental conditions, can weaken within short periods of time—days, weeks, months.

2. The strength of a rock mass can decrease very rapidly once creep gets under way.

3. Evidence of active creep should be considered as a warning that warrants immediate technical assessment, since acceleration to collapse can occur quickly.

Engineering Implications

Speed of sliding movement. Rock masses are capable of translatory movement as fast as quick clays or at a liquid-like speed.

Strain energy: its influence on a rock mass, its release and associated movement are critical. The interplay between wetting of a rock mass, buoyancy effect, and the lightening of a rock mass (1.0 or less in density) allows an accelerated release of the inherent strain energy, thereby creating more release fractures and the cycle is repeated. The net result is the increase in amount of water and a stronger buoyancy effect—both aided by the energy release phenomenon.

Potential for landslides at reservoir sites that have been in operation for many years must be studied; new projects must be evaluated for landslide potential in a more critical manner.

The tremendous amount of potential energy that is stored in a rock mass undergoing creep on an incline (as at Vaiont). With the increasing displacements, gliding friction factor drops and the velocity of mass increases. This means that a sliding mass has a potential to increase from slow creep to a fantastically high rate of movement in a brief time of seconds or minutes. The energy goes into momentum and not into deforming the interior of the sliding mass as in the typical rotational slide.

Two techniques assist the engineer today in this connection. First, he has and can use the most improved methods for observing and measuring the changes of strain within a rock mass. And second, he can use a forewarning system in case this phenomenon acts quickly and the failure of a rock is imminent.

References

Anon. 1958. Some SADE developments: *Water Power*, vol. 10, nos. 3–6, Mar.–June 1958.

Anon. 1961. Italy builds more dams: *Engineering News-Record*, vol. 167, no. 18, pp. 30–36.

Anon. 1963. Vaiont Dam survives immense overtopping: *Engineering News-Record*, vol. 171, no. 16, pp. 22–23, Oct. 17, 1963.

Boyer, G. R. 1913. Etude géologique des environs de Longarone (Alpes venitiennes): *Soc. Géologique de France, Bulletin*, vol. 13, no. 4, pp. 451–485.

Muller, L. 1963. Differences in the characteristic features of rocks and mountain masses: 5th Conference of the International Bureau of Rock Mechanics, *Proceedings*. Leipzig, Germany, Nov. 1963.

Muller, L. 1963. Rock mechanics considerations in the design of rock slopes, in *State of Stress in the Earth's Crust*, International Conference, Rand Corp., Santa Monica, Calif., June 1963.

Pancini, M. 1961. Observations and surveys on the abutments of Vaiont Dam. *Geologie und Bauwesen*, vol. 26, no. 1, pp. 122–141.

Pancini, M. 1962. Results of first series of tests performed on a model reproducing the actual structure of the abutment rock of the Vaiont Dam. *Geologie und Bauwesen*, vol. 27, no. 1, pp. 105–119.

14 Permafrost

U.S. Geological Survey

In 1577, on his second voyage to the New World in search of the Northwest Passage, Sir Martin Frobisher reported finding ground in the far north that was frozen to depths of "four or five fathoms, even in summer," and that the frozen condition "so combineth the stones together that scarcely instruments with great force can unknit them." This permanently frozen ground, now termed *permafrost*, underlies perhaps a fifth of the earth's land surface. It occurs in Antarctica, but is most extensive in the northern hemisphere. In the lands surrounding the Arctic Ocean its maximum thickness has been reported in terms of thousands of feet—as much as 5,000 feet in Siberia and 2,000 feet in northern Alaska.

For almost 300 years after Frobisher's discovery, little attention was paid to this frost phenomenon, but in the 19th Century during construction of the Trans-Siberian Railroad across vast stretches of frozen tundra, permafrost was brought to the attention of the Russians because of the engineering problems it caused. In North America the discovery of gold in Alaska and the Yukon Territory near the turn of the century likewise focused the attention of miners and engineers on the unique nature of permafrost.

Published by the U. S. Government Printing Office, 1973. Based on material provided by Louis L. Ray.

Frozen ground poses few engineering problems if it is not disturbed. But changes in the surface environment—such as the clearing of vegetation, the building of roads and other construction and the draining of lakes—lead to thawing of the permafrost, which in turn produces unstable ground susceptible to soil creep and landslides, slumping and subsidence, icings, and severe frost heaving. The many environmental problems stemming from the rapid expansion of human activities in areas underlain by frozen ground during and following World War II, have demonstrated that a thorough understanding of the nature of permafrost is of prime importance for wise land-use planning. Not only is it an economic requirement that the least possible disturbance be made of the frozen ground, but it is a practical necessity if the future land-use potential of the vast areas underlain by the frozen ground is to be preserved.

In the Northern Hemisphere, permafrost decreases in thickness progressively from north to south. Two major zones are distinguished: a northern zone in which permafrost forms a continuous layer at shallow depths below the ground surface and, to the south, a discontinuous zone in which there are scattered permafrost-free areas. Land underlain by continuous permafrost almost completely circumscribes the Arctic Ocean. A less clearly defined zone of so-called sporadic permafrost occurs along the southern margin of the discontinuous

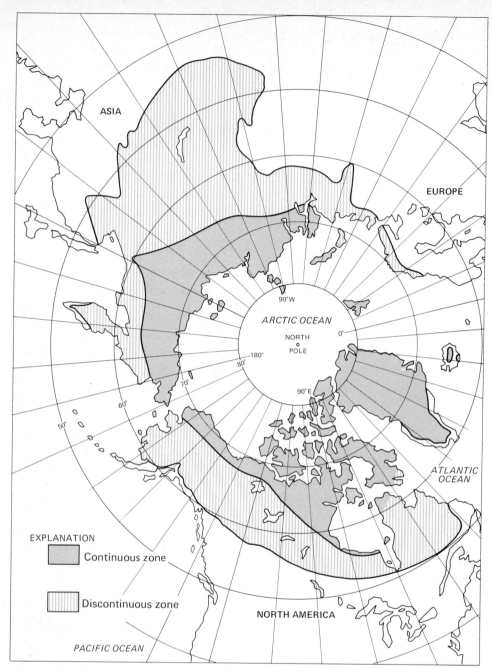

FIG. 1. Distribution of permafrost in the Northern Hemisphere.

zone where widespread permafrost-free ground contains scattered small, isolated masses of the frozen ground.

The term *permafrost*, a contraction of permanently frozen ground, was proposed in 1943 by Siemon W. Muller, of the U. S. Geological Survey, to define a thickness of soil or other superficial deposit, or even of bedrock beneath the surface of the earth in which a temperature below freezing has existed continuously for 2 or more years. When the average annual air temperature is low enough to maintain a continuous average surface temperature below 0°C, the depth of winter freezing of the ground exceeds the depth of summer thawing, and a layer of frozen ground is developed. Downward penetration of the cold will continue until it is balanced by the heat flowing upward from the earth's interior. In this manner, permafrost hundreds of feet thick can form over a period of several thousand years. Distribution and thickness of permafrost depends, however, on many factors that control surface temperatures and cold penetration. Some of these factors are: geographic position and exposure, character of seasonal and annual cloudiness, precipitation, vegetation, drainage, and the properties of the earth materials that underlie the ground surface.

Although thawing of the surface and near surface layers may occur quickly when summer thawing exceeds winter freezing, it would require long periods of time, estimated as high as tens of thousands of years, for thawing air temperatures to penetrate and melt the thickest known permafrost. Thus, the distribution and variations of subfreezing temperatures recorded at depth in the thicker permafrost layers reflect the ancient cold temperatures of the past that are commonly assigned to the Pleistocene Epoch, the so-called Great Ice Age, which began about 3 million years ago.

The ground above the permafrost which thaws in summer and refreezes in winter is known as the *active layer*. Its thickness, like that of the underlying permafrost, depends on the many factors that influence the flow of heat into and out from the Earth's surface.

The upper surface of the permafrost layer is known as the *Permafrost Table*. When winter freezing does not penetrate to the permafrost table, an unfrozen layer remains between the base of the frozen active layer and the permafrost table. This unfrozen material, as well as the rare isolated masses of unfrozen ground present within the permafrost itself, are called *talik*, a term adopted from the Russian. The talik, like the ground ice, is a major consideration in any appraisal of the permafrost environment because unfrozen ground water, commonly concentrated within the talik, may be highly mineralized and under hydrostatic pressure. Frequently the water

FIG. 2. Typical section of permafrost terrain.

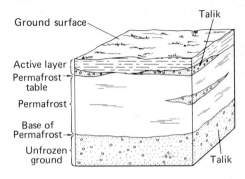

bursts forth to the ground surface under pressure where a point of issue has been opened either by natural or artificial means. On reaching the surface the water may freeze, producing a thick and perhaps widespread ice sheet or an ice mound. Such *icings* may pose serious problems in areas of concentrated human activity.

In the upper part of the permafrost layer, large masses of ground ice of various shapes and origins may be present. In places in Alaska it is estimated that more than half of the volume of the upper 10 feet or so of the permafrost layer consists of ice. The ice may occur as coatings, as individual grains, veinlets and lenses, and as ice wedges that extend downward from the permafrost table.

The presence of ice-rich permafrost is readily apparent whenever the insulating effect of vegetation is modified or destroyed. Such changes trigger the thawing of the underlying permafrost and ground ice so that distinctive changes in the landscape are produced. Especially important from the standpoint of human activities is the ground subsidence that results from thawing of the ground ice and the solifluction or gravity controlled mass movement of thawed, water-saturated surficial sediments that produce a variety of land forms.

Artificial stripping of the insulating vegetation mat from the active layer or the removal or destruction of the active layer while preparing the land for agricultural use or construction projects such as roads, railroads, airfields, and buildings, disturbs the delicate thermal balance which may, unless preventive measures are taken, result in thawing of the ice masses in the

underlying permafrost with consequent irregular subsidence of the land surface. Even the casual crossing of tundra areas on foot or a single traverse by a wheeled vehicle may so upset the thermal balance through slight changes in the insulating properties of the vegetation mat that thawing of the underlying ground may result. Once thawing starts, its control may be difficult or even impossible.

Experience has shown that ice-rich permafrost is generally present where the active layer is relatively thin, the drainage poor, and the frozen materials are fine-grained sediments. These conditions are well developed on the Arctic Slope of Alaska where ground ice is commonly present as a honeycomb-like, polygonal network of vertical ice wedges that extends downward from the base of the active layer. Where the thermal balance has been so modified that thawing is initiated, the normal thickness of the active layer is increased, and the surface of the permafrost table is depressed below the upper part of the ground ice. Consequent melting of the tops of the ice wedges produces subsidence in the overlying ground that is reflected on the land surface by a network of shallow-to-deep interconnected furrows that form a polygonal pattern.

Where conditions are such that no polygonal ground has developed, the presence of the underlying network of polygonal ice wedges may be reflected in the character of the stream courses. The concentrated heat in the water flowing across terrain underlain by the ice-wedge polygons thaws them to produce a series of subsidence pools along the stream course at ice-wedge intersections. When viewed from the air, such stream courses with their series of pools give the impression of a string of beads, hence the name *beaded drainage*.

When masses of ground ice thaw, subsidence may produce isolated depressions, *thermokarst pits*, or basins occupied by *thaw lakes*. Once initiated, thawing and consequent calving along the thaw shores tends to increase the lake area. If wind directions are relatively constant in the summer season, thawing may be concentrated in directions controlled by the wind and wave action along the lake shores, produc-

ing features such as the well-known "oriented lakes" of Alaska and Russia.

Curious, rounded, ice-cored hills called *pingos* are present both in areas of tundra and boreal forest. Rising above the surrounding landscape of unconsolidated, fine-grained sediments to heights as much as several hundred feet, the largest pingos may be as much as several thousand feet in diameter. Pingos are relatively ephemeral features of the landscape, believed to result from concentrations in the talik of unfrozen water under hydraulic pressure that bows up the overlying

FIG. 3. Building located south of Fairbanks is subsiding because of thawing permafrost. (Photo by T. L. Péwé, Courtesy of O. J. Ferrians, Jr., U. S. Geological Survey.)

sediments and freezes. The summits of pingos may contain crater-like lakes fed by springs of fresh water that may flow throughout the year.

On the tundra of the Arctic Slope of Alaska and Canada, pingos are generally present in old lake basins that may be partly filled with swamp deposits (muskeg). In the forested areas to the south, pingos form in areas of valley-bottom alluvium adjacent to the base of steeply sloping, south-facing valley walls. Pingos in areas of boreal forests may commonly be recognized by the vegetation which grows profusely on the well-drained soils of their steeply sloping sides.

Solifluction, the slow mass movement of surficial unconsolidated water-saturated sediments downslope, is another common-place phenomenon in regions underlain by permafrost. Commonly, during the summer season of thawing, the surficial sediments become supersaturated because melt-waters are unable to percolate into the impervious permafrost below. At the interface between the frozen and unfrozen materials, these melt-waters provide a lubricant to the frozen surface over which the mobile unfrozen mass can readily slide by gravity. Surficial materials may move as sheets or lobes over fronts ranging from a few to several hundreds of feet wide. Where there is a well-developed vegetation mat above the fluidlike mass of supersaturated debris, a *solifluction sheet* may move downward as a well defined sheet, or lobe, or as a series of partially overriding folds. At times the vegetation mat may rupture, and the slurry of water-saturated sediments may produce sudden destructive mud flows. Because solifluction can be trig-gered on slopes as low as 3°, and movement may be increased by disturbance of the normal ground conditions, it is a serious hazard at construction sites, especially those underlain by permafrost at shallow depth.

Frost heaving, commonplace in all environments where there is marked freezing and thawing of the ground, results from an upward or expanding force occasioned by swelling of the ground during freezing. The effects of this process are magnified in the colder climates where the land is underlain by permafrost, although in most areas there is generally little evidence of frost heaving under natural conditions. If surface conditions are disturbed, however, by construction or other activities permitting an increase in summer thawing with a consequent thickening of the active layer to which frost heaving is confined, it expectedly becomes a serious problem. It is necessary, therefore, to preserve the insulating value of the ground surface as much as possible so that the thickness of the active layer will be retained at a minimum, in order that the effects of frost heaving may also be kept at a minimum. This can be accomplished by not removing the vegetation mat and by increasing the surficial insulation of the ground at the construction site, generally by the addition of a coarse gravel fill.

It is readily apparent that man's uninhibited and careless use of land underlain by permafrost can produce serious problems today and in some cases, can cause lasting detrimental effects. Likewise, natural changes such as variations in climate, erosion by shifting streams, increased precipitation, unusually severe storms, earth-

FIG. 4. Ground subsidence along abandoned Copper River Railroad. (Photo by L. A. Yehle, Courtesy of O. J. Ferrians, Jr., U. S. Geological Survey.)

quakes, landslides, forest fires, and many other natural phenomena can also modify the environment, producing similar adverse effects. Each modification, whether naturally or artificially induced, must be carefully evaluated, and if necessary controlled in order to minimize the detrimental ef-

fects if the vast regions underlain by permafrost are to remain continuously serviceable to man.

Suggested Reading

Ferrians, O. J., Jr. Kachadoorian, Reuben, and Greene, G. W., 1969. Permafrost and related engineering problems in Alaska: *U. S. Geol. Survey Prof. Paper 678*, 37 p.

Lachenbruch, A. H., 1968, Permafrost, *in* Fairbridge, R. W., ed., *The Encyclopedia of Geomorphology:* New York, Reinhold Publishing Corp., p. 833–839.

Muller, S. W., 1943, Permafrost or permanently frozen ground and related engineering problems: *U. S. Army, Office of Chief of Engineers, Military Intelligence Div. Strategic Eng. Study 62*, 231 p.; reprinted with corrections, 1945; also, 1947, Ann Arbor, Mich., Edwards Bros.

Ray, L. L., 1956, Perennially frozen ground, an environmental factor in Alaska: *Internatl. Geogr. Congr., 17th, and 8th Gen. Assembly, Washington, D. C., 1952 Proc.*, p. 260–264.

15   Expansive Soils—The Hidden Disaster

D. Earl Jones, Jr., and Wesley G. Holtz

Expansive soils cause more damage in the U.S. than earthquakes, tornados, hurricanes and floods combined. Much of Nature's violence is vented on vacant lands, as only 3% of our nation is occupied by cities and only 13% is agriculturally tilled. Within the average American's lifetime, 14% of our land will be lashed by earthquakes, tornados and floods—but over 20% will be affected by expansive soil movements.

When dry, expansive soils are hard and strong, like a rock. But when water is added, these soils swell and soften. In pure form, montmorillonite clays (bentonite) may swell to over 15 times their dry volume. Most soils contain only small amounts of montmorillonite, so not many swell to more than 1½ times dry volume. When fully expanded and saturated with water, the soil loses much of its strength. Such water-saturated soils become soft, with the consistency and slipperiness of lubricating grease; they will squeeze out from beneath a load—as toothpaste squeezes from a tube.

Expansive Soils: Big Problem

Expansive soils, found in every state, cover one-fourth the U.S. They often occur on moderately sloped lands, most attractive for human occupancy. Irregular movements of expansive soils put unexpected pressures on man's works, or remove support from them. Expansive soil movements may occur slowly, may continue for years. Continued movements caused by seasonal ground moisture changes or by wet-dry cycles of several years' duration are common.

Expansive plastic clay soils shrink on drying, often badly cracking structures built on them. While expansion is more of a problem in arid and semiarid areas, shrinkage caused by dessication can affect structures in areas with sizable precipitation. In many cases trees are planted near houses and along streets underlain by montmorillonite clays. As the trees mature, they withdraw more soil moisture

Reprinted from *Civil Engineering*, v. 43, n. 8, pp. 49–51, with permission of the authors and the American Society of Civil Engineers, © 1973 by ASCE. D. Earl Jones, Jr., was Chief Civil Engineer for the Department of Housing and Urban Development and Wesley Holtz a consulting engineer when this article was published.

than replenished by Nature, thereby drying out the soil—and shrinking it. The resulting soil movements disrupt houses, sidewalks, driveways, streets, utilities.

Severe structural damage to homes often stems from expansive soil heaving, triggered by water from a leaking sewer. Damage ranges from sticking doors and hairline plaster cracks to totally destroyed homes. Homeowners must spend $ thousands to correct damage.

Over 250,000 new homes are built on expansive soils each year. 60% will experience only minor damage during their useful lives, but 10% will experience significant damage—some beyond repair. Owners have been known to patch and redecorate badly cracked and moving walls, sell the house quickly, and pass the problem to someone else.

Commercial Building Damage

One-half the nation's light construction involves service and light commercial facilities, usually constructed less carefully than homes and built of materials more easily damaged by soil

Table 1 Structural damage from soil movements

single family homes	$300,000,000.
commercial buildings	360,000,000.
multi-story buildings	80,000,000.
walks, drives & parking areas	110,000,000.
highways and streets	1,140,000,000.
buried utilities and services	100,000,000.
airport installations	40,000,000.
involved in urban landslides	25,000,000.
other	100,000,000.
Total annual damage in U.S.	$2,255,000,000.

movements. Damages unacceptable in a home are negligible in many commercial facilities, where appearance is unimportant. But broken and irregular floors and broken and rotated walls and foundations are common, calling for costly reconstruction.

Damage to Multi-story Buildings

Expansive soil movements also damage multi-story buildings. Walls may be heavy enough to resist damage from soil swelling (uplifting). But the same heavy walls are most vulnerable if supporting soil drys out and shrinks, removing wall support.

Wall loads above the first floor are carried by the building frame. But first floor wall loads are ground supported. Many building codes permit all wall loads up to three or four stories to be ground supported—thus susceptible to soil-shrinkage damage.

Wall breakage in multi-story buildings, then, can be a major problem in warm climates and arid areas—unless first floor wall loads are carried by the building frame and the walls isolated from the soil. Even in cooler climates and non-arid areas, extended drought causes soil shrinkage—and thus wall damage.

Even multi-story buildings on caissons and piles have suffered severe damage when expansive clay subsoils increased in moisture. Clay clings to the pile or caisson and lifts it and the building—unless the pile is anchored in a lower, non-reactive soil zone. Also, pile reinforcing steel must be able to resist tensile forces generated.

Ground supported floors in multi-story buildings are often badly broken by swelling. Small soil movements may induce undesirable floor tilting, unsightly cracks and even tripping hazards. Frequent repairs are needed until stable moisture conditions prevail in the underlying soil.

Building Site Damage

Walks, driveways, patios and porches often are more badly broken by expansive soil movements than buildings. Slabs are usually light, move easily, have low flexural strength. Many concrete slabs are poured directly on dry clay soils. Subsoils are then protected from drying; so soils gather moisture and expansion begins. So it's important, during site grading, to assure runoff will drain rapidly away rather than saturate subsoils.

During the rush to complete construction, parking lots on expansive soils often are graded and paved without careful design or construction. Such pavements are frequently destined to rapid deterioration, high maintenance and early replacement.

The Short-Lived Highway

Highway pavements built on expansive soils usually break up long before pavements on stable soils. On expansive soils, pavement repair and replacement costs are high, and the public often inconvenienced. Hundreds of "one car" auto accidents each year may be caused by loss of driver control on wavy pavements. Several states now use construction procedures on primary roads to lessen the undesirable effects of expansive clay subgrades—at considerable additional cost.

Underground Utilities and Services

Expansive soil movements are greatest near the ground surface and generally diminish to nothing between 5 and 30 ft underground. A light building supported on the ground surface moves up and down more than underground-sewer, water and gas pipes connected to it. This can stretch or bend pipes, causing them to break and leak. Sewer pipes with packed bell and spigot joints are particularly susceptible to movement-caused leakage. Leaking water or sewer pipes trigger localized soil expansion. The more the swell, the more the leakage—thus more soil expansion. "Flexible" pipes, such as thin-walled plastic or bitumized fiber, are unusually vulnerable to crushing in expansive soils, because of external pressures.

Urban Landslides

Even the land man builds on can be damaged by expansive soil movements. On hillsides, alternate shrinking and swelling of expansive soils induces a downhill creep of the soil mantle. Fences, walks and even utilities may move considerably. Clay hillsides may develop deep shrinkage cracks during dry weather, permitting the next rain to penetrate deeply and perhaps trigger slippage.

Highway builders can "take a chance" and make cut or fill slopes steep. Most highway landslides can be repaired quickly; repair costs will be less than wider rights-of-way with flatter, more stable slopes. In cities, however, landslides may damage structures: urban designers cannot afford to "take a chance." Slopes near buildings should be much less steep than the usual highway side slopes.

Concrete and other rigid slabs in canal or reservoir linings and in swimming pools are very vulnerable to expansive soil damage. A slight leak releases water into the supporting soil, triggering damaging localized expansion. For 20 years, the concrete lining of a California water-moving canal has been plagued by expansive soil movements. Its reconstruction, at $500,000 per mile, is now being considered.

Electric power substations built on expansive soils have suffered damages. Movements of slabs supporting electrical equipment have been known to break switch bars, insulator stacks and other equipment, resulting in power disruptions. Some dams on expansive soils have experienced sliding failures similar to landslides.

The Losses

Average annual losses this century from floods, earthquakes, hurricanes and tornados are less than one-half of damages from expansive soils (see Table 1).

The expansive soil problem is diffi-cult. Laboratory tests identify how much a soil can shrink or swell—but not how much it will shrink or swell beneath a particular building. Engineers must also evaluate how the soil will react to future environmental conditions—difficult. People working with expansive soils have been able to choose only between taking a risk or constructing unusually strong foundations. They are reluctant to strengthen foundations to resist *all* possible soil movements, as such strengthening could cost more than the damages prevented.

To stimulate research for preventing expansive soil damages, ASCE recently established a research council on expansive behavior of earth materials.

In sum, our country has not recognized the losses from expansive soils. We may be able to prevent much damage. The country has spent $ billions to control floods—but virtually nothing to mitigate expansive soil damages, which are far greater than flooding losses. One person in 10 is affected by floods; but one in five by expansive soils. The civil engineering profession has an obligation to solve the expansive soil problem—the hidden disaster.

Editor's note: D. Earl Jones, Jr., writing in *Underground Space* (v. 3, n. 5, 1979) notes that, as of 1979, a most conservative estimate of annual damages by expansive soils in the United States was $8 billion. In 1981, the estimate was $10 billion (personal communication).

Erosion and Sedimentation

Lake Michigan storm waves crash against a condominium on Sheridan Road, Chicago. Reproduced with permission from *Chicago Daily News*. Photograph by Fred Stein.

*" Along the coastlines of the world, numerous engi-
neering works in various states of disintegration
testify to the futility and wastefulness of disregarding the tremendous destructive forces of the
sea."*

M. P. O'Brien

Nature of the Problem

Erosion and sedimentation, both natural geologic proccesses, are two
of the most pervasive environmental problems. They also deplete
valuable topsoil, introduce suspended sediment into streams and
lakes, and cause siltation of reservoirs and harbors. Average annual
losses in the United States from sediment damage exceed $1 billion,
and more than $15 billion has been spent on soil conservation since
the mid-1930s. Figure 1 illustrates the magnitude of soil erosion in the
United States.

Fluvial erosion (stream or river erosion), is responsible for 96 percent of the sediment eroded from the world's land surfaces and is the
principal subject of this chapter; wind (*eolian*) erosion accounts for 2
percent, and *glacial* erosion 1 percent. Shoreline erosion erodes less
than 1 percent of the land area, but may have a dramatic impact on a
local scale. In the United States, 40 percent of the stream sediment
load is eroded from agricultural lands; 26 percent from streambanks;
12 percent from pastures and rangeland; and 7 percent from forest
lands.

Most of the sediment reaching stream channels is the product of
sheet, rill, or gully erosion. *Sheet erosion* is most pervasive and occurs
when broad continuous sheets of running water wash away thin
layers of topsoil (Fig. 2). *Rill erosion* is the removal of soil along small
but visible channels. *Gully erosion* takes place in distinct, narrow channels which are larger and deeper than rills but carry water only during or immediately after storms or snowmelt. Rills can be removed by
normal tilling; gullies cannot.

Causes of Erosion

The principal factors affecting the erosion and transportation of sediments are climate, soil character, topography, and soil cover (Table 1).

It is important to note that human activities may accelerate the rates

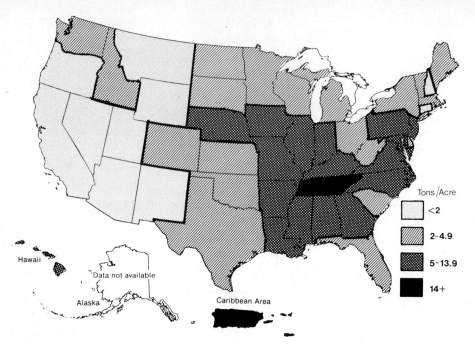

FIG. 1. Average annual sheet and rill erosion on cropland. The national average is 4.66 tons per acre. (From U.S. Soil Conservation Service.)

of erosion to a significant degree. Deforestation, cultivation, and construction activities will rob soil of its protective vegetative cover and expose it to maximum erosion. Grading building sites may increase the length and degree of a slope, thereby encouraging gullying and slope failure. The U.S. Department of Agriculture estimates that only about 30 percent of the eroded sediment in the United States is the result of natural erosion, and that 70 percent is the result of human activities which have accelerated erosion.

Control of Erosion and Sedimentation

The U.S. Soil Conservation Service was created in 1935 to combat soil erosion, which by that time had become a serious problem in many of the nation's agricultural areas. Initial efforts were directed at preserving the topsoil of prime agricultural land. Later efforts were directed at preventing soil erosion at construction sites and at keeping sediments out of streams and lakes.

The most common techniques for controlling erosion in agricultural areas are

FIG. 2. Uncontrolled sheet and rill erosion depleted topsoil at the rate of 150 tons per acre on this slope in Sherman County, Oregon, in the 1950s. (Photo courtesy U.S. Department of Agriculture.)

1. Contour plowing. Furrows that follow the contour of the topography trap runoff and sediments. In rolling terrain, contour plowing may reduce erosion by 50 percent.
2. Contour strip cropping. Dense, erosion-resistant crops are planted in strips between row crops such as corn in order to retard runoff and trap sediment (Fig. 3).
3. Terracing. Terraces reduce runoff and trap sediments.
4. Construction of check dams and retention ponds. These also reduce runoff and trap sediments.
5. Conservation tillage. "No till—plant in residue" reduces exposure of soil to the forces of erosion. New crops are planted in the residue left after harvesting without tilling the ground.
6. Crop rotation. Cover crops are rotated with crops that expose fields to erosion.
7. Pasture management. This may be employed to prevent overgrazing.

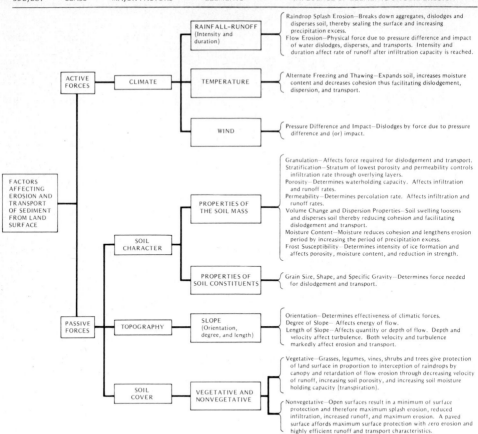

Table 1   Chart of the principal factors affecting erosion and transportation of sediment from the land surface [From *U.S. Geological Survey Professional Paper 462-E*, 1964]

In urban areas, the most common techniques used for preventing erosion are

1. Identification of erosion-prone areas. Detailed soil surveys are used to identify steep terrain and soils of high erodibility, which are set aside for open space.
2. Reducing soil exposure. Grading and removal of native vegetation is held to a minimum. Large tracts are developed sequentially, and exposed areas are covered with mulch to reduce exposure during construction. Permanent cover is established as soon as possible.

FIG. 3. Contour strip croping has brought erosion under control on land near Zumbro Falls, Minnesota, and the sediment load of streams has been reduced. (Photo courtesy U.S. Soil Conservation Service.)

3. Scheduling construction. Grading and land clearing is conducted during periods of minimum rainfall.
4. Controlling runoff. Control structures include terraces, diversion ditches, and water bars (barriers which divert runoff to roadside ditches).
5. Stabilizing stream channels. Channels are lined with riprap or concrete to reduce erosion.
6. Controlling sediments. Small check dams and settling ponds prevent sediments from entering streams and rivers.

Predicting Soil Loss and Sediment Yield

Rainfall Erosion Losses

Soil conservation not only requires a knowledge of the factors that cause soil erosion and the techniques that help to reduce soil loss, it also requires an understanding of the complex interrelationships between these factors. A certain amount of soil loss is inevitable even when people are not involved in agricultural or construction activities. The term *soil loss tolerance* denotes the maximum level of soil erosion that will permit a high level of crop productivity to be sustained economically and indefinitely. Widespread experience in the United States has shown that annual soil loss tolerances ranging from 1814 to 4536 kg per 0.4 ha (2 to 5 tons per acre) are feasible and generally adequate for sustaining high productivity levels indefinitely. The *Universal Soil Loss Equation* (USLE) is an erosion model designed to predict the longtime average annual soil losses caused by sheet and rill erosion at specific sites. It is an empirical tool based on widespread field and plot data collected over more than two decades. Use of the USLE allows prediction of soil losses under current conditions and under changes in crops or such "support" practices as contouring and terracing. The soil loss equation is expressed as

$$A = R \cdot K \cdot L \cdot S \cdot C \cdot P$$

where

A   is the average annual soil loss per unit area (usually expressed in tons per acre per year),
R   is the measure of the erosion forces of rainfall and runoff,
K   is the soil erodibility factor,
L   is the slope-length factor,
S   is the slope-steepness factor,
C   is the crop cover and management factor, and
P   is the support practice factor (e.g., reflects benefits of contouring and strip cropping).

The USLE, therefore, relates upland soil loss to the complex interrelationships between rainfall, runoff, soil erodibility, topography, vegetative cover, and support practices.

Numerical values for each of the six factors of the model have been derived from analyses of the research data on test plots and from the National Weather Service precipitation records. Representative values for the factors may be obtained from various sources (see Wischmeier and Smith, 1978). Widespread field use has substantiated the usefulness of the USLE in a variety of agricultural conditions and in construction sites. It should be noted, however, that the predicted soil losses are only estimates, and that they have been averaged over a long period of time. In some watersheds, because of the complexity of the interrelationships between the factors, the averaging of values might detract from the accuracy of the results (Wischmeier, 1976).

Soil scientists and engineers have demonstrated that remote sensing techniques can be used to predict soil loss on cropland in Wisconsin (Morgan *et al.*, 1978). Their estimates compare favorably with those of the USLE, and the method has the advantage of providing a continuous flow of information about land-use practices that influence soil loss.

Watershed Sediment Yield

Predictions of watershed sediment yields can be used in water quality studies and in determining reservoir sedimentation. One method of estimating sediment yield is the gross erosion-sediment delivery method. The equation is

$$Y = E \, (DR) \, _{Ws}$$

where

Y   is sediment yield per unit area,

E   is the gross erosion,

DR   is the sediment delivery ratio, and

Ws   is the area of the watershed above the point for which the sediment yield is being computed.

Gross erosion includes both sheet and rill erosion, which may be determined by using the USLE, and channel-type erosion which must be estimated by other means. The amount of sediment delivered to a stream will be less than the amount produced through gross erosion because some sediment will be deposited on the more level and densely vegetated slopes remote from the stream channel. The ratio of sediment delivered at a given location in the stream basin to gross erosion from the drainage area above that location is called the *sedimentary delivery ratio*. If the runoff drains directly into the stream with no intervening obstructions or flattening of the topographic slope, the sedimentary delivery ratio will approach one. On the other hand, a substantial area of flat ground or dense vegetation may cause deposition of almost all the sediment before it reaches the channel. The decrease in stream load may result in a compensatory increase in channel erosion.

Case Histories and Readings

Erosion of the Land, or What's
Happening to Our Continents?

We have seen that erosion and sedimentation affect the entire surface of continents and have a significant impact on human beings. What information is available concerning the erosion of continents? How does one quantify the role of people in accelerating soil erosion? Is it possible to determine erosion rates on a global scale? These questions are considered by Sheldon Judson in Reading 16.

Shoreline Structures as a Cause of
Shoreline Erosion: A Review

From a geologic point of view, the zone where land and sea meet is one of the most dynamic and most vulnerable to erosion and sedimentation. Yet this zone is also one of the most desirable settings for recreation, residence, and commercial development. Measuring rates of coastal erosion is simpler and more direct than measuring rates of erosion on land. Topographic maps, charts, and aerial photographs provide useful baseline data.

Some of the more significant factors which influence coastal erosion are

1. Changes in water levels. There have been periodic changes in sea level (*eustasy*) throughout geologic time. Sea level has been slowly rising since the retreat of the continental ice caps began about 10,000 years ago; this has been an important factor in shoreline erosion. In large inland lakes, water levels change in response to changes in precipitation.
2. Composition of bluff material.
3. Steepness of bluffs.
4. Vegetative cover on bluff.
5. Ground-water flow. Generally, flow is from land toward the body of water, exerting a hydraulic force in that direction.
6. Interference with littoral drift and supply of shoreline sediments.
7. Storm frequency, intensity, and associated waves.

Coastal zone management programs generally prohibit development of hazardous areas. But what about areas that have already been developed or that lack management programs? What structural protection devices are available and what are the consequences of employing these structures? In Reading 17, James Rosenbaum reviews the role of structural devices in stabilizing the shoreline environment.

Sinkhole

A more subtle aspect of erosion is the solution of bedrock that commonly takes place in humid areas underlain by limestone. *Sinkholes* or *solution craters* often are the result of such erosion. The hazard is often intensified during extended drought or when large volumes of underground water are removed by pumping, since the water may have stabilized the rocks by floatation. Philip LaMoreaux and William Warren report a spectacular example of the sudden formation of a large sinkhole in Shelby County, Alabama. Extensive areas throughout the southeastern United States are subject to this hazard. For example, a recent cave-in in Winter Park, Florida, created a large sinkhole (more than 115×100 m; 377×328 ft) within minutes, causing extensive property damage. The event was probably triggered by a combination of a long drought and ground-water pumping to meet the needs of area residents (Foose, 1981).

Sediment

Early soil conservation efforts were directed at holding productive soils in place. Although that need still exists, the current emphasis is on the effects of waterborne and deposited sediments in the aquatic environment. Sediment that adversely affects the aquatic environment originates from a number of sources, and its physical, economic, and aesthetic impact is great. A. R. Robinson presents some insights on the dual role of sediment as both a pollutant and a scavenger of pollutants in the concluding article of this section.

The deposition of sediments in a stream channel limits the capacity of the channel so that in times of high water, flooding may occur. Additional causes of flooding are examined in the following section.

16 Erosion of the Land, or What's Happening to Our Continents?

Sheldon Judson

Not quite two centuries ago James Hutton, Scottish medical man, agriculturalist, and natural scientist—now enshrined as the founder of modern geology—and Jean André de Luc, Swiss emigré, scientist, and reader to England's Queen Charlotte, carried on a spirited discussion concerning the nature and extent of erosion of the natural landscape. De Luc believed that once vegetation had spread its protective cloak across the land, erosion ceased. Not so, in Hutton's opinion. He argued (Hutton, 1795):

According to the doctrine of this author (de Luc) our mountains of Tweed-dale and Tiviotdale, being all covered with vegetation, are arrived at the period in the course of times when they should be permanent. But is it really so? Do they never waste? Look at rivers in a flood—if these run clear, this philosopher has reasoned right, and I have lost my argument. [But] our clearest streams run muddy in a flood. The great causes, therefore, for the degradation of mountains never stop as long as there is water to run; although as the heights of mountains diminish, the progress of their diminution may be more and more retarded.

Judson, S., 1968, "Erosion of the Land, or What's Happening to Our Continents?" *Amer. Scientist*, vol. 56, pp. 356–74. Lightly edited by permission of the author and reprinted by permission of The Society of Sigma Xi. Copyright 1969 by The Society of Sigma Xi.

Dr. Judson is Professor of Geology at Princeton University, Princeton, New Jersey.

We know today, of course, that vegetation plays an important role in the preparation of material for erosion. We know also that although vegetation may slow the removal of material from a slope it does not stop it completely. Hutton's view is overwhelmingly accepted today. Erosion continues in spite of the plant cover, which in fact is conducive to certain aspects of erosion. The discussion now centers on the factors determining erosion, the nature of the products of this process, how these products are moved from one place to another, and at what rates the products are being produced. Hutton, in his day, had no data upon which to make a quantitative estimate of the rates at which erosion progressed. Today we, unlike Hutton, measure rates of erosion for period of a fraction of a man's lifetime, as well as for periods of a few hundreds or thousands of years of human history. In addition, radioactive dating and refined techniques of study in field and laboratory allow us to make some quantitative statements about the rates at which our solid lands are wasted and moved particle by particle, ion by ion, to the ocean basins.

This report sets forth some of what we know about these erosional rates. We will understand that erosion is the process by which earth materials are worn away and moved from one spot to another. As such, the action of water, wind, ice, frost-action, plants and animals, and gravity all play their roles. The destination of material eroded is eventually the great world ocean, although there are pauses in the journey and, as we will see later, the material delivered to the ocean must be in some way reincorporated into the continents.

Some Modern Records

Let us now examine some modern records of erosion of various small areas on the earth's crust, essentially determinations of rates at specific points. There is a large amount of information to be gleaned from agricultural, forestry, and conservation studies as well as from some studies by geologists.[1]

Even a casual inspection of our cemeteries demonstrates that some rock goes to pieces at a measurable rate and that rocks have differing resistance to destruction. Four marble headstones photographed in 1968 in the Princeton, N. J., cemetery indicate what can happen to marble in the 172 years involved. The marker erected in 1898 was still easily legible 70 years later, but the crisp, sharp outline of the stone carver's chisel was gone. The headstone erected 70 years earlier was still partially legible in 1968, but the stone put up in 1796 was completely illegible. In this instance the calcite ($CaCO_3$), which makes up the marble, was attacked by a carbonic acid formed by rain water and the CO^2 of the atmosphere. In general, marble headstones become illegible in the humid northeastern states after 150 to 175 years of exposure.

In contrast to the marble headstones is a marker in the Cambridge, Massachusetts Burying Ground, that was erected in 1699 and photographed in 1968. It is made of slate, often used as a headstone material in many New England cemeteries until marble became fashionable at the turn of the nineteenth century. Unlike marble it is resistant to chemical erosion. Nearly 270 years after the stone was erected the inscription stands out clearly.

Graveyards do most certainly provide examples of the impermanence of rock material as well as of the relative resistance of different rock types. The earliest study in such an environment that I have seen was by Sir Archibald Geikie, in Edinburgh, published in 1880. More recent studies have been made of the rates at which erosion proceeds on tombstones. Thus, in an area near Middletown, Connecticut, it is estimated that tombstones of a local red sandstone are weathering at the rate of about 0.006 centimeters per year (Matthias, 1967). In general, however, a graveyard does not present the best conditions for the accumulation of quantitative data.

More reliable data seem to come from agricultural stations. Here is an example. A summary of measurements has been made at 17 different stations on plots measuring 6 by 72.6 ft and under differing conditions of rainfall, soil, slope, and vegetative cover (Musgrave, 1954). Periods of record in this instance vary between 4 and 11 years. On the average, erosion from plots with continuous grass cover annually lost 75 tons per square kilometer, a lowering of about 3 meters per 1000 years. This is a dramatic demonstration of the role of plants in affecting erosion. In this instance the rate of erosion increased 100 times between grass-covered plots and well-tilled rowcrop plots.

Obviously climate will also affect the rate of erosion. For example, recent studies by Washburn (1967) in eastern Greenland show that seasonal freeze and thaw in a nearly glacial climate produce erosion rates ranging between 9 and 37 meters per thousand years. This contrasts with the rates in more temperate climates cited previously. In semiarid lands, where vegetation is discontinuous and rainfall low (± 25 cm per year) and unpredictable, the erosion rates are high but not as high as those in the rigorous climate of northeastern Greenland. Studies of bristlecone pines in Utah and California have allowed an estimate of erosion rates on a time base of hundreds and even thousands of years (Eardley, 1967). Thus the pines, which may reach 4000 years in age, betray the amount of erosion during their lifetime by the amount of exposure of their root systems. The depth of exposed roots on living trees is a measure of the amount the land surface has been reduced since the tree began to grow. Rates of lowering in general vary with exposure (greater on north-facing slopes) and with declivity of slopes (greater on steeper slopes). On the average, the rate varies between about 2 cm per 1000 years on slopes of 5 degrees and 10 cm on slopes of 30 degrees. A total of 42 observations indicate a direct relation between the erosion rate and the sine of the slope.

A different sort of study, this one in the rain forest of New Guinea Mountains, has yielded the estimate that between 1 and 2 cm per 1000 years is lost from the area by landslides alone (Simonett, 1967). How much addi-

tional material is lost through the agency of other processes is not known.

Archaelogical sites may yield information on erosional rates and have, as in the case of the bristlecone pines, a fairly long time base. Data collected in Italy show that for the sites studied the range in rates is 30 to 100 cm per 1000 years (Judson, 1968).

These are but a sample of the type of information that abounds in the literature on the rate of erosion. They are enough, however, to indicate how variable the rates can be when, as in the examples cited, the observation is for a single spot or limited area. Not only are they highly variable but they can hardly be representative of rates of erosion over large areas. It is apparent that the material eroded in one spot may be deposited nearby, at least temporarily, and thus the net loss to an area may be little or nothing. Erosion is more rapid at some spots than others for any one of many different reasons. Material removed from its position at any single spot on the landscape follows a slow, halting, devious course as natural processes transport it from the land to the ocean.

River Records

When we ask now how much material is being lost by the continents to the ocean, the spot measurements such as those reported above are of little help. We need some method of integrating these rates over larger areas. One way to do this is to measure material carried by a stream from its drainage basin at the point where the stream leaves the basin. Alternatively, the amount of sediment deposited in a

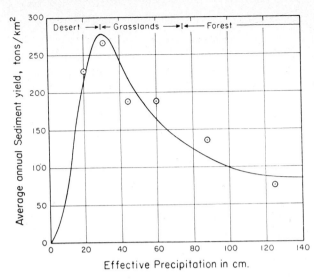

FIG. 1. Variation of the yield of sediments with precipitation. Effective precipitation is defined as precipitation necessary to produce a given amount of runoff. (After Langbein and Schumm, 1958.)

reservoir or in a natural lake over a specific length of time is indicative of the rate at which the land has been worn away in the basin lying upstream. The mass of sediments accumulated in unit time can be averaged out over the area of the contributing drainage basin to produce an erosion rate. Of course the erosion rate is not uniform over the entire basin, but it is convenient for our purposes here to assume that it is.

If we examine the solid load of a stream carried in suspension past a gauging station we discover that the amount of material per unit area of the drainage basin varies considerably according to a number of factors. But, if we hold the size of the drainage basin relatively constant, we find pronounced correlation between erosion and precipitation. Figure 1 is based on data presented by Langbein and Schumm (1958) from about 100 sediment gauging stations in basins averaging 3900 sq km. It suggests that a maximum rate of erosion is reached in

areas of limited rainfall (± 25 cm per year) and decreases in more arid as well as in more humid lands.

Considering small drainage basins (averaging 78 km²), Langbein and Schumm also show a similar variation in erosion with rainfall, but at rates which are 2 to 3 times as rapid as for the larger basins. In still smaller basins erosion rates increase even more. A small drainage basin in the Loess Hills of Iowa, having an area of 3.4 km² provides an extreme example. Here sediments are being removed at a rate which produces a lowering for the basin of 12.8 m per 1000 years.

We have data based on river records for larger areas. Judson and Ritter (1964) have surveyed the regional erosion rates in the United States and have shown that, on the average, erosion is proceeding at about 6 cm² per 1000 years. Here too, as shown in Table 1, there are variations. These appear to be related to climate as in the smaller areas already discussed. Greatest erosion occurs in the dry Colorado

Table 1 Rates of regional erosion in the United States (After Judson and Ritter, 1964)

Drainage region	Drainage[1] Area Km² × 10³	Runoff m³/sec	Load tons Km²/yr			Erosion cm/1000 yr	% Area sampled	Avg. years of record
			Dissolved	Solid	Total			
Colorado	629	0.6	23	417	440	17	56	32
Pacific Slopes, California	303	2.3	36	209	245	9	44	4
Western Gulf	829	1.6	41	101	142	5	9	9
Mississippi	3238	17.5	39	94	133	5	99	12
S. Atlantic & Eastern Gulf	736	9.2	61	48	109	4	19	7
N. Atlantic	383	5.9	57	69	126	5	10	5
Columbia	679	9.8	57	44	101	4	39	<2
Totals	6797	46.9	43	119	162	6		

[1]Great Basin, St. Lawrence, Hudson Bay drainage not considered.

River basin. In examining the rates of regional erosion we note that although erosion rates increase with decrease in discharge per unit area, they do not increase quite as rapidly as of the major component, the detrital load, increases. This is so because the absolute dissolved load decreases with decreasing discharge per unit area. This inverse relation between solid and dissolved load is shown in Fig. 2.

These data suggest that on the average the United States is now being eroded at a rate which reduces the land surface by 6 cm each 1000 years. Actually the rate is somewhat less when we consider that the area of the Great Basin, with no discharge to the sea, is not included in these figures— and that for all practical purposes the

FIG. 2. Relation by regions in the United States between solid load and dissolved load in tons/km²/yr. (After Judson and Ritter, 1964.)

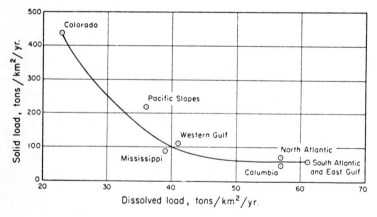

net loss from this area is presently close to zero.

Effect of Man

What effect does man's use of the land have on the rate at which it is destroyed by natural forces? Three examples are cited here:

Bonatti and Hutchinson have described cores from a small volcanic crater lake, Lago di Monterosi, 41 km north of Rome. (See Judson, 1968, note 3.) An archaeological survey of the environs of the lake indicate that intense human activity dates from approximately the second century B.C. when the Via Cassia was constructed

through the area. At this moment the cores indicate a sudden increase of sedimentation in the lake. The rate varies somewhat but continues high to the present. Extrapolation of the sedimentation rate in the lake as to the surrounding watershed shows that prior to intensive occupation by man (that is, prior to the second century B.C.) the erosion rate was 2 to 3 cm per 1000 years. Thereafter it rose abruptly to an average of about 20 cm per 1000 years.

Ursic and Dendy (1965) have studied the annual sediment yields from individual watersheds in northern Mississippi. The results of their data are shown in Fig. 3. These indicate that, when the land is intensively cultivated,

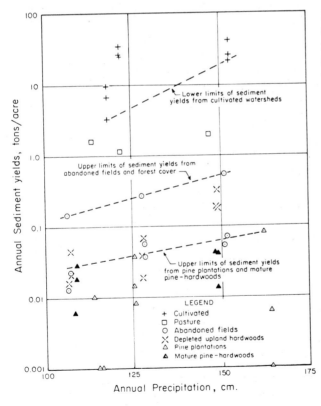

FIG. 3. Variation in sediment yields from individual watersheds in northern Mississippi under different types of land use and changing amounts of precipitation. One ton/acre equals 224 tons/km². (After Ursic and Dendy, 1965.)

the rate of sediment production and hence the rate of erosion is three orders of magnitude or more above that experienced from areas with mature forest cover or from pine plantations.

Wolman (1967) has described the variation of sediment yield with land use for an area near Washington, D. C. These data are summarized in Fig. 4. They show that, under original forest conditions, erosion proceeded at the low rate of about 0.2 cm per 1000 years. With the rapid increase of farmland in the early nineteenth century the rate increased to approximately 10 cm per 1000 years. With the return of some of this land to grazing and forest in the 1940's and 1950's this high rate of erosion was reduced perhaps by one-half. Areas undergoing construction during the 1960's show yields which exceed 100,000 tons per square kilometer for very small areas, which approximate a rate of lowering of 10 m per 1000 years. For completely urban areas the erosion rates are low, less than 1 cm per 1000 years.

There is no question that man's occupancy of the land increases the rate of erosion. Where that occupation is intense and is directed toward the use of land for cultivated crops the difference is one or more orders of magnitude greater than when the land is under a complete natural vegetative cover such as grass or forest. The intervention of man in the geologic processes raises questions when we begin to consider the rates of erosion for the earth as a whole and to apply modern rates to the processes of the past before man was a factor in promoting erosion.

Ian Douglas (1967) postulates that man's use of the landscape has so increased the rates of erosion that they far exceed those of the past before man became an important geologic agent. He presents persuasive data and arguments to suggest that any computation of present-day erosion rates on a world-wide basis are unrepresentative of those that pre-date man's tampering with the landscape. So, as we turn to

FIG. 4. A sequence of land use changes and sediment yield beginning prior to the advent of extensive farming and continuing through a period of construction and subsequent urban landscape. Based on experience in the middle Atlantic region of the United States. (After Wolman, 1967.)

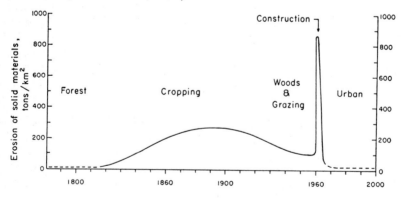

Table 2 Rates of erosion for the Amazon River Basin, United States and Congo River Basin

Drainage region	Drainage area Km^2 × 10^6	Load, tons × 10^6/yr			Tons Km^2/yr	Erosion cm/1000 yr
		Dissolved	Solid	Total		
Amazon River[1] Basin	6.3	232	548[2]	780	124	4.7
United States[3]	6.8	292	248[3]	540	78	3.0
Congo River[4] Basin	2.5	99	34[2]	133	53	2.0
Totals	15.6	623	830	1453	93	3.6

[1]From Gibbs, 1967.
[2]Solid load increased by considering bed load as 10% of suspended load.
[3]From Judson and Ritter, 1964. Solid load reduced to adjust for increased erosion because of man's activity.
[4]From Spronck, 1941, quoted in Gibbs 1967.

the question of world-wide erosion, we will want to distinguish between present-day rates which are profoundly affected by man's activity and those of the immediate past before man introduced grazing, agriculture, and other activities.

Let us first attempt an estimate of erosion before man began to affect the process. It is estimated that approximately one fourth of the United States is in cropland. If this area is now undergoing a rate of erosion ten times that of its natural rate then, for the United States as a whole, the increase of rate of erosion because of man's use of the land increases the rate of the removal of solid particles from the earth's crust by a factor of a little over three times. Assuming that this is correct and that the dissolved load does not change appreciably, then, as a first approximation, the present rates of erosion listed in Table 1 for the United States would be decreased to approximately 3 cm per 1000 years, which is about 78 tons per square kilometer per year. This figure would

apply then to the area of the United States before the intervention of man with intensive agricultural practices.

Rates for Entire Earth

What can we say now about the rate of erosion for the entire earth? Presented in Table 2 are data for approximately 10 per cent of the earth's surface. The table includes erosional data for the drainage basins for the Amazon, the world's largest river; the Congo; and for that part of United States covered in Table 1. Here, however, the data for the United States have been adjusted to account for the increased rates of erosion presumed to have occurred because of man's cultivation of the land. Neither the Congo nor the Amazon basin are significantly affected by man. For the 15 million square kilometers of these three areas the average rate of erosion is 3.6 cm per 1000 years, or 93 tons per square kilometer annually.

Let us accept the figures just given as representative of erosion rates prior to man's intervention in the process and use them to extrapolate to erosion rates for the whole area of the earth. The earth's land surface has approximately 151 million square kilometers, but much of this area has no streams which drain directly to the ocean. For example, a large area of western United States is without direct drainage to the sea, as is a large percentage, about 50 per cent, of Australia. Areas of little or no drainage to the sea are estimated to occupy approximately one third of the earth's surface. So for our purposes we estimate that 100 million kilometers of the earth's surface are contributing sediments directly to the sea by running water. In addition to this there is a certain amount of wind erosion, and part of the materials eroded by the wind are delivered to the sea. It is even more difficult to find data on the amounts of regional erosion by wind than it is by running water. We have some preliminary estimates for the amount of eolian material which has been dumped into the oceans. These lie between 1 and 0.25 mm per 1000 years.[2] Whatever the figure, wind erosion of the land is volumetrically unimportant when compared with the amount of material carried by the streams.

We can estimate, then, the amount of sediment carried as solids and as dissolved material from the continents each year to the ocean basins as 9.3×10^9 tons. This figure is based on the assumption that on the average, 3.6 cm per 1000 years are eroded from the 100 million square kilometers of land which are estimated to drain into the oceans. Further, the figure attempts to eliminate the effect on the erosion rate of man's activity. If we include an estimate for the amount of erosion by wind action then this figure increases by an amount approximating 10^8 tons. Glacier ice may add a similar amount.

We can now compare this estimate of the tonnage of eroded materials with other estimates in the following paragraphs and Table 3.

Barth (1962) presents data on some geochemical cycles indicating that weathering of the land produces on the average of 2.5 kg per cm^2 per million years. From this figure we calculate that the average tonnage per year of all material, dissolved and solid, would be 3.8×10^9 tons which seems low. Strakhov (1967) quotes Lopatin (1950) to the effect that annual dissolved and solid loads of the rivers total 17.5×10^9 tons of which 4.9×10^9 tons are dissolved material. Two other estimates on dissolved loads should be quoted. Clarke (1924) estimates 2.7×10^9 tons per year and Livingstone (1963) 3.9×10^9 tons per year. This last figure can be duplicated by extrapolation of the data in Table 3. Livingstone indicates that the figure might be high. Indeed new figures on the salinity and discharge of the Amazon River by Gibbs (1967) indicate that Livingstone's figure should be adjusted downward by 5 per cent.

MacKenzie and Garrels (1966) estimate that the rivers of the world carry 8.3×10^9 tons of *solid material alone* to the oceans each year. In arriving at this figure they adopted from Livingstone an average annual world-wide runoff of 3.3×10^{16} liters and an average suspended sediment concentration equal to that of the Mississippi River. If man's occupancy has indeed increased erosion rates as we have sug-

Table 3 Estimates of world-wide erosion rates by various authors. All material assumed to reach the oceans

	10^9 metric tons/yr
Carried by rivers	
Dissolved load	
Livingstone (1963)	3.9
Clarke (1924)	2.7
Solid load[1]	
Fournier (1960) as calculated by Holeman (1968)	58
Kuenen (1950)	32.5
Schumm (1963) as calculated by Holeman (1968)	20.5
Holeman 1968	18.3
MacKenzie and Garrels (1967)	8.3
Combined solid and dissolved loads	
Lopatin (1950)[1]	17.5
Judson (this paper)[2]	9.3
Barth (1962)[2]	3.8
Carried by wind from land	
Calculated from various sources	0.06–0.36
Carried by glacier ice	
Estimated	0.1

[1]Does not include bed load.
[2]Solid load includes both suspended and bed load.

gested, then this figure is high. Kuenen (1950) gives an estimate for solid load of 32.5×10^9 tons per year, a high estimate, the basis for which is not clear.

Even higher is the estimate of 58×10^9 tons of suspended load calculated by Holeman (1968) from data in Fournier (1960). Douglas (1967) points out that the data presented by Fournier seem to be strongly influenced by man's activity. Holeman also extrapolates data of Schumm (1963), from selected drainage basins in central United States to obtain a figure of 18.3×10^9 tons of suspended sediment per year. These data, too, are affected by man. Holeman, himself (1968), presents suspended sediment data for rivers draining 39 million square kilometers of the earth's surface, and extrapolates this to the approximately

100 million square kilometers of land surface draining to the ocean. He obtains a figure of 18.3×10^9 tons per year of suspended sediments carried annually to the oceans. The figure is strongly affected by data from the Asiatic rivers, particularly those of China, India, and the Southeast. These provide 80 per cent of the total sediment from 25 per cent of the land area in Holeman's figures. These are the same areas where the world's greatest population is concentrated and where the largest areas of intensive agriculture are located.

Let us now estimate the present rate of erosion. In this the major component is the suspended load carried by rivers. Of the data available, Holeman's appear to be the most inclusive and reliable. Allowing the bed load to be 10 percent of suspended load and

Table 4 Mass of material estimated as moved annually by rivers to the ocean before and after the intervention of man

	10^9 metric tons
Before man's intervention	9.3
After man's intervention	24

adding these two figures to the dissolved load as calculated by Livingstone, then the total material delivered annually to the sea by rivers at the present is 24×10^9 metric tons. This is about two and one half times the rate that we estimated existed before man started tampering with the landscape on a large scale (See Table 4).

Returning now to our estimate of the material produced by erosion before the serious intervention by man, we should be able to check our figure by comparing it with the amount of material deposited annually in the oceans. Thus far our only way of determining annual sedimentation rates over large areas is to average them out over the last several thousand years. Because man has only recently become a world-wide influence on erosion, this averaging serves to curtail his impact on the rate of accumulation of the sedimentary record.

What figures do we have on sedimentation in the oceans? Large areas of the ocean floor and the rates at which sedimentation takes place there are but dimly known at the present. We have data from coring of the ocean bottom but our data are scanty at best. In considering the tonnage which settles annually to the ocean floors we should distinguish between the deep oceans and the shallower oceans. As far as sedimentation goes there is probably a difference between those ocean floors lying below 3000 m and those above 3000 m. For the deep seas—those below 3000 m—current figures suggest something like 4.2×10^{-4} gm per cm^2 per year.[3] Spread over the nearly 280,000,000 km^2 of area for the deep sea, this amounts to 1.17×10^9 tons of sediments per year. Estimates for the shallower waters are probably less reliable than for the deep waters. For those waters shallower than 3000 m, about 72,000,000 km^2, I have assumed that between 10 and 20 cm of sediment accumulates every thousand years. Given a density of 0.7, there would be approximately 7 to 14×10^{-3} gm deposited for each square centimeter per year. This is equivalent to a total tonnage of between 5 and 10 $\times 10^9$ tons per year. Totaling the tonnage for the deep and shallow waters, we have a range of 6.2 to 11.2 $\times 10^9$ tons. Most of this is provided by the rivers. Wind provides an estimated 10^8 tons per year. The contribution of ice is also estimated as 10^8 tons. Extraterrestrial material is estimated by various authors as between 3.5×10^4 to 1.4×10^8 tons per year (Barker and Anders, 1968). Table 5 compares the estimate of the amount of material deposited each year in the oceans with the estimate of the amount delivered by various agents annually to the oceans. In both estimates we have tried to eliminate the effect of man.

Whether we use the rate of erosion prevailing before or after man's advent, our figures pose the problem of why our continents have survived. If we accept the rate of sediment production at 10^{10} metric tons per year

Table 5 Estimated mass of material deposited annually in the oceans compared with estimated mass of material delivered annually to the oceans by different agents[1]

	10⁹ metric tons/year
Estimated mass of material deposited in ocean	
Oceans shallower than 3000 meters	5–10
Oceans deeper than 3000 meters	1.17
Total	6.2–11.2
Estimated mass of material delivered to oceans	
From continents	
By rivers	9.3
By wind	0.06–0.36
By glacier ice	0.1
From extraterrestrial sources	0.00035–0.14
Total	~9.6

[1]Man's influence on rates of erosion is excluded from estimates.

(the pre-human intervention figure) then the continents are being lowered at the rate of 2.4 cm per 1000 years. At this rate the ocean basins, with a volume of 1.37×10^{18} m³, would be filled in 340 million years. The geologic record indicates that this has never happened in the past, and there is no reason to believe it will happen in the geologically foreseeable future. Furthermore, at the present rate of erosion, the continents, which now average 875 m in elevation, would be reduced to close to sea level in about 34 million years. But the geologic record shows a continuous sedimentary history, and hence a continuous source of sediments. So we reason that the continents have always been high enough to supply sediments to the oceans.

Geologists long ago concluded that the earth was a dynamic system, being destroyed in some places and renewed in others. Such a state would help resolve the problem of what happens to the sediments and why continents persist. Thus, although the sediments are carried from continents to oceans to form sedimentary rocks, we know that these rocks may be brought again to the continental surface. There they are in turn eroded and the products of erosion returned to the ocean. These sedimentary rocks may also be subjected to pressures and temperatures which convert them from sedimentary rocks to metamorphic rocks. If this pressure and temperature is great enough, the metamorphic rocks in turn will melt and become the parent material of igneous rock. These relationships are the well known rock cycle which has been going on as long as we can read the earth's rock record.

Inasmuch as we have been talking about the sedimentary aspects of the rock cycle, we should ask how much time it takes to complete at least the sedimentary route within the whole cycle. Poldervaart (1954) gives the total mass of sediments (including the sedimentary rocks) as 1.7×10^{18} tons. Taking the annual production of sediments as 10^{10} tons, then one turn in the sedimentary cycle approximates 1.7×10^8 years. At the present rates then we could fit in about 25 such cycles during the 4.5 billion years of earth history.

Accepting Poldervaart's figure of 2.4×10^{10} tons as the mass of the earth's crust then there has been time enough for a mass equivalent to the earth's crust to have moved two times through the sedimentary portion of the cycle.

We began this review with a brief examination of the homely process of erosion. As we continued we found that man has appeared on the scene as an important geologic agent, increasing the rates of erosion by a factor of two or three. We end the review face to face with larger problems. Regardless of the role of man, the reality of continental erosion raises anew the question of the nature and origin of the forces that drive our continents-above sea level. In short, we now seek the mechanics of continental survival.

Notes

[1]Data on erosion are expressed in metric tons per square kilometer and as centimeters of lowering either per year or per thousand years. A specific gravity of 2.6 is assumed for material eroded from the land.

[2]Although data are very incomplete the interested reader will find some specific information in Bonatti and Arrhenius (1965); Delany, et al. (1967); Folger and Heezen (in press); Goldberg and Griffin (1964); Rex and Goldberg (1958, 1962); and Riseborough, et al. (1968).

[3]I use data from deep sea cores as reported by Ku, Broecker and Opdyke, 1968. In calculating weights of sediments from rates of sedimentation I have used a density of 0.7 per cm³ (Ku, personal communication, 1968) and sedimentation rates which include original $CaCO_3$ content.

References

Barker, John L., Jr. and Edward Anders, 1968. Accretion rate of cosmic matter from iridium and osmium contents of deep-sea sediments. *Geochimica et Cosmochimica Acta, 32,* p. 627–645.

Barth, T. F. W., 1962. *Theoretical Petrology.* 2nd edition. John Wiley & Sons, Inc.: New York and London, 416 pp.

Bonatti, E. and G. Arrhenius, 1965. Eolian sedimentation in the Pacific off northern Mexico. *Marine Geology, 3,* p. 337–348.

Clarke, F. W., 1924. Data of geochemistry, 5th edition, *U. S. Geological Survey, Bulletin 770,* 841 p.

Delany, A. C. et al., 1967. Airborne dust collected at Barbados. *Geochimica et Cosmochimica Acta, 31,* p. 885–909.

Douglas, Ian, 1967. Man, vegetation and the sediment yields of rivers. *Nature, 215,* Pt. 2, p. 925–928.

Eardley, A. G., 1967. Rates of denudation as measured by bristlecone pines, Cedar Breaks, Utah. *Utah Geological and Mineralogical Survey, Special Studies, 21,* 13 p.

Folger, D. W. and B. C. Heezen. (in press), Trans Atlantic sediment transport by wind. (abstract) *Geological Society of America.* Special paper.

Fournier, F., 1960. *Climat et Erosion,* Presses Universitaires de France.

Geikie, Archibald, 1880. Rock-weathering as illustrated in Edinburgh church yards. *Proceedings, Royal Society, Edinburgh, 10,* p. 518–532.

Gibbs, R. J., 1967. The geochemistry of the Amazon River system: Part I, *Bulletin, Geological Society of America, 78,* p. 1203–1232.

Goldberg, E. D. and J. J. Griffin, 1964. Sedimentation rates and mineralogy in the South Atlantic, *Jour. of Geophysical Research, 69,* p. 4293–4309.

Holeman, John N., 1968. The Sediment Yield of Major Rivers of the World. *Water Resources Research, 4,* No. 4, p. 737–747.

Hutton, James, 1795. *Theory of the earth.* Vol. 2, Edinburgh.

Judson, Sheldon, 1968. Erosion rates near Rome, Italy. *Science, 160,* p. 1444–1446.

Judson, Sheldon and D. F. Ritter, 1964. Rates of regional dunudation in the United States. *Journal of Geophysical Research, 69,* p. 3395–3401.

Ku, Teh-Lung, W. S. Broecker, and Neil Opdyke, 1968. Comparison of sedimentation rates measured by paleomagnetic and the ionium methods of age determinations. *Earth and Planetary Science Letters, 4,* p. 1–16.

Kuenen, Ph. H., 1950. *Marine Geology.* John Wiley and Sons: New York and London, 551 p.

Langbein, W. B. and S. A. Schumm, 1958. Yield of sediment in relation to mean annual precipitation. *Transactions, American Geophysical Union, 39,* p. 1076–1084.

Leet, L. Don and Sheldon Judson, 1965. *Physical Geology.* 3d edition. Prentice-Hall, Inc.: Englewood Cliffs, N. J., 406 p.

Livingstone, D. A., 1963. Chemical Composition of Rivers and Lakes. U. S. *Geological Survey Professional Paper* 440-G., 64 p.

Lopatin, G. V., 1950. Erosion and detrital discharge. *Priroda,* No. 7. (Quoted by Strakhov, 1967.)

MacKenzie, F. T. and R. M. Garrels, 1966. Chemical mass balance between rivers and oceans. *American Journal of Science, 264,* p. 507–525.

Matthias, George F., 1967. Weathering rates of Portland arkose tombstones. *Journal of Geological Education, 15,* p. 140–144.

Musgrave, G. W., 1954. Estimating land erosion-sheet erosion. *Association Internationale d' Hydrologie Scientifique, Assemblée Générale de Rome, 1,* p. 207–215.

Poldervaart, Arie, 1954. Chemistry of the earth's crust, in *Crust of the earth.* Edited by A. Poldervaart. Geological Society of America Special Paper 62, p. 119–144.

Rex, R. W. and E. D. Goldberg, 1958. Quartz content of pelagic sediments of the Pacific Ocean. *Tellus, 10,* p. 153–159.

———, 1962. Insolubles. *in* M. Hill, ed. *The Sea,* Interscience: New York, vol. 1, p. 295–312.

Riseborough, R. W., R. J. Huggett, J. J. Griffin, and E. D. Goldberg, 1968. Pesticides: Transatlantic movements in the northeast trades. *Science, 159,* p. 1233–1236.

Schumm, S. A., 1963. The disparity between present rates of denudation and orogeny, *U. S. Geological Survey,* Prof. Paper 454H, p. 1–13.

Simonett, David S., 1967. Landslide distribution and earthquakes in the Bewani and Torricelli Mountains, New Guinea, in *Landscape Studies from Australia and New Guinea,* Edited by J. N. Jennings and J. A. Mabbutt, Australian National University press: Canberra, p. 64–84.

Spronck, R. 1941. Measures hydrographique *effectuées* dans la region divagante du Bief Maritime du Fleuve Congo. *Brussels, Institute Royale Colonial Belge Memoire,* 156 p. (quoted by Gibbs, 1967).

Strakhov, N. M. 1967, *Principles of Lithogenesis,* vol. 1. Translated from the 1962 Russian edition by J. P. Fitzsimmons. Oliver and Boyd: Edinburgh and London, 245 p.

Ursic, S. J. and F. E. Dandy, 1965. Sediment yields from small watersheds under various land uses and forest covers. *Proceedings of the Federal Inter-Agency Sedimentation Conference, 1963, U. S. Department of Agriculture,* Miscellaneous Publications 970, p. 47–52.

Washburn, A. L., 1967. Instrumental observations of mass-wasting in the Mesters Vig district, northeast Greenland. *Meddeleser om Gronland, 166,* No. 4, p. 1–296.

Wolman, M. G., 1967. A cycle of sedimentation and erosion in urban river channels. *Geografiska Annaler, 49-A,* p. 385–395.

17 Shoreline Structures as a Cause of Shoreline Erosion: A Review

James G. Rosenbaum

Introduction

Shoreline erosion is an increasingly serious worldwide problem. In the United States alone, about one fourth of the 84,000 miles (140,000 km.) of shoreline are presently subject to erosion (U.S. Army, Corps of Engineers, 1971).

Along ocean coastlines, increased erosion is often attributed to a slight rise in world sea level since the turn of the century, coupled with unusually severe storms. Similarly, persons living along the Great Lakes of the United States often attribute erosion to high lake levels. In fact, it is man's activities that have disrupted fundamental shoreline processes, creating a potential for erosion which is realized whenever nature is less than benign. According to Inman and Brush (1973), two major types of man-made disruptions are:

1. Dams. Artificial impoundments behind dams act as settling basins for sediment which would otherwise nourish beaches on the coastline.

2. Shoreline structures such as harbor breakwaters, jetties, groins, and landfills. The present report will discuss how these shoreline structures cause erosion.

Mr. Rosenbaum is a consulting geologist in Milwaukee, Wisconsin.

The Littoral Drift

Littoral drift is common to virtually all coastlines of the world and is basic to an understanding of the mechanism of erosion. Littoral drift is the term applied to sediment which moves laterally along the shore. This movement is induced by the action of waves breaking at an angle to the shore, and operates principally in the narrow zone from the breaker line to the beach (Zenkovitch, 1960; Ingle, 1966; Komar, 1971; Inman and Brush, 1973). Within this zone, sand transport is greatest where waves break in conjunction with topographic highs, as at the offshore bar and the seaward face of the beach (Fig. 1). On a given coastline individual storms may move shore material in one direction or another, but over a longer period of time there is a well defined net movement in one direction. This net movement is the subject of the present discussion. The direction towards which material moves is termed downdrift. The direction from which material moves is termed updrift. The movement is an entirely normal process, and does not lead to a loss of shore material. Relative to a fixed observer, material moving downdrift is replaced by material which had been updrift. On a stable shoreline, sediment in the littoral zone is essentially a steady state collection

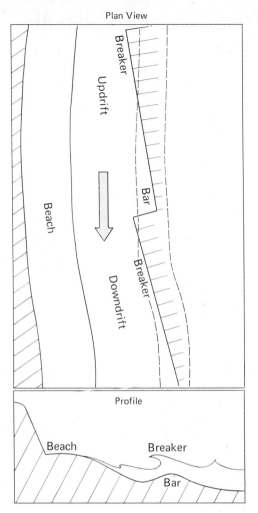

Plan View

FIG. 1. Normal shoreline with an offshore bar (after Ingle, 1966). Arrow indicates the direction of the net littoral drift. Sand transport is greatest where waves break in conjunction with topographic highs; i.e., at the offshore bar and at the seaward face of the beach (Zenkovitch, 1960).

of drift. Man-made structures upset this equilibrium (Fig. 2).

Consideration of the Action of a Single Structure

For many people, the presence of shoreline harbor breakwaters, jetties, groins, and landfills may be reassuring, since these engineering works signify that man has "made a stand against the sea." Ironically, such structures are actually responsible for much erosion. They alter the natural distribution of shoreline material. Areas subject to severe erosion do not occur uniformly along the shorelines, as one might expect if high water levels or storms were the principle cause of erosion (Davis et al., 1973). Instead, severely eroding shorelines are frequently downdrift of man-made structures.

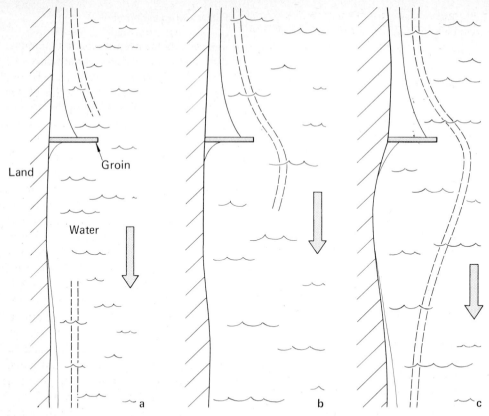

FIG. 2. Successive stages of shoreline condition after emplacement of a groin. Although not shown here, the land area downdrift of the groin may recede during the time of a and b. In Fig. 2c, a limited region downdrift of the groin remains deprived of sediment nourishment, and may continue to erode indefinitely.

Field studies and reviews of case histories, (Spring, 1914; Caldwell, 1950; Johnson, 1957; Hartley, 1964; and Dyhr-Nielsen and Sorensen, 1968), laboratory tank models (Kressner, 1928; Johnson, 1948; Dyhr-Nielsen and Sorensen, 1968; Sato and Irie, 1970), and mathematical models (Bakker, 1968; Dean and Jones, 1974) have considered the effects of various man-made structures and have determined that these effects are quite predictable.

A groin is a solid, occasionally per-meable, narrow structure, which projects seaward approximately perpendicular to the beach. It usually rises several feet above water level. Groins are built to trap littoral drift, and by doing so cause an abnormally wide accumulation of beach material updrift, which can thus provide protection for a limited section of shoreline.

The sequence of shoreline changes following placement of a groin (or other structures attached to the shore, such as harbor breakwaters or jetties) suggests that there are two distinct phases in the erosion history of the region downdrift.

Phase 1

During an initial phase, the groin acts as a complete barrier to littoral drift (Fig. 2a, 2b). Immediately after construction, material will accumulate updrift of the groin. As drift material continues to encounter the groin, it will also migrate seaward into water that is deep relative to the depths at which it would otherwise move along the shore. Flow velocities near the bottom at these greater depths are insufficient to maintain transport of most drift material, so deposition occurs. Such deposition will cause the bottom to shoal until wave action is again able to move all drift material.

The volume of material comprising the accretion updrift and offshore of the groin does not reach the shoreline farther downdrift, where it would have traveled prior to construction of the groin. Downdrift beaches will thus have a negative sand budget. The material they lose to beaches even farther downdrift will not be replaced until the impoundment capacity of the groin or jetty is exceeded.

A well known case history, in which a large structure acted as a complete littoral barrier, is the shore-attached breakwater at Santa Barbara, California (Wiegel, 1964). This breakwater was constructed in 1930. Until a dredging program was initiated to move material downdrift past the harbor, beaches were affected for up to 23 miles (38 km) downdrift (W. C. Krumbein, personal communication).

Phase 2

Beach and bar accumulation will eventually stabilize updrift and offshore of the groin. After sufficient offshore shoaling has occurred, sediment is able to continue moving past the groin (Fig. 2b) and gradually approaches the coast, as shown in Fig. 2c (U. S. Army Coastal Engineering Research Center, 1973, pp. 5–34). On barred shorelines, this transport will take place along a rebuilt outer bar (Fig. 2c), which had earlier disappeared (Fig. 2a, 2b) due to lack of sediment nourishment (Dyhr-Nielsen and Sorensen, 1970). Because this material will not reach its normal position on the shoreline for a certain distance downdrift, a section of shoreline downdrift of the groin will remain deprived of its normal sand nourishment, and will continue to erode indefinitely. In the author's experience, the length of continually eroding shoreline is frequently from 3 to 5 times the distance by which the structure projects seaward from the shore. Beaches at a greater distance downdrift, which were starved during phase 1, may return to normal (Inman and Brush, 1973). The time before onset of this recovery varies widely, depending on the size of the structure, the volume of littoral drift, and offshore topography. On Lake Michigan near Milwaukee, Wisconsin, one or two years might elapse before phase 2 is reached downdrift of a moderate size groin.

Downdrift and inshore of singlearm breakwaters, shoreline changes are commonly caused by wave refraction and diffraction (Johnson and Eagleson, 1966), as well as by the more general nourishment problems already discussed. The altered wave regime may cause a local reversal of the littoral drift, which then removes sediment from an area downdrift and deposits it in the lee of the breakwater. Deposition will continue until the shoreline has aligned itself with the new wave front, at which time the reversed littoral drift will be eliminated. However, downdrift beaches appear to remain undernourished (Johnson and Eagleson, 1966, Fig. 9.43). A somewhat similar process takes place at detached breakwaters.

Self-Propagating Nature of Shoreline Structures

Severe erosion downdrift of shoreline structures often prompts construction of additional structures in an attempt to stop that erosion. These second generation structures almost invariably cause erosion downdrift in a region which previously had been unaffected by the first generation structure. Third generation structures might then be built, causing additional erosion, and so forth (Schijf, 1959; Inman and Brush, 1973). Lawsuits brought by parties suffering erosion attributable to structures updrift may become an additional problem (Lillevang, 1965). Federal, state and local authorities may allow construction, but this will not be complete defense to such suits.

Attention has been drawn to the temporary erosion which occurs during phase 1 on shorelines far downdrift of a new groin (Fig. 2b). Installation of additional structures is a common response to this erosion. This is unfortunate, since beach nourishment on such

shorelines often returns to normal during phase 2 (Inman and Brush, 1973). The additional structures may be needless, and may themselves cause further downdrift erosion.

The history of some major shoreline projects in Milwaukee, Wisconsin, typifies the way in which shoreline structures propagate (Fig. 3). The federal government constructed the main harbor breakwater in stages from 1881 to 1929. Construction progressed from north to south. By 1916, downdrift erosion had become severe enough to cause the city of Milwaukee to construct a rubble mound breakwater approximately 1000 feet offshore, and extending southward parallel to the shore. It terminated at a position opposite the small breakwater of a local utility. The city's rubble breakwater was built from 1916 to 1931. It was observed that the shoreline opposite the end of the structure experienced severe erosion, and that "as the breakwater was extended from year to year, the point of greatest erosion on the shore kept pace" (Milwaukee County, 1945).

The utility's landfill and small breakwater, built in 1920, probably compounded erosion to the south until 1930–1931, when its effects would have been largely masked by the overlapping city breakwater.

In 1933–1934 a system of eleven permeable groins was placed to protect the downdrift third of a 1½ mile section of eroding parkland that extended south of the termination of the new rubble breakwater. In turn, the groin system is currently responsible for accelerated erosion for at least ¼ mile to the south. To protect this and other eroding areas a citizen task force

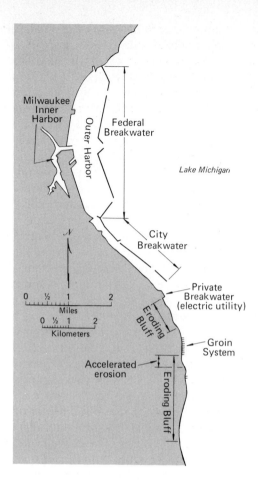

FIG. 3. Lake Michigan shoreline structures at Milwaukee, Wisconsin. The longshore drift is from north to south. The Federal breakwater was built in stages from 1881 to 1929. The city breakwater was constructed in stages from 1916 to 1930. The utility company's breakwater was built in 1920 and the groins in 1933–34.

on lakefront planning has recently proposed construction of a series of offshore islands, each one being several miles long, up to a mile wide, and 3000–5000 feet offshore. If ever built, these islands, which function similar to a detached breakwater, would undoubtedly cause extremely severe erosion for many miles downdrift.

Solutions

Proposals for avoiding downdrift erosion caused by shoreline structures include building new and different structures, removing structures, bypassing sediment around structures, and zoning to forbid structures.

Artificial filling of the accretion zone updrift from a new groin is partially effective in reducing downdrift erosion (U. S. Army Coastal Engineering Research Center, 1973). It is thought that such fill allows an uninterrupted movement of drift past the structure. To be most effective, fill should not only cover the subaerial part of the new shoreline, but should also extend a considerable distance offshore. This practice is usually ignored. However, even an elaborate program of artificial filling should be followed by periodic renourishment, otherwise "starved" beaches would be expected to occur along a region downdrift of the structure (U. S. Army, Corps of Engineers, 1973, p. 5–45). The mechanism of sediment loss would be the same as that of phase 2, outlined above (Fig. 2c).

Of course, artificial filling requires a source of fill that is hydrodynamically stable. At the present time, sources of suitable beach fill are often not readily available, partly because they have been exhausted, and partly because the process of obtaining the fill would cause physical disturbance in other sectors of the environment.

Detached offshore breakwaters parallel to the shoreline have been proposed as a solution, the concept being that they will provide shelter from wave action, but will not interfere with the littoral drift. Johnson (1957) has shown that the latter assumption is erroneous. Unless the breakwater is separated from the shore by a distance of three to six times its length, the lack of turbulence in its lee will cause sediment deposition (Inman and Frautschy, 1965). Deposits have been known to extend all the way from the shoreline out to the breakwater, forming a tombolo. Erosion downdrift will accompany this deposition. However, as discussed below, this ability of detached offshore breakwaters to trap sand may be used to great advantage.

Realization of the damaging effect of groins often prompts efforts to build seawalls, the idea in this case being that one can at least "hold the line" against the sea. However, it is recognized that because they reflect rather than dissipate wave energy, vertical impermeable walls cause a net erosion of the nearshore profile fronting the wall (Dean and Jones, 1974). This erosion may ultimately lead to undermining and collapse of the structure. Downdrift of these structures, erosion may also accelerate (U. S. Army Coastal Engineering Research Center, 1973, pp. 5-3, 5-4). Permeable revetments, such as rubble fill, cause somewhat less erosion of the nearshore profile (A. M. Wood, 1969). On a pebble or cobble beach, a low permeable stockade projecting perhaps 1 foot (.3 m) above beach level, and aligned parallel to the shore is frequently able to trap material (Dobbie, 1946). Of course, this material is trapped at the expense of downdrift regions.

The most fruitful method of stopping erosion is to identify and treat the first cause of erosion (Schijf, 1959). Very often this will be a harbor structure, which subsequently causes other relatively minor structures to be constructed downdrift. The damaging effects of these large structures can be greatly reduced by sediment bypassing. Bypassing consists of mechanically or hydraulically moving sediment across a shoreline structure to the vulnerable region downdrift, rather than allowing the structure to shunt sediment away from shore into deep water. This approach was suggested as early as 1913 by Berridge, who commented on the disastrous accretion and erosion experienced at the shoreline harbor at Madras, India.

Most harbor breakwaters require maintenance dredging to remove littoral drift that has settled in the relatively deep, quiet water of the harbor entrance. An elementary form of bypassing could be achieved by depositing this dredged material onto the adjacent downdrift shoreline, rather than following the common practice of dumping it out at sea (Johnson and Eagleson, 1966; Watts, 1968).

The first fixed hydraulic pumping station intended for sediment bypassing was built on the updrift jetty at South Lake Worth Inlet, Florida, in 1937 (Fig. 4). Downdrift erosion was stopped from 1938 to 1942 (Caldwell, 1950; Watts, 1962). A shut down during World War II coincided with renewed erosion, which however, was stopped when pumping was resumed in 1945.

A different method of bypassing has been applied at Port Hueneme and Ventura, California. A detached offshore breakwater and small boat harbor at Ventura was completed in 1962, updrift of an older jetty system at Port Hueneme (Fig. 5). The jetty system had been deflecting sediment into the Hueneme submarine canyon, causing shoreline erosion for up to 8 miles downdrift (Watts, 1962; Tornberg, 1968). The offshore breakwater is able to trap the sediment which would otherwise be deflected into the submarine canyon. It also provides shelter for a floating hydraulic dredge which periodically pumps the trapped sand to a disposal area downdrift of the Port Hueneme jetties. This system has been notably successful in restoring the downdrift beaches. Also, because most sand is trapped before reaching the Ventura County harbor entrance, relatively little maintenance dredging of this entrance is necessary. In 1966, the U. S. Army Corps of Engineers,

Coastal Engineering Research Center, reported that "This general method of bypassing is considered to provide greater assurance of effectiveness than any other thus far considered."

Bypassing systems are expensive to construct and maintain (Middleton, 1958), which apparently is why they are not used more widely. However, their cost should be compared to the ultimate cost of the entire proliferation of shoreline structures which would otherwise occur downdrift of the principle structure, keeping in mind that additional shoreline erosion will be caused by such a proliferation. Long sections of coastlines cluttered with engineering works also lose much of their aesthetic appeal. An important financial consideration is that bypassing systems can greatly reduce maintenance dredging of harbor entrance channels.

Once the main structure has been bypassed, consideration should be given to removal of smaller structures. Otherwise, an abundance of smaller structures might by themselves starve downdrift regions. The sequence of removal should proceed from the most undrift structure to progressively more downdrift structures.

Many cases exist in which the first cause of erosion is not an essential commercial or navigational structure, and in which the structure is not large enough to warrant a sediment bypassing system. For example, storm sewer and industrial outfalls are often built as groins. Park commissions commonly construct groins to augment the natural beach, usually at the expense of downdrift beaches. Landfills are often extended offshore and used as parking

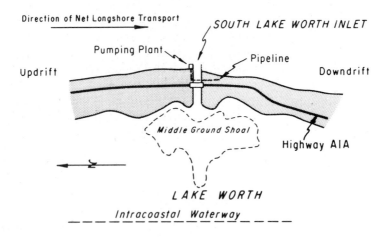

FIG. 4. Fixed bypassing plant, South Lake Worth Inlet, Florida (Figure courtesy U. S. Army Corps of Engineers).

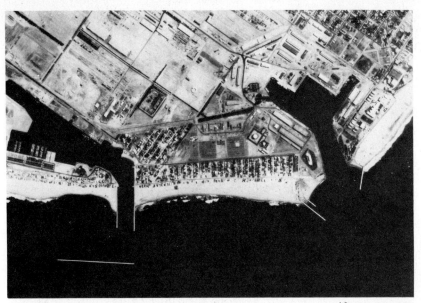

(September 1965

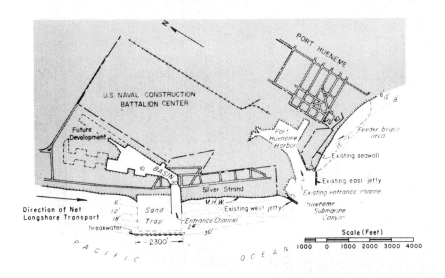

FIG. 5. Sand bypassing, Channel Islands Harbor, California. The photo was taken just after 3 million cubic yards had been dredged from the trap. (U. S. Army Corps of Engineers).

lots. These types of structures might be removed or modified. If subsequent downdrift erosion were included in the initial cost/benefit analysis of these structures, it is doubtful that they would be so attractive to public officials responsible for shoreline management.

Probably the best way of avoiding future problems in yet undeveloped areas is to require a wide buffer zone separating any future development from the shoreline (Schijf, 1959), and to forbid the introduction of shoreline structures. This policy would greatly reduce the chance of future erosion, and would make it unnecessary to use public funds to assist those affected by erosion. Condemnation of erosion prone properties, converting them to public ownership, and forbidding construction of protective works on such properties is another possibility. In these cases, shoreline management might parallel certain aspects of flood plain management. Such programs would likely be a far wiser use of public funds than construction of additional engineering works.

Conclusions

In ignorance of the consequences, well intentioned governmental bureaus, planning agencies, and park commissions frequently recommend a new groin system, landfill, or even offshore islands as a solution to erosion problems. Although erosion may be stopped or slowed along the areas which such structures are designed to protect, erosion will merely appear at a new location downdrift.

The shore drift along Milwaukee County is lean and part of the problem becomes one of stabilizing and holding what beach now exists (Milwaukee County, 1945).

This statement is typical of the thinking that has often justified groin construction. However, it is precisely on such relatively narrow beaches that groin construction should be avoided. It is not usually appreciated that the amount of beach accumulation updrift of a groin is smaller than the amount of beach erosion and land loss which will occur downdrift as a consequence of groin construction (Schijf, 1959; Bruun and Lackey, 1962). Net erosion occurs because the scour hole downdrift of a groin continues to enlarge even after accretion updrift has stablized.

During the past several decades, sediment bypassing systems have demonstrated their effectiveness in stopping erosion caused by major shoreline structures. Except for the possibility of more efficient and varied bypassing systems (Inman and Brush, 1973; U. S. Army Corps of Engineers, 1973), it is unlikely that any technical breakthrough will occur in the field of preventing shoreline erosion. Instead, breakthroughs must be made in public policy (Schijf, 1959; Dolan, 1972), and in an increased level of technical awareness of those responsible for shoreline management. An understanding of the continuing liability for erosion caused by shoreline structures might do much to dampen the enthusiasm of those who would advocate additional structures (Lillevang, 1965).

Agencies and individuals responsible for planning shoreline structures are slow to learn from past mistakes. M. P. O'Brien's preface to the Proceedings of the First Conference on Coastal Engineering, 1950, is as appropriate today as when first published:

Along the coastlines of the world, numerous engineering works in various states of disintegration testify to the futility and wastefulness of disregarding the tremendous destructive forces of the sea. Far worse than the destruction of insubstantial coastal works has been the damage to adjacent shorelines caused by structures planned in ignorance of, and occasionally in disregard of, the shoreline processes operative in the area.

References Cited and Selected Bibliography

Bajorunas, L., and Duane, D. B., 1967, Shifting offshore bars and harbor shoaling: *Journal of Geophysical Research*, v. 76, p. 6195–6205.

Bakker, W. T., 1968, The dynamics of a coast with a groyne system: *Proceedings of the Eleventh Conference on Coastal Engineering, American Society of Civil Engineers*, p. 492–517.

Berridge, H., 1913, Correspondence on coastal sand travel near Madras harbour, *in* Spring, F. J. E., 1914, Coastal sand travel near Madras harbour: *Min. Proc. Institution of Civil Engineers*, v. 194, p. 190.

Bruun, P., and Lackey, J. B., 1962, Engineering aspects of sediment transport including a section on biological aspects, *in Reviews in Engineering Geology, v.* 1, ed. by Thomas Fluhr and Robert F. Leggett, Geological Society of America, p..39–103.

Bruun, P., 1962, Review of beach erosion and storm tide conditions in Florida 1961–1962: *Engineering Progress at the University of Florida, Technical Progress Report No. 13,* v. XVI, n. 11, 104 p.

Bruun, P., 1973, *Port Engineering:* Gulf Publishing Company, Houston, 436 p.

Caldwell, J. M. 1950, By-passing sand at South Lake Worth Inlet, Florida: *Proceedings of First Conference on Coastal Engineering, American Society of Civil Engineers,* p. 320–325.

Collinson, C., 1974, Sedimentological studies: a basis for shoreland management in southwestern Lake Michigan (abs.): *Geological Society of America, Abstracts with Programs,* v. 6, n. 1, p. 14.

Davis, R. A., Jr., Siebel, E., and Fox, W. T., 1973, Coastal erosion in eastern Lake Michigan—causes and effects, *Proceedings of the Sixteenth Conference Great Lakes Research, Internat. Assoc. Great Lakes Res.,* p. 404–412.

Dean, R. G., and Jones, D. F., 1974, Equilibrium beach profiles as affected by seawalls: *Transactions, American Geophysical Union, v.* 55, n. 4, p. 322.

Dobbie, C. H., 1946, Some sea defence works for reclaimed lands: *Journal of the Institution of Civil Engineers,* v. 22, n. 4, p. 257–274.

Dolan, R., 1972, Barrier dune systems along the outer banks of North Carolina: a reappraisal: *Science,* v. 176, p. 286–288.

Dyhr-Nielson, M., and Sorensen, T., 1970, Some sand transport phenomena on coasts with bars; *Proceedings of the Twelfth Coastal Engineering Conference, American Society of Civil Engineers,* p. 855–865.

Hartley, R. P., 1964, Effects of large structures on the Ohio shore of Lake Erie: *Ohio Division of Geological Survey, Report of Investigation No. 53,* 30 p.

Ingle, J. C., Jr., 1966, *The Movement of Beach Sand: An Analysis Using Flourescent Grains: Developments in Sedimentology 5:* Elsevier, 221 p.

Inman, D. L., and Brush, B. M., 1973, The coastal challenge: *Science,* v. 181, p. 20–32.

Inman, D. L., and Frautschy, J. D., 1965, Littoral processes and the development of shorelines: *Coastal Engineering Santa Barbara Specialty Conference, American Society of Civil Engineers,* p. 511–536.

Johnson, J. W., 1948, The action of groins on beach stabilization: *University of California Department of Engineering, Fluid Mechanics Laboratory, Berkeley, California, Navy Department—Bureau of Ships—Contract NObs2490, Technical Report He-116-283,* 21 p.

Johnson, J. W., 1957, The littoral drift problem at shoreline harbors: *Journal of the Waterways and Harbors Division, American Society of Civil Engineers,* v. 83, n. WW1, paper 1211, 37 p.

Komar, P. D., 1971, The mechanics of sand transport on beaches: *Journal of Geophysical Research*, v. 76, p. 713–721.

Kressner, B., 1928, Tests with scale models to determine the effect of currents and breakers upon a sandy beach, and the advantageous installation of groins: *Bautechnik*, v. 25, translated, 1930 by G. P. Specht, U. S. Army Coastal Engineering Research Center, 25 p.

Larsen, C. E., 1973, Variation in bluff recession in relation to lake level fluctuations along the high bluff Illinois shore: *Illinois Institute for Environmental Quality*, Chicago, 73 p.

Lee, C. E., 1953, Filling pattern of the Fort Sheridan groin system: *Proceedings of the Fourth Conference on Coastal Engineering, American Society of Civil Engineers*, p. 227–248.

Lillevang, O. J., 1965, Groins and effects—minimizing liabilities: *Coastal Engineering Santa Barbara Specialty Conference, American Society of Civil Engineers*, p. 749–754.

Middleton, S. R., 1958, Financing of sand bypassing operations: *Journal of the Waterways and Harbors Division, Proceedings of the American Society of Civil Engineers*, v. 84, n. WW5, paper 1875, 8 p.

Milwaukee County Committee on Lake Michigan Shore Erosion, 1945, Report: Courthouse, Milwaukee, Wisconsin, 33 p.

O'Brien, M. P., 1950, Preface: *Proceedings of the First Conference on Coastal Engineering, American Society of Civil Engineers*.

Sato, S., and Irie, I., 1970, Variation of topography of sea-bed caused by the construction of breakwaters: *Proceedings of the Twelfth Coastal Engineering Conference, American Society of Civil Engineers*, p. 1301–1319.

Schijf, J. B., 1959, Generalities on coastal processes and protection: *Journal of the Waterways and Harbors Division, American Society of Civil Engineers*, v. 85, n. WW1, p. 1–12.

Spring, F. J. E., 1914, Coastal sand-travel near Madras harbour: *Min. Proc. Institution of Civil Engineers*, v. 194, p. 153–246.

State of Illinois, 1958, *Interim Report for Erosion Control: Illinois Shore of Lake Michigan:* Department of Public Works and Buildings, Division of Waterways, 108 p.

Tornberg, G. F., 1968, Sand bypassing systems: *Shore and Beach*, v. 36, n. 2, p. 27–33.

United States Army, Corps of Engineers, Coastal Engineering Research Center, 1966, Shore Protection Planning and Design, *Technical Report #4*.

United States Army, Corps of Engineers, 1971, *Report on the National Shoreline Study*, Washington, D.C.

United States Army, Corps of Engineers, Coastal Engineering Research Center, 1973, *Shore Protection Manual*, 3 v.

Watts, G. M., 1962, Mechanical bypassing of littoral drift at inlets: *Journal of the Waterways and Harbors Division, American Society of Civil Engineers*, v. 88, n. WW1, p. 83–99.

Watts, G. M., 1968, Field inspection of erosion problems in India, *Shore and Beach*, v. 36, n. 2, p. 34–60.

Wiegel, R. L., 1964, *Oceanographical Engineering:* Prentice-Hall Inc., 532 p.

Wood, A. M. Muir, 1969, *Coastal Hydraulics:* Gordon and Breach Science Publishers, 187 p.

Wood, S. M., 1944, Erosion of our coastal frontiers—part II: *The Illinois Engineer*, v. XX, n. 5, p. 5–34.

Zenkovitch, V. P., 1960, Flourescent substances as tracers for studying the movement of sand on the sea bed; experiments conducted in the U.S.S.R.; *Dock Harbor Authority*, v. 40, p. 280–283.

18   Sinkhole

Philip E. LaMoreaux and William M. Warren

Last Dec. 2, Hershel Byrd, a resident of rural southern Shelby County, Alabama, was startled by a rumble that shook his house, followed by the distinct sound of trees snapping and breaking. Two days later, hunters in nearby woods found a crater about 140 m long, 115 m wide, and 50 m deep. Those events mark the time of formation and discovery of the largest recent sinkhole in Alabama and possibly one of the largest in the United States.

Aerial reconnaissance indicates that in approximately 16 square km about 1,000 sinkholes, other areas of subsidence, and internal drainage features have formed. Most of the sinks are in a limestone valley where solution activity and subsurface soil erosion are most pronounced. . . . Although the sink discovered by the hunters is on a hillside at a higher altitude, it is believed to be hydraulically connected with the flanking valleys.

Deeply weathered Cambrian dolomites of the Knox Group underlie the low eroded ridge where the large collapse occurred, south of Birmingham. The dolomites crop out on the eastern flank of an eroded northeast-plunging anticline. The collapse occurred in white and orange residual clays; bedrock is not exposed. Steep sides are usually found in recent col-

Reprinted from *Geotimes*, v. 18, p. 15, by permission of the authors and the American Geological Institute. Copyright 1973.

Mr. LaMoreaux is the State Geologist for Alabama and Mr. Warren is on the staff of the Geological Survey of Alabama.

lapses but not in this sinkhole, probably because of its large size and unstable walls. Slumping is active on all sides (see Fig. 1), indicating that the sinkhole will continue to grow.

Two smaller sinks, one more than 15 m in diameter and 4 to 5 m deep, were discovered later about 100 m south of the large collapse. Water is standing in all 3 sinks, but the brilliant azure water in the largest is about 30 m lower than that in the smaller sinks, and its level apparently represents the altitude of the water table.

Investigations by the Alabama Geological Survey and the U. S. Geological Survey in Alabama indicate that many areas underlain by carbonate rocks are prone to subsidence. Sinkhole collapses are related to natural phenomena such as heavy rainfall, seasonal fluctuations in the water table, earthquakes, or other changes in the hydrogeologic regime affecting residuum stability. Man-imposed effects such as artificial drainage, dewatering, seismic shocks, breaks in water or sewage pipes, or even over-watering of gardens may result in collapse.

Formation of sinkholes often results from collapse of cavities in residual clay that are caused by 'spalling', or downward migration of clay through openings in underlying carbonate rocks. The spalling and formation of cavities is caused by (or may be accelerated by) a lowering of the water table resulting in a loss of support to clay overlying openings in bedrock, fluctuation of the water table against the base of residual

FIG. 1. View of sinkhole formed December 2, 1972, in Shelby County, Alabama. The sinkhole measures 425 feet long, 350 feet wide, and 150 feet deep. (Photo by U. S. Geological Survey.)

clay, downward movement of surface water through openings in the clay, or an increase in water velocity in cones of depression to points of discharge. Collapses have occurred where spalling and resulting enlargement of cavities has progressed upward until overlying clay could not support itself, and where sufficient vibration, shock, or loading over cavities caused the clay to be jarred loose or forced down.

There are no large ground-water withdrawals from the Knox Group in the area of the December collapse. However, about 1.5 km to the east the Newala and Longview Limestones are being dewatered for limestone extraction at 1 underground and 2 recessed quarries. Small sinkholes have been a common phenomenon around the quarries for about 5 years but few collapses have been observed outside

this valley. 2 km west of the collapse a municipal well has been pumped for 13 years in an adjacent valley underlain by Ketona Dolomite without development of subsidence.

Apparently the large collapse is in an area that may be affected by extensive ground-water withdrawal but the history of water levels in the area currently is unknown. Rain fell almost daily last November and concentrations of surface water were observed in December. Otherwise very little data is available and the hydrogeologic factors responsible for the collapse may only be speculated upon and can be confirmed or disproved by future investigations.

The Geological Survey of Alabama, in cooperation with the U. S. Geological Survey, has published 3 reports on investigations of similar subsidence problems on limestone terranes. We are considering plans for a long-term study on Shelby County, including use of thermal imagery and multispectral photography as well as conventional aerial photography.

19 Sediment

A. R. Robinson, P.E.

A pollutant has been defined as a resource out of place. Sediment, with its two-fold effect upon the environment, is a perfect example. It depletes the land resource from which it is derived, and it impairs the quality of the water in which it is entrained and deposited (5).

Sediment is by far our nation's largest single water pollutant, exceeding the sewage load by some 500 to 700 times (2). Since about half of all sediment originates on agricultural land, agriculture has a major role in maintaining environmental quality (8). Improved agricultural practices have greatly reduced sediment damage and enhanced environmental quality, but new practices are needed to cope with the excessive erosion resulting from rapidly changing land use and cultural patterns in many areas.

Sediment is an extremely complicated material with untold physical, chemical, and biological implications. It becomes a pollutant when it fills reservoirs, lakes, and ponds; clogs stream channels; settles on productive land; destroys aquatic habitats; creates turbidity; degrades water needed for consumptive or other uses; or impairs water distribution systems. It may also carry other pollutants, such as plant nutrients, pesticides, toxic metals, and perhaps bacteria and viruses.

In assessing the role of sediment as a pollutant, one must keep in mind a couple of points that are often overlooked. First, any stream is a dynamic body which has a certain inherent ability or energy to transport sediment and other materials. Unless a stream flows in a nonerodible channel, such as concrete, it will attempt to transport sediment up to its ability, and it may erode its banks or bed to obtain the sediment. On the other hand, if the sediment load exceeds a stream's ability to carry the load, sediment will be deposited. The challenge is to control a stream's energy and create nonerodible boundaries.

The second point is that sediment may act as a scavenger. Chemically active surfaces on fine sediment give it the ability to adsorb or desorb other elements (4,6). It may receive chemicals from solution and serve as a trap or sink to remove chemicals from the stream. In nutrient-rich water, sediment may help prevent eutrophication. But it may also serve as a medium to promote the growth of aquatic

Reprinted from *Journal of Soil and Water Conservation*, v. 26, pp. 61–62, with permission of the author and the Soil Conservation Society of America, © 1971 by scsa. Mr. Robinson is a consulting engineer.

plants. In either event, sediment must be considered a major factor in determining the quality of surface waters.

Sediment as a Pollutant

Any discussion of sediment as a pollutant and its effect upon the environment can logically be divided into five parts: (1) sediment properties, (2) sediment yield, (3) sediment transport, (4) sediment deposition, and (5) stream channel systems.

Sediment Properties

The physical, chemical, and biological properties of sediment all influence its effect on the environment. Physical properties include size, density, and shape. The erosion, transport, and deposition processes are selective since coarse sediment moves differently than fine sediment. Fine sediment is composed of silts, clays, and organic materials which may have chemically active properties. It may sorb ions from solution or release ions to solution depending on the chemical environment. Reactions between chemicals and colloidal sediment determine the relative concentration of pollutants in solution and suspension. In general, coarse sediment tends to buffer the erosive potential of streamflow while fine sediment tends to buffer the dissolved and suspended chemical load. It is primarily coarse sediment that is more readily controlled with available technology. We do not know yet how to control the amounts of clay and colloidal fractions which constitute the bulk of our sediment problems. This is true both at the sediment source and in the final disposition of the material.

In its role as a scavenger, sediment may sorb chemicals from solution and then deposit them in stream channels or reservoirs. The deposited pollutants may or may not stay in place. They may desorb or react to re-enter the stream in another form. Reactions between chemicals and colloidal sediment may determine the relative concentration of other pollutants that remain in solution or suspension.

Little is known about the chemistry of sediment in field situations, and much study is needed. The following are examples of recent research findings: Radioactive materials from fallout selectively erode with soil particles and may result in higher concentrations of radioactivity in reservoir deposits. A phosphorus-deficient soil, at sediment concentrations of 10,000 parts per million, was capable of reducing the phosphorus concentration in solution from sewage effluent from 6.6 to 4.3 ppm. At relatively high sediment concentrations—60,000 ppm—a soil low in organic material but high in clay content depleted the dissolved oxygen in water from 8 to 4 ppm in a period of 40 hours.

Sediment Yield

The detachment of a sediment particle, whether it be from a cultivated field, pasture, forest roadbank, urban area under development, gully, or streambank, is an important link in the chain of environmental quality. Unfortunately, we have not yet perfected methods for predicting sediment yield at points downstream from such erosion sources. Current methods are empirical and urgently need refinement.

Since complete elimination of erosion is virtually impossible, guidelines are needed to define acceptable levels of erosion and sedimentation in watersheds and river basins. Not all sediment is detrimental. Some may even be necessary for stability. The question then becomes how much can sediment in streamflow be reduced without destroying stream channel stability?

The basic principles of erosion processes that we have learned over the years are still sound, but we must continually update and modernize specific practices and erosion control systems to keep abreast of changing times and conditions. Present-day farm machinery, along with current cropping and tillage practices, requires new erosion control systems that are compatible with today's farming methods. Urbanization and other developmental activities present new erosion control problems also.

Soil and water conservation researchers must foresee future changes in land use and develop suitable practices to meet these needs. For example, chemical control of surface and channel erosion is needed. There is also special need for a means of quick, temporary erosion control on bare areas that will last until permanent vegetation or other control measures can be established (7).

Sediment Transport

Sediment in transport enters the environmental quality picture in many ways. Streamflow is the primary transport vehicle. Sediment in transport discolors and degrades the quality of water for many consumptive uses. A muddy river is not very picturesque either. Sediment concentrations for rivers in the United States range from 200 to 50,000 ppm, with an occasional concentration as high as 600,000 ppm (2).

Despite decades of study, the mechanics of sediment transport are not well known. Sediment moves in a number of ways. Large material rolls or jumps along a stream bed. Coarse material is transported in suspension with a high concentration near the stream bed. Fine material, which may concentrate near the stream bed, generally is uniformly distributed throughout a stream's depth. This sediment distribution pattern is usually unstable and changing so that deposition and aggradation or scouring and degradation take place over small spaces and times in a flowing stream.

Sediment Deposition

Sediment deposition is one of the important facets of environment quality. It creates unsightly and troublesome accumulations of material along streams, in valleys, and in reservoirs. These deposits may contain large amounts of chemical, organic, and biological materials which represent a potential source of pollution. A recent survey along tributaries of the lower Mississippi River revealed isolated bottom sediment deposits containing as much as 0.5 ppm DDT (1).

Reservoirs in the United States have an approximate storage capacity of 500 million acre-feet. An average, measured depletion rate from sediment deposition of 0.2 percent represents an annual loss in capacity of one million acre-feet (3). The cost of removing this sediment, if it were possible, would approximate $1 billion.

Reservoir sites are an exhaustible natural resource. Because good sites are becoming scarce, preservation of existing reservoir capacity is critical. Therefore, bypassing or removing sediment will be needed in the future if the advantages of water storage and flood control are to be continuously enjoyed.

Among the aspects of sediment deposition needing continuing study are watershed delivery rates, deposition in valleys immediately above structures, distribution of deposited materials within reservoirs, means of bypassing or flushing sediment from reservoirs, sediment and chemical trap efficiencies, and changes in water quality upon impoundment as a result of sediment.

Stream Channel Systems

The morphology of stream channel systems is highly complicated. A meandering stream with wooded banks and riffled flow may be aesthetically pleasing but highly inefficient for carrying large flows without extensive flooding. Straightened, improved channels may erode, meander, and degrade, presenting stabilization problems.

Channel stabilization works are often needed to protect farmsteads and other valuable lands from stream meandering. Sediment from streambank erosion contributes heavily to deposition in reservoirs and harbors in many parts of the country. The present emphasis on stream channel alterations in many watershed improvement programs makes stream channel design an important part of planning. Stream channel alterations involve environmental considerations when streams are straightened, timber is removed, and lowlands are drained.

The engineering designs for adequate, economical streambank stabil-

ization are only beginning to be understood. Works of improvement must be based on proper design methods. These methods differ for each channel condition to be corrected. All are complex, and the basic streamflow and channel data needed in design are often difficult to obtain.

The Challenge

Sediment, as a pollutant, has a two-fold effect on the environment. It depletes the land resource from which it comes and impairs the quality of the water resource in which it is entrained and deposited. Sediment may also act as a scavenger, sorbing other pollutants, such as agricultural fertilizers and pesticides, from solution and depositing them in stream channels or reservoirs. In some cases this may be beneficial; in others it may concentrate pollutants with harmful repercussions.

With improved technology, sediment yields can be reduced, but it is virtually impossible to reduce erosion to the extent that sediment in streams, lakes, reservoirs, and valleys would no longer be a problem. Unless a stream flows in a nonerodible channel, the stream will attempt to transport sediment up to its energy ability, and it may erode its bed or banks to obtain this material. Man's challenge is to reduce erosion, control the energy of streamflow, and create, where possible, nonerodible stream boundaries.

References Cited

[1] Barthel, W. F., J. C. Hawthorne, J. H. Ford, G. C. Bolton, L. L. McDowell, E. H. Grissinger, and D. A. Parsons. 1969. *Pesticide residue in*

sediments of the lower Mississippi River and its tributaries. Pesticide Monitoring J. 3(1): 8–66.

[2] Glymph, L. M., and C. W. Carlson. 1966. Cleaning up our rivers and lakes. Paper No. 66-711. Am. Soc. Agr. Eng., St. Joseph, Mich. 43 pp.

[3] Glymph, L. M., and H. C. Storey. 1967. Sediment—its consequences and control. Pub. 85. Am. Ass. Advance. Sci., Washington, D. C. pp. 205–220.

[4] Grissinger, E. H., and L. L. McDowell. 1970. Sediment in relation to water quality. Water Resources Bul. 6(1): 7–14.

[5] Joint Task Force of the U. S. Department of Agriculture and the State Universities and Land Grant Colleges. 1968. A national program of research for environment quality-pollution in relation to agriculture and forestry. Washington, D. C. 111 pp.

[6] McDowell, L. L., and E. H. Grissinger. 1966. Pollutant sources and routing in watershed programs. Proc., 21st Annual Meeting, Soil Cons. Soc. Am.: 147–161.

[7] Robinson, A. R. 1969. Technology for sediment control in urban areas. Proc., Nat. Conf. on Sediment Control: 41–47.

[8] Wadleigh, C. H. 1968. Wastes in relation to agriculture and forestry. Misc. Pub. No. 1065. U.S. Dept. Agr., Washington, D. C. 112 pp.

Floods

> *"And the waters prevailed and were greatly increased upon the earth."*
>
> *Genesis*

Nature of the Problem

Floods are a natural and inevitable part of life along most rivers and many seashores. Some floods are seasonal, such as those that occur when spring rains and melting snow fill river channels, or when the storm waves and rain that accompany tropical hurricanes inundate coastal communities. Other floods (flash floods) are sudden, the result of torrential rain or cloudbursts, failure of dams, or glacial outbursts (jökulhaups). There are three classes of floods: riverine, coastal, and flash floods. Riverine floods are caused by precipitation over large areas or by the melting of winter accumulation of snow or both. The large volume of water overflows the river channel and inundates the adjacent floodplain. Riverine floods are of greater areal extent and duration than flash floods. The emphasis in this chapter is on riverine floods—probably the most ubiquitous and costly of all natural hazards in loss of life, property, and land.

There are about 1.9 million km (3 million mi) of streams in the conterminous United States, and about 6 percent of the land area is prone to flooding. More than 20,000 communities have flood problems, and more than 6000 of these have populations greater than 2500. Each year, an average of 75,000 Americans are driven from their dwellings and 90 persons lose their lives. Property damage now exceeds $1 billion a year, and the loss of both lives and property is increasing. Table 1 and the map (Fig. 1) that accompany it show the distribution of major floods in the United States since the latter part of the nineteenth century. The impact of single disasters can be devastating. In 1972, one flood alone was responsible for 237 deaths in the United States (Table 1). Worldwide, losses are staggering.

Besides the loss of lives and property associated with direct inundation, water supplies often are polluted, creating health hazards; erosion and sedimentation are accelerated; wildlife habitat is destroyed;

Flood of March 12, 1963, on North Fork Kentucky River at Hazard, Kentucky. A graphic example of people and nature competing for a river floodplain. (Photograph by Billy Davis, *Courier-Journal and Louisville Times*.)

Table 1 Great floods in the United States since May 1889

Number[a]	Type of flood	Date	Location	Lives lost	Estimated damages (millions of dollars)
1	b	May 1889	Johnstown, Pennsylvania dam failure	3,000	—
2	c	September 8, 1900	Hurricane—Galveston, Texas	6,000	30
3	d	May–June 1903	Kansas, Lower Missouri, and Upper Mississippi River	100	40
4	d	March 1913	Ohio River and Tributaries	467	147
5	c	September 14, 1919	Hurricane—south of Corpus Christi, Texas	600–900	22
6	b,e	June 1921	Arkansas River, Colorado	120	25
7	d	September 1921	Texas rivers	215	19
8	d	Spring of 1927	Mississippi River valley	313	284
9	d	November 1927	New England rivers	88	46
10	b	March 12–13, 1928	St. Francis Dam failure, southern California	450	14
11	f	September 13, 1928	Lake Okeechobee, Florida	1,836	26
12	d	May–June 1935	Republican and Kansas Rivers	110	18
13	d	March–April 1936	Rivers in Eastern United States	107	270
14	d	January–February 1937	Ohio and Lower Mississippi River basins	137	418
15	d	March 1938	Streams in southern California	79	25
16	d	September 21, 1938	New England	600	306
17	e	July 1939	Licking and Kentucky Rivers	78	2
18	d	May–July 1947	Lower Missouri and Middle Mississippi River basins	29	235
19	d	June–July 1951	Kansas and Missouri	28	923
20	d	August 1955	Hurricane Diane floods—Northeastern United States	187	714
21	d	December 1955	West coast rivers	61	155
22	d	June 27–30, 1957	Hurricane Audrey—Texas and Louisiana	390	150
23	d	December 1964	California and Oregon	40	416
24	d	June 1965	South Platte River basin, Colorado	16	415
25	c	September 10, 1965	Hurricane Betsy—Florida and Louisiana	75	1,420
26	d	January–February 1969	Floods in California	60	399
27	c,d	August 17–18, 1969	Hurricane Camille—Mississippi, Louisiana, and Alabama	256	1,421
28	c	July 30–August 5, 1970	Hurricane Celia—Texas	11	453
29	b	February 1972	Buffalo Creek, West Virginia	125	10
30	e	June 1972	Black Hills, South Dakota	237	165
31	c,d	June 1972	Hurricane Agnes floods—Eastern United States	105	4,020
32	d	Spring 1973	Mississippi River basin	33	1,155
33	d	June–July 1975	Red River of the North basin	<10	273
34	c,d	September 1975	Hurricane Eloise floods—Puerto Rico and Northeastern United States	50	470

35	b	June 1976	Teton Dam failure, southeast Idaho	11	1,000
36	e	July 1976	Big Thompson River, Colorado	139	30
37	e	April 1977	Southern Appalachian Mountains area	22	424
38	b,e	July 1977	Johnstown-western Pennsylvania	78	330
39	d	April 1979	Mississippi and Alabama	<10	500
40	c	September 12–13, 1979	Hurricane Frederic floods—Mississippi, Alabama, and Florida	13	2,000

[a]Number corresponds to those in Fig. 1.   [d]Riverine flood.
[b]Dam break flood.   [e]Flash flood.
[c]Tidal flood.   [f]Flood wave generated in Lake Okeechobee by hurricane.
Adapted from Climatological Data, National Summary, 1977, National Oceanic and Atmospheric Administration, vol. 28, no. 13, and by information furnished from the Federal Disaster Assistance Administration
From *U.S. Geological Survey Prof. Paper 1240-B*, 1981

and commerce and governmental services are disrupted. Expenditures for flood protection and relief are extremely high.

In view of this, why do people occupy flood-prone areas? Floodplains, where the most severe losses occur, are the relatively flat areas adjacent to a river channel. Historically, they were developed because of their high fertility and proximity to water supplies, hydroelectric power, transportation, and a waste-disposal medium. The flat topography facilitated building and agriculture. Furthermore, because of the hazard, the land was significantly cheaper than adjacent high land, and river-front lots often afforded scenic views. Once a floodplain was developed, a certain historical momentum encouraged its continued use. Today, large parts of many major communities are located on floodplains.

Physical Setting

A *floodplain* is the normally dry land area adjacent to a stream channel or ocean shoreline that is likely to be flooded. The erosive power of a stream carves a channel large enough to carry the most frequent flow; the floodplain receives stream flow in excess of channel capacity. The areal extent of inundation is related to the magnitude of flood discharge and the physical characteristics of the stream valley. The term *floodway* refers to the part of the floodplain that provides the avenue for most floods. In humid climates, this area may be flooded every two to three years. The *flood fringe* is located between the floodway and valley walls and is inundated only by exceptionally high floodwaters.

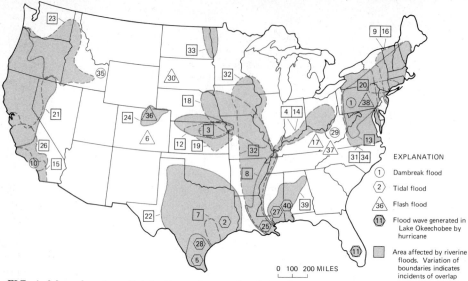

FIG. 1. Map showing distribution of great floods in the conterminous United States since 1889. Numbers correspond to those in Table 1. (From *U.S. Geological Survey Professional Paper 1240-B, 1981.*)

Factors Affecting the Severity of Floods

The frequency and duration of flooding and the size of the area inundated depend on both the natural features of a watershed and the land-use practices.

Natural Features

Natural features may be viewed as *physical*—topography, soils, bedrock geology, vegetation, and drainage pattern—and *hydrologic*—amount of moisture in soil before rainfall and the quantity, intensity, and geographical distribution of rainfall and snowmelt. High antecedent moisture conditions, steep slopes, and impermeable soils or bedrock will combine to produce a high rate of runoff, which may exceed the carrying capacity of a stream even when rainfall is moderate (Fig. 2).

Land-Use Practices

Most floods stem from natural causes, but land-use practices may greatly increase the severity of flooding by increasing the rate of runoff. In urban areas, buildings and pavement seal off water-absorbing

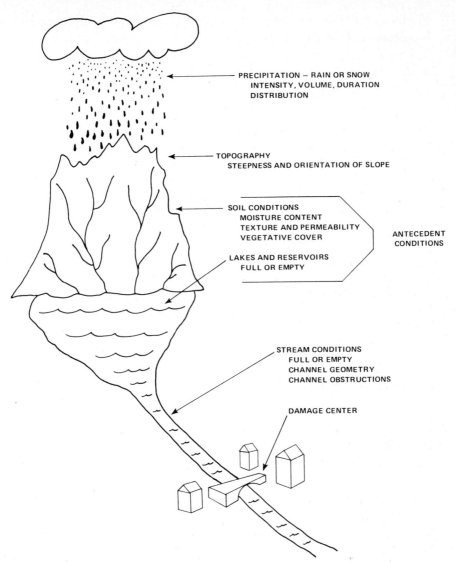

PRECIPITATION – RAIN OR SNOW
INTENSITY, VOLUME, DURATION
DISTRIBUTION

TOPOGRAPHY
STEEPNESS AND ORIENTATION OF SLOPE

SOIL CONDITIONS
MOISTURE CONTENT
TEXTURE AND PERMEABILITY
VEGETATIVE COVER

LAKES AND RESERVOIRS
FULL OR EMPTY

ANTECEDENT
CONDITIONS

STREAM CONDITIONS
FULL OR EMPTY
CHANNEL GEOMETRY
CHANNEL OBSTRUCTIONS

DAMAGE CENTER

FIG. 2. Flood-flow factors (Adapted from Wisconsin Dept. of Natural Resources.)

ground, or construction activities may leave land bare and subject to runoff for months at a time. Because storm sewers often lead directly to stream channels, large volumes of water can be emptied into a stream in a short period of time. In rural areas, ill-conceived lumbering or agricultural practices remove protective vegetation, or mining activities may leave the land bare for long periods of time.

The damage potential of floods may also be increased by inadequately designed dams, which in times of peak flow may fail and produce catastrophic floods. Also, buildings and other structures may be floated off their foundations, becoming destructive weapons against other property. Or, if buildings and structures do not float away, they can interfere with flow, causing a backup of water and an increase in the size of the area inundated.

Riverine Flood-Hazard Areas

The problems resulting from floodplain occupancy have led to regulations aimed at reducing flood damage while promoting beneficial use. A basic feature of floodplain regulation is the establishment of the standard project flood and the regulatory flood limits. *Standard project flood* (SPF) is the high-water stage expected from the most severe combination of meteorologic and hydrologic conditions that are reasonably characteristic of a stream valley. The most common recurrence interval of an SPF is 100 years. The *regulatory flood limit* is the outer limit of the floodway fringe and accommodates smaller discharges. The *regulatory floodway* is that area which is to be kept open to carry floodwater. In the adjacent *regulatory floodway fringe*, development is permissible, but only when building density is regulated to prevent backup of floodwater, and structures are floodproofed. Figure 3 shows the relationships between riverine flood-hazard areas.

Risk Assessment

Hydrologic Variables

Some of the hydrologic variables of interest to planners and engineers in evaluating flood risks are (1) frequency, (2) magnitude, (3) rate of rise, (4) flood peak, (5) lag time, and (6) duration.

Frequency and Magnitude. Flood flows can occur annually in temperate climates or as infrequently as once in a 1000 years in arid climates. The magnitude of a flood can be measured by *rate of flow* (cubic meters or feet per second) or by *river stage* (meters or feet above some reference point). The frequency, or *recurrence interval*, of a flood of a particular magnitude can be calculated if sufficient stream-flow data are available (Fig. 4). Large floods have a low frequency of occurrence; smaller floods a higher frequency. Because stream flow cannot be predicted, frequency is expressed as the *probability* of a given discharge being equaled or exceeded in any one year. The lower the frequency the less are the chances of that occurring, but the odds are never zero. Even a flood with a return period of 100 years (100-year flood) may occur twice in the same year. The odds that it will occur

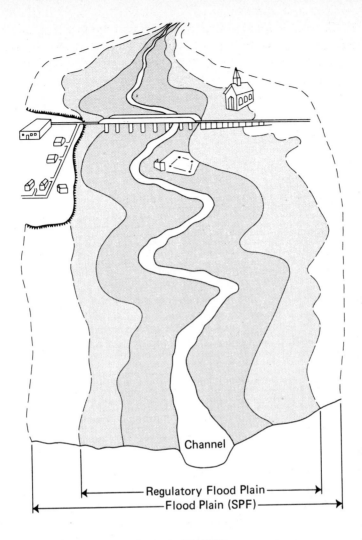

Channel

|← Regulatory Flood Plain →|

|← Flood Plain (SPF) →|

EXPLANATION

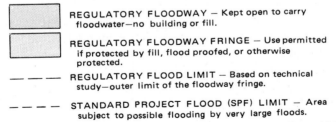

REGULATORY FLOODWAY — Kept open to carry floodwater—no building or fill.

REGULATORY FLOODWAY FRINGE — Use permitted if protected by fill, flood proofed, or otherwise protected.

— — — REGULATORY FLOOD LIMIT — Based on technical study—outer limit of the floodway fringe.

— — — STANDARD PROJECT FLOOD (SPF) LIMIT — Area subject to possible flooding by very large floods.

FIG. 3. Riverine flood-hazard areas of a regulatory floodplain. (From *U.S. Geological Survey Professional Paper 942, 1977.*)

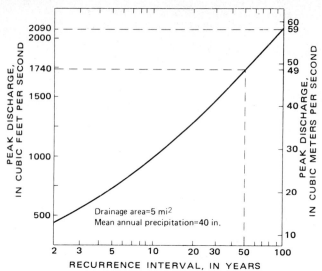

FIG. 4. Flood-frequency curve. This curve represents a cumulative distribution relating the magnitude of a flood to frequency of occurrence. The 50-year flood discharge has a 2 percent (1 in 50) chance of being exceeded in *any* given year. Its discharge or magnitude, in this example, is 49.3 m³/sec (1740 ft³/sec). (Figure from U.S. *Geological Survey Professional Paper 942*, 1977.)

again in the following year remain the same; that is, there is a 1 percent chance *every* year of a 100-year flood.

Rate of Rise. The time from flood stage to flood peak is a measure of the rate of rise. In upstream areas the rate of rise is more rapid than downstream.

Flood Peak and Lag Time. Flood peak is the time when the quantity of water draining to a gauging station has reached its maximum, and is generally coincident with the greatest water-surface elevation during a rise in the stream. Lag time is the interval between peak rainfall and peak discharge, or flow. These relationships can be illustrated in a flood hydrograph (Fig. 5).

Duration. The time of inundation can vary from a few minutes to weeks or even months. Upstream areas are inundated for shorter periods than downstream areas.

Floodplain Maps

Reports on floods and proposed flood-control projects commonly include maps of the probable extent of the areas subject to inundation from floods of designated recurrence intervals. Such maps permit detailed evaluation of flood hazards and are useful in developing land-use plans and policies (see Fig. 2, Reading 21).

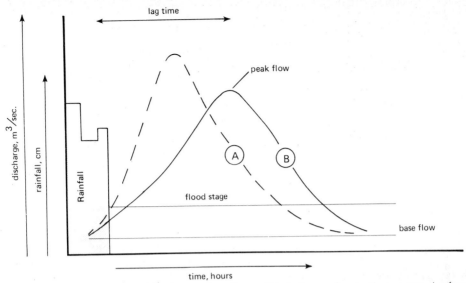

FIG. 5. Hypothetical flood hydrographs. Curve B is before urbanization; curve A after urbanization. Note how urbanization causes a decrease in lag time and an increase in peak flow. This is a result of an increase in the percentage of rainfall which becomes runoff and the more rapid delivery of runoff to the stream channel. Base flow represents the sustained or fair-weather flow of the stream.

Floodplain maps, which in the United States are produced for the most part by the Army Corps of Engineers and the U.S. Geological Survey, are based on the 100-year (1 percent) flood. The data required to map this flood come from various sources. Historical records of rainfall and floods are helpful, but may not be available for some parts of a drainage basin or for extended periods of time. The use of historical data must also be tempered by the fact that extensive urbanization, agricultural development, and climatic changes can significantly alter runoff and stream-flow characteristics. If changes have been extensive, some data may even be irrelevant.

Areas subject to inundation may also be identified on the basis of such geomorphic features as meander scars, oxbow lakes, natural levees, terraces, and scarps (Fig. 6). Certain soils and vegetation (such as cottonwood, black willow, and honey locust in the Upper Mississippi Valley) are also characteristic of floodplains and may serve as useful indicators. Some of these features can be identified on air photos or LANDSAT (satellite) imagery.

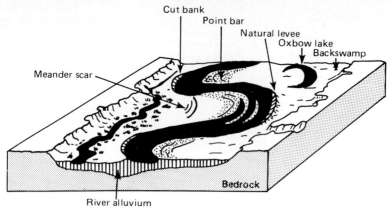

FIG. 6. Characteristic features of the floodplain of a meandering river. A cut bank is an erosional feature produced by lateral erosion of the outside bank of a meander. A point bar is a depositional feature on the inside of a meander. Natural levees are low embankments adjacent to a river channel. Crescent-shaped lakes (oxbow lakes) and scars (meander scars) are formed when a meander is cut off from the main stream. Low-lying ground between natural levees and the river bluff is known as backswamp.

Flood Forecasting

The National Oceanic and Atmospheric Administration (NOAA) collects and analyzes data on rivers and rainfall to provide forecasts of river levels and flood warnings. The flood warnings, of course, allow the temporary evacuation of low-lying areas, and forecasts of river levels are essential for navigation, hydroelectric power production, reservoir operation, pollution control, industrial activities, and fish and wildlife management.

The goal of flood forecasting is to predict the amount of water that will find its way into rivers and the time it will take to reach them under different conditions. Techniques must be adapted to the size, terrain, climate, and structural features, such as dams and reservoirs, of an individual basin. Success in prediction is related to the quality and relevance of historic data and the length of lag time. The most useful historic data are those that show the relationship between precipitation, snowmelt, soil conditions, and river levels. Computer technology has led to the development of sophisticated conceptual models of the factors determining stream flow.

Flood warnings can be issued hours to days in advance of the flood peak on major tributaries and weeks in advance on main streams. Forecasts and warnings include crest stage (maximum elevation of a flood) expected and its time of occurrence, the areas to be affected, and duration. Flood potential outlooks and forecasts of seasonal water supplies are often made months in advance.

Flash-Flood Warnings. Flash floods, local floods of relatively high magnitude and short duration, are most common in semiarid regions, the headwaters of mountain streams, and urbanized areas. Heavy precipitation combined with rapid runoff results in a rapid rate of rise, high peak flows, and short lag times. There are several flash-flood warning systems. Some are highly automated while others rely heavily on individual observations.

A vital part of any prediction system is the education of people who occupy the floodplain. They must be apprised of the nature of the hazard, how warnings will be issued, and how to respond to the warnings. NOAA has established a set of flood safety rules which are available to residents of floodplain areas (Table 2).

Reducing Losses

A variety of approaches have been suggested to reduce flood-related losses. For purposes of discussion, it is convenient to classify these approaches as either structural or nonstructural (administrative).

Structural Controls and Methods

The most common structural controls are dams, reservoirs, diversion channels, dikes, and levees; and channel dredging, straightening, and widening. Dams and reservoirs temporarily store floodwaters while the other controls increase the capacity of a channel to transport flood flows.

The feasibility of control structures is frequently the subject of controversy. In some instances they have proved to be both structurally sound and cost effective. In other instances construction costs have outweighed the benefits, or the improvements have caused increased erosion and flooding downstream. Further, structural controls may also encourage further development of floodplains in anticipation of the elimination of flood hazards. Accelerated development is frequently followed by demands for greater protection, which in turn encourages more development. This "snowball effect" explains, in part, the intensive development of many floodplains and rising property losses. For example, since the passage of the Federal Flood Control Act in 1936, the Federal government has spent an average of $1 billion annually on structural controls while annual losses have quadrupled (Costa, 1978).

Structural methods also include techniques designed to reduce or delay surface runoff and erosion. Agricultural practices such as contour plowing, strip cropping, terracing, and reforestation of clear-cut

Table 2 Flood safety rules [NOAA]

Before the flood:

1. Keep on hand materials like sandbags, plywood, plastic sheeting, and lumber.
2. Install check valves in building sewer traps, to prevent flood water from backing up in sewer drains.
3. Arrange for auxiliary electrical supplies for hospitals and other operations which are critically affected by power failure.
4. Keep first aid supplies at hand.
5. Keep your automobile fueled; if electric power is cut off, filling stations may not be able to operate pumps for several days.
6. Keep a stock of food which requires little cooking and no refrigeration; electric power may be interrupted.
7. Keep a portable radio, emergency cooking equipment, lights and flashlights in working order.

When you receive a flood warning:

8. Store drinking water in clean bathtubs, and in various containers. Water service may be interrupted.
9. If forced to leave your home and time permits, move essential items to safe ground; fill tanks to keep them from floating away; grease immovable machinery.
10. Move to a safe area before access is cut off by flood water.

During the flood:

11. Avoid areas subject to sudden flooding.
12. Do not attempt to cross a flowing stream where water is above your knees.
13. Do not attempt to drive over a flooded road—you can be stranded, and trapped.

After the flood:

14. Do not use fresh food that has come in contact with flood waters.
15. Test drinking water for potability: wells should be pumped out and the water tested before drinking.
16. Seek necessary medical care at nearest hospital. Food, clothing, shelter, and first aid are available at Red Cross shelters.
17. Do not visit disaster area; your presence might hamper rescue and other emergency operations.
18. Do not handle live electrical equipment in wet areas; electrical equipment should be checked and dried before returning to service.
19. Use flashlights, not lanterns or torches, to examine buildings; flammables may be inside.
20. Report broken utility lines to appropriate authorities.

FLASH FLOODS

Flash flood waves, moving at incredible speeds, can roll boulders, tear out trees, destroy buildings and bridges, and scour out new channels. Killing walls of water can reach 10 to 20 feet. You won't always have warning that these deadly, sudden floods are coming.

WHEN A FLASH FLOOD WARNING IS ISSUED FOR YOUR AREA OR THE MOMENT YOU FIRST REALIZE THAT A FLASH FLOOD IS IMMINENT, ACT QUICKLY TO SAVE YOURSELF. YOU MAY HAVE ONLY SECONDS.

Get out of areas subject to flooding. Avoid already flooded areas. Do not attempt to cross a flowing stream on foot where water is above your knees. If driving, know the depth of water in a dip before crossing. The road may not be intact under the water. If the vehicle stalls, abandon it immediately and seek higher ground—rapidly rising water may engulf the vehicle and its occupants and sweep them away. Be especially cautious at night when it is harder to recognize flood dangers.

tracts are particularly effective. The use of small ponds and check dams are also helpful. In urban areas, drainage tunnels are constructed or the capacity of natural drainage channels can be expanded.

Non-structural Measures

In recent years, public policy on flood control in the United States has encouraged the use of non-structural or administrative solutions to flood hazards. These measures include risk assessment, flood warnings, public awareness programs, evacuation, tax incentives, floodplain zoning, and insurance requirements. Permanent evacuation is an extreme and seldom-used approach. The residents of Soldiers Grove, Wisconsin, however, did elect to abandon the floodplain of the Kickapoo River after a disastrous flood in 1978, and evacuation is being implemented at Prairie du Chien, Wisconsin (see Reading 22).

Floodplain Zoning. Land use is frequently regulated through local zoning ordinances, which divide a jurisdiction into districts or zones, establish for each district requirements for use of the land, and specify codes for the construction of buildings, roads, sanitary facilities, and so on. Floodways are frequently designated as open space (greenbelts) for parks, golf courses, or hiking and riding trails, or they may be zoned for grazing or parking. Development may be permitted on the floodway fringe if buildings are floodproofed and well anchored. Special storm sewers and sanitary codes may be invoked to reduce damage and health hazards. Legal authority for zoning is derived from a state's "police power", which can be defined as the power inherent in the state to prescribe reasonable regulations necessary to preserve the "health, safety, morals, and general welfare" of the community.

Rational floodplain zoning may be difficult because existing development is frequently protected by a "grandfather clause." This means that zoning may be restricted to future development and is not retroactive. Of course, governments can acquire private property, using their power of eminent domain to establish proper land-use practices in floodplains, but this is an expensive process and not feasible in most areas. Figure 7 shows a zoning map and a floodplain zoning map of the same area.

Insurance. Flood insurance policies can reduce individual losses and encourage wise management of floodplains. The National Flood Insurance Program (1968) provides public subsidies of flood insurance premiums for residents of communities that agree to adopt and enforce minimum floodplain management regulations. The Federal Insurance Administration administers the program and is responsible

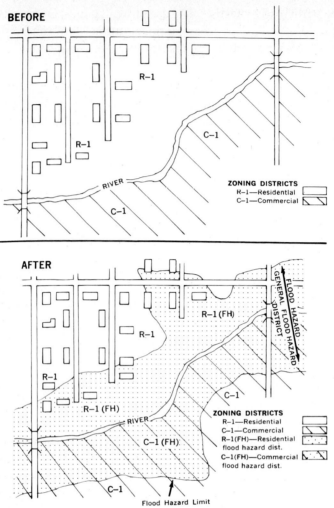

BEFORE

R-1

C-1

ZONING DISTRICTS
R-1—Residential
C-1—Commercial

RIVER

C-1

AFTER

R-1 (FH)

R-1

GENERAL FLOOD HAZARD DISTRICT

FLOOD HAZARD GENERAL FLOOD HAZARD

R-1

C-1

ZONING DISTRICTS
R-1—Residential
C-1—Commercial
R-1(FH)—Residential
flood hazard dist.
C-1(FH)—Commercial
flood hazard dist.

R-1 (FH)

RIVER

C-1 (FH)

C-1

Flood Hazard Limit

FIG. 7. Typical zoning map before and after addition of flood regulations, using a single district approach. (From Water Resources Council, *Regulation of Flood Hazard Areas,* vol. 1, 1971.)

for identifying the nations' floodplains and promulgating minimum regulations. Purchase of flood insurance is voluntary, but it is required when federal assistance or loans are needed for construction on floodplains. In 1973, the Flood Disaster Protection Act extended the program to cover landslides and shoreline erosion associated with storms.

Case Histories and Readings

Global Summary of Human
Response to Natural Hazards:
Floods

In Reading 20, Jacquelyn Beyer presents an overview of human response to floods from a global perspective. She notes that even though there is, from the global and national institutional viewpoints, adequate sensitivity to the nature of the hazard, the range of adjustments to the hazard is broad. Local conditions of economic development and financial structure, individual attitudes, government policies and philosophies, and historical precedents clearly influence the choices in adjusting to the problem.

Flood-Hazard Mapping in
Metropolitan Chicago

As urban pressures encourage expanded development in fringe areas, it becomes necessary to develop new techniques to identify potentially hazardous settings and to reduce the hazards associated with their development. The concept and application of flood-hazard mapping are reviewed in Reading 21.

Flood-Plain Management Must Be
Ecologically and Economically
Sound

Floods exact a costly toll in lives and property damage, while measures to control floods often result in heavy ecological losses. Ecologically acceptable solutions must be found. James Goddard, in Reading 22, presents several case histories where floodplain management incorporated solutions that are both ecologically and economically sound.

20 Global Summary of Human Response to Natural Hazards: Floods

Jacquelyn L. Beyer

Definition

All streams are subject to flooding in the hydrological sense of inundation of riparian areas by stream flow which exceeds bank full capacity. In arid regions the channel itself, not usually filled with water, is "flooded" at times of high runoff. The point at which the channel discharges an overbank surplus is the flood stage. This may not, however, coincide with the amount of water outside the normal channel which will cause damage to human works. It is also possible to calculate the stage of high water which is the threshold for damage to property or dislocation of human activities. Frequently the use of the term "flood stage" is based on such a perception of the event, and is therefore a definition subject to change as conditions of floodplain occupance change. This paper is not concerned with coastal flooding.

Spatial Extent

Floods are the most universally experienced natural hazard, tend to be larger in spatial impact, and involve greater loss of life than do other hazards. Floods can occur on both perennial and ephemeral stream beds or in an area

where no defined channel exists, such as in an arid region subject to cloudburst type storms. The problem is compounded for human adjustment by the fact that few other hazards present the ambivalent Januslike aspect of good and evil. Humans are attracted to settlement in flood hazard areas by the very characteristics—water supply and floodplain terrain—which contribute to the damage potential.

For this reason it is not surprising to find that historic attempts have been made to resolve the conflict between the need for riparian occupance and the inevitable damage, as Wittfogel describes it: "Thus in virtually all major hydraulic civilizations, preparatory (feeding) works for the purpose of irrigation are supplemented by and interlocked with protective works for the purpose of flood control" (Wittfogel, 1957, p. 24). Less elaborately organized preindustrial societies have also worked out ecological adjustments to flooding. Familiar examples of peasant adaptation to periodic flooding include the traditional agricultural organization along the lower Nile, now altered by the construction of the high Aswan dam, and the village rice culture of the lower Mekong, which will eventually be affected by flood-control components of the Mekong basin development. Another such example is the people of Barotseland in northwest Zambia where migration to higher ground is the organized response to the annual seasonal inundation of the reaches of the upper Zambezi which mark the coreland of Barotse occupance. Changes in socioeconomic patterns as such societies industrialize will undoubtedly accelerate the damage from floods. Familiar adjustments such as migration will fall outside the range of choice. Alternative workable adjustments may be inhibited by lack of knowledge, technology, and/or capital.

For industrial societies the twentieth-century concept of multiple-purpose river basin planning, now widely diffused (United Nations, 1969a), involves the consideration of flood damage reduction along with planning for beneficial use of water.

In summary, the potential for flooding is global in nature and can occur, with the proper combination of factors, whenever there is precipitation. This precipitation may range from uniform and general to sporadic and highly localized. Adjustments to hazards must be made in the context of both universality and randomness. In addition there needs to be sensitivity to beneficial uses for floodplains and water courses.

Damage Potential

Types of floods are so varied in origin, duration, strength, timing, volume, depth, and seasonality that it is difficult to identify damage potential except in the most general terms. The amount of damage and the damage potential in any flood-hazard area is very closely related to the nature of occupance and to the stage of economic development as well as to the physical parameters. There also seems

to be an inverse relationship between property damage as measured in monetary terms and loss of life. Societies which have much to lose in terms of structures, utilities, transportation facilities, etc., also have the technological sophistication to ensure better monitoring, warnings, evacuations, and rehabilitation—all of which contribute to the lowering the human costs. Conversely, preindustrial societies, especially with dense rural populations, do not suffer large property losses but are less well equipped to provide preventative or rescue measures for people.

Clearly the main damage agent is the water itself, overflowing normal channels and inundating land, utilities, buildings, communications, transportation facilities, equipment, crops, and goods which were never meant to operate in or withstand the effects of water. In addition, high velocity of running water operates as a damage agent either directly or indirectly. In the latter case, debris carried by the water or dislodged materials batter structures, people, and goods. Debris and silt carried by the water and left behind as the water recedes operate as further damage agents.

Damage is considered to be either direct or indirect (primary or secondary). Such a classification is useful in any assessment which attempts to clarify the benefits of a damage reduction program. Loss of human life is the most dramatic and certainly the easiest to identify as a direct result of flood events. Loss of livestock may be especially costly in rural zones.

In agricultural areas, damage involves inundation of land accompanied by erosion and/or loss of crops. It is in such cases that the season of flooding is especially significant. Water damages farm equipment, stored materials (seed, fertilizer, feed), disrupts irrigation systems and other water supply, and disrupts communication.

Urban facilities are all subject to water and force damage—buildings of all kinds, public facilities, utilities, transportation, waterway facilities, and open space. Machinery, manufacturers, goods in retail establishments, household furniture can all be damaged by water, debris, and silt.

Indirect damages are generally associated with health and general welfare although such amenities as scenic values, recreational services, and wilderness preservation may also be taken into account. Normal public-health services are subject to greater pressures in the face of disruption of transportation and utilities, especially water supply. Contamination and pollution are more probable, epizootics emerge, stagnant water is left as a flood legacy, and general morbidity increases. Flooding affects the normal sources of food and shelter and hence adversely affects health conditions. Opposed to these considerations is the possibility that emergency relief operations might provide better health care and food than is normally available to some communities counter-acting, to some extent, the effects of both direct and indirect damages (White, 1945).

Benefits

Rivers in flood clearly are hazardous for many kinds of human use, but a complication in planning for amelioration of the hazard is presented by the benefits of naturally flowing rivers, including overbank flooding. "Control" of flooding by protective works, especially, may negate benefits from soil nutrient renewal and fisheries. Restrictions on use of floodways will provide for damage reduction and also enhance community values through preservation of open space. In some parts of the world, exemplified by India and Bangladesh, the rhythms of agricultural production are dependent upon water brought by major storms and renewal of fertility through siltation. In such cases serious weather modification efforts should proceed only after careful assessment of the total benefit-cost pattern.

Damage Assessment

Assessment includes the costs of repair, including temporary repairs; replacement and cleanup; and loss of improvements and inventories. Emergency costs are also involved. Loss of business and employment during flood disruption is also a direct cost associated with the physical damage. Similarly the transfer of public economic development funds to flood emergency and control programs represents a deferred opportunity which may be especially significant as a social cost in developing nations.

There are no comprehensive calculations of global damage from floods. In many cases, local flooding in areas remote from communications may not even be recorded. There are immense difficulties for any inventory of flood consequences including those of cost, technical expertise, comparability of data collected, allocation of losses to proper causes, and time. A pilot survey of global natural disasters by Sheehan and Hewitt (1969) provides some information for the 20 years 1947–67 for most of the world—the USSR is the major country excluded. During this period Asia (excluding the USSR) led in loss of life, with 154,000 deaths. Europe (excluding the USSR) followed with 10,540 deaths. Africa, South America, and the Caribbean area each recorded 2,000–3,000 deaths. During the same period 680 lives were lost to floods in North America, and 60 in Australia, totaling for all these regions 173,170 deaths.

This total can be compared with the 269,635 deaths attributed to all . . . other hazards . . . and if it is considered that many of the other categories—e.g., tornadoes, typhoons, hurricanes, and tidal waves—also involve flooding, the death loss is the most impressive comparison. Table 1 will give some idea of the magnitude of the hazard in terms of area affected, people involved, and property damage.

Damage Factors

The factors which should be taken into account in assessing the damage potential for flooding in any basin include the following.

Frequency

Flood flows can occur on the average as often as once every 2 years in temperate climates and as infrequently as once in 1,000 years elsewhere. The recurrence of a particular flood flow can be predicted with reasonable accuracy over a long time span if sufficient stream-flow data are available over a long period. This emphasizes the certainty of the event, not its timing, which is not predictable since flood flows are assumed to be random events.

Frequency is a physical parameter clearly related both to perception and adjustments. The greater the frequency the more accurate is the perception of the hazard by floodplain occupants and the greater is their willingness to consider a wider range of adjustments, including alternative sites for their activities. This is demonstrated by variations in community decision making which can be correlated with frequency of flooding (Kates, 1962).

Magnitude (depth)

Magnitude of a flood may be expressed in physical, or probabilistic terms. The physical measures are rate of flow measured in cusecs, m^3/sec (cubic meters per second), or river stage in meters (or feet) above some datum (reference) point. Both of these require carefully established measuring installations to obtain reliable data. Flow can be graphically plotted versus stage to give a stage-discharge relationship which can then be useful in predicting damages. This relationship can be quite complex, and considerable care should be exercised in its use.

The probabilistic measure is a statistical method of ordering various magnitudes of flow and stating the probability that a given flow will be exceeded. Under this procedure, a 5-year flood is a flow or stage which will be equaled or exceeded 20 times, on the average, in a 100-year time span. The statistical method has validity only in areas where good flow records over a long period of time (at least 30 years) are available, although some simulation can be done. The ability to determine probability of flood magnitudes is not the same as the ability to state when such floods will occur. Some of the parameters upon which magnitude depends and which are useful in classifying the flood characteristics of a given area include the type, intensity, duration, areal extent, and distribution of precipitation; basin size and shape; floodplain topography; surface conditions of soil; and land use. There is an effort in Japan, for example, to elaborate a system of flood-hazard classification through the use of landform analysis. It is suggested that such analysis will provide for predictability of flood current, ranges of submersion, depth of stagnant water, and length of period of stagnation (Oya, 1969).

Depth can be important in terms of both the kinds of damage and possible adjustments, e.g., floodproofing of structures.

Rate of Rise

This is the time from flood stage (or zero damage stage) to flood peak and is a measure of the intensity of the flood. As Sheaffer (1961) notes, this time between flood stage and flood peak represents an adjustment time during which persons affected by flooding can engage in activities to lessen the damage. Generally, people will not respond to the danger of flooding until at least flood stage, so this time period is critical. It is clear that there is a relationship between the nature of the drainage system and the rate of rise—upstream areas will have a more rapid rise and shorter duration of flooding than will downstream areas.

Seasonality

This is one of the more significant factors for agricultural damage and probably the main basis for the adjustments made by preindustrial riverine societies. Clearly the hazard increases where the growing season is limited and coincides with the season of flooding. Winter floods might also account for increased loss and disruption in urban areas where heating and sanitary facilities are needed to guard against increase in disease and discomfort.

Duration

The time of inundation for flood flows can vary from a few minutes to more than a month. The duration is highly correlated to the rate of rise and fall of flood crests except where drainage of land area is impeded by obstructions. Flood duration is dependent upon such parameters as source of runoff; runoff characteristics including slope and surface conditions; nature of obstructions impeding recession of waters; and man-created controls such as reservoirs, levees, and channelization.

Nature of Floodplain Occupance

This includes the density of settlement; types of facilities; extent of fixed facilities, buildings, and equipment; and value of facilities. Obviously every increase in such occupance will increase the potential damage and call for some kind of adjustment, whether protective works, warning systems, and public relief capabilities, or a willingness to accept the losses.

Efficacy of Forecasting and Warning Systems

The ability to forecast the occurrence of overbank flooding is limited to a time

Table 1 Significant historical flood events

Date	Place	Deaths	Property damage
June 1972[a]	Eastern U. S.	100+	$2 billion
June 1972[a]	Rapid City, S. D.	215 (est.)	$100 million
May 11–23, 1970[b]	Oradea, Rumania	200	225 towns destroyed
January 25–29, 1969	Southern California	95	
July 4, 1969	Southern Michigan and Northern Ohio	33	
August 23, 1969	Virginia	100	
May 29–31, 1968	Northern New Jersey	8	$140 million
August 8–14, 1968	Gujarat, India	1,000	
January–March 1967	Rio de Janeiro and São Paulo states	600+	
November 26, 1967	Lisbon	457	
January 11–13, 1966	Rio de Janeiro	300	
November 3–4, 1966	Arno Valley, Italy	113	Art treasures in Florence and elsewhere destroyed
June 18–19, 1965	Southwest U. S.	27	
June 8–9, 1964	Northern Montana	36	
December 1964	Western U. S.	45	
October 9, 1963	Belluno, Italy	2,000+	Vaiont Dam overtopped
November 14–15, 1963	Haiti	500	
September 27, 1962	Barcelona	470+	$80 million
December 31, 1962	Northern Europe	309+	
May 1961	Midwest U. S.	25	
December 2, 1959	Frejus, France	412	Malpasset Dam collapsed
October 4, 1955	Pakistan and India	1,700	5.6 million crop acres at loss of $63 million
August 1, 1954	Kazvin District, Iran	2,000+	
January 31–February 1, 1953	Northern Europe	2,000+	Coastal areas devastated
July 2–19, 1951	Kansas and Missouri	41	200,000 homeless, $1 billion
August 28, 1951	Manchuria	5,000+	
August 14, 1950	Anhwei Province, China	500	10 million homeless; 5 million acres inundated
July–August 1939	Tientsin, China	1,000	Millions homeless
March 13, 1928	Santa Paula, Calif.	450	St. Francis Dam collapsed
March 25–27, 1913	Ohio and Indiana	700	
1911	Yangtze River, China	100,000	
1903	Heppner, Ore.	250+	Town destroyed
May 31, 1889	Johnstown, Pa.	2,000+	
1887	Honan, China	900,000+	Yellow River overflowed; communities destroyed
1642	China	300,000	

[a]Press reports.
[b]Adapted from Table of Disasters/Catastrophes, *New York Times Encyclopedic Almanac (1970), p. 1228;* (1972), pp. 322–33.

span in which the hydrologic conditions necessary for flooding to occur have begun to develop. The formulation of a forecast for flood conditions requires information on current hydrologic conditions such as precipitation, river stage, water equivalent of snowpack, temperature, and soil conditions over the entire drainage basin as well as weather reports and forecasts.

In small headwater regions a forecast of crest height and time of occurrence is all the information required to initiate effective adjustments since the relatively rapid rate of rise and fall makes the period of time above flood stage relatively short. In lower reaches of large river systems where rates of rise and fall are slower it is important to forecast the time when various critical stages of flow will be reached over the rise and fall. Reliability of forecasts for large downstream river systems is generally higher than for headwater systems.

Warning time for peak or overbank conditions can range from a few minutes in cloudburst conditions to a few hours in small headwater drainages to several days in the lower reaches of large river systems. As with forecasting, the time and reliability of the warning increase with distance downstream where adequate knowledge of upstream conditions exists.

Clearly the amount of information required, the data collection network necessary for collecting the information, the technical expertise required for interpretation, and the communication system needed to present the information in time to potential victims are such as to preclude many poor and developing nations from having an adequate service. The World Meteorological Organization of the United Nations, through its World Weather Watch and Global Data Processing System, hopes to coordinate efforts to improve forecasting. A recent report (Miljukov, 1969) notes that quantitative precipitation forecasts for 24–49 hours in advance are provided in parts of Australia, Byelorussian SSR, Cambodia, Canada, Czechoslovakia, France, Federal Republic of Germany, Hong Kong, India, Iraq, Japan, Mauretania, Norway, Pakistan, Philippines, Rhodesia, Romania, Sweden, Ukrainian SSR, the USSR, and the U. S. Precipitation forecasts for hydrological purposes are provided in Australia, Canada, Czechoslovakia, France, Federal Republic of Germany, India, Iraq, Japan, the USSR, and the U. S. The report also notes: "Precipitation forecasts are not accurate and reliable enough, in the present state of meteorological science, for use in the preparation of quantitative forecasts of river discharges" (Miljukov, 1969, p. 10). Most developing nations will have to rely on much less data than are ideally needed for forecasting and warning, which in turn will lessen the effectiveness of this factor with respect to flood losses.

Efficacy of Emergency Services
The helplessness of many small and poor societies is exemplified by the situation in Bangladesh during the tropical cyclone of 1970. Where resources are not available for planning, for the physical effort of relief and evacuation, and for coordination with other activities, little will be done outside of contributions of international aid agencies. Even in more developed areas, local conditions of

transportation and public attitudes will lessen the usefulness of emergency aid, e.g., the disaster at Buffalo Creek, West Virginia, in 1972. The more elaborate and dependable such services are, however, the more there is a tendency to rely on such aid as a major adjustment and to reject consideration of less costly and more effective adjustments. There clearly must be provisions for first-level emergency aid where settlement already exists and where alternative moderations of the hazard are difficult to implement or costly, but these should normally not supplant other measures to reduce losses.

The Range of Adjustments and Their Adoption

The accompanying table (Table 2) suggests that the cumulative experience of centuries provides for any society or group wishing to alleviate the social and economic costs of flood losses a choice of methods, to be used singly or in strategic combinations. Much of the wisdom with respect to the need for such strategies, adapted to local conditions of basin hydrography, settlement characteristics, and economic capabilities, has been gained in industrial nations after painful and costly trial and error. It has been suggested (Goddard, 1969) that developing nations need not repeat the errors of the past, that they have models for actions and policies which would provide a much more coherent and suitable response to the flood hazard. It is clear, even with such models, that adaptation to local circumstances in any society will not be a simple matter. It is increasingly evident that various combinations of individual psychology, institutional inertia, costs, governmental policies and philosophies, and historical precedents help to condition the choice of adjustments.

Modify the Flood

This category includes engineering works affecting the channel which represent the most widely accepted feasible adjustment with the possible exception of bearing the loss. Such protective works are justified where benefits exceed the costs of implementation and especially where high damage potential exists for relatively intensive settlement in urban and industrial situations. In such cases the high value of fixed facilities will justify levees, dams, and channelization even when 100 percent protection cannot be guaranteed. While benefits accrue to both private and public sectors, the costs are necessarily largely public. Partly for this reason this adjustment strongly tends to encourage persistent settlement and even attracts, through a false sense of security, further floodplain encroachment. On the other hand, such engineering works are important components of multiple purpose projects which are directed toward comprehensive land and water planning and they can be complemented, for floodplain management, with other measures. A major problem is to encourage engineers and officials to think in terms of nonstructural alternatives or supplements to protective works.

Watershed treatment practices have more subtle implications with respect to flood control and are frequently more significant for their contribution

Table 2 Adjustments to the flood hazard

Modify the flood	Modify the damage susceptibility	Modify the loss burden	Do nothing
Flood protection (channel phase)	Land-use regulation and changes	Flood insurance	Bear the loss
Dikes	Statutes	Tax write-offs	
Floodwalls	Zoning ordinances	Disaster relief	
Channel improvement	Building codes	volunteer	
Reservoirs	Urban renewal	private	
River diversions	Subdivision regulations	activities	
Watershed treatment (land phase)	Government purchase of lands and property	government aid	
Modification of cropping practices	Subsidized relocation	Emergency measures	
Terracing	Floodproofing	Removal of persons and property	
Gully control	Permanent closure of low-level windows and other openings	Flood fighting	
Bank stabilization	Waterproofing interiors	Rescheduling of operations	
Forest-fire control	Mounting store counters on wheels		
Revegetation	Installation of removable covers		
Weather modification	Closing of sewer valves		
	Covering machinery with plastic		
	Structural change		
	Use of impervious material for basements and walls		
	Seepage control		
	Sewer adjustment		
	Anchoring machinery		
	Underpinning buildings		
	Land elevation and fill		

Source: Adapted from Sewell (1964), pp. 40–48; and Sheaffer, Davis, and Richmond (1970).

to improved *in situ* land management. All such measures have their limitations with respect to major flood events. There may be some contribution to lowering the depth in small floods and to lengthening the flood-to-peak interval, but essentially the appeal of such practices is lower costs. About 90 percent of the costs for such practices are public while benefits accrue largely—about 85 percent—to private land users. Land treatment measures often complement and make more effective protective measures but will also tend to encourage continued settlement for flood-prone areas.

Weather modification is a fairly recent technique with respect to flood control and too little is known about its effectiveness. One major problem, even given scientific certainty about effectiveness, will be the necessity to allay public fears that tampering with weather processes will increase rather than lessen floods. Recent news stories of the use of weather modification techniques in the Indochina war will not make this task easier (Shapley, 1972). The immediate postflood news reports from the 1972 Rapid City, South Dakota, flood suggest that this has already become an issue. Costs for weather modification are entirely public while benefits are about equally divided between private users and the public.

Modify the Damage Susceptibility

Given that there may be a need to encroach on floodplains or to accept present settlement patterns, certain measures are possible which are either less costly than protective works or bearing the loss, or which will lessen the actual damages even more. There is also a greater shift of cost bearing to private interests, especially in the case of floodproofing, with resultant increased awareness of the need for flood adjustments.

Land-use regulation, including changes in occupance, is especially suitable where there is competition for floodplain land for uses other than agricultural or recreational. The legislative and police powers of the state can be used to control and guide development of floodplains. According to Goddard, in the United States "about 35 states have adopted regulations and 500 additional places in 41 states have them in adoption process" (Goddard, 1971). Encouraging this is the 1969 Federal Flood Insurance Act which provides for governmental flood insurance subsidy to individuals in communities which agree to adopt floodplain regulation guidelines. These measures tend to encourage more efficient and less costly use of floodplains and there is a greater shared responsibility between floodplain users and authorities. Strong leadership and a commitment to long-range planning and rational allocation of land uses are also prerequisites to widespread adoption of such measures.

Structural changes and floodproofing (including land elevation) provide for even larger shifts of costs as individual users may bear all the costs and share benefits with the public on an equal basis. Such adjustments are most appropriate where flooding is not intense either in velocity or depth and where some warning time is possible. Floodproofing especially requires a network of forecasting and warning facilities along with a flood-hazard information program which will encourage preflood adjustments. Structural modifications are possible for existing structures as well as for new structures although this will increase costs. In many cases it would be too expensive to modify old buildings. Some types of buildings are better suited to modification than others but clearly damage reduction is related to size of structures and costs of modification. These adjustments tend to encourage persistent occupance and lose effectiveness where flood frequency is low. At the same time they place more responsibility on the user and thus heighten sensitivity to and knowledge about the flood hazard.

Modify the Loss Burden

There is much more emphasis in this category on humanitarian responses rather than calculated economic rationale, based on the inevitability of flooding and the unlikely possibility of preventing all damage by eliminating floodplain occupance. Losses will thus occur even in the face of widespread use of appropriate adjustments. When people suffer trauma and loss there can be little question of a social obligation to provide assistance. The dilemma for rational flood damage reduction, however, is that relief measures and emergency assistance unless properly designed tend to encourage persistent occupance and reluctance to accept more rational adjustments.

Insurance and tax write-offs will not decrease flood losses but there will be a spreading of loss over time and a shift of some costs to the general public. As the flood insurance program has been worked out in the United States, the insurance subsidy by the government to private carriers must be coupled with community planning for land-use regulation and other adjustments to lessen potential damage. In Hungary, where levees and flood fighting are the principal adjustments, agricultural insurance was extended in 1968 to cover flood damages (Bogardi, 1972). Whether this works as an incentive for private adjustments is not clear. There is obviously a sensitive line between encouraging further encroachment or private irresponsibility and alleviating the damages to those who must occupy floodplains. Purchases of insurance, according to recent reports after the June 1972 floods in the eastern U. S., have not been commensurate with the danger nor with the benefits to eligible individuals. Problems resulting from a hazard insurance program which was not thoroughly planned have been noted.

Disaster relief is a necessary adjustment in order to lessen the immediate impact of a flood event and to ease the implementation of rehabilitation efforts. Whether government or private, the major disadvantage is that such measures, necessary though they may seem when disaster strikes, strongly encourage the belief that nothing else need be done.

The effectiveness of emergency measures depends largely upon the nature of the flood hazard (ideal combination of high flood frequency, low velocity and depth, long flood-to-peak interval, and short duration) and the quality of forecasting. The immediate governmental obligation is generally seen to be the removal of persons and property from flood threatened areas.

Do Nothing

Bearing the loss is still the major adjustment for large numbers of floodplain occupants in developing countries (Ramachandran and Thakur, chap. 5), and is frequently modified in developed nations only by the widespread expectation of relief and emergency measures. In all cases, however, it is clear that an increasing effort to clarify public interest in floodplain situations will restrict the choice of doing nothing and management strategies will become more common (Sheaffer, Davis, and Richmond, 1970).

Reduction of Loss

One element in the acceptance by any group of decision makers of a particular mix of components in a flood damage reduction program is the assessment of the comparative return from each possible choice of adjustment or combination of adjustments. If damage assessment after the fact of flooding is extremely difficult, it is even more difficult to predict what the damage will be under a set of assumptions about responses of various kinds. White and Burton have suggested methods whereby maximum damage reduction and minimum cost can be calculated for particular situations (White, 1964; Burton, 1969). Such methods may hopefully provide an additional planning tool in those situations where encroachment onto the floodplain is neither as intensive as in some industrial countries nor necessary. Some such tool is essential also to ensure the most efficient allocation of scarce resources, whether of materials, man-power, or money.

The relative contributions of each possible adjustment to reduction of potential damage can only be crudely measured at present. Such measurement is further complicated by the fact that only infrequently is a single adjustment adopted. Clearly any protective works which provide for 100 percent security under any feasible flood condition will provide 100 percent loss reduction although costs of providing such protection are likely to be unacceptable. Such security is highly improbable, both because of costs and imperfect knowledge of potential floods. The damage reduction to structures may range from 40 to 100 percent, dependent upon the size of the flood experienced and the nature of structures (White, 1964).

Watershed treatment data are inconclusive with respect to damage reduction and there are no data available for weather modification. Land-use regulation and change can provide for up to 90 percent damage reduction dependent upon the effectiveness of the regulations and the speed of application.

Data from one United States town (White, 1964) suggest that even minimal floodproofing of present structures under conditions of frequent but shallow flooding can be very effective, reducing damages by 60–85 percent. Great depths and/or high velocities would call for consideration of flood-proofing as part of building design.

Emergency action increases in effectiveness where there is a long flood-to-peak interval, high flood frequency, low depths, short duration, and low velocity. Where such conditions prevail, and assuming adequate warning facilities plus personnel and equipment, emergency action can reduce damages by 15 to 25 percent. A lower range of 5 to 10 percent is more probable.

"Adequate" warning would seem to be a minimal requirement for communities subject to flood hazard, but it is not simple nor inexpensive to provide a good system. Meteorological services and communications are part of the costs. Even given an excellent network of knowledge about the physical event, it may be difficult to convey that information to persons who will have to make adjustment decisions. Factors involved in a less than optimal warning system include:

1. Reluctance of officials to give false alarms.
2. Lack of complete coverage of median used to transmit warnings (radios, telephones, etc.)—communities and individuals may not be able to afford facilities.
3. Reluctance of people to see themselves affected by distant events (storms, runoff).
4. Individual interpretation of warning messages, especially where several messages may be contradictory or the messages may be incomplete.
5. Failure to provide exact information about what recipients of warnings are to do.

6. Impossibility of warning in time for much else than rapid evacuation.
7. Dramatic warning signals triggering an influx of the curious which negates warning advantages.

Flood insurance and tax subsidies spread the burden through time and shift much of the loss to the general public but do not reduce damage.

Another indication of the relative efficacy of various adjustments in reducing damage is the importance placed on them in national and regional plans. A recent report on Hungary (Bogardi, 1972), for example, suggests that reduction of damages is to be achieved largely through levee construction and maintenance, flood fighting, and, to some extent, insurance. Recommendations for Malaysia (Flood Control, 1968) are for a flood-control program involving better data collection and improved organization for relief and evacuation, combined with structural controls, land-use regulations, and flood-resistant crops. It is estimated that these measures would reduce anticipated damage from presently known levels of flooding by 50 percent. This report does not consider dams in catchment areas justifiable for flood control alone. Engineering works are still considered primary tools for India although some attention is being given to catchment area management and weather modification. The Japanese have extended their management approach to include regulatory measures (Oya, 1969). A comprehensive summary of national efforts to cope with floods as one of many natural hazards would probably justify a comment in a report from the United Nations (1969b): "Although there is still a considerable gap in many countries between the needs for governmental action and the actual institutional framework, new administrative patterns have evolved in others which responded to the need for a more coordinated and system oriented approach to resource administration."

Perception of Hazard and Adjustments

The global nature of the flood hazard is suggested not only by maps of large floods and by tables of deaths, but also by referece to international interest in the problem. The special agencies of the United Nations are involved in a wide spectrum of activities, including hydrological and meteorological data collection, flood forecasting methods, world catalogue of large floods, problems of health due to floods, and relief and aid to victims. Agencies involved include the Economic Commission for Africa, Economic Commission for Asia and the Far East, World Meteorological Organization, World Health Organization, and UNESCO. In many cases small nations will have to rely on technical help and assistance through United Nations channels. There is a discernible diffusion of efforts to plan and implement comprehensive programs including Canada, Japan, United Kingdom, and the United States. Because river basin management is so popular as an economic development tool, this opens the door for widespread consideration of comprehensive flood control as a component of such programs. From the global and national institutional viewpoints there is probably adequate sensitivity to the nature of the problem, if not to the possible range of adjustments. At the individual level, it is more difficult to judge whether the knowledge gained in recent years about perception of the hazard in the United States (Kates, 1962; Burton and Kates, 1964; James, Laurent, and Hill, 1971) is applicable to individual perception in developing or industrializing nations—or even industrial nations with different social and political conditions. A summary of some of the findings of these hazard perception studies, especially of floodplain occupants in Georgia, may be listed in the form of planning guidelines (adapted from James, Laurent, and Hill, 1971):

1. It cannot be assumed that accurate knowledge of the flood hazard will inhibit all persons from moving onto the floodplain.

2. The flood hazard itself will process people over time in terms of perception of the hazard and willingness to make adjustments. Management programs can short-circuit the unhappy experiences of those who remain unaware of the hazard and reluctant to adopt adjustments by preventing their settlement (e.g., through insurance programs).

3. Prospective floodplain occupants who are initially unaware cannot be swayed by large amounts of technical information; they also tend to be people who avoid contact with public officials and are not observant with respect to natural features.

4. In contrast, people who are knowledgeable about the flood hazard and settle anyway on floodplains will be responsive to more sophisticated information than is usually presented.

5. Delineation of flood-hazard areas on a map is ineffective as a form of communication.

6. Officials who disapprove of settlement on floodplains or who think in technical terms about risk will not be effective with those who are unaware of the risk.

7. Those who know about the flood hazard will be sensitive to depths, if not to frequency, and will therefore be open to flood proofing and possibly insurance as adjustments.

8. Time reduces awareness of the hazard, especially for those moving into a hazard area where indications of past flood events are not evident.

9. The wave of concern for environmental issues has brought with it evidence that those who are unaware of the flood hazard, but who have a concern for environmental damage, may respond more to appeals that land-use regulations are ecologically sound than to information about potential property damage.

10. Flood damage sufferers who contact, or who are contacted by, officials are a biased sample in terms of response to flood hazard. Frequently this bias is associated with speculation as a motive for owning floodplain property.

11. Upstream development frequently becomes the scapegoat for downstream floodplain users threatened by floods.

12. Floodplain users who are alien-

ated from government or authority because of other contacts are poor candidates for participation in floodplain management programs.

13. Extended delays in programs to reduce flood losses will increase alienation and make user participation more unlikely.

14. Encouragement of particular users should be part of policy, e.g., it should be made easy for those who are unaware of the hazard and/or reject adjustments to leave and be replaced by those who know something about the hazard and will be willing to adopt reasonable adjustments, including insurance and flood proofing or structural change. Where even these adjustments are too costly in the light of potential damage the policy should be to consider purchase and reversion to open space and recreational use.

It is hard to believe that persons would vary much with respect to a number of factors involved in determining the degree of knowledge about flood events, anticipation of future events, and willingness to consider various possible adjustments. Confirmation of this belief awaits further investigations of human response in diverse societies.

References

Baroyan, O. V. (1969) "Problems of health due to floods." Tbilisi, USSR: United Nations Inter-regional Seminar on Flood Damage Prevention Measures and Management.

Bogardi, I. (1972) "Floodplain control under conditions particular to Hungary." International Commission on Irrigation and Drainage, 8th Congress.

Burton, Ian. (1969) "Methods of measuring urban and rural flood losses." Tbilisi, USSR: United Nations Inter-regional Seminar on Flood Damage Prevention Measures and Management.

———, and Kates, Robert W. (1964) "The perception of natural hazards in resource management." Natural Resources Journal 3 (2):412–41.

Flood Control, Report of the Technical Subcommittee for (1968) Government of Malaysia: Director of Drainage and Irrigation.

Goddard, James E. (1969) "Comprehensive flood damage prevention management." Tbilisi, USSR: United Nations Inter-regional Seminar on Flood Damage Prevention Measures and Management.

———. (1971) "Flood-plain management must be ecologically and economically sound." Civil Engineering—ASCE 000:81–85.

James, L. Douglas, Laurent, Eugene A., and Hill, Duane W. (1971) The Flood Plain as a Residential Choice: Resident Attitudes and Perceptions and Their Implication to Flood Plain Management. Atlanta: Georgia Institute of Technology, Environmental Resources Center.

Kates, Robert W. (1962) Hazard and Choice Perception in Flood Plain Management. Chicago: University of Chicago, Department of Geography, Research Paper No. 78.

Miljukov, P. I. (1969) "Review of research and development of flood forecasting methods." Tbilisi, USSR: United Nations Inter-regional Seminar on Flood Damage Prevention Measures and Management.

Oya, Masahiko (1969) "Flood plain adjustments, restricted agricultural uses, zoning, and building codes as damage prevention measures." Tbilisi, USSR: United Nations Inter-regional Seminar on Flood Damage Prevention Measures and Management.

Sewell, W. D. F. (1964) Water Management and Floods in the Fraser River Basin. Chicago: University of Chicago, Department of Geography, Research Paper No. 100.

Shapley, Deborah. (1972) "News and comment." Science 176: 1216–20.

Sheaffer, John. R. (1961) "Flood-to-peak interval." In Gilbert F. White, ed., Papers on Flood Problems. Chicago: University of Chicago,

Department of Geography, Research Paper No. 70.

——, Davis, George W., and Richmond, Alan P. (1970) *Community Goals—Management Opportunities: An Approach to Flood Plain Management*. Chicago: University of Chicago, Center for Urban Studies. Report by Institute for Water Resources, Department of the Army, Corps of Engineers.

Sheehan, Lesley, and Hewitt, Kenneth. (1969) "A pilot survey of global natural disasters of the past twenty years." Toronto: University of Toronto, Natural Hazards Research Working Paper No. 11.

United Nations. (1969a) *Integrated River Basin Development*. New York: rev. reprinting.

——. (1969b) "Some institutional aspects of adjustments to floods." Tbilisi, USSR: Resources and Transport Division, Department of Economic and Social Affairs, United Nations Inter-regional Seminar on Flood Damage Prevention Measures and Management.

White, Gilbert F. (1945) *Human Adjustment to Floods: A Geographical Approach to the Flood Problem in the United States*. Chicago: University of Chicago, Department of Geography, Research Paper No. 29.

——. (1964) *Choice of Adjustment to Floods*. Chicago: University of Chicago, Department of Geography, Research Paper No. 93.

Wittfogel, Karl A. (1957) *Oriented Despotism: A Comparative Study of Total Power*. New Haven, Conn.: Yale University Press.

21   Flood-Hazard Mapping in Metropolitan Chicago

John R. Sheaffer, Davis W. Ellis, and Andrew M. Spieker

Introduction

The effective management of flood plains consists of more than building detention reservoirs and levees. As urban pressures are forcing more and more developments on flood plains, such devices as flood-plain regulations and flood proofing are coming into wider use. These devices, however, require information as to what areas are likely to be flooded. The need for floodplain information is further intensified by Federal legislation such as the National Flood Insurance Act of 1968

Sheaffer, J. R., Ellis, D. W., and Spieker, A. M., 1969, "Flood-Hazard Mapping in Metropolitan Chicago" *U. S. Geol. Survey Circ. 601–C,* 14 pp. Mr. Sheaffer is affiliated with the Center for Urban Studies, The University of Chicago. Davis Ellis and Andrew Spieker are on the staff of the U. S. Geological Survey.

(Title XIII, Public Law 90–448) and recent Federal policies on use of flood plains (U. S. Congress, 1966; Executive Order 11296).

The present report describes how these needs are being met in the Chicago SMSA (Standard Metropolitan Statistical Area) by a cooperative program involving the six counties of the metropolitan area—Cook, Du Page, Kane, Lake, McHenry, and Will—the Northeastern Illinois Planning Commission, the State of Illinois, and the U. S. Geological Survey. This unique flood-mapping program, in progress since 1961, has resulted in coverage of nearly the entire six-county metropolitan area by maps showing the flood hazard. Figure 1 is a map of the area showing the extent of coverage in June 1969. Quadrangles showing an HA (U. S. Geological Survey Hydrologic Investigations Atlas) number are published and available for sale at the Northeastern Illinois Planning Commission, or the U. S. Geological Survey, Washington, D. C. Quadrangles without an HA designation are in progress or are scheduled for future mapping. At present this coverage is about 85 percent complete. Metropolitan Chicago is the only large metropolitan area in the United States for which this information is so widely available.

The purpose of this report is to describe how the program originated and is being carried out, the outlook for improving this program to meet the changing needs of the rapidly urbanizing metropolitan area, and the various ways flood maps can be used by individuals and public and private institutions.

Flooding in Metropolitan Chicago

Floodflows in the rivers and waterways of Metropolitan Chicago have periodically spilled from their channels and inundated the adjacent lowlands or flood plains. The earliest recorded flood in the Chicago area occurred on March 29, 1674, when the explorer priest Marquette and his companions were driven from their camp near Damen Avenue by high water coming through Mud Lake, from the Des Plaines River. However, such overflows did not become hazards, except possibly to navigation, until development of the flood plains gave the floods something to damage.

In retrospect, it is conceivable that if adequate land-use planning, based on sound hydrologic data in conjunction with regulatory and flood-proofing measures, had guided the development of our flood plains, there would be little, if any, improper use today and no major flood problems would exist.

Flood damages have been steadily increasing as urban sprawl has engulfed many flood plains and subdivisions have been located on sites subject to flooding. The absence of accurate information on these areas subject to flooding has been a limitation on efforts to formulate a comprehensive flood damage reduction program. The need for this information is particularly acute in Metropolitan Chicago and other topographically similar regions of flat terrain and poorly developed drainage, where the flood plains are not readily perceptible to the human eye.

The Concept of Flood-Hazard Mapping

Flood-hazard mapping is a means of providing flood-plain information for planning and management programs. Such information should be designed to assist officials and private interests in making decisions and alternative plans concerning the development of specific lands subject to flooding. Proper use of flood-hazard mapping will help to:

1. Prevent improper land development in flood-plain areas.
2. Restrict uses that would be hazardous to health and welfare and which would lead to undue claims upon public agencies for remedy.
3. Encourage adequate stream channel cross-section maintenance.

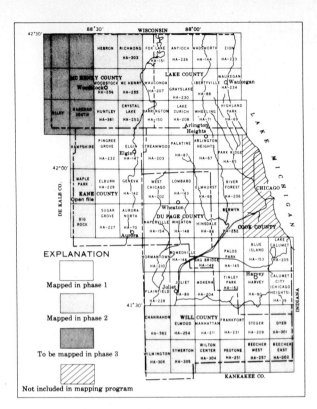

FIG. 1. The Chicago metropolitan area showing location of quadrangles included in flood-hazard mapping program.

4. Protect prospective home buyers from locating in flood-prone areas.
5. Preserve potential for natural ground-water recharge during flood events.
6. Guide the purchase of public open space.
7. Avoid water pollution resulting from the flooding of sewage treatment plants and solid waste disposal sites that were located on flood plains.

What Is a Flood-Hazard Map?

A flood-hazard map uses as its base a standard U. S. Geological Survey topographic quadrangle which includes contours that define the ground elevation at stated intervals. Each of the quadrangles covers an area 7½ minutes of longitude wide by 7½ minutes of latitude deep, or approximately 57 square miles. The scale of the flood maps is 1:24,000, or 1 inch equals 2,000 feet. The area inundated by a particular "flood of record" is superimposed in light blue on the map to designate the "flood-hazard area." Also marked on the flood-hazard map are distances (at ½-mile intervals) along and above the mouth of each stream and the locations of gaging stations, crest-stage gages, and drainage divides. Figure 2 shows part of the Elmhurst quadrangle, a typical flood map.

Profiles and Probabilities

Accompanying the flood-hazard map are explanatory texts, tables, and

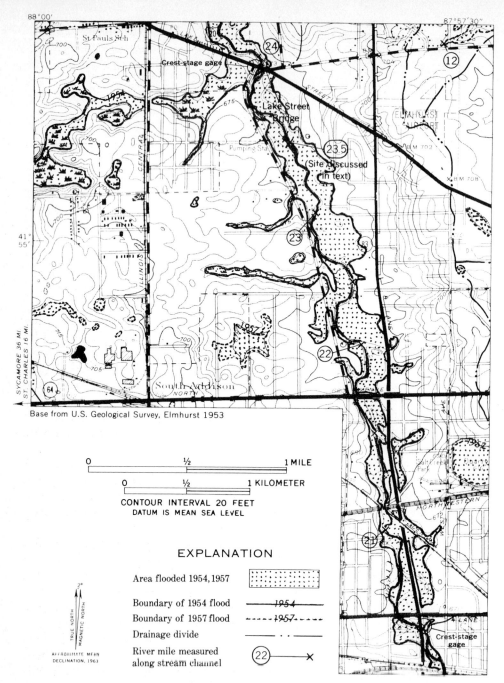

FIG. 2. Flood-hazard map of part of the Elmhurst quadrangle. (Adapted from Ellis, Allen, and Noehre (1963).)

graphs, which facilitate their use. One set of graphs shows the linear flood profiles (see Fig. 3) of the major streams in the quadrangle; from them, the user can tell how high the water rose at any given point during one or more floods.

Another valuable tool (Fig. 4) is a set of graphs showing probable frequency of flooding at selected gaging stations. These charts indicate the average interval (in years) between floods that are expected to exceed a given elevation. Frequencies can also be expressed as probabilities which make it possible to express the flood risk or "flood hazard" for a particular property; for example, a given area may have a 5-percent chance of being inundated by flood waters in each year.

How to Use a Flood-Hazard Map

To illustrate the use of a flood-hazard map, assume that you own property along Salt Creek, near Elmhurst and about half a mile south of Lake Street, Perhaps you plan to build there, and you want to know the risk of being flooded. You examine the Elmhurst quadrangle flood-hazard map (Fig. 2) and note that your property is located at a point 23.5 miles above the mouth of Salt Creek. (The river miles are shown on the map.)

One of the graphs accompanying the map is a flood-frequency curve for Salt Creek (Fig. 4). This curve, however, is for a particular point on Salt Creek—the Lake Street Bridge. To apply the flood-frequency relationship to your own property will require an adjustment for the water-surface slope between the two points. So you consult another graph, the one which shows profiles or high-water eleva-

tions, of floods along Salt Creek (Fig. 3). There, you find that at the Lake Street Bridge (river mile 24) the 1954 flood crested at 671.5 feet, while at the point you are interested in (river mile 23.5) the crest was at 671 feet.

Returning now to the flood-frequency curve, you find that the 1954 flood has an 8-year "recurrence interval," meaning that, over a long period of time, floods can be expected to reach or exceed that level on an average of once every 8 years. That level, you have already found, is 671.5 feet at Lake Street and 671 feet at your property. Another way of thinking of it is this: if you were to erect a building on your property at 671-foot elevation, the chances of a flood reaching the structure in any given year would be approximately one in eight. These are only odds—probabilities—and the actuality may be better or worse. But the odds are poorer than most property owners are willing to accept, so you will probably want to seek better odds at higher ground.

Suppose you were willing to accept a flood risk of one every 25 years: What is the ground elevation at which a building should be situated to enjoy that much security?

The flood-frequency curve indicates that, at the Lake Street Bridge, an elevation of 672.3 feet corresponds to the 25-year recurrence interval,. You now plot this elevation at river mile 24 on the flood-profile chart and draw a straight line through the point you have plotted and parallel to the 1954 flood profile. You now have the profile for a flood with a 25-year recurrence interval and it shows that the elevation reached by such a flood at your property would be 671.8 feet. Using the line

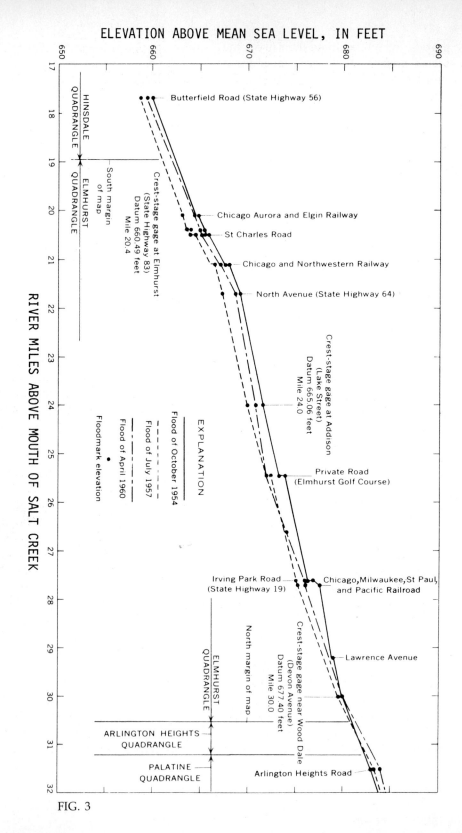

FIG. 3

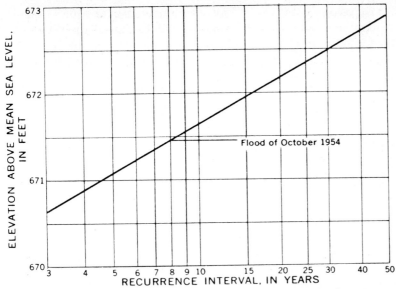

FIG. 4. Frequency of floods on Salt Creek at Addison (Lake Street).

you have drawn, you can determine corresponding elevations (for the same recurrence interval) at other points along Salt Creek. And, of course, you can use the method outlined to approximate the elevation at your property for other recurrence intervals—up to 50 years.

The Metropolitan Chicago Flood-Mapping Program

It was determined that flood-hazard mapping could meet some of the needs that had become evident in Metropolitan Chicago. However, it was also recognized that flood-hazard mapping of such a large area could not be accomplished overnight. It would require financing, time, careful planning, and data. The flood-mapping program is a cooperative effort, financed jointly by the six counties of Metropolitan Chicago, the Northeastern Illinois Planning Commission, the State of Illinois, and the U. S. Geological Survey. Funds offered by the six counties through the Planning Commission which serves in an administrative and coordinating role, are matched on a one-to-one basis with Federal funds. The actual mapping is done by personnel of the U. S. Geological Survey. In 1968 the State of Illinois entered into a separate, though similar, cooperative agreement with the U. S. Geological Survey to assist with part of the financing. The flood-mapping program was carried out in phases. The first phase extended from July 1, 1961 to June 30, 1966. Phase 2 extended from July 1, 1966 to June 30, 1969. The formulation of a phase 3 is currently (1969) being discussed among the principal agencies involved.

Phase 1

In phase 1 of the program, flood maps were prepared for 43 7½-minute quadrangles in the six-county area. One flood atlas, U. S. Geological Survey Hydrologic Investigations Atlas HA-39, "Floods in the Little Calumet River basin near Chicago Heights, Ill.," had been prepared previously by the Geological Survey as a prototype for the program. Each quadrangle is given the name of a principle city or prominent geographic feature located on the map. The location of these quadrangles is shown in Fig. 1.

The scope of phase 1 is shown in Table 1 and Fig. 1. The average cost of preparing a flood map initially was estimated as $6,250, or a cost to the local agencies of $3,125, and the initial agreements between local agencies and the Planning Commission were prepared on this basis. Early in the program, however, it was found that, for several quadrangles, particularly in Lake County, there was need for supplemental contours on the flood-plain areas. These were provided under a supplemental agreement among the appropriate agencies. Partly because of this change, and partly because of steadily rising costs throughout the 5-year period of the program, the total expenditure for phase 1, including the supplemental contours, the prepara-

tion of inundation maps for 43 quadrangles, and the installation and operation to June 30, 1966, of the initial 229 creststage gages, was $299,860, or about $6,975 per quadrangle.

Another item of possible interest to those who may plan similar programs is the expenditure of manpower. All operations were conducted from the Survey subdistrict office at Oak Park, near the geographic center of the area. A total of 37,372 direct man-hours were required to complete phase 1; this indicates an average of 869 direct man-hours per quadrangle. There was, however, considerable variation for the individual quadrangles, ranging from a maximum of 1,455 man-hours to a minimum of 520 man-hours. Man-hours required for providing the supplemental contours are not included in these figures, as this part of the work was performed under a contractual arrangement with the Topographic Division of the Geological Survey.

Because of insufficient hydrologic data in much of the area, it was necessary to establish 229 crest-stage gages to record instantaneous flood peaks so that flood profiles and flood-plain limits could be better defined along the approximately 1,000 miles of streams located in the 43 quadrangles.

Preparation for phase 2 of the flood-mapping program involved the installation of an additional 165 crest-stage gages in McHenry, Kane, and Will Counties in 1963. The installation of these gages was necessary because the hydrologic events on many of the streams in southern Will County and western Kane and McHenry Counties had never been recorded. These gages are located in 19 quadrangles which

Table 1   Scope of phase I of flood-mapping program, July 1961 to June 1966

Counties	Quadrangles mapped
Cook	[1]13
Du Page	6
Kane	6
Lake	10
McHenry	1
Will	8
Total	44

[1] Includes Calumet City quadrangle. Chicago Heights, which was prepared as a pilot project, was not included as part of program.

were scheduled for mapping during phase 2. The costs were covered by supplementary cooperative agreements with the affected counties.

A flood-hazard mapping program can lead to other related hydrologic studies. A study of the role of flood-plain information and related water resource management concepts in comprehensive land-use planning (Spieker, 1969) was made by the Geological Survey, at the request of the Planning Commission, in 1965–67. This study used the Salt Creek basin in Cook and Du Page Counties as a demonstration area to illustrate principles which govern the effects of alternative land-use practices, particularly uses of the flood plains, on the overall water resources of the area. Emphasis was placed on the interrelationship of the various components of the hydrologic system, particularly the interrelationship between surface water and ground water.

Phase 2

Phase 2 involved the preparation of 19 additional flood maps. (See Fig. 1.) In addition, the 394 crest-stage gages, including those located in areas already mapped, were kept in operation as part of phase 2 to extend the hydrologic records. The completion of phase 2 will make flood maps available for the entire metropolitan area with the exception of the western part of McHenry County and the completely urbanized area of Chicago and the close-in suburbs in Cook County. This area, which comprises four quadrangles, was not mapped because urbanization has obliterated nearly all the natural flood plains and overbank flooding is generally not a problem.

Table 2 Scope of phase 2 of flood-mapping program, July 1966 to June 1969

Counties	Quad-rangles mapped	Total number of gages	Gages having peak dis-charge data
Cook	2	70	8
Du Page		36	12
Kane	3	58	11
Lake		40	11
McHenry	6	52	8
Will	8	138	19
Total	19	394	69

The scope of phase 2 is presented in Table 2. A proportionally larger share of the local cost was allocated to McHenry and Will Counties because a major part of the work was done in those two counties. In 1967, the State of Illinois, through the Division of Waterways of the Department of Public Works and Buildings entered into a separate cooperative agreement with the Geological Survey to assume part of Will County's share of phase 2 mapping. The cost of mapping in phase 2 was $174,600, of which $86,200 was provided by local agencies and $88,400, by the Geological Survey. (The difference of $2,200 was due to supplemental allotments of Survey funds, unmatched by local funds, to partially cover interim increases in Federal salary rates.) At the completion of phase 2, the total cost of flood mapping in the metropolitan area was $474,460. (This cost includes operation of the entire network of 394 crest-stage gages to June 30, 1969.)

The Formulation of Phase 3

Providing adequate flood information in an urban area is a continuing activity. Floods wil continue to occur and will provide new and additional information. Spreading urbanization can alter both the frequency and the patterns of flooding. Paving and covering of the land tends to accelerate storm runoff and increase flood peaks. Man-made changes in the channel cross section can alter flooding patterns. Examples of such changes are bridges, culverts, fill on the flood plain, and building on the flood plain. These changes take place at a rapid pace in a fast growing area such as Metropolitan Chicago.

To keep up with these changes will require periodic revision of the flood-hazard maps. Many of the maps are based on information which is 8 years old. Additional flooding and a great deal of urbanization has taken place during these 8 years. The crest-stage gage network has provided a wealth of data to document these flood events and to help in analyzing the changes resulting from urbanization.

Phase 2 of the flood-mapping program terminated in 1969. In continuing the program into its third phase, the following four activities should be considered.

1. Continued operation of the existing network of crest-stage gages. The crest-stage gage network is believed to be the densest such network in the country. About 8 years of record will be available at the completion of phase 2 of the program. As urban development continues, the continued availability of flood-stage information will be increasingly important. Such data would be valuable in determining rates if a flood-insurance program became operational.

2. Evaluation of the crest-stage gage network for adequacy and relevance. Although the existing network is one of the most comprehensive in the country, there exists a need for its review to eliminate redundant gages and to add new ones where needed. The 8 years of record would be useful in this evaluation.

3. Extension of the program to unmapped areas. At the completion of phase 2 all of the six-county metropolitan areas except the completely urbanized central city and the western 40 percent of McHenry County will be mapped. The remainder of McHenry County is already planned for inclusion in phase 3. Before the metropolitan area expands into Kankakee and Kendall Counties, the flood-mapping program should be extended there to provide a part of the basis for orderly growth.

4. Periodic and systematic revision of the flood maps prepared in phases 1 and 2. All existing maps should be evaluated as to their adequacy and a systematic program should be planned for updating the maps where urbanization and additional flood data warrant it. This should be a continuing process. Examples of maps greatly in need of revision are the Calumet City (HA-39) and Arlington Heights (HA-67) quadrangles. In addition to mapping floods of record, consideration should be given to defining floods of given frequencies: for example, at 25–, 50–, and 100-year recurrence intervals. Even though the cost of such a mapping program would be considerably greater than that of mapping historical floods, the maps would provide a more sound and consistent basis for considering the element of

risk in planning and decision making. Profiles at the selected frequencies also should be included in future mapping. This kind of flood information would be especially useful in determining premium rates under the National Flood Insurance Act of 1968. It has been agreed,[1] for example, that the area inundated by the 100-year flood should define the regulatory area under the Flood Insurance Act.

Continuation of the cooperative flood-hazard mapping program along these lines will assure that local governmental bodies, industries, utilities, developers, and citizens of Metropolitan Chicago will have more and better flood information which can be used in furthering the region's orderly development.

The Crest-Stage Gage Network

The Northeastern Illinois Planning Commission and the U. S. Geological Survey's cooperative flood-mapping program required the establishment of a network of crest-stage gaging stations.

A crest-stage gage is a rather simple device that records the maximum elevation of floods. These gages are mounted on wingwalls or piers of highway bridges and culverts or anchored in concrete along stream banks. After the gages are mounted, levels are run from nearby benchmarks to establish datum (zero) of the gages referred to mean sea level, datum of 1929. The base of these gages is set above normal water levels so that they record only flood elevations. The sketch in Fig. 5 illustrates how the gage functions. Water enters the gage through specially designed holes at

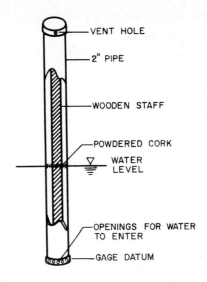

FIG. 5. Typical crest-stage gage.

the bottom of the pipe. Finely ground cork at the bottom of the gage floats on the water surface and comes in contact with the wooden staff located inside of the pipe. As the water recedes, the cork adheres to the staff and provides a record of the maximum stage of the flood.

After a flood, the crest-stage gage is serviced by opening the gage, withdrawing the staff, and measuring the distance from the base of the staff to the top of the cork line. After the measurements are made, the cork is removed from the staff, any debris that has collected is removed from the holes at the base of the gage, new cork is added, and the gage is reassembled. It is then ready to record the elevation of the next flood. By adding the depth of water recorded on the staff (gage height) to the gage datum, the elevation of the flood in feet above mean sea level is determined.

Usefulness of Flood-Hazard Maps in Urban Development

The main purpose of flood-hazard mapping, as stated previously, is to make available information which can be used to bring about the orderly and beneficial use of areas subject to flooding. A wide range of institutions and devices exists through which this information can be put to use. The following outline presents the general categories of flood-plain information use:

1. Regulation of private development:
   a. By public institutions:
      (1) Building, subdivision, and zoning regulations.
      (2) Sewer connection permits.
      (3) Public financial institutions (that is, Federal Housing Administration, Veterans Administration).
      (4) Land management and use criteria of the flood insurance program.
   b. By private institutions:
      (1) Financial institutions.
      (2) Private utilities (that is, gas, electric).

2. Purchase of property for public use:
   a. Forest preserve districts (county).
   b. Parks and recreation facilities.
   c. Municipal parking lots.

3. Development of public facilities:
   a. Highways and streets.
   b. Sewer extensions, treatment plant locations.

4. Guidelines (planning) for future development.

Following is a résumé of how the flood-hazard maps have actually been used to give direction to urban development in the Chicago metropolitan area.

Regulation of Private Development

One of the most frequently employed devices of flood-plain management is flood-plain zoning. The zoning authority is usually delegated to local governments, villages, and cities. County governments may exercise zoning powers in the unincorporated areas.

The Northeastern Illinois Planning Commission (1964) has prepared a model flood-plain zoning ordinance for the assistance of county and local governments. This model is the basis for many of the flood-plain zoning ordinances that have been adopted by Metropolitan Chicago communities.

Progress in the adoption of flood-plain zoning ordinances by county and municipal governments has been varied. As of late 1968 three of the six metropolitan counties—Cook, Du Page, and Lake—had adopted such ordinances. Kane County does not have a flood-plain zoning ordinance as such, although its zoning ordinance and subdivision regulations set forth conditions for subdivision development in flood-hazard areas. As of October 1968, 94 of the 117 Cook County municipalities located in the Metropolitan Sanitary District had adopted flood-plain zoning ordinances. There are 20 Cook County municipalities within the Sanitary District which do not have recognized flood hazards. An additional three are revising ordinances which have been rejected as unsatisfactory. Outside of

Cook County, only a few municipalities have adopted flood-plain zoning ordinances.

One of the reasons for the large number of Cook County municipalities taking action is the policy adopted in 1967 by the Metropolitan Sanitary District regarding the issuance of sewer permits. The policy states that: "No permits shall be issued by the Metropolitan Sanitary District for sewers to be constructed within a flood-hazard area, as delineated on the maps prepared by the United States Geological Survey in cooperation with the Northeastern Illinois Planning Commission, until the local municipality has adopted a flood-plain zoning ordinance which meets the approval of the Sanitary District."

"Permits in undeveloped areas will not be approved until Cook County adopts flood-plain zoning regulations."

"The ordinance shall include but not be limited to the following:

1. Restrictions on residential development.

2. Provisions for establishing permanent flood way channels through acquisition of rights-of-way, including easements for maintenance and improvements.

3. Requirements for flood proofing buildings within the flood-hazard areas. The ordinance shall be adopted before September 1, 1967."

This policy has proved highly effective in encouraging municipalities to adopt flood-plain zoning ordinances. In addition, Cook County has adopted a flood-plain ordinance which applies to all its unincorporated areas.

Financial institutions, public and private, can exert a powerful influence over the location of private urban development. Where flood-plain information is available, these institutions are generally reluctant to finance housing development in flood-hazard areas. The financing of housing in flood-prone areas is a risk that financial institutions would rather not assume, provided that there exists a knowledge of this risk. In the Chicago metropolitan area the Veterans Administration and the Federal Housing Administration routinely check the location on the U. S. Geological Survey's flood-hazard maps of new housing developments which they are considering financing. These agencies as a matter of policy will not finance developments in areas known to be subject to flooding. A large number of private financial institutions (banks, savings and loan companies) make similar use of the flood-hazard maps.

Private utility companies can influence urban development by where they choose to extend—or not to extend—gas and electric lines. By recognizing that development on flood plains is not wise, utility companies are in an excellent position to prevent their development by refusing to service them. Flood-hazard maps thus can be useful to utility companies by helping them to identify those areas where they might wish to discourage development.

Purchase of Property for Public Use

The public development of flood-hazard areas for recreation or aesthetic purposes has long been recognized as a technique of flood-plain management. Green belts, or undeveloped areas along streams, can provide breaks in the monotony of urban sprawl. The construction of municipal parking lots is another example of public use of flood plains. Identification on flood-hazard maps of those areas subject to flooding can assist public officials in acquiring these lands at a reasonable price.

The Du Page County Forest Preserve District is now engaged in a long-range program of land acquisition whose purpose is to develop a major green belt along the West Branch of the Du Page River. The U. S. Geological Survey's flood-hazard maps have been extremely useful in providing guidelines for land purchase. Also, they have been helpful in negotiations for public open space acquisition in Cook and Lake Counties.

Development of Public Facilities

Public facilities frequently lead urban development into flood-hazard areas. The State Division of Highways has made frequent use of the flood-plain maps in their highway planning process. Proper planning of access can tend to discourage improper flood-plain development.

The location of sewage treatment plants and sanitary landfills is also influenced by flood-plain information. An example of such use is found in "Rules and Regulations for Refuse Disposal Sites and Facilities" (Illinois Department of Public Health, 1966, p. 1). This document states that: "sites subject to flooding should be avoided . . . or protected by impervious dikes and pumping facilities provided." Thus, flood-plain information becomes involved in all decisions regarding the establishment of disposal sites and facilities and is cited by the State Geological Survey in their site-evaluation reports.

Guidelines (Planning) for Future Development

Planning, or the formulation of guidelines for future development, provides the overall framework in which the previously discussed uses of flood-hazard maps are implemented. It is in the planning process that the broad, long-range goals and objectives are set out. These objectives can be attained by alternate tactics.

The importance of wise management of the flood plains has been recognized by the Regional Planning Agency for Metropolitan Chicago, Northeastern Illinois Planning Commission, almost from its inception. Examples of how the flood-hazard maps are influencing long-range planning can be found in the two following policy statements taken from the Northeastern Illinois Planning Commission's comprehensive general plan for Metropolitan Chicago (Northeastern Illinois Planning Commission, 1968, p. 7): "Lands unsuited for intensive development due to flooding, unstable soil conditions, or where the provision of essential public services and facilities is difficult, should be maintained in suitable open space use." And on page 11: "Intensive urban development should be directed so as to avoid flood plains, protect ground water deposits, and preserve lands particularly suited for multi-purpose resources management programs."

Notes

[1] Consensus of Seminar on Flood Plain Management held at the Center for Urban Studies, University of Chicago, December 16–18, 1968, at the request of the U. S. Department of Housing and Urban Development.

References

Bue, C. D., 1967, Flood information for flood-plain planning: U. S. Geological Survey Circular 539, 10 p.

Ellis, D. W., Allen, H. E., and Noehre, A. W., 1963, Floods in Elmhurst quadrangle, Illinois: U. S. Geol. Survey Hydrol. Inv. Atlas HA-68.

Illinois Department of Public Health, 1966, Rules and Regulations for Refuse Disposal Sites and Facilities: Illinois Dept. Public Health, Springfield, Ill., 7 p.

Mitchell, W. D., 1964, Some problems in flood mapping in Illinois: Natl. Acad. Sci.—Natl. Research Council Highway Research Board, Highway Research Rec. 58, p. 42–43.

Northeastern Illinois Planning Commission, 1964, Suggested flood damage prevention ordinance with commentary: Northeastern Illinois Planning Comm., Chicago, Ill., 28 p.

—— 1968, A regional armature for the future: The comprehensive general plan for the development of the northeastern Illinois counties: Northeastern Illinois Planning Comm., Chicago, Ill., 12 p.

Sheaffer, J. R., 1964, The use of flood maps in northeastern Illinois: Natl. Acad. Sci.—Natl. Research Council Highway Research Board, Highway Research Rec. 58, p. 44–46.

Sheaffer, J. R., Zeizel, A. J., and others, 1966, The water resources in northeastern Illinois—Planning its use: Northeastern Illinois Planning Comm. Tech. Rept. 4, 182 p.

Spieker, A. M., 1969, Water in metropolitan area planning: U. S. Geological Survey Water-Supply Paper 2002. (In press.)

U. S. Congress, 1965, A unified national program for managing flood losses: U. S. 89th Cong., 2d sess., House Doc. 465, 47 p.

22 Flood-Plain Management Must Be Ecologically and Economically Sound

James E. Goddard

Over the past 35 years in the U. S., some $8 billion has been spent for dams, levees and channelization to limit flood losses. Yet over that same 35 years, losses due to floods have

Reprinted from *Civil Engineering*, v. 41, pp. 81–85, by permission of the author and the American Society of Civil Engineers. Copyright 1971 ASCE.

Mr. Goddard was on the staff of the Tennessee Valley Authority and the Corps of Engineers. He is currently a consulting engineer in Tucson, Arizona.

risen, as shown in the reports "Changes in Urban Occupance of Flood Plains," by Gilbert F. White & Associates, Univ. of Chicago, 1958; "Types of Agricultural Occupance of Flood Plains," by Ian Burton, Univ. of Chicago, 1962; and House Document 465, "A Unified National Program for Managing Flood Losses," 1966. The average annual potential flood losses are now well over $1 billion.

Such data as these led to the conclusion that flood-control works are not the entire answer. Better agricultural practices and land control over development in flood plains, to hold the rainwater and snowmelt on the land longer and to wisely limit occupation and investment in flood plains must be part of the answer, though difficult to achieve.

In and near the cities, especially, a judicious blend of flood-control measures—and regulatory measures—is required. A few cities are finding that their work in tackling flood problems has had broader results, with the effort leading to redevelopment of the entire town.

Following are case histories of flood-damage prevention illustrating positive ecological and environmental effects.

Channelization

In "Crisis on our Rivers," in *Reader's Digest*, December 1970, James Nathan Miller charged that federal agencies are blindly channelizing rivers and streams in order to foster development and stop flooding, but that for ecological and other reasons it would be better to forbid channelization and stop flood-plain development. There may be cases where Mr. Miller's advice should be followed; however, in many instances development either cannot be stopped or need not and there will be little ecological price to pay. Following are examples to support this statement.

Bear Creek—Alabama and Mississippi
Bear Creek winds through fertile bottomlands, from northwest Alabama into northeast Mississippi before it empties into Pickwick Lake on the Tennessee River. Trees crowd its banks and, in places, almost choke off its normal flow. Most of the time, stream and neighbors remain at peace. But several times each year, heavy rains swell the flow and a smoldering controversy is resumed.

It happens that farmers and owners of flood-prone bottomland propose channelization, for it would mean relief from flooding of their lands. But to the sportsmen and fish and wildlife biologists, channelization would mean uprooting the creek's complex system of plant and animal life which nature has established over many years. They cite studies which indicate reductions of up to 90 percent in both the weight and number of game fish per acre following stream channelization.

A solution has been found which largely satisfies both interests. Floodways are being built by the Tennessee Valley Authority in areas where channelization is required to provide flood protection to some 17,600 acres of potentially productive land. These wide and shallow floodways cut diagonally across the meandering river channel, at a higher level than the existing streambed. Low flows will continue to be handled by the old streambed, with its ecology largely undisturbed. But during floods, the surplus flow will be handled by the shallower but larger-capacity floodways. (See Fig. 1.) When not in use these grassed flood-ways will be dry and hardly noticeable as their broad, flat bottoms and gently sloping sides provide pasture and walkways for livestock.

A similar application is the separate, auxiliary floodway built by the Corps of Engineers to bypass floods on the

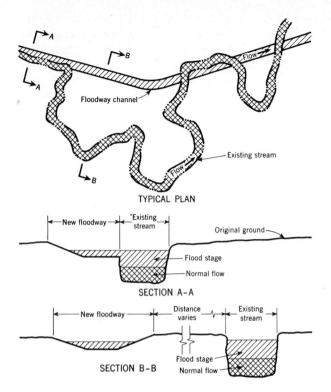

FIG. 1. In Bear Creek, both ecology and development have been served. Most of the time, all flow is through the deeper but constricted natural channel, thus preserving the river ecology. During floods, the overflow passes through the new channel, thereby avoiding flooding of adjacent lands and making them developable.

TYPICAL PLAN

Floodway channel

Existing stream

Flow

New floodway — Existing stream — Original ground

Flood stage

Normal flow

SECTION A–A

New floodway — Distance varies — Existing stream

Flood stage

Normal flow

SECTION B–B

Arkansas River around Wichita, Kansas, and thus preserve the waterfront through the city.

Coastal Marsh near Houston

On the low coastal lands east of Houston, about 75 percent of the 36,000-acre East Bay Bayou watershed was in a rice-then-pasture rotation. Rice production requires intricate irrigation and water-disposal systems. July and August are not only the high-rainfall months, but also rice-harvest time when floods can ruin a crop. Drainage is into the narrow, shallow East Bayou (see Fig. 2) that twisted through the marsh to the coast. During heavy rains the stream overflowed, dumping excessive fresh water into the marsh.

However, the brackish water, with salinity one-fourth to one-half that of sea water due to occasional high tidewaters lapping over a high rim near the coast, is excellent habitat for muskrats, ducks and geese. It also supports some of the best livestock grazing to be found. But during years of heavy rainfall, the excessive fresh water often diluted the brackish water. This damaged wildlife habitat, dumped sediment into the marsh, and reduced livestock forage from plants thriving under brackish conditions.

Watershed planners and biologists agreed on a plan to channelize East Bay Bayou and add spoil banks along both sides. To permit regulating the amount of fresh or saline water going into or out of the marsh, and thus controlling brackishness and flood

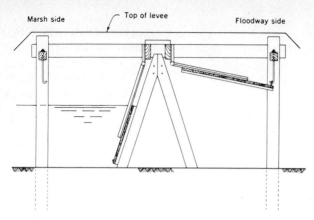

FIG. 2. East Bay Bayou floodway, near Houston; ten of these semi-automatic, two-way gates in spoil banks on either side of floodway maintain proper levels and mixtures of brackish-fresh water to benefit both agriculture and wildlife. When the right-hand gate is lowered, water will flow to the right, but not to the left.

heights, 10 semi-automatic two-way gates and 16 other flap gates were installed in the spoil banks (see Fig. 2).

As a result of this regulation, duck and geese hunting flourishes with 134,000 man-days annually in the region; muskrat trapping provides more than 3,500 pelts; commercial fishing in the bayou exceeds 20,000 lb annually; range grasses in the marsh have been improved and flood damages to the rice crop have been sharply reduced, nearly doubling the income. This was a cooperative effort of the U. S. Agriculture Department's Soil Conservation Service, with the U. S. Bureau of Sport Fisheries and Wildlife, and the Texas Parks and Wildlife Department.

Danger if Flood Problem Is not Tackled

It is unfortunate that many cities and counties are permitting bad floodplain practices to continue.

This seems to be happening in the Upper Mill Creek Watershed that lies 10 miles north of Cincinnati in Ohio, and within the influence of that major city. Of the 50-sq-mi watershed, about three sq mi are subject to one to three floods per year.

Despite the flood hazard, farming continues and industrial development is accelerating. Private levees built to protect private lands restrict flows because of their haphazard location. When one looks at the situation as a watershed-wide problem, they are a bad partial solution. Cost of industrial flood damages will triple from $200,000 to $678,000 annually in a few years if nothing is done.

A plan to alleviate the situation, as prepared by the Soil Conservation Service, included four flood-retarding and one multipurpose structure and some channel improvement—but there have been delays in implementation. The proposed project would cost $4.6 million, with average annual benefits of $0.9 million, and average annual costs of $0.2 million, for a benefit-cost ratio of about 4 to 1.

Landowners and governments are aware of the danger of inaction. Local sponsors know that delays will bring increased costs, and thus threaten the economic feasibility of the program, because of increasing encroachments of new construction in the flood plain, of housing and utility lines and industrial plants. This construction should not be permitted because of the negative effect on the economy and ecology. Three of five sites proposed for flood-control structures have already been seriously affected by the developments.

Unfortunately, this watershed is just one of hundreds of such cases in which unwise practices are continuing. Engineers must increase their efforts in arousing greater public understanding that will spark appropriate action.

Sometimes, Relocation Is the Answer

Prairie du Chien, a town of about 6,000, is on the Mississippi River in Wisconsin. The river valley at this point is some 8,000 ft wide between high cliffs, and the Mississippi normally is confined to two channels, each 1,000 ft wide. The town is in the valley, and is occasionally flooded, usually by the combination of spring snow melt and spring rains. (See Fig. 3, showing flood plain.)

Thorough investigation determined that levees, walls and upstream reservoirs were neither technically sound nor economically justified. The most feasible solution was found to be evacuation and flood-plain regulations. Such a plan has been recommended by the Board of Engineers for Rivers and Harbors, and by the Chief of Engineers (Corps of Engineers).

The proposed solution provides for local ownership of project lands and for federal flowage easements. Implementation of flood-plain regulations and other land management is to be by nonfederal governments. About

FIG. 3. Map of Prairie du Chien, Wi., showing flood plain. It is proposed that the area shown be evacuated and that 205 buildings be removed from the flood plain. Most would be relocated to higher ground.

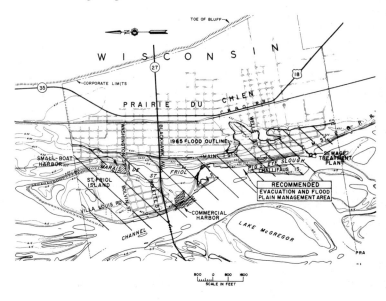

48 buildings would be purchased and torn down, 157 relocated, 33 raised above 100-year flood level, and seven flood-proofed. Relocation sites would be graded and landscaped and utilities installed. Sites of razed buildings would be cleared of debris, filled, and graded as necessary to make them safe and attractive. Historical sites (this is an early fur-trading post) and business-complementing open-space use can remain, if they are floodproofed. City and state interests plan to develop the evacuated flood plain for recreation and tourism uses, during periods of other than flood flow, as demand for those uses warrants.[1]

Flood Proofing of Buildings in the Flood Plain

Sometimes it is not feasible or not desirable to move buildings out of the flood plain. And there are often major advantages or needs for new buildings in flood plains. An alternative is to make them water-tight to above flood level.

A good example is found in St. Bernard Parish (county), east of New Orleans, La. Most of the land is low marshland, typically about 1.5 ft. above sea level. The lowlands extend inland for varying distances of about 20 to 50 miles. These Louisiana coastal areas have experienced many severe hurricanes and lesser storms. Perhaps even more damaging than the winds are the hurricane-induced flood waters, which push inland from the Gulf. Thus flood protection is required.

There are flood levees in portions of St. Bernard Parish, but extending them around additional areas is not economically justified. A recent Corps of Engineers study indicated that the most practical approach would be to flood-proof structures, including constructing buildings with their lowest floor at least 12 ft above mean sea level (see Fig. 4). An alternative, if a building is not so constructed, would be early evacuation to protected areas.

Land Filling

Land can be filled and buildings constructed thereon.

An example is found in Waterloo, Iowa (pop. 74,000) on the Cedar River. Some 4,200 acres are in the flood plain; this land holds most of the city's commercial development, 46 industrial plants, more than 5,000 homes, water and sewage plants, municipal buildings, and many schools, churches, etc.

Fill available from river-channelization dredging will be placed behind a new levee-road, and this filled land will be available for flood-safe construction. Behind about two miles of the berm, some 1,500 acres of filled land will be available.

Flood-Proofing

Individual buildings, both new and existing, can be flood proofed. Details are contained in the report, "Introduction to Flood Proofing," by John R. Schaeffer, Center for Urban Studies, University of Chicago, 1967.

Other Management (non-structural) Solutions

By habit, training, and historic precedent, civil engineers and officials tend to think first of structural solutions to a problem. In the case of flood problems, there are important non-structural alternatives that must also be

FIG. 4. New Sabastian Roy School replaced one destroyed by Hurricane Betsy south of New Orleans, La. It was flood proofed by elevating the lowest floor 10 ft to level of 100-year flood. The open ground level, paved, will be used for recreation. (Photo courtesy U.S. Corps of Engineers.)

considered. Generally, the optimum solution will be a combination of several approaches, both structural and non-structural. Three of the many non-structural techniques or management tools are flood-plain regulations, flood insurance, and flood forecasting.

Flood-Plain Regulations

One of the outstanding programs, which by its very nature provides and insures opportunities for ecological and environmental-quality enhancement and preservation is that of flood-plain regulations. In this type of program, a community or state uses its police powers to guide and control use and development of flood-plain lands.

Generally zoning ordinances, subdivision regulations, building codes, and similar ordinances are enacted which will: (1) provide for open areas or floodways which will permit floods to pass without so constricting the flow as to unduly raise the flood level and thus worsen the problem, and (2) require floors of buildings in the remainder of the flood plain (outside the open floodway) to be higher than selected flood elevations, or that buildings be flood-proofed to this elevation.

The benefits of such regulations are seen in the case of Lewisburg, Tn. Because it had not adopted flood-plain regulations, the construction of some 20 houses was halted (the city wanted to wait to see the regulations before issuing a building permit). Had these homes been built, they would have been under 6 to 10 ft of water in the flood that occurred four months later.

Several states have active, effective statewide programs of this type, requiring regulation, including Nebraska, Connecticut, Iowa, Minnesota, and Wisconsin. Wisconsin also has a shorelands program to preserve nature and to guide development along the shores of its many lakes and streams. (See "Wisconsin's Shoreland Management Program," December 1970 issue, p. 80.) California, Washington, Oregon and New Jersey also have programs designed to lead to wise use. Tennessee, North Carolina, and others have still different approaches. But many of these programs need to be strengthened and the remaining two-thirds of the states need to act. One acre in 16 in the U. S. is in a flood plain, and most states have not done enough.

Over the past 15 years, about 360 communities throughout 35 states have adopted regulations and 500 additional places in 41 states have them in adoption process. This is heartening but this is a discouragingly small fraction of the 5,200 significant-size towns and cities wholly or partially in flood plains.[2]

Help is available to the cities and states. Reports containing hydrologic and hydraulic data outlining the local flood situation are prepared by the Corps of Engineers, TVA, and others. These are the basis for designing the needed regulations.

Flood Insurance

A Federal Flood Insurance Program, authorized by Congress in 1968, for the first time makes flood damage insurance available at premiums which large numbers of owners are willing to pay. Already nearly 100,000 policies have been written.

As intended, the federal law is proving to be a spur to responsible development (or non-development, if indicated) of the flood plain. For example, availability of the insurance has led scores of cities and counties, such as Alexandria, Va. and Santa Cruz County, Az., to adopt their first flood-plain regulations. More than 500 others, from Fairbanks, Al. to Maricopa County, Az. to Palm Beach, Fl., are in the process of adopting regulations. If the cities and counties do not pass such regulations, their landowners cannot continue to get the insurance.

Flood Forecasting

The forecasting of flood stages, and then temporary evacuation, permit continued occupation of innumerable areas where other measures that may upset ecology have not been taken.

In recent years the U. S. Weather Service has improved its techniques, more of the data required for reliable forecasting are available, organization for implementing disaster and relief plans are improved, and public awareness is greater.

A good example of benefits to be derived from flood forecasting was Operation Foresight of the Corps of Engineers, Associated General Contractors, and others. This was in the spring of 1970 for the upper Mississippi River Valley when record flood stages hit cities on the Mississippi, Red, and Souris rivers causing more than $100 million in damage. Relying on accurate stage forecasts, levees were raised by earthmoving equipment provided at cost by regional contractors, other remedial structural action was taken, many structures flood-proofed, and some areas evacuated in an effort estimated to have saved $190 million in flood damages (CE, June 1969, p. 99).

Conclusion

Today "ecology" is a popular watch-word in the mass media and the federal government. "Development" is currently out of favor in much of the land. Clearly, in the past too little attention has been paid to conservation and ecology. Some construction projects were undertaken which should not have been, or which should have been modified. But today, to some engineers the danger seems to be that some construction projects are being held up or cancelled, despite their being ecologically acceptable on balance; they are held up because there is political capital to be gained in so doing, or because politically there is no choice but to kill the projects.

Stopping all river channelization, or all flood-plain and water-resource development, is not the answer. In some individual cases, yes; in others, no. Needed today is a team effort by development interests and conservation interests.

Human beings also are part of the environment. America has never needed the decent middle, honest center ground of discussion and compromise as it does now. Hopefully, knowing that man cannot live by ecology alone, public-spirited members of environmental groups will curb harmful pursuits of some of their leaders in the interest of more constructive approaches. And, hopefully, the engineering profession will take the steps necessary to be the effective catalyst.

Today all federal agencies have the same worthy objectives. The differences and misunderstandings rise from the methods of reaching these goals. Through close cooperation of all parties and better public understanding of both sides of the issues, channelization, flood-plain regulations, and other flood-plain management measures can be used, but wisely.

The federal government is moving in the right direction: the federal Water Resources Council was established in 1965 to coordinate the actions of federal agencies; a Presidential Task Force report of 1966 on managing flood losses is gradually being implemented; and in 1970 Congress instructed the Secretary of the Army (Corps of Engineers) to prepare and promulgate by 1972 appropriate guidelines for an ecological impact assessment and statement with each federal project.[3]

But the goals are not yet reached. Today developers are asked to fully consider alternative ways to reach goals; this costs more money, manpower, and time than have been available in the past. Government must meet this need without further delay.

Public understanding is needed. In public discussions by the media and government officials and representatives, reason should replace emotion, and both sides of a question should be aired, not just the conservation or the construction side. Engineers and ecologists and environmentalists must work more closely together as partners, and must develop evaluation measures in addition to economic (the dollar is not the only measure of well being).

Notes

[1] The floodplain management plan for Prairie du Chien was authorized by Congress in the Water Resources Development Act of 1974. Construction funding was authorized in October 1978, and the project is scheduled to be completed by September 1983. As of September 1981, a floodplain regulation was in effect and floodplain insurance was available. 130 buildings were to be relocated or torn down and 175 were identified for optional flood-proofing.

[2] As of September 30, 1981, a total of 10,139 communities were participating in the Federal Emergency Management Agency's Emergency Phase of the National Flood Insurance Program. An additional 6956 communities adopted and were enforcing more comprehensive floodplain management measures which enabled them to participate in the Agency's Regular Program. All of these communities are eligible for insurance coverage at subsidized rates.

[3] Environmental Impact Statements are required by Section 102 (2)(c) of the National Environmental Policy Act of 1969 (Public Law 91-190) and prepared according to Council on Environmental Quality Guidelines for Statements on Proposed Federal Actions Affecting the Environment, dated August 1, 1973. The requirement applies to all planning, design, construction, management, and regulation of civil works activities of the Corps of Engineers. [Ed.]

Supplementary Readings

General

Bolt, B., Horn, W., Macdonald, G., and Scott, R., 1975, *Geological Hazards*, Springer-Verlag, New York.

Burton, I., Kates, R., and White, G., 1978, *The Environment as Hazard*, Oxford Univ. Press, New York.

Hays, W., ed., 1981, Facing Geologic and Hydrologic Hazards—Earth-Science Considerations, *U.S. Geological Survey Prof. Paper 1240-B*, 109 pp.

Leet, L. D., 1948, *Causes of Catastrophe*, McGraw-Hill, New York.

Olson, R. A., and Wallace, M. M., 1969, *Geologic Hazards and Public Problems, Conference Proceedings* (May 27–28, 1969), U.S. Office of Emergency Preparedness.

Radbruch-Hall, D., 1979, Environmental Aspects of Engineering Geological Mapping in the United States, *Bull. of International Assoc. of Engineering Geologists*, n. 19, pp. 351–58.

Robinson, G., and Spieker, A., eds., 1978, Nature To Be Commanded, *U.S. Geological Survey Prof. Paper 950*, 95 pp.

White, G. F., 1974, *Natural Hazards: Local, National, Global*, Oxford Univ. Press, New York.

Volcanism

Bullard, F., 1976, *Volcanoes of the Earth*, Univ. of Texas Press, Austin, 576 pp.

Crandell, D., and Mullineaux, D., 1978, Potential Hazards from Future Eruptions of Mount St. Helens Volcano, Washington, *U.S. Geological Survey Bull. 1383-C*, 26 pp.

Crandell, Dwight R., 1971, Postglacial Lahars from Mount Rainier Volcano, Washington, *U.S. Geological Survey Prof. Paper No. 677*.

Crandell, Dwight R., and Mullineaux, Donal R., 1967, Volcanic Hazards at Mount Rainier, *U.S. Geological Survey Bull. 1238*, 26 pp.

Decker, R., 1978, State of the Art in Volcano Forecasting, in *Geophysical Predictions*, National Academy of Sciences, Washington, D.C., pp. 47–57.

Eaton, J. P., Richter, D. H., and Ault, W. V., 1961, The Tsunami of May 23, 1960, on the Island of Hawaii, *Seismological Soc. of Amer. Bull.* v. 51, pp. 135–57.

Gorshkov, G., 1971, Prediction of Volcanic Eruptions and Seismic Methods for Locations of Magma Chambers—a review, *Bull. Volcanologique*, v. 35, pp. 198–211.

Hyde, J., and Crandell, D., 1978, Postglacial Volcanic Deposits at Mount Baker, Washington, and Potential Hazards from Future Eruptions, *U.S. Geological Survey Prof. Paper 1022-C*, 17 pp.

Lansford, H., 1981, Vulcan's Chimneys: Subduction-zone Volcanism, *Mosaic*, v. 12, n. 2, pp. 46–53.

Lear, J., 1966, The Volcano That Shaped the Western World, *Saturday Review*, v. 49 (Nov. 5, 1966), pp. 57–66.

MacDonald, G., 1962, The 1952 and 1960 Eruptions of Kilauea Volcano, Hawaii, and the Construction of Walls to Restrict the Spread of the Lava Flows, *Bull. Volcanologique*, v. 24, pp. 248–94.

MacDonald, G. A., 1972, *Volcanoes*, Prentice-Hall, Englewood Cliffs, N.J.

Maiuri, A., Bianchi, P. V., and Battaglia, L. E., 1961, Last Moments of the Pompeians, *Natl. Geog.*, v. 120, pp. 651–69.

Mason, A. C., and Foster, H. L., 1953, Diversion of Lava Flows at Oshima, Japan, *Amer. Jour. Sci.*, v. 251, pp. 249–58.

Moore, J., Nakamura, K., and Alcaray, A., 1966, The 1965 Eruption of Taal Volcano, *Science*, v. 151, pp. 955–60.

Richter, D. H., et al., 1970, Chronological Narrative of the 1959–60 Eruption of Kilauea Volcano, Hawaii, *U.S. Geological Survey Prof. Paper 537-E*.

Rittman, A., 1962, *Volcanoes and Their Activity*, John Wiley & Sons, New York, 305 pp.

Sheets, P., and Grayson, D., 1979, *Volcanic Activity and Human Ecology*, Academic Press, New York, 672 pp.

Wexler, Harry, 1952, Volcanoes and World Climates, *Sci. Amer.*, v. 186, n. 4, pp. 74–80.

Williams, Howell, 1951, Volcanoes, *Scientific Amer.*, v. 185, n. 5, pp. 45–53.

Earthquakes and Tectonic Movements

Algermissen, S., 1969, Seismic-risk Studies in the United States, *Proc. 4th World Conf. on Earthquake Engineering*, Santiago, Chile, pp. 19–27.

Allen, C., 1975, Geological Criteria for Evaluating Seismicity, *Geological Soc. Amer. Bull.*, v. 86, pp. 1041–57.

Anderson, D. L., 1971, The San Andreas Fault, *Scientific Amer.*, v. 225, n. 5, pp. 52–67.

Bernstein, J., 1954, Tsunamis, *Sci. Amer.*, v. 191, n. 2, pp. 60–63.

Blair, M., and Spangle, W., 1979, Seismic Safety and Land-Use Planning—Selected Examples from California, *U.S. Geological Survey Prof. Paper 941-B*, 82 pp.

Bolt, B., 1978, *Earthquakes, A Primer*, W. H. Freeman and Co., San Francisco, 241 pp.

Carder, D., 1970, Reservoir Loading and Local Earthquakes, in, *Engineering Geology Case Histories No. 8*, Adams, W., ed., Geological Soc. Amer., pp. 51–61.

Dickinson, W. R., and Grantz, A., eds., 1968, Proceedings of Conference on Geologic Problems of San Andreas Fault System, *Stanford Univ. Pubs. Geol. Sci.*, v. 11, pp. 70–82.

Eckel, E., 1970, The Alaska Earthquake March 27, 1964: Lessons and Conclusions, *U.S. Geological Survey Prof. Paper 546*, 57 pp.

Fairbridge, R., 1958, Dating the Latest Movements of the Quaternary Sea Level, *Trans. N.Y. Acad. Sci., Ser. II.*, v. 20, n. 6, pp. 471–82.

Fairbridge, R., 1960, The Changing Level of the Sea, *Sci. Amer.*, v. 202, n. 5, pp. 70–79.

Flint, R. F., 1971, *Glacial and Quaternary Geology* (chap. 12, Fluctuation of Sea Level, and chap. 13, Glacial-Isostatic Deformation), John Wiley & Sons, New York, 892 pp.

Fuller, M. L., 1914, The New Madrid Earthquake, *U.S. Geological Survey Bull. 494*.

Guttenburg, B., 1941, Changes in Sea Level, Postglacial Uplift and Mobility of the Earth's Interior, *Bull. Geol. Soc. Amer.*, v. 52, n. 5, pp. 721–22.

Hagiwara, T., and Rikitake, T., 1967, Japanese Program on Earthquake Prediction and Control, *Science*, v. 166, no. 3912, pp. 1467–74.

Hammond, A. L., 1973, Earthquake Prediction (II): Prototype Instrumental Networks, *Science*, v. 180, pp. 940–41.

Hauf, H., 1968, Minimizing Earthquake Hazards II, Architectural Factors, *Amer. Inst. Archt. Jour.*, v. 19, pp. 65–72.

Hays, W., 1980, Procedures for Estimating Earthquake Ground Motions, *U.S. Geological Survey Prof. Paper 1114*, 77 pp.

Hess, H. H., 1946, Drowned Ancient Islands of the Pacific Basin, *Amer. Jour. Sci.*, v. 244, pp. 772–91.

Hodgson, J. G., 1964, *Earthquakes and Earth Structure*, Prentice-Hall, Englewood Cliffs, N.J., 166 pp.

Kisslinger, C., 1974, Earthquake Prediction, *Physics Today*, v. 27, pp. 36–42.

Lomnitz, C., 1970, Casualties and Behavior of Populations During Earthquakes, *Bull. Seismological Soc. Amer.*, v. 60, pp. 1309–13.

Lomnitz, Cinna, 1973, Global Tectonics and Earthquake Risk, *Developments in Geotectonics*, v. 5, Elsevier, Amsterdam, 334 pp.

Lyons, J., and Snellenburg, J., 1971, Dating Faults, *Geol. Soc. Amer. Bull.*, v. 82, pp. 1749–52.

National Academy Sciences, 1971, *The San Fernando Earthquake of February 9, 1971: Lessons Learned from a Moderate Earthquake on the Fringe of a Densely Populated Region*, Washington, D.C.

Pakiser, L., Eaton, J., Healy, J., and Raleigh, C., 1969, Earthquake Prediction and Control, *Science*, v. 166, pp. 1467–74.

Press, F., 1975, Earthquake Prediction, *Sci. Amer.*, v. 232, pp. 14–23.

Press, F., and Brace, W. F., 1966, Earthquake Prediction, *Science*, v. 152, n. 3729, pp. 1575–84.

Radbruch, D., et al., 1966, Tectonic Creep in the Hayward Fault Zone California, *U.S. Geological Survey Circular 525*.

Raleigh, C., Healy, J., and Bredehoeft, J., 1976, An Experiment in Earthquake Control at Rangely, Colorado, *Science*, v. 191, pp. 1230–37.

Reasenberg, P., 1978, Unusual Animal Behavior Before Earthquakes, *Earthquake Info. Bull.*, v. 10, n. 2, pp. 42–50.

Reid, H. F., 1914, The Lisbon Earthquake of November 1, 1755, *Seismol. Soc. of Amer. Bull.*, v. 4, pp. 53–80.

Rogers, T. H., 1969, A Trip to an Active Fault in the City of Hollister, *Mineral Information Service*, v. 22, n. 10, pp. 159–64.

Russell, R. J., 1957, Instability of Sea Level, *Amer. Scientist*, v. 45, pp. 414–30.

Ryall, A., Slemmons, D. B., and Gedney, L. D., 1966, Seismicity, Tectonism, and Surface Faulting in the Western United States During Historic Time, *Seismol. Soc. Amer. Bull.*, v. 61, n. 12, pt. 2, pp. 1529–30.

Scholz, C. H., Sykes, L. R., and Aggarwal, Y. P., 1973, Earthquake Prediction: A Physical Basis, *Science*, v. 181, pp. 803–10.

Steinbrugge, K. V., 1968, *Earthquake Hazard in the San Francisco Bay Area: A Continuing Problem in Public Policy*, Univ. of California Press, Berkeley.

Wallace, R. E., 1974, Goals, Strategy, and Tasks for the Earthquake Hazard Reduction Program, *U.S. Geological Survey Circular 701*, 26 pp.

Wesson, R., Helley, E., Lajoie, K., and Wentworth, C., 1975, Faults and Future Earthquakes, in Borcherdt, R., ed., Studies for Seismic Zonation in the San Francisco Bay Region, *U.S. Geological Survey Prof. Paper 941-A*, pp. A5–A30.

Whitcomb, J. H., Garmany, J. D., and Anderson, D. L., 1973, Earthquake Prediction: Variation of Seismic Velocities before the San Francisco Earthquake, *Science*, v. 180, pp. 632–35.

Mass Movement

Arora, H. S., and Scott, J. B., 1974, Chemical Stabilization of Landslides by Ion Exchange, *California Geology*, v. 27, pp. 99–107.

Black, R. F., 1954, Permafrost—A Review, *Geol. Soc. Amer. Bull.*, v. 65, pp. 839–56.

Campbell, R., 1975, Soil Slips, Debris Flows, and Rainstorms in the Santa Monica Mountains and Vicinity, Southern California, *U.S. Geological Survey Prof. Paper 855*, 51 pp.

Davies, W., Bailey, J., and Kelly, D., 1972, West Virginia's Buffalo Creek Flood: A Study of the Hydrology and Engineering Geology, *U.S. Geological Survey Circular 667*, 32 pp.

Eckel, E. B., ed., 1958, Landslide and Engineering Practice: Washington, D.C., *Highway Research Board Spec. Rept. 29, NAS-NRC 544*, 232 pp.

Fleming, R., and Taylor, F., 1980, Estimating the Cost of Landslide Damage in the United States, *U.S. Geological Survey Circular 832*, 21 pp.

Hansen, W., 1966, Effects of the Earthquake of March 27, 1964, at Anchorage, Alaska, *U.S. Geological Survey Prof. Paper 542-A*, 68 pp.

Kerr, Paul F., 1963, Quick Clay, *Sci. Amer.*, v. 209, n. 5, pp. 132–42.

Krohn, J., and Slosson, J., 1976, Landslide Potential in the United States, *California Geology*, v. 29, n. 10, pp. 224–31.

Leighton, F., 1972, Origin and Control of Landslides in the Urban Environment of California, *Proceedings 24th Session, International Geological Congress, section 13*, pp. 89–96.

Leighton, F., 1976, Urban Landslides: Targets for Land-Use Planning in California, in

Urban Geomorphology, Coates, D., ed., *Geol. Soc, Amer. Spec. Paper 174,* pp. 37–60.

McDowell, B., and Fletcher, J., 1962, Avalanche! 3,500 Peruvians Perish in Seven Minutes, *Natl. Geog.,* v. 121, pp. 855–80.

Schuster, R., and Krizek, R., eds., 1978, Landslides, Analysis and Control, *Natl. Research Council, Transportation Research Board Special Report 176,* pp. 1–10.

Sharpe, C. F. S., 1938, *Landslides and Related Phenomena,* Columbia Univ. Press, New York.

Shreve, R. L., 1968, The Blackhawk Landslide; *Geol. Soc. Amer., Spec. Paper No. 108.*

Terzaghi, K., 1950, Mechanism of Landslides, *Geol. Soc. Amer., Berkey Volume,* pp. 83–123.

Tourtelot, H., 1974, Geologic Origin and Distribution of Swelling Clays, *Assoc. of Engineering Geologists Bull.,* v. 11, n. 4, pp. 259–75.

Varnes, D. J., 1958, Landslide Types and Processes, chap. 3 in *Landslides and Engineering Practice, Highway Research Board, Spec. Rept. 29.*

Zaruba, A., and Mencl, V., 1968, *Landslides and Their Control,* Elsevier, Amsterdam, 205 pp.

Erosion and Sedimentation

American Society of Agricultural Engineers, 1977, Soil Erosion and Sedimentation, *Amer. Soc. Agr. Eng., Pub. No. 4-77,* St. Joseph, Mich., 151 pp.

Carter, L., 1977, Soil Erosion: The Problem Exists Despite the Billions Spent on It, *Science,* v. 196, pp. 409–11.

Dendy, F. E., 1968, Sedimentation in the Nation's Reservoirs, *Jour. Soil and Water Conservation,* v. 23, n. 4, pp. 135–37.

Dolan, R., 1973, Man's Impact on the Barrier Islands of North Carolina, *Amer. Scientists,* v. 61, pp. 152–62.

Dow Chemical Company, 1972, *An Economic Analysis of Erosion and Sediment Control Methods for Watersheds Undergoing Urbanization,* PB-209-212, Nat. Tech. Info. Serv., U.S. Dept. Commerce.

El-Ashry, M. T., 1971, Causes of Recent Increased Erosion along United States Shorelines, *Geol. Soc. Amer. Bull.,* v. 82, pp. 2033–38.

Foose, R., 1981, Sinking Can Be Fast or Slow, *Geotimes,* v. 26, pp. 21–24.

Gorsline, D. S., 1966, Dynamic Characteristics of West Florida Gulf Beaches, *Marine Geology,* v. 4, pp. 187–206.

Guy, H., 1976, Sediment-control Methods in Urban Development: Some Examples and Implications, in *Urban Geomorphology,* Coates, D., ed., *Geol. Soc. Amer. Spec. Paper 174,* pp. 21–35.

Guy, H. P., 1970, Sediment Problems in Urban Areas, *U.S. Geological Survey Circular 601-E.,* 8 pp.

Guy, H. P., and Ferguson, G. E., 1970, Stream Sediment: An Environmental Problem, *Jour. Soil and Water Conservation,* v. 25, pp. 217–20.

Hester, N., and Fraser, G., 1973, Sedimentology of a Beach Ridge Complex and Its Significance in Land-Use Planning, *Env. Geol. Notes No. 63,* Illinois State Geological Survey, 24 pp.

Ippen, A. T., ed., 1966, *Estuary and Coastline Hydrodynamics,* McGraw-Hill, New York.

Johnson, J. W., 1956, Dynamics of Nearshore Sediment Movement, *Amer. Assoc. Petroleum Geologists Bull.,* v. 40, p. 2211–32.

Krumbein, W. C., 1950, Geological Aspects of Beach Engineering, in *Application of Geology to Engineering Practice,* Paige, S., ed., Geol. Soc. Amer.

Legrand, H. E., and Stringfield, V. T., 1973, Karst Hydrology—A Review, *Jour. of Hydrology,* v. 20, pp. 97–120.

Martinez, J. D., 1972, Environmental Geology at the Coastal Margin, *Proceedings, XXIV Int. Geol. Congress, Symposium 1,* pp. 45–58, Montreal, Canada.

Moore, W., and Smith, C., 1968, Erosion Control in Relation to Watershed Management, *Amer. Soc. Civil Engr. Proc.,* v. 94, n. 1R3, pp. 321–31.

Morgan, K., Lee, G., Kiefer, R., Daniel, T., Bubenzer, G., and Murdock, J., 1978, Prediction of Soil Loss on Cropland with Remote Sensing, *Jour. Soil and Water Conservation,* v. 33, n. 6, pp. 291–93.

Newton, J. G., and Hyde, L. W., 1971, Sinkhole Problem in and near Roberst Industrial Subdivision, Birmingham, Alabama, recon. *Geological Survey Alabama Bull. 68,* 42 pp.

Schwartz, M. L., 1967, The Bruun Theory of Sea Level Rise as a Cause of Shore Erosion, *Jour. of Geology,* v. 75, n. 1, pp. 76–92.

Shepard, F. P., 1973, *Submarine Geology,* 3d ed., Harper & Row, New York.

Shepard, F. P., and Wanless, H. R., 1970, *Our Changing Coastlines*, McGraw-Hill, New York, 539 pp.

Soil Conservation Society of America, 1977, *Soil Erosion: Prediction and Control*, Ankeny, Iowa, 393 pp.

Stall, J. B., 1966, Man's Role in Affecting Sedimentation of Streams and Reservoirs, pp. 79–95, in *Proceedings, 2nd Ann. Amer. Water Resources Cong.*, Bowder, K. L., ed., 465 pp.

Trask, P. D., ed., 1950, *Applied Sedimentation*, John Wiley & Sons, New York, 707 pp.

Warren, W. M., and Wielchowsky, C. C., 1973, Aerial Remote Sensing of Carbonate Terranes in Shelby County, Alabama, *Ground Water*, v. 11, pp. 14–26.

Wischmeier, W., 1976, Use and Misuse of the Universal Soil Loss Equation, *Jour. Soil and Water Conservation*, v. 31, pp. 5–9.

Wischmeier, W., and Smith, D., 1978, Predicting Rainfall Erosion Losses, *U.S. Dept. Agric. Handbook No. 537*, 58 pp.

Wolman, M., and Schick, A., 1967, Effects of Construction on Fluvial Sediment, Urban and Suburban Areas of Maryland, *Water Resources Research*, v. 3, pp. 451–64.

Floods

American Meteorological Society, 1978, Flash Floods—A National Problem, *Bull. Amer. Meteorological Soc.*, v. 59, pp. 585–86.

Baker, V., 1976, Hydrogeomorphic Methods for the Regional Evaluation of Flood Hazards, *Environmental Geology*, v. 1, pp. 261–81.

Benson, M., 1964, Factors Affecting the Occurrence of Floods in the Southwest, *U.S. Geological Survey Water Supply Paper 1580-D*, 72 pp.

Boffey, P., 1977, Teton Dam Verdict: A Foul-up by the Engineers, *Science*, v. 195, pp. 270–72.

Brahtz, J. F., 1972, *Coastal Zone Management: Multiple Use with Conservation*. Wiley Interscience, New York, 352 pp.

Bue, C. D., 1967, Flood Information for Flood-Plain Planning, *U.S. Geological Survey Circular 539*, 10 pp.

Cain, J., and Beatty, M., 1968, The Use of Soil Maps in the Delineation of Flood Plains, *Water Resources Research*, v. 4, pp. 173–82.

Costa, J., 1974, Stratigraphic, Morphologic, and Pedologic Evidence of Large Floods in Humid Environments, *Geology*, v. 2, pp. 301–303.

Costa, J., 1978, The Dilemma of Flood Control in the United States, *Environmental Management*, v. 2, n. 4, pp. 313–22.

Davies, W., Bailey, J., and Kelly, O., 1972, West Virginia's Buffalo Creek Flood—Study of the Hydrology and Engineering Geology, *U.S. Geological Survey Circular 667*, 32 pp.

Dingman, S., and Platt, R., 1977, Flood Plain Zoning: Implications of Hydrologic and Legal Uncertainty, *Water Resources Research*, v. 13, pp. 519–23.

Dougal, M. D., ed., 1969, *Flood Plain Management, Iowa's Experience*, Iowa State Univ. Press, Ames, 270 pp.

Emerson, J. W., 1971, Channelization: A Case Study, *Science*, v. 173, pp. 325–26.

Fisk, N. H., 1952, Geological Investigations of the Atchafalaya Basin and the Problem of Mississippi River Diversion, *Corps of Engineers, U.S. Army Waterways Exp. Station, Vicksburg, Miss.*

Hinson, H. G., 1965, Floods on Small Streams in North Carolina, Probable Magnitude and Frequency, *U.S. Geological Survey Circular 517*, 7 pp.

Hoyt, W. G., and Langbein, W. B., 1955, *Floods*, Princeton Univ. Press, Princeton, N.J.

Judge, J., 1967, Florence Rises from the Flood, *Natl. Geog.*, v. 132, pp. 1–43.

Knox, G., Parker, O., and Milfred, C., 1972, *Development of New Techniques for Delineation of Flood Plain Hazard Zones*, Water Resources Center, Univ. Wisc., Madison, 15 pp.

Leopold, L. B., 1962, Rivers, *Amer. Scientist*, v. 50, pp. 511–37.

Leopold, L. B., 1968, Hydrology for Urban Land Planning, *U.S. Geological Survey Circular 554*.

Post, A., and Mayo, L., 1971, Glacier Dammed Lakes and Outburst Floods in Alaska, *U.S. Geological Survey Hydrologic Investigations Atlas HA-455*, 10 pp.

Rantz, S. E., 1970, Urban Sprawl and Flooding in Southern California, *U.S. Geological Survey Circular 601-B*.

Schneider, W. J., and Goddard, J. E., 1974, Extent and Development of Urban Flood Plains, *U.S. Geological Survey Circular 601-J*.

U.S. Geological Survey Water-Supply Papers (series includes descriptions of severe floods which occur each year), U.S. Govt. Printing Office, Washington, D.C.

U.S. Water Resources Council, 1976, *A Unified National Program for Flood Plain Management*, U.S. Govt. Printing Office, Washington, D.C.

U.S. Water Resources Council, 1971, *Regulation of Flood Hazard Areas to Reduce Flood Losses*, v. 1.

Waananen, A., et al., 1977, Flood-prone Areas and Land-use Planning—Selected Examples from the San Francisco Bay Region, California, *U.S. Geological Survey Prof. Paper 942*, 75 pp.

Wolman, M., 1971, Evaluating Alternative Techniques of Floodplain Mapping, *Water Resources Research*, v. 7, pp. 1383–92.

Films

General

When the Earth Moves (U.S. Geological Survey, 1980: 22 min.)

Volcanism

Days of Destruction (Icelandic Film Distributors, 1973: 27 min.)

Case History of a Volcano (National Educational Television, 1966: 30 min.)

Eruption of Kilauea, 1959–60 (U.S. Geological Survey, 1960: 25 min.)

The Heimaey Eruption: Iceland 1973 (Univ. of Waterloo, 1974: 28 min.)

Volcano: The Birth of a Mountain (Encyclopedia Britannica Educational Corp., 1977: 24 min.)

Volcanoes: Exploring the Restless Earth (Encyclopedia Britannica, 1973: 18 min.)

Earthquakes and Tectonic Movements

The Alaska Earthquake, 1964 (U.S. Geological Survey, 1966: 22 min.)

Earthquakes and Technology (British Broadcasting Co., 1977: 23 min.)

The San Andreas Fault (Encyclopedia Britannica, 1973: 21 min.)

San Francisco—The City That Waits to Die (Time-Life, 1971: 58 min.)

Tomorrow's Quake: Earthquake Prediction (American Education Films, 1977: 16 min.)

The Trembling Earth (National Educational Television, 1968: 30 min.)

Tsunami (National Ocean Data Center, 1965: 28 min.)

Mass Movement

Mud (National Association of Conservation Districts, 1968: 20 min.)

Erosion and Sedimentation

Barrier Beach (ACI Films, 1971: 20 min.)

The Beach—A River of Sand (Encyclopaedia Britannica, 1965: 20 min.)

Erosion—Leveling the Land (Encyclopaedia Britannica, 1964: 14 min.)

The New Jersey Shoreline (Environmental Films, Inc., 1971: 18 min.)

Waterbound—The Carolina Barrier Islands (North Carolina State Univ., 1974: 20 min.)

Floods

Planning for Floods (Environmental Defense Fund, 1974: 30 min.)

Storm—Tropical Storm Agnes (U.S. Defense Civil Preparedness Agency, 1974: 29 min.)

II
MINERAL RESOURCES AND THE ENVIRONMENT

Part I dealt with the geologic hazards that limit our use of the physical environment. We have seen that there are environmental limits that function, or should function, as constraints on the activities of humans. Part II deals with mineral resources—the geology of mineral deposits, the limits of these resources, and the environmental impact of their intensive exploitation.

The Rock Cycle

The rock cycle depicts the cyclical relationships between natural processes and materials in the earth's crust (Fig. 1). The fundamental processes are volcanism, gradation, metamorphism, and diastrophism. The fundamental materials are minerals which combine to form the three common rock groups: igneous, metamorphic, and sedimentary rocks.

The crust of the earth consists largely of igneous rocks, which owe their origin to the processes of igneous intrusion and volcanism. Igneous intrusion begins with the melting of earth materials to produce a mass of molten material (*magma*) and concludes with the crystallization of the magma to produce igneous rocks. Crystallization may take place in the subsurface during the intrusive phase to form coarse-grained igneous rocks such as granites or pegmatites, or it may take place at the surface during the extrusive phase to form basalt or other types of lava rock (Appendix A).

When igneous rocks are exposed on the earth's surface, they are subject to agents of gradation (leveling of land through erosion and deposition), such as chemical and physical weathering, and erosion, transportation, and deposition by running water, wind, landslides, and glacial ice. The gradational agents will produce either sedimentary particles or ions in solution or both. Sedimentary particles may be lithified by compaction or cementation to form such sedimentary rocks as sandstone and shale, or the ions in solution may be chemically precipitated to form limestone, gypsum, or salt (Appendix A).

Sedimentary particles and rocks may be reworked by gradational agents or they may be metamorphosed by heat, pressure, or chemically active fluids to produce such metamorphic rocks as gneiss, schist, quartzite, marble, or slate (Appendix A). If metamorphic rocks are exposed to gradational agents, sediments will again be produced.

Bingham Canyon Mine, Utah.

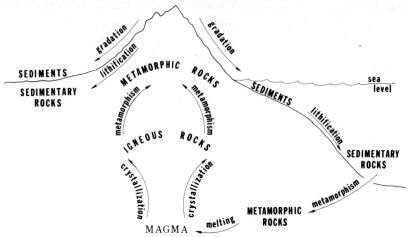

FIG. 1. Diagram of the rock cycle, illustrating the relationships between the major types of rocks and earth processes.

If, on the other hand, metamorphic rocks are subjected to intense heat, they will melt to form magma, and the cycle will have been completed. The cycle suggests that some earth materials have an extremely complex history.

The process of diastrophism includes the folding, faulting, fracturing, and warping of the earth's crust. Diastrophic forces generate heat and may be responsible for metamorphism. They may also uplift and expose rocks to gradational agents, or downwarp the crust, which will lead to the burial and preservation of earth materials.

Reserves and Resources

The crust of the earth ranges in thickness from 10 to 50 km (6.2 to 31 mi) and contains at least trace amounts of 88 chemical elements. Table 1 shows that the 8 most abundant elements account for 98.5 percent of the mass of the continental crust; the remaining 80 elements account for only 1.5 percent of the mass.

The geochemically abundant metals that are widely used in industry are aluminum, iron, and magnesium. Like most metals, however, they do not occur in an elemental state but are present in inorganic compounds called minerals. The feasibility of recovering elements from their mineral compounds depends, in part, on the nature of the compounds. Carbonates and oxides, for example, require far less en-

Table 1 Relative abundance
of elements in the
earth's crust

Element	Percentage by weight
Oxygen	46.6
Silicon	27.7
Aluminum	8.1
Iron	5.0
Calcium	3.6
Sodium	2.8
Potassium	2.6
Magnesium	2.1
	98.5

ergy to process than do silicates and are therefore desired mineral forms.

But more than the desired mineral form is necessary if a deposit is to be minable. In addition, concentrations must be far above average crustal abundance. Table 2 shows that even aluminum, the most abundant metal on earth, must usually be concentrated by a factor of 3 and that the rarer elements must be concentrated by much larger factors (e.g., mercury × 62,500). We will look at some of the ways in which high concentrations are achieved in the next section.

Table 2 Approximate concentration factors of some
metals [Concentration factors vary with
changes in economic conditions and technology. *Mining and Minerals Policy, 1977 Annual
Report of the Secretary of the Interior*]

Metal	Crustal abundance (parts per million)	Concentration factor
Aluminum	83,000	3
Iron	48,000	6
Zinc	81	300
Chromium	77	4,300
Nickel	61	160
Copper	50	100
Lead	13	2,570
Uranium	2.2	680
Molybdenum	1.1	2,275
Mercury	0.08	62,500
Silver	0.065	1,500
Gold	0.0035	2,280

The feasibility of a mining operation is influenced not only by the anomalously high concentration of elements in desirable chemical forms but also by government policy and such economic factors as world market price and costs of transportation, labor, and equipment. Economic factors explain why such rare metals as silver and gold may be profitably mined even if they are not found in high concentrations. Only a mineral deposit that can be mined and processed at a profit under current economic and technological conditions is considered an *ore deposit*—a term applied to metals, nonmetals, or fossil fuels.

Classification and Genesis of Ore Deposits

There are numerous approaches to the classification of ore deposits but those schemes based on the probable mode of origin (*genetic classification*) are perhaps the most instructive and useful. We will, therefore, look at some examples of ore deposits and how they are related to the fundamental earth processes of volcanism, metamorphism, gradation, and diastrophism (Table 3).

Magmatic Deposits and Pegmatites

The process of igneous intrusion is responsible for magmatic deposits and pegmatites. When a magma intrudes country rock (surrounding rocks) and cools, there is a definite order to the crystallization pro-

Table 3 General classification of ore deposits

Major geologic process	Type of deposit	Typical ores
Volcanism	Early magmatic segregation	Platinum, nickel, chromite
	Late magmatic segregation	Magnetite
	Disseminated ores	Diamond
	Pegmatites	Beryllium, lithium, tin, tungsten
Metamorphism	Hydrothermal (may originate from magmatic or metamorphic source)	Lead, zinc, copper, gold, silver, uranium
	Contact metamorphic	Iron, copper, lead, zinc, silver, tin
Gradation	Placer	Gold, tin, uranium, platinum, diamond
	Residual concentrate	Aluminum, iron, phosphate, manganese
	Supergene enrichment	Copper, iron, silver
	Primary chemical precipitate	Salt, potash, borates, iron

cess. This may lead to a segregation and anomalously high concentration of early and late magmatic deposits which may be of ore quality. On the other hand, minerals may crystallize over a long time span and may be disseminated through a large volume of rock. Even though the disseminated deposits are usually not as highly concentrated, they may also be of economic interest. In the diamond mines of South Africa for example, disseminated deposits are being worked. The chromite deposits at Stillwater, Montana, are early magmatic segregations, and the iron ores in the Adirondack Mountains of New York are late magmatic deposits.

Pegmatites are coarse-grained rocks which may occur in a variety of shapes and contain a variety of minerals. Most pegmatites are produced during the late stages of the crystallization process and include a history of the migration of fluids, rich in the volatile constituents of the magma, through fractures in the surrounding bedrock. A high volatile content and high temperatures and pressures allow for the growth of large crystals and for chemical reactions between the migrating fluids and the country rock. Some of the elements associated with pegmatites are beryllium, lithium, tin, and tungsten.

Metamorphic and Hydrothermal Deposits

The crystallization of a magmatic body is associated with the liberation of great quantities of heat, water vapor, and other volatiles. Because rocks are relatively poor conductors, the effect of liberating heat is most pronounced at the contact between the magmatic body and the country rock. The effect of fluids is also most pronounced near the contact zone, but if the country rock is permeable, the effect of the fluids will extend well beyond the contact zone. The combined effects of the heat and fluids acting on the country rock may result in a variety of textural and compositional changes which depend upon the temperature, pressure, nature of the country rock, and the composition of the fluids. The ore minerals are, however, not generally confined to veins, but may be widely distributed in the contact zone. These deposits are called *contact metamorphic deposits*. The most typical ores found in them are copper, lead, zinc, and tin.

If the country rock is highly faulted and fractured, or contains bedding planes or other pronounced openings, hot mineral-bearing solutions known as *hydrothermal solutions* will migrate great distances from their magmatic source and form vein-type hydrothermal deposits. Not all hydrothermal solutions are related to a magmatic source.

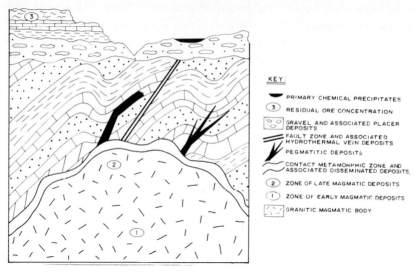

KEY:

⬛ PRIMARY CHEMICAL PRECIPITATES

③ RESIDUAL ORE CONCENTRATION

GRAVEL AND ASSOCIATED PLACER DEPOSITS

FAULT ZONE AND ASSOCIATED HYDROTHERMAL VEIN DEPOSITS

PEGMATITIC DEPOSITS

CONTACT METAMORPHIC ZONE AND ASSOCIATED DISSEMINATED DEPOSITS

② ZONE OF LATE MAGMATIC DEPOSITS

① ZONE OF EARLY MAGMATIC DEPOSITS

GRANITIC MAGMATIC BODY

FIG. 2. Hypothetical cross section showing the spatial relationships between the most common types of ore deposits.

Some are related to chemically active liquids or gases contained within the pore spaces of rocks and are said to be of metasomatic origin. The liquids and gases exchange ions with the rocks, and in some instances concentrations may reach ore-body proportions. This type of replacement process is known as *metasomatism*. Characteristic ores include lead, zinc, copper, gold, silver, and uranium.

The spatial relationships between the magmatic body and magmatic deposits, pegmatites, contact metamorphic deposits, and hydrothermal deposits are shown in Fig. 2. Examples of typical ores associated with these deposits are given in Table 3.

Sedimentary Deposits

Sedimentary mineral deposits may be classified as 1. placer deposits, 2. residual concentrations, 3. supergene enrichments, and 4. primary chemical precipitates. All of these deposits owe their origin to the weathering of pre-existing rock, which facilitates the segregation of the ore into an anomalously high concentration. Placer deposits are deposits in which certain heavy minerals have been concentrated by the selective sorting and transportation action of currents and waves. The ore material is chemically and physically resistant and has a higher density than associated sedimentary materials. As this material is transported, heavy minerals tend to settle out and accumulate where the velocity of the stream decreases—at obstructions in the channel, along inner bends of meandering streams, or in plunge pools at the base of waterfalls. The deposits are usually dominated by coarse-grained particles and are discontinuous. The most common ores are gold, diamond, platinum, and tin oxide. Placer deposits may be associated with modern or ancient streams and under certain conditions with the action of waves on beaches.

Residual Concentrations

Residual concentrations represent the stable end products of extreme and protracted chemical weathering. Modern residual concentrates are found at the surface while ancient concentrates may be deeply buried. Such deposits are generally associated with climates that facilitate chemical weathering (humid, tropical, or subtropical) and with topographical features such as elevated plateaus, which, because they have limited potential for runoff, retard the mechanical removal of the residual products of weathering. The most important ores occurring as residual concentrates are bauxite, which is an ore of aluminum, and hematite, which is an ore of iron; valuable deposits of manganese and phosphate are also known.

Supergene Enrichments

Supergene enrichments are precipitates of ore minerals from percolating ground water. Surface waters dissolve the ore minerals from the weathered zone, carry the dissolved material downward, and reprecipitate it just below the water table. Although the precipitated material is confined to a narrow zone associated with the water table, it may have a great areal extent. The most common ores associated with supergene enrichment deposits include copper, iron, silver, and uranium.

Primary Chemical Precipitates

Primary chemical precipitates of economic importance are most commonly formed in arid environments. Chemical weathering may dissolve ions, which are transported in solution to inland lakes or restricted marine basins like the Mediterranean Sea. Under extreme aridity, the waters evaporate and a large variety of salts may precipitate. Vast deposits of sodium chloride (common salt), gypsum, potash, borates, nitrates, and other salts may form in this way.

Under other unusual and extreme environmental conditions, unstable materials may be produced, preserved, and incorporated in sedimentary rocks in large volumes. Tropical and semitropical swamp environments, for example, produce large volumes of plant material which may be rapidly buried by sediments and gradually converted to coal. Other environments like deltas, reefs, and restricted marine basins may generate large volumes of organic material which may be buried by sediments and converted to petroleum through complex geochemical processes.

Our discussion has been concerned with ore deposits and the fossil fuels. It is important to note that the processes that produce these minerals are also responsible for producing the stone used in the construction industry. Granite, marble, slate, and limestone are all popular building stones; they are also used as crushed rock. The largest nonfuel mineral industry in the United States is concerned with crushed rock, sand, and gravel.

Diastrophism

Our discussion of the genesis of ore deposits would not be complete if we did not consider the role of diastrophism and the complex interrelationships between it and the processes we have reviewed.

Diastrophism exerts an indirect but significant effect upon the gen-

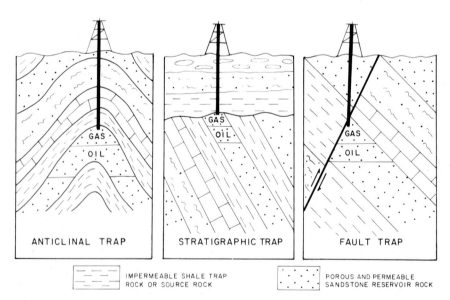

FIG. 3. Typical petroleum traps.

eration of some ore deposits. Uplift exposes the deeply buried pegmatites, magmatic segregations, and contact metamorphic and hydrothermal deposits, bringing them within the range of mining technology and gradational processes which may further enrich the deposits.
Folding, faulting, and fracturing produce pathways for the migration
of hydrothermal solutions and volatiles. These movements also generate frictional heat, which may facilitate metamorphic reactions. Downwarping of the crust may lead to the accumulation and preservation
of ore minerals and organic matter in sedimentary basins.

Folding and faulting also cause the migration of oil and gas and
may cause the trapping of these hydrocarbons in commercial accumulations. Fluids migrate in response to the density differences, and the
barrier to migration to the surface is the presence of an impermeable
rock. Figure 3 illustrates the more common petroleum traps that are
related to diastrophic features.

Outlook for the Future

Solar energy facility near Odeillo, France. (Courtesy of Combustion Engineering Incorporated.)

"Better methods for estimating the magnitude of potential mineral resources are needed to provide the knowledge that should guide the design of many key public policies."

V. E. McKelvey

The Roman scholar Pliny the Elder described the earth as "gentle and indulgent, ever subservient to the wants of man." Today, however, we often express concern about whether the earth can provide sufficient materials, energy, and food for an increasing population. Questions concerning the magnitude of resources arise at local, national, and international levels. Decisions concerning price structure, import-export policies, research and development programs, exploration activities, population control, rationing, conservation, land use, and so on, are influenced, in part, by our estimates of the magnitude of our resources. We must know not only what exists now and can be produced under present economic and technological conditions, but we must also know the potential represented by undiscovered deposits and known deposits that cannot be profitably produced at present.

Classifying Mineral Resources

The U. S. Geological Survey and the U. S. Bureau of Mines have collaborated on a joint statement on the classification of mineral resources (Fig. 1). The distinction between the various classes is based on current geologic and economic factors. A *resource* is a concentration of naturally occurring solid, liquid, or gaseous materials in or on the earth's crust in such form that economic extraction of a commodity is currently or potentially feasible. A *reserve* is that portion of an identified resource from which a usable mineral or energy commodity can be economically and legally extracted at the time of determination. The term *ore* is used for reserves of metalliferous material. Resources thus include, in addition to reserves, mineral deposits that eventually may become available—known deposits that cannot be profitably mined at present because of economics, technology, or legal restraints—and unknown deposits that may be inferred on the basis of geologic evaluation.

As shown in Fig. 1, total resources are divided into two major

TOTAL RESOURCES

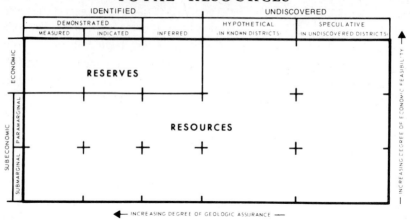

FIG. 1. Classification of mineral resources.

categories, identified and undiscovered; they in turn are subdivided. Definitions of the most frequently used terms follow.

Identified resources are specific bodies of mineral-bearing material, the location, quality, and quantity of which are known from geologic evidence. If they can be supported by engineering measurements, they are called *demonstrated resources.*

Reserves are that portion of the identified resource from which a usable commodity can be economically and legally extracted at present.

Identified subeconomic resources are known deposits not now economically minable. *Paramarginal* resources border on being economically producible or are not commercially available because of legal restrictions or political circumstances. The market price of *submarginal* deposits must be substantially higher (more than 1.5 times the price at the time of determination) or there must be a major reduction in the cost of recovery through technological advances.

Measured, indicated, and inferred are terms applicable to both reserves and identified subeconomic resources. *Measured* resources are those for which tonnage is computed from dimensions revealed in outcrops, trenches, workings, and drill holes, and for which grade is computed from the results of detailed sampling. The computed tonnage and grade are judged to be accurate within stated limits, and no such limit is judged to be different from the computed tonnage or grade by more than 20 percent. *Indicated* resources are those for which tonnage and grade are computed partly from specific measurements, samples, or production data, and partly from projection of geologic data. The sites available for inspection, measurement, and sampling are too widely or otherwise inappropriately spaced to permit the mineral bodies to be outlined completely or the grade to be established throughout. *Inferred* resources are those for which quantitative estimates are based largely on broad knowledge of the geologic character of the deposit and for which there are few samples or measurements.

Inasmuch as the distinction between the various classes is based on *current* geologic and economic factors, a resource inventory can change with technological advances, exploration, and production activities. Mineral reserves and resources are dynamic quantities and must be constantly appraised.

Undiscovered resources are bodies of mineral-bearing material surmised to exist on the basis of broad geologic knowledge and theory. *Hypothetical resources* are undiscovered resources that may reasonably be expected to exist in a known mining district under known geologic conditions, *Speculative resources* may exist either as known types of deposits in a favorable geologic setting where no discoveries have been made, or as types of deposits that remain to be recognized. Exploration that confirms the existence of undiscovered resources and reveals quantity and quality will permit their reclassification as reserves or as identified subeconomic resources.

Domestic Reserves and Resources

The general domestic outlook for reserves and resources for the remainder of the twentieth century is summarized in Table 1.

It can be seen that known United States reserves of some minerals represent only a few years' supply and that we must frequently rely on foreign sources for many of our minerals (Table 2). Increasing our domestic supply through discovery of new deposits is imperative, but exploration that leads to the discovery of an ore deposit is exceptional and the economic risks are great. Furthermore, it frequently takes decades to complete geologic mapping, physical exploration, and economic evaluation. Mine development and plant construction requires additional time. There are, however, more immediate ways to alleviate the problem of scarce domestic supplies. The recovery and recycling of scrap and used materials already provides significant amounts of some metals, such as iron, lead, and nickel, while the demand for other scarce metals, such as chromium, manganese, and cobalt, may be reduced by substituting more abundant materials and by reducing waste.

Table 1 General outlook for domestic reserves and resources through A.D. 2000
[From *U.S. Geological Survey Professional Paper 940, 1975*]

Within each group, commodities are listed in order of relative importance as determined by dollar value of U.S. primary demand in 1971.

Group 1: RESERVES in quantities adequate to fulfill projected needs well beyond 25 years.

Coal	Clays	Argon
Construction stone	Potash	Diatomite
Sand and gravel	Magnesium	*Barite
Nitrogen	Oxygen	Lightweight
Chlorine	Phosphorus	aggregates
Hydrogen	Silicon	Helium
Titanium (except	Molybdenum	Peat
rutile)	Gypsum	*Rare earths
Soda	Bromine	*Lithium
Calcium	Boron	

Group 2: IDENTIFIED SUBECONOMIC RESOURCES in quantities adequate to fulfill projected needs beyond 25 years and in quantities significantly or slightly greater than estimated UNDISCOVERED RESOURCES.

Aluminum	Vanadium
*Nickel	*Zircon
Uranium	Thorium
Manganese	

Group 3: Estimated UNDISCOVERED (hypothetical and speculative) RESOURCES in quantities adequate to fulfill projected needs beyond 25 years and in quantities significantly greater than IDENTIFIED SUBECONOMIC RESOURCES; research efforts for these commodities should concentrate on geologic theory and exploration methods aimed at discovering new resources.

Iron	*Silver	*Cadmium
*Copper	*Fluorine	*Bismuth
*Zinc	Platinum	Selenium
Gold	Tungsten	*Niobium
*Lead	*Beryllium	
Sulfur	*Cobalt	

Group 4: IDENTIFIED-SUBECONOMIC and UNDISCOVERED RESOURCES together in quantities probably not adequate to fulfill projected needs beyond the end of the century; research on possible new exploration targets, new types of deposits, and substitutes is necessary to relieve ultimate dependence on imports.

Tin	*Antimony
Asbestos	*Mercury
Chromium	*Tantalum

*those commodities which may be in much greater demand than is now projected because of known or potential new applications in the production of energy

Table 2 U.S. dependence on foreign sources for some of its minerals [*Mining and Mineral Policy, Report of the Secretary of the Interior,* 1973]

a. Less than half imported from foreign sources:

Copper	Magnesium	Cement
Iron	Molybdenum	Salt
Titanium	Vanadium	Gypsum
(ilmenite)	Antimony	Barite
Lead	Tellurium	Rare earths
Silicon	Stone	Pumice

b. One-half to three-fourths imported from foreign sources:

Zinc	Nickel
Gold	Cadmium
Silver	Selenium
Tungsten	Potassium

c. More than three-fourths imported from foreign sources:

Aluminum	*Titanium	*Strontium
*Manganese	(rutile)	Asbestos
Platinum	*Niobium	*Sheet mica
Tin	Tantalum	Mercury
*Cobalt	Bismuth	
*Chromium	Fluorine	

*Commodities more than 90 percent imported.

Readings

Replenishing Non-Renewable Mineral Resources—A Paradox

In Reading 23 Richard P. Sheldon explores the intriguing paradox of replenishing nonrenewable mineral resources. His essay is an instructive example of the application of the concept of the dynamic nature of resources and reserves in the context of the mineral supply system of the United States. Even though mineral resources may be replenished by scientific and technological advance, the mineral supply system is subject to serious constraints. Mr. Sheldon recognizes these constraints and presents various points of view about their significance.

Realities of Mineral Distribution

Many scientists, engineers, and economists have expressed grave doubts about the adequacy of our resources, while others are more optimistic. The optimists, or "cornucopians," rely on the market mechanism and technology for transforming "infinite" resources into almost infinite reserves. Almost invariably they rely upon a drastically altered, energy-intensive world in which a cheap and limitless supply of energy replaces raw-material-intensive processes. They envision contained nuclear fusion or breeder reactors as the energy source and look to the electrolytic production of hydrogen to extend the mineral resource base. Coal could be hydrogenated to make liquid fuel or gas, and metalliferous ores could be cheaply reduced, thereby making available an abundant supply of low-grade ores. Sea water could be desalinated at a fraction of the present cost.

The "doomsters" see a catastrophic exhaustion of critical resources in the foreseeable future. They focus their attention on "reserves" and suggest that, although the reserve base can be expanded as price rises and technical advances make lower-grade resources more easily available, the additional inventory would be won at unacceptable social, environmental, and economic costs. These costs, they claim, set a limit to the potential for transforming resources into reserves.

Preston Cloud suggests in Reading 24 that "the situation calls neither for gloomy foreboding nor facile optimism, but for positive and imaginative realism." He suggests that "the technological fix is not a panacea but an anaesthetic," and he challenges other fundamental premises underlying the cornucopian concept of unlimited mineral resources. While acknowledging the role of economics and technology in evaluating current and potential resources, he is careful to demonstrate that these are not the sole factors governing the availability of mineral resources. They may, however, buy the time in which to find better solutions.

Global Energy Resources

Reading 25 reviews energy resources from a global perspective. Conventional resources (coal, petroleum, natural gas, and uranium) and unconventional resources (enhanced oil recovery, tar sands, oil shale, tight gas formations, and geopressured brines) are summarized in terms of location, estimated amounts, and availability in the context of the next 20 years. Although supplies of coal appear to be adequate, the same cannot be said of conventional sources of petroleum or natural gas. The resource base for uranium is also adequate, but only if breeder reactors are developed. The countries of the world will have to develop a strategy for moving from conventional fuels to unconventional sources of energy that will minimize increases in energy costs as well as social, environmental, and health problems associated with their exploitation. The resource base appears to be sufficient to allow an orderly transition from present patterns of energy use to the new pattern that will be required in the future.

Is Nuclear Energy Necessary?

In Reading 26 Alvin Weinberg argues that it is highly desirable, if not absolutely necessary, to maintain the nuclear option at least in some parts of the world. In reaching this conclusion, he explores three issues which dominate the long-term (beyond the year 2030) outlook for fission: (1) the availability and acceptability of alternatives, (2) the long-term energy demand, and (3) the possible constraints on fossil fuels imposed by accumulation of carbon dioxide in the atmosphere. He recognizes, however, that if people are to live with fission over the long term they must significantly reduce the *a priori* probability of accidents.

23 Replenishing Non-Renewable Mineral Resources—A Paradox

Richard P. Sheldon

In 1922 a joint committee of petroleum geologists from the American Association of Petroleum Geologists and the U.S. Geological Survey estimated that the United States had only 9 billion barrels of oil left in the ground either as reserves or as resouces to be discovered (U.S. Geological Survey, 1922). Eleven years later, the 9 billion barrels had been produced and an additional 13 billion barrels had been discovered.

In 1952 the President's Materials Policy Commission estimated the Nation's foreseeable copper resource (as of 1950) to be 25 million tons. Twenty-five years later, 31 million tons of copper had been produced and an additional 57 million tons of reserve had been discovered.

These are two of many examples of carefully reasoned mineral resource predictions by credible highly qualified geologists and engineers that have been overtaken in a few tens of years by additional production and discovery.

Mineral resource estimates ordinarily are requested by national planners when they perceive possible future shortages. The 1922 oil estimate was made during the "John Bull" oil shortage scare, and the 1952 copper estimate was undertaken during the post-World

Reprinted from *U.S. Geological Survey Yearbook, Fiscal Year 1977*, U.S. Government Printing Office.

War II period when the United States was thought to be "outgrowing its resource base." The engineers and geologists responsibly furnish these estimates, usually qualifying them as conservative, particularly in regard to minerals expected to be added by additional discovery. Unfortunately, such qualifiers are quite often dropped by many of those who use the estimates. These estimates generally deepen the concern over impending shortage, but become irrelevant when the period of shortage gives way to a period of adequate or even over supply.

Behavior of these non-renewable mineral resources over time is opposite to what we intuitively expect. Geologists and engineers measure and report resource abundance. The resources they estimate are then depleted at ever increasing rates that foreseeably should exhaust the resource. Concern about shortages grows. Yet when the time approaches when we should have run out of the resource, we find paradoxically—almost alchemically—that we have more than we started with. At best we distrust the forecaster's ability and at worst his motives. What has gone wrong? Are such underestimates going to continue to be made in the future? To answer these questions, the nature and dynamics of resources need to be understood.

Nature of Resources

In Webster's dictionary, a resource is defined as a fresh or additional stock or store of something available at need. Thus, in the short term, we think of resources as a stockpile of inventoried material with immediate availability. If we consider long-term demand, the question of future availability becomes important, so that undeveloped resources, that is mineral resources awaiting discovery in the ground or living resources yet to be born or planted, must be considered. Thus resources have two essential characteristics: (1) a demand, and (2) an availability. Depending on the time frame being considered, the demand and the availability are either immediate or potential.

Resource Demand

Potential resources are based on a projection of future demands, a process that carries some risk. For example, in view of the threatened deforestation in France in the late seventeenth century, the King planted an oak forest near Paris as a reserve to supply, some 200 years hence when the trees matured, oak logs for masts and timbers for warship construction. His foresight created a beautiful forest that still stands, but did little to meet the needs of the modern French Navy.

The changing nature of mineral resource demand and its effect on resources over time can be seen by considering the mineral resources of the State of Montana at two times: a thousand and ten years ago and ten years ago.

On the one hand, the stone-age Indian living in 968 in what was to become Montana had a very small but highly specialized need for stones. Each year he used a few pounds of flint and obsidian for tools, arrowheads, and axe heads, a few pounds of sandstone for mortars, a little salt and mineral dye. Their value at today's prices would be a few cents, or at his prices, a few belts of wampum. The total resources available in his shallow quarries would be worth perhaps a few thousands of dollars in our terms. Of course, he used so little of his mineral resource that their eventual depletion was of little if any concern to him.

On the other hand, the industrial-age Montanan living in 1968 had tremendous needs for minerals, and huge production facilities and mineral resources to meet them. Along with his fellow citizens from the other 49 states, the Montanan used, on the average, 20 tons of minerals a year. The value of the cumulative mineral production of Montana from 1880 up to 1960 was over 4 billion dollars, showing that Montana lives up to its name, the Treasure State. In 1968, Montana contained 52 varieties of significant mineral deposits ranging from asbestos to vermiculite (U.S. Geological Survey, 1968). The most important of these were oil, natural gas, coal, copper, phosphate rock, and chromium. Their reserves at that time were worth 1.17 trillion dollars.

The mineral needs of the stone-age Montanan were low, and his assessed resources were correspondingly low. On the other hand, the mineral needs of the industrial-age Montanan are large and varied, but so are his mineral resources. The difference in mineral needs, supplies and resources between the two ages is staggering. It is no matter that the stone-age Montanan was standing on vast deposits of minerals that were to become highly valuable to Montanans a millenium later. To him they were rocks to walk on, not resources to use. The separation of ages is complete when one realizes that the industrial-age Montanan does not even include among his vast mineral resources the small deposits of flint, obsidian and sandstone used by his predecessor a millenium earlier.

It is clear from considering this example that even though most mineral deposits are permanent and unchanging on the human time frame, mineral resources are temporal and changing. V. E. McKelvey pointed out (1972, p. 20) that, "Defining resources as materials usable by man, a little reflection reveals that whereas it is God who creates minerals and rocks, it is man who creates resources." One can reflect further that a mineral *deposit* can be characterized as nonrenewable, but mineral *resources* are another thing entirely. By additional effort by man new mineral resources can be "created," not in the sense of creation by the Almighty, but in the sense that a body of rock is identified for the first time as useable. Within the limits of geologic availability, one can conclude that the character, variety and size of mineral resources depend on the technology that needs them and the technology for developing them.

Resource Availability

To be anything other than a wishful thought, resources must be available or potentially so. A mineral deposit that has been found, measured, and determined to be economically mineable at the current price using current mining and extraction technology is available and clearly a resource. If the deposit is known and measured, but no process is known or foreseen by which the material can or could be recovered economically, it is not available for use and is not a resource. However, if it seems to qualified engineers technologically feasible to develop in the future a process to recover the material economically, the deposit would be potentially available for use and would be called a sub-economic resource (U.S. Bureau of Mines, and the U.S. Geological Survey, 1976). For example, in 1950, when the 6.2 billion tons of U.S. iron ore reserve included no taconite, the low-grade taconite deposits were sub-economic and were foreseen to be only potentially available. The developing of the technology to drill, mine and concentrate taconite made it economic and, in fact, the preferred ore, which in 1975 made up most of the U.S. iron ore reserve of 17 billion tons.

Another factor of availability is the knowledge of the existence of a deposit. It is obvious that a deposit must be identified to be available and that an undiscovered deposit is unavailable. Exploration and resource geologists can identify areas where undiscovered deposits might occur and then can make a knowledgeable guess about how many deposits exist there, and of these how many might be discoverable. They also can make knowledgeable guesses about the size of such undiscovered deposits. Such deposits can then be considered potentially available; that is we have the potential to discover them with current exploration techniques. Nearly all of our known deposits that now make up our past production and present mineral reserve were once a part of the undiscovered but discoverable resource.

There is no way in which the ultimate amount of the undiscovered resources can be determined even though some portions of the ultimate amount can be estimated. We can predict a discoverable portion of undiscovered resources using well supported *hypotheses* of the occurrence of deposits, as well as an additional discoverable portion using poorly supported *speculations* on the occurrence of deposits. These portions make up the *hypothetical* and *speculative* categories of undiscovered resources used by the Geological Survey and the Bureau of Mines. However, a still further portion of undiscovered ultimate resources cannot be predicted because it is undiscoverable using either current or foreseeable future exploration technology. For example, some rocks of the western United States are mineralized where they are exposed, but in large areas where they are covered by younger lava flows, they cannot be prospected for by anything other than the too-expensive drill or shaft. Geologists can confidently predict that many deposits exist beneath the lava flows but the deposits are not economically discoverable with present or foreseeable future technology and cannot be counted as a part of our resources. A still further portion of undiscovered ultimate resources cannot be predicted because of lack of scientific evidence of the existence of the deposits. Such deposits probably exist but are unsuspected by geologists. A clear hindsight example of such a deposit is the Red Sea metalliferous mud. On February 17, 1965, marine geologists on the oceanographic research vessel, *R. V. Atlantis, II,* were astonished to find that a core of mud taken in the central part of the Red Sea was enriched in zinc, copper, lead, silver, and gold (Degens and Ross, 1969). Subsequent surveys showed that the metalliferous muds in the Red Sea are widespread and fairly thick and contain large quantities of scarce metals. These deposits now are a part of the world's sub-economic resource, but there was no reason whatever before their chance discovery to suspect that they existed. They were totally unconceived and were certainly not visualized as a part of undiscovered resources.

Resource Flow

A common but incorrect way of viewing mineral resources is to regard them as the sum of the known and predicted economic deposits of commodities in current use, and from that to conclude that mineral resources in general are fixed and non-renewable. As seen in the discussion in the previous section, mineral resources consist of known and suspected mineral deposits that are counted as resources by virtue of industrial needs for them and subsequently are categorized according to knowledge of their existence, the economics and technology of their discovery, and the economics and technology of their mining and extraction. These factors change over time, causing the make-up and magnitude of resources to change. Recognizing that resources are so heavily influenced by these temporal economic factors, economists David Brooks and P. W. Andrews pointed out in 1974 that in matters of long-term supply, minerals should be treated not as a fixed stock, but as a flow that responds to demand.

The misconception of resources as a fixed stock answers part of the question raised at the start of this paper, "What has gone wrong with our mineral forecasting?" At a time of concern over threatening shortages of minerals, resource geologists and engineers are asked to join forces and estimate the known mineral resources and predict the unknown mineral resources. They would like to estimate once and for all the total or ultimate resources of the country, but they cannot. Their problem is this—both geologists and engineers, no matter how technically liberal they may be, must stay within the confines both of their data and their technical understanding and methodology. They come up with estimates, but each one is outdated the day it is published, because continuing exploration and study generate new data, and new basic research sparks new ideas of occurrence or recovery. In the past, most estimates have turned out to be too low, which is expectable. Regardless of the liberalism of the estimater, the methodologic conservatism that must be followed insures that the estimate will exclude deposits that are unrecoverable with foreseeable technology as well as deposits that are undiscoverable—as were the mineral deposits beneath basalt flows—or are unpredictable—as were the Red Sea metalliferous muds. Over time with the accumulation of more knowledge, significant amounts of such deposits will become recoverable, discoverable or predictable and add to the total resources.

This is not to say that rocks, minerals, and their natural concentrations are not finite or that geologic availability is not a limiting factor in resource magnitude, but only that the conception and perception of resources at any given time are likely to be limited.

Mineral Supply System

The mineral supply system of the United States yields this flow of most mineral materials from one resource category to another progressively from speculative-undiscovered resources to refined material production. As we have seen, the system is driven by the demands of the industrial age.

To understand how this supply system works and the factors influencing it, one must look at its components. It is commonly conceived to have three major phases: research, exploration, and exploitation; however, each of these phases is divided into two parts. Research consists of conception and assessment of undiscovered resources; exploration consists of discovery and delineation of mineral deposits; and exploitation consists of extraction and processing of ores. Table 1 shows this breakdown along with the actual activity carried out, the mineral resource category developed, and the institutions with the prime responsibility.

The flow of material is initiated by research organizations in the government, academic, and private sectors conducting basic research on geologic processes of rock and mineral formation and distribution. Originally all resources were unconceived, and only by such basic study and thinking was each kind of deposit conceived. Once conceived, the magnitude, location and character of the deposits are

Table 1 Phases of mineral supply system.

Major phases	Detailed phases	Activity		Mineral resource category developed	Prime responsibility
RESEARCH	CONCEPTION	Research in geologic processes, i.e. plate tectonics, formation of mineral deposits, etc.	UNDISCOVERED RESOURCES	SPECULATIVE	Universities, Government, research organizations, private institutes
	ASSESSMENT	Geologic, geophysical, and geochemical mapping, geostatistical analysis		HYPOTHETICAL	Government Industry
EXPLORATION	DISCOVERY	Prospecting	RESERVES	INFERRED	Industry
		Research on prospecting techniques			Government and Industry
	DELINEATION	Exploration		INDICATED AND MEASURED	Industry
		Research on exploration techniques			Industry and Government
EXPLOITATION	EXTRACTION	Mining and land reclamation		Produced raw material	Industry
		Research and development on extraction		Reserves	Industry and Government
	PROCESSING	Beneficiation reduction and refining		Produced refined material	Industry
		Research and development on processing		Reserves	Industry and Government

speculated on and reported as a *speculative* undiscovered resource, generally by government and academic research organizations.

In the next phase, mineral resources are further defined by government resource agencies and to a lesser degree (and mainly for its own purposes) by the exploration sector of industry. They conduct geologic, geophysical, and geochemical mapping of areas of speculative resources. Application of well-supported hypotheses concerning the occurrence of mineral deposits to these regional data allows estimation of *hypothetical undiscovered resources.* In this way the certainty of actual existence of the undiscovered resource is increased to the point that the hypothetical resource estimates have sufficient reliability for national planning in government or exploration planning in industry.

At this point exploration is initiated by industry. The regional maps produced at the assessment stage are used to plan a prospecting program. More detailed field studies are carried out to narrow the target areas, and finally drilling or tunneling is undertaken to search for the deposit. This activity, when successful, develops *reserves* of the *inferred* class. Further detailed exploration improves the accuracy of the reserve by better delineating the extent and shape of the deposit as well as its grade and mineralogy. This activity develops *indicated* and *measured reserves* which have the degree of certainty necessary for the investment by industry of large amounts of capital needed for exploitation of the deposit.

The mineral supply system consists of a series of sequential steps, each one necessary for the initiation of the succeeding step, and each one designed

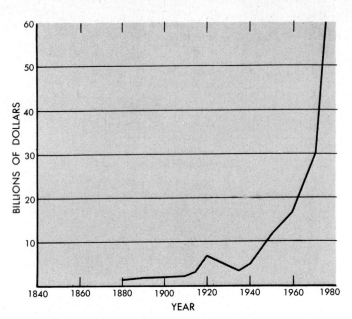

FIG. 1. Growth of U.S. annual production of principal raw material mineral commodities. (From U.S. Bureau of Mines, 1975.)

to improve the effectiveness and economic efficiency of the total system. The demand for minerals drives the resource flow. The overall economic efficiency of this system is set by the technologic level and is improved by research and development at all phases. That is to say, the estimated magnitude of undiscovered resources is increased by improved basic concepts of mineral deposits and mapping and resource assessment of potentially mineralized areas. Reserves are increased by prospecting, which is made more effective by improvement of prospecting, extraction and processing techniques. The increase in resources over time is directly related to the amount of effort put into improving the technology as well as to the amount of exploration effort. That is to say, we replenish, expand and diversify our "non-renewable" mineral resources by technologic advance through research and development effort.

The relationship between mineral resource supply and mineral science and technology can be shown by comparing the growth in value of U.S. annual production of principal raw material mineral commodities (Fig. 1) with several indices of the growth of scientific and technologic effort in the minerals area. The first is the cumulative numbers of scientific articles on the geology of North America (Fig. 2). The second is the growth of the yearly number of earth scientists in the U.S. (Fig. 3). The strength of the geoscience educational system is shown by the growth of the number of degree-granting geoscience departments in the U.S. Universities (Fig. 3). Finally, increasing sophistication of geoscience is indicated in part by the fields of specialization, which are in turn measured by the increasing number of National geoscience societies (Fig. 4). If the cost to support a geoscientist in terms of salary, administrative and

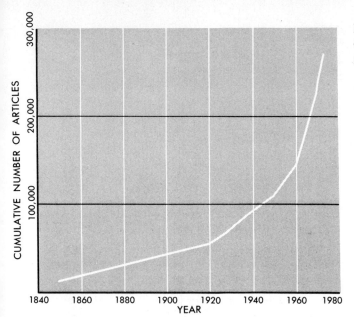

FIG. 2. Cumulative number of scientific articles on the geology of North America.

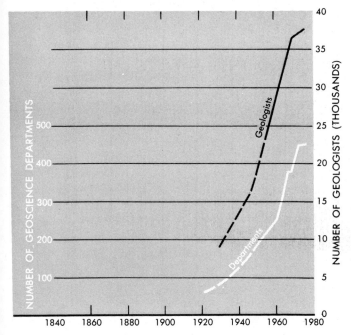

FIG. 3. Number of geologists and number of degree-granting geoscience departments.

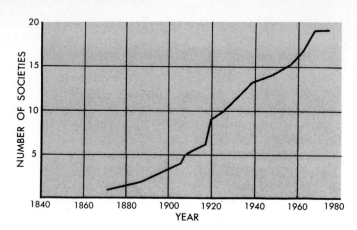

FIG. 4. Growth of national societies of the American Geological Institute.

scientific support is taken at $50,000, a conservative figure, nearly 2 billion dollars would have been spent in 1975 to support this geoscience effort. Of course, not all of this geologic effort would be directed at the mineral supply system, but a major portion would be, either directly or indirectly. Also the geologic effort is by no means the whole of the effort going into resource studies, as much effort is made by mineral and mining engineers.

Long-Term Mineral Supply

Our mineral resources are replenished by scientific and technologic advance, but how long can this keep up? Even with replenishment, will we eventually outgrow our mineral resource base? Much thought has been given to this question and much diversity of opinion exists. Economic geologist B. J. Skinner, in an article titled "A second iron age ahead?" (1976), predicts that the day when "we will have to come to grips with the way in which the earth offers us its riches . . . is less than a century away, perhaps less than a half century. When it dawns we will have to learn to use iron and other

abundant metals for all our needs." On the other side, resource economist J. F. McDivitt (1974) believes that "if the peoples of the world continue to work closely together and to move towards an ever more efficient pattern of resource use, . . . mineral shortages will continue to be only a faint cloud on the world's horizon."

The mineral supply system certainly will have to deal in the long term with serious constraints if it is to keep up with demand. First of all, demand itself has been growing exponentially and, of course, that sort of growth cannot continue. Brooks and Andrews (1974) have suggested that "relative demands (for minerals) decline after a point with increase in per capita income. Indeed . . . the relative growth is sufficiently damped that it was suggested in one study that mineral production will need to grow less fast in the future than it has grown in the recent past, exactly the opposite of most conclusions based on trend analysis." This is the same sort of hopeful sign that in recent years we have witnessed in some population growth in reaction to increased per capita income. Another constraint to the long-term mineral supply system is a threatened shortage of energy. Much of the technologic advance that replenishes mineral resources is energy intensive, so this could be a serious future constraint and, in fact, the rise in energy costs in the last years has been severely felt in most parts of the mining industry. Another present major constraint that could increase even more is the accommodation of the mineral supply system to the regulatory controls and costs concerned with environmental pollution and degradation as well as to the public desire to withdraw from mineral entry all public lands judged better used as a wilderness or an ecologic reserve. Another possible constraint is what B. J. Skinner calls the "mineralogical barrier." Scarce metals in the earth's crust, such as copper, lead, nickel, tin, and tungsten, are mostly disseminated within the atomic structure of host minerals that make up common rock. In that way they are accessible to recovery only by chemically breaking down the host minerals, a feat which requires very large amounts of energy. Such scarce metals are now recovered from the geologically rare deposits in which they occur as a principal component. Geologist Skinner believes that such deposits will soon be depleted. Economists Brooks and Andrews argue against this concept by holding that "every bit of evidence we have indicates the existence of mineral resources (at lower grades) that could be mined and, further, that either as their price goes up or as their cost goes down (which is to say, as technology of extraction improves), the volume of mineable material increases significantly—not by a factor of 5 or 10 but by a factor of 100 or 1,000." The arguments on this critical issue could be better focused by additional scientific information on the amounts of mineralized rock available at different grades, because existing data are too limited for definite conclusions. Thus, it is not certain that the geologic availability of lower grade resources to the mineral resource supply system is assured so a potential barrier to the resource system remains.

Whether the mineral resource supply system can overcome these restraints in the long run depends ultimately on the magnitude of those mineral resources that we cannot now assess. They are the resources that we are unable to conceive or for which we are unable to foresee the discovery or recovery techniques. There is no question that the resources we know about are a fixed stock and eventualy will run out. But, can they be replaced?

Another way of posing this problem is to ask the question whether or not an equilibrium of mineral use can be established that will last indefinitely or nearly so. If the real world of supply and demand is likened to a model where mineral resources are fixed and demand is dynamic and expanding, minerals will run out. But this model is not like the real world where supply and demand are dynamically interrelated and mineral shortages set off a whole set of changes including higher prices, reduced demand, increased conservation, increased substitution, increased recycling, increased technologic and scientific research, increased exploration and increased exploitation of lower grades ores. In the real world, mineral use has evolved to overcome such shortages as firewood from depleted forests, or copper from mined out high grade veins, and such evolution will continue to operate as new shortages arise. This is not to say that such minerals as petroleum in conventional fields, and presently economic deposits of mercury, helium, platinum, and other such geologic rarities will not be exhausted eventually. But our technology likely will evolve to accommodate to a lesser or more expensive supply of such minerals, much as it has developed without abundant supplies of the metals praeseodymium, neodymium, promethium, gadolinium, terbium, and the other rare earths, which were they abundant, probably would be used widely.

Mineral economist John E. Tilton in his excellent book, *The Future of Nonfuel Minerals*, concluded:

In the more distant future—the twenty-first century and beyond—depletion could become a more pressing problem. It is important to stress this possibility, for the consequences to industrial societies could be most severe. At the same time, it should be noted that the arsenal available to mankind for dealing with this threat is not empty. As pointed out above, public policies that support research in minerals and reduce their consumption increase the likelihood that technological progress will continue to offset the adverse effects of depletion. Other policies, such as programs to encourage smaller families, to slow the growth of population, may be desirable for other reasons as well. Finally, even in the absence of such policies, one cannot be certain that depletion will ultimately overwhelm the cost-reducing impact of new technology. For as depletion starts to push mineral prices up, it unleashes forces that stimulate the substitution of cheaper and more abundant materials for the increasingly scare minerals, encourages the search for new and unconventional sources of supplies, and promotes the development of more cost-reducing technologies. Conceivably, these forces could by themselves keep the specter of depletion at bay indefinitely.

It seems clear that our long-term mineral supply system is much stronger than believed by the analysts who regard mineral resources as a fixed and essentially known and fully conceived stock. At the same time, we have the responsibility of keeping the system healthy, and some steps should be taken to do so. The workings of the mineral resource system in its full complexity needs examination to better understand the factors that affect it. A statistical monitoring series that would give early warning of a weakening in the resource replenishment process in any part of the system should be devised and set up. Finally, adequate research and development should be carried out in order to strengthen weak parts of the mineral supply system.

References Cited

Brooks, David B. and P. W. Andrews, 1974, Mineral Resources, economic growth, and world population: Science, v. 185, no. 4145, p. 13–19.

Degens, Egan and David A. Ross, 1969, Hot brines and recent heavy metal deposits in the Red Sea: Springer Verlag New York Inc., New York, 600 p.

McDivitt, James F., 1974, Minerals and Men: Published for Resources for the Future, Inc. by The Johns Hopkins Press, Baltimore, 175 p.

McKelvey, V. E., 1972, Mineral Resources, Environmental Quality and the Limits to Growth: Intermet Bulletin, no. 2, vol. 2, p. 17–21.

President's Materials Policy Commission, W. S. Paley, chm, 1952, Resources for freedom: Washington, D.C., U.S. Govt. Printing Office, 818 p, 5 vols.

Skinner, Brian J., 1976, A second iron age ahead?: American Scientist, v. 64, p. 258–269.

Tilton, John E., 1977, The Future of Nonfuel Minerals: The Brookings Institution, Washington, D.C., 113 p.

U.S. Bureau of Mines and U.S. Geological Survey, 1976, Principles of the mineral resources classification system of the U.S. Bureau of Mines and the U.S. Geological Survey: U.S. Geol. Surv. Bull. 1450-A, 5 p.

U.S. Geological Survey, 1922, The oil supply of the United States: Bull. Amer. Assoc. of Petrol. Geologists, vol. 6, no. 1, p. 42–46.

U.S. Geological Survey, 1968, Mineral and water resources of Montana: 90th Cong. Senate Document no. 98, U.S. Govt. Printing Office, 166 p.

24 Realities of Mineral Distribution

Preston E. Cloud, Jr.

Introduction

. . . Optimism and imagination are happy human traits. They often make bad situations appear tolerable or even good. Man's ability to imagine solutions, however, commonly outruns his ability to find them. What does he do when it becomes clear that he is plundering, overpopulating, and despoiling his planet at such a horrendous rate that it is going to take some kind of a big leap, and soon, to avert irreversible degradation?

Dr. Weinberg, with his marvelous conception of a world set free by nuclear energy sees man at this juncture in history as comparable to the frog who was trying, unsuccessfully, to jump out of a deep rut. A second frog came along, and, seeing his friend in distress, told him to rest awhile while he fetched some sticks to build a

Reprinted from, *Texas Quarterly*, v. 11, pp. 103–26 by permission of the author and the publisher.

Mr. Cloud is Professor of Biogeology at the University of California, Santa Barbara.

platform from which it would be but a short leap to the top of the rut. When frog number two returned, however, his friend was nowhere to be seen. A glance around soon revealed him sitting at the top of the rut. "How did you get up there?" the second frog exclaimed. "Well," said the first frog, "I had to—a hell of a big truck came down the road." In this story we are the first frog, the truck is overpopulation, pollution, and dwindling mineral resources, and the extra oomph that gets us out of the rut is nuclear power—specifically the breeder reactor, and eventually contained fusion.

The inventive genius of man has got him out of trouble in the past. Why not now? Why be a spoil-sport when brilliant, articulate, and well-intentioned men assure us that all we need is more technology? Why? Because the present crisis is exacerbated by four conditions that reinforce each other in a very undesirable manner: (1) the achievements of medical technology which have brought on the run-away imbalance between birth and death able national dream of an ever-increasing real Gross National Product based on obsolescence and waste; (3) the finite nature of the earth and particularly its accessible mineralized crust; and (4) the increased risk of irreversible spoilation of the environment which accompanies overpopulation, overproduction, waste, and the movement of ever-larger quantities of source rock for ever-smaller proportions of useful minerals.

Granted the advantages of big technological leaps, therefore, provided they are in the right direction, I see real hope for permanent long-range solutions to our problems as beginning with the taking of long-range views of them. Put in another way, we should not tackle vast problems with half-vast concepts. We must build a platform of scientific and social comprehension, while concurrently endeavoring to fill the rut of ignorance, selfishness, and complacency with knowledge, restraint, and demanding awareness on the part of an enlightened electorate. And we must not be satisfied merely with getting the United States or North America through the immediate future, critical though that will be. We must consider what effects current and proposed trends and actions will have on the world as a whole for several generations hence, and how we can best influence those trends favorably the world over. Above all, we must consider how to preserve for the yet unborn the maximum flexibility of choices consistent with meeting current and future crises.

Rhetoric, however, either cornucopian or Malthusian, is no substitute for informed foresight and rational action or purposeful inaction.

What are the problems and misconceptions that impede the desired progress? And what must we invest in research and action—scientific, technological, and social—to assure a flexibility of resource options for the long range as well as for the immediate future? Not only until 1985, not only until the year 2000, not only even until the year 2050, but for a future as long as or longer than our past. In the nearly five billion years of earth history is man's brief stay of now barely a million years to be only a meteoric flash, and his industrial society less than that? Or will he last with reasonable amenities for as long as the dinosaurs?

Nature and Geography of Resources

Man's concept of resources, to be sure, depends on his needs and wants, and thus to a great degree on his locale and place in history, on what others have, and on what he knows about what they have and what might be possible for him to obtain. Food and fiber from the land, and food and drink from the waters of the earth have always been indispensable resources. So have the human beings who have utilized these resources and created demands for others—from birch bark to beryllium, from buffalo hides to steel and plastic. It is these other resources, the ones from which our industrial society has been created, about which I speak today. I refer, in particular, to the nonrenewable or wasting resources— mineral fuels which are converted into energy plus carbon, nuclear fuels, and the metals, chemicals, and industrial materials of geological origin which to some extent can be and even are recycled but which tend to become dispersed and wasted.

All such resources, except those that are common rocks whose availability and value depend almost entirely on economic factors plus fabrication, share certain peculiarities that transcend economics and limit technology and even diplomacy. They occur in local concentrations that may exceed their crustal abundances by thousands of times, and particular resources tend to be clustered within geochemical or metallogenic provinces from which others are excluded. Some parts of the earth are rich in mineral raw materials and others are poor.

No part of the earth, not even on a continent-wide basis, is self-sufficient in all critical metals. North America is relatively rich in molybdenum and poor in tin, tungsten, and manganese, for instance, whereas Asia is comparatively rich in tin, tungsten, and manganese and, apparently, less well supplied with molybdemun. The great bulk of the world's gold appears to be in South Africa, which has relatively little silver but a good supply of platinum. Cuba and New Caledonia have well over half the world's total known reserves of nickel. The main known reserves of cobalt are in the Congo Republic, Cuba, New Caledonia, and parts of Asia. Most of the world's mercury is in Spain, Italy, and parts of the Sino-Soviet bloc. Industrial diamonds are still supplied mainly by the Congo.

Consider tin. Over half the world's currently recoverable reserves are in Indonesia, Malaya, and Thailand, and much of the rest is in Bolivia and the Congo. Known North American reserves are negligible. For the United States loss of access to extra-continental sources of tin is not likely to be offset by economic factors or technological changes that would permit an increase in potential North American production, even if present production could be increased by an order of magnitude. It is equally obvious that other peculiarities in the geographical distribution of the world's geological resources will continue to encourage interest both in trading with some ideologically remote nations and in seeking alternative sources of supply.

Economic geology, which in its best sense brings all other fields of geology to bear on resource problems, is concerned particularly with questions of how certain elements locally attain geochemical concentrations that greatly exceed their crustal abundance and with how this knowledge can be applied to the discovery of new deposits and the delineation of reserves. Economics and technology play equally important parts with geology itself in determining what deposits and grades it is practicable to exploit. Neither economics, nor technology, nor geology can *make* an ore deposit where the desired substance is absent or exists in insufficient quantity.

Estimated Recoverable Reserves of Selected Mineral Resources

Consider now some aspects of the apparent lifetimes of estimated recoverable reserves of a selection of critical mineral resources and the position of the United States with regard to some of these. The selected resources are those for which suitable data are available.

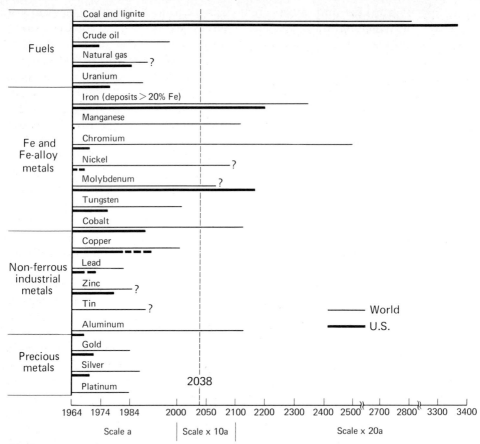

FIG. 1. Lifetimes of estimated recoverable reserves of mineral resources at current mineable grades and rates of consumption (no allowance made for increasing populations and rates of consumption, or for submerged or otherwise concealed deposits, use of now submarginal grades, or imports). (Data from Flawn, 1966).

Figure 1 shows such lifetimes for different groups of metals and mineral fuels at *current* minable grades and rates of consumption. No allowance is made for increase of populations, or for increased rates of consumption which, in the United States, tend to increase at twice the rate of population growth. Nor is allowance made for additions to reserves that will result from discovery of submarine deposits, use of submarginal grades, or imports—which may reduce but will not eliminate the impact of growth factors. Data are from the U.S. Bureau of Mines compendia *Mineral Facts and Problems* and its *Minerals Yearbooks*, as summarized by Flawn (*Mineral Resources*, Rand McNally, 1966). The thin lines represent lifetimes of world reserves for a stable population of roughly 3.3×10^9 at current rates of use. The heavy lines represent similar data for a United

States population of about 200 million. Actual availability of some such commodities to the United States will, of course, be extended by imports from abroad, just as that of others will be reduced by population growth, increased per capita demands, and perhaps by political changes. The dashed vertical line represents the year 2038. I have chosen this as a reference line because it marks that point in the future which is just as distant from the present as the invention of the airplane and the discovery of radio-activity are in the past.

The prospect is hardly conducive to unrestrained optimism. Of the nineteen commodities considered, only fourteen for the world and four or five for the United States have assured lifetimes beyond 1984; only ten for the world and three for the United States persist beyond the turn of the century; and only eight for the world and three for the United States extend beyond 2038. I do not suggest that we equate these lines with revealed truth. Time will prove some too short and others perhaps too long. New reserves will be found, lower-grade reserves will become minable for economic or technological reasons, substitutes will be discovered or synthesized, and some critical materials can be conserved by waste control and recycling. The crucial questions are: (1) how do we reduce these generalities to specifics; (2) can we do so fast enough to sustain current rates of consumption; (3) can we increase and sustain production of industrial materials at a rate sufficient to meet the rising expectations of a world population of nearly three and one-half billion, now growing with a doubling time of about thirty to thirty-five years, and for how long; and (4) if the answer to the last question is no, what then?

A more local way of viewing the situation is to compare the position of the United States or North America with other parts of the world. Figs. 2 to 4 show such a comparison for sixteen commodities with our favorite measuring stick, the Sino-Soviet bloc. Figure 2 shows the more cheerful side of the coin. The United States is a bit ahead in petroleum, lignite, and phosphate, and neither we nor Asia have much chromium—known reserves are practically all in South Africa and Rhodesia. Figure 3, however, shows the Sino-Soviet bloc to have a big lead in zinc, mercury, potash, and bauxite. And Fig. 4 shows similar leads in tungsten, copper, iron, and coal.

Again there are brighter aspects to the generally unfavorable picture. Ample local low grade sources of alumina other than bauxite are available with metallurgical advances and at a price. The United States coal supply is not in danger of immediate shortage. Potassium can be extracted from sea water. And much of the world's iron is in friendly hands, including those of our good neighbor Canada and our more distant friend Australia.

No completely safe source is visible, however, for mercury, tungsten, and chromium. Lead, tin, zinc, and the precious metals appear to be in short supply throughout the world. And petroleum and natural gas will be exhausted or nearly so within the lifetimes of many of those here today unless we decide to conserve them for petrochemicals and plastics. Even the extraction of liquid fuels from oil shales and "tar sands," or by hydroge-

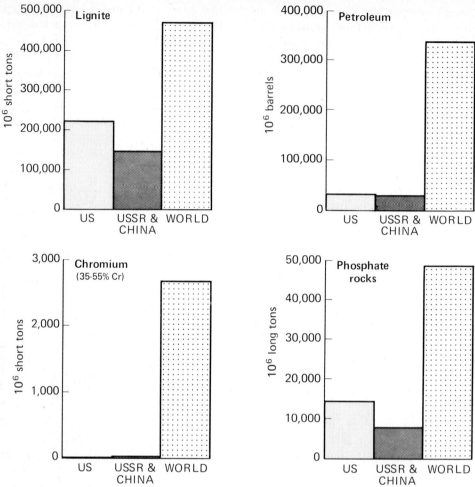

FIG. 2. Estimated recoverable reserves of minerals (above sea level) for which U.S. reserve estimates exceed, equal, or fall only slightly below those of the U.S.S.R. plus Mainland China. (Data from Flawn, 1966).

nation of coal, will not meet energy requirements over the long term. If they were called upon to supply all the liquid fuels and other products now produced by the fractionation of petroleum, for instance, the suggested lifetime for coal the reserves of which are probably the most accurately known of all mineral products, would be drastically reduced below that indicated in Fig. 1—and such a shift will be needed to a yet unknown degree before the end of the century.

The Cornucopian Premises

In view of these alarming prospects, why do intelligent men of good faith seem to assure us that there is nothing to be alarmed about? It can only be

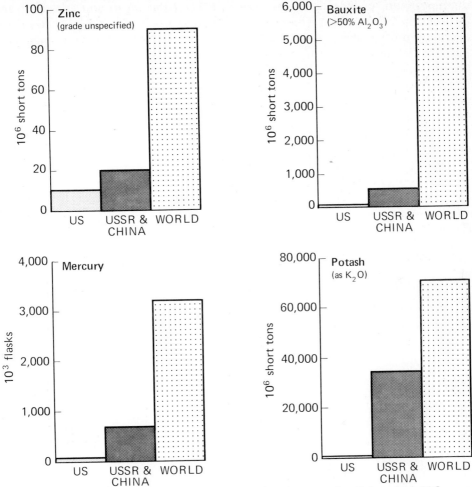

FIG. 3. Estimated recoverable reserves of minerals (above sea level) for which U.S. reserve estimates are less than those of the U.S.S.R. plus Mainland China. (Data from Flawn, 1966).

because they visualize a completely nongeological solution to the problem, or because they take a very short-range view of it, or because they are compulsive optimists or are misinformed, or some combination of these things.

Let me first consider some of the basic concepts that might give rise to a cornucopian view of the earth's mineral resources and the difficulties that impede their unreserved acceptance. Then I will suggest some steps that might be taken to minimize the risks or slow the rates of mineral-resource depletion.

The central dilemma of all cornucopian premises is, of course, how to sustain an exponential increase of anything—people, mineral products, industrialization, or solid currency—on a finite resource base. This is, as everyone must realize, obviously impossible in the long run and will become increasingly difficult in the short run. For great though the mass of the earth is, well under 0.1 per cent of that mass is accessible to us by any imaginable means (the entire crust is only about 0.4 per cent of the total mass of the earth) and this relatively minute accessible fraction, as we have seen and shall see, is very unequally mineralized.

But the cornucopains are not naive or mischievous people. On what grounds do they deny the restraints and belittle the difficulties?

The five main premises from which their conclusions follow are:

FIG. 4. Estimated recoverable reserves of minerals (above sea level) for which U.S. reserve estimates are less than those of the U.S.S.R. plus Mainland China. (Data from Flawn, 1966).

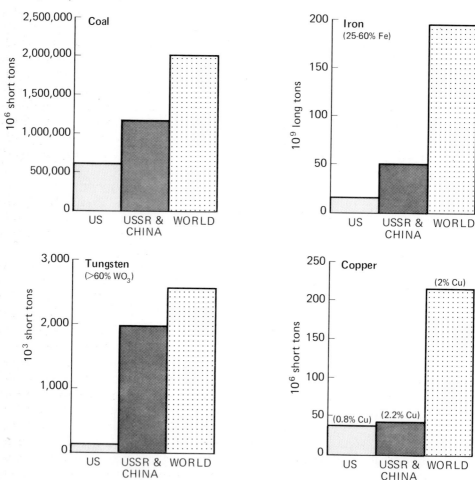

Premise I—the promise of essentially inexhaustible cheap useful energy from nuclear sources.

Premise II—the thesis that economics is the sole factor governing availability of useful minerals and metals.

Premise III—the fallacy of essentially uninterrupted variation from ore of a metal to its average crustal abundance, which is inherent in Premise II; and from which emanates the strange and misleading notion that quantity of a resource available is essentially an inverse exponential function of its concentration.

Premise IV—the crucial assumption of population control, without which there can be no future worth living for most of the world (or, worse, the belief that quantity of people is of itself the ultimate good, which, astounding as it may seem, is still held by a few people who ought to know better—see, for instance, Colin Clark, *Population Growth and Land Use*, Macmillan, 1967).

Premise V—the concept of the "technological fix."

Now these are appealing premises, several of which contain large elements of both truth and hope. Why do I protest their unreserved acceptance? I protest because, in addition to elements of truth, they also contain assumptions that either are gross oversimplifications, outright errors, or are not demonstrated. I warn because their uncritical acceptance contributes to a dangerous complacency toward problems that will not be solved by a few brilliant technological "break-throughs," a wider acceptance of deficit economy, or fall-out of genius from unlimited expansion of population. They will be solved only by intensive wide-ranging, and persistent scientific and engineering investigation, supported by new social patterns and wise legislation.

I will discuss these premises in the order cited.

Premise I

The concept of essentially inexhaustible cheap useful energy from nuclear sources offers by all odds the most promising prospect of sweeping changes in the mineral resource picture, as Dr. Weinberg so brilliantly argued. We may be on the verge of developing a workable breeder reactor just in time to sustain an energy-hungry world facing the imminent exhaustion of traditional energy sources. Such a development, it has been persuasively stated, will also banish many problems of environmental pollution and open up unlimited reserves of metals in common crustal rocks. There are, unhappily, some flaws in this delightful picture, of which it is important to be aware.

As Dr. Weinberg has told you, uranium 235 is the only naturally occurring spontaneously fissionable source of nuclear power. When a critical mass of uranium is brought together, the interchange of neutrons back and forth generates heat and continues to do so as long as the U^{235} lasts. In the breeder reactor some of the free neutrons kick common U^{238} over to plutonium 239, which is fissionable and produces more neutrons, yielding heat and accelerating the breeder reaction. Even in existing reactors some breeding takes place, and, if a complete breeding system could be produced, the amount of energy available from uranium alone would be increased about 140 fold. If thorium also can be made to breed, energy generated could be increased about 400 fold over that now attainable. This would extend the lifetime of visible energy resources at demands anticipated by 1980 by perhaps 1000 to 3000 years and gain time to work on contained nuclear fusion.

The problem is that it will require about 275,000 short tons of $6.00 to $10.00 per pound U_3O_8 (not ore, not uranium) to fuel reactors now on order to 1980, plus another 400,000 tons to sustain them until the turn of the century, burning only U^{235} with currently available enrichments from slow breeding (Charles T. Baroch, U.S. Bureau of Mines, oral comment). Only about 195,000 of the 675,000 tons of uranium needed is known to be available at this price, although known geologic conditions indicate the possibility of another 325,000 tons. Thus we now appear to be about 155,000 tons short of the U_3O_8 needed to produce the hoped-for 150,000 megawatts of nuclear energy on a sustained basis from 1985 until the end of the century without a functioning breeder reactor. Unless we find a lot more uranium, or pay a lot more money for it, or get a functioning complete breeder reactor or contained nuclear fusion within ten or fifteen years, the energy picture will be far from bright. There is good reason to hope that the breeder will come, and after it contained fusion, *if* the U^{235} and helium hold out—but there is no room for complacency.

If and when the breeder reactor or contained fusion does become avail-able as a practicable energy source, however, how will this help with mineral resources? It is clear immediately that it will take pressure off the fossil "fuels" so that it will become feasible, and should become the law, to reserve them for petrochemicals, polymers, essential liquid propellants, and other special purposes not served by nuclear fuels. It is also clear that cheap massive transportation, or direct transmittal of large quantities of cheap electric power to, or its generation at, distant resources will bring the mineral resources of remote sites to the market place—either as bulk ore for processing or as the refined or partially refined product.

What is not clear is how this very cheap energy will bring about the extraction of thinly dispersed metals in large quantity from common rock. The task is very different from the recovery of liquid fuels or natural gas by nuclear fracturing. The procedure usually suggested is the break-up of rock in place at depth with a nuclear blast, followed by hydrometallurgical or chemical mining. The problems, however, are great. Complexing solutions, in large quantity also from natural resources, must be brought into contact with the particles desired. This means that the enclosing rock must be fractured to that particle size. Then other substances, unsought, may use up and dissipate valuable reagents. Or the solvent reagents may escape to ground waters and become contaminants. Underground electrolysis is no more promising in dealing with very low concentrations. And the bacteria that catalyze reactions of metallurgical interest are all aerobic, so that, in addition to having access to the particles of interest, they must also be provided with a source of oxygen underground if they are to work there.

Indeed the energy used in breaking rock for the removal of metals is not now a large fraction of mining cost in comparison with that of labor and capital. The big expense is in equipping and utilizing manpower, and, although cheap energy will certainly reduce manpower requirements, it will probably never adequately substitute for the intelligent man with the pick at the mining face in dealing with vein and many replacement deposits, where the sought-after materials are irregularly concentrated in limited spaces. There are also limits to the feasible depths of open pit mining, which would be by all odds the best way to mine common rock. Few open pit mines now reach much below about 1,500 feet. It is unlikely that such depths can be increased by as much as an order of magnitude. The quantity of rock removable decreases exponentially with depth because pit circumference must decrease downward to maintain stable walls.

It may also not be widely realized by non-geologists that many types of ore bodies have definite floors or pinch-out downward, so that extending exploitative operations to depth gains no increase in ore produced. Even where mineralization does extend to depth, of course, exploitability is ultimately limited by temperature and rock failure.

Then there is the problem of reducing radioactivity so that ores can be handled and the refined product utilized without harm—not to mention heat dispersal (which in some but not all situations could itself be a resource)

and the disposal of waste rock and spent reagents.

Altogether the problems are sufficiently formidable that it would be foolhardy to accept them as resolved in advance of a working efficient breeder reactor plus a demonstration that either cheap electricity or nuclear explosions will significantly facilitate the removal of metals from any common rock.

A pithy comment from Peter Flawn's recent book on *Mineral Resources* (Rand McNally, 1966, p. 14) is appropriate here. It is to the effect that "average rock will never be mined," It is the uncommon features of a rock that make it a candidate for mining! Even with a complete nuclear technology, sensible people will seek, by geological criteria, to choose and work first those rocks or ores that show the highest relative recoverable enrichments in the desired minerals.

The reality is that even the achievement of a breeder reactor offers no guarantee of unlimited mineral resources in the face of geologic limitations and expanding populations with increased per capita demands, even over the middle term. To assume such for the long term would be sheer folly.

Premise II
The thesis that economics is the sole, or at least the dominant, factor governing availability of useful minerals and metals is one of those vexing part-truths which has led to much seemingly fruitless discussion between economists and geologists. This proposition bears examination.

It seems to have its roots in that interesting economic index known as the Gross National Product, or GNP.

No one seems to have worked out exactly what proportion of the GNP is in some way attributable to the mineral resource base. It does, however, appear that the dollar value of the raw materials themselves is small compared to the total GNP, and that it has decreased proportionately over time to something like 2 per cent of the present GNP, as I recall. From this it is logically deduced that the GNP could, if necessary, absorb a several-fold increase in cost of raw materials. The gap in logic comes when this is confused with the notion that all that is necessary to obtain inexhaustible quantities of any substance is either to raise the price or to increase the volume of rock mined. In support of such a notion, of course, one *can* point to diamond, which in the richest deposit ever known occurred in a concentration of only one to twenty-five million, but which, nevertheless, has continued to be available. The flaw is not only that we cannot afford to pay the price of diamond for many substances, but also that no matter how much rock we mine we can't get diamonds out of it if there were none there in the first place.

Daniel Bell (1967, Notes on the Post-industrialist Society II: *in* the *Public Interest,* no. 7, p. 102–118) comments on the distorted sense of relations that emerges from the cumulative nature of GNP accounting. Thus, when a mine is developed, the costs of the new facilities and payroll become additions to the GNP, whether the ore is sold at a profit or not. Should the mine wastes at the same time pollute a stream, the costs of cleaning up the stream or diverting the wastes also become additions to the GNP. Similarly if you hire someone to wash the dishes this adds to GNP, but if your wife does them it doesn't count.

From this it results that mineral raw materials and housework are not very impressive fractions of the GNP. What seems to get lost sight of is what a mess we would be in without either!

Assuming an indefinite extension of their curves and continuance of access to foreign markets, economists appear to be on reasonably sound grounds in postulating the relatively long-term availability of certain sedimentary, residual, and disseminated ores, such as those of iron, aluminum, and perhaps copper. What many of them do not appreciate is . . . that the type of curve that can with some reason be applied to such deposits and metals is by no means universally applicable. This difficulty is aggravated by the fact that conventional economic indexes minimize the vitamin-like quality for the economy as a whole of the raw materials whose enhancement in value through beneficiation, fabrication, and exchange accounts for such a large part of the material assets of society.

In a world that wants to hear only good news some economists are perhaps working too hard to emancipate their calling from the epithet of "dismal science," but not all of them. One voice from the wilderness of hyperoptimism and overconsumption is that of Kenneth Boulding, who observes that *"The essential measure of the success of the economy is not production and consumption at all, but the nature, extent, quality, and complexity of the total capital stock, including in this the state of the human bodies and minds included in the system"* (p. 9 in K. E. Boulding, 1966, "The economics of the coming spaceship Earth," p. 3–14 in *Environmental Qual-*

ity in a Growing Economy, Resources of the Future, Inc., The Johns Hopkins Press). Until this concept penetrates widely into the councils of government and the conscience of society, there will continue to be a wide gap between the economic aspects of national and industrial policy and the common good, and the intrinsic significance of raw materials will remain inadequately appreciated.

The reality is that economics per se, powerful though it can be when it has material resources to work with, is not all powerful. Indeed, without material resources to start with, no matter how small a fraction of the GNP they may represent, economics is of no consequence at all. The current orthodoxy of economic well-being through obsolescence, over-consumption, and waste will prove, in the long term, to be a cruel and preposterous illusion.

Premise III

Premise III, the postulate of essentially uninterrupted variation from ore to average crustal abundance is seldom if ever stated in that way, but it is inherent in Premise II. It could almost as well have been treated under Premise II; but it is such an important and interesting idea, whether true or false, that separate consideration is warranted.

If the postulated continuous variation were true for mineral resources in general, volume of "ore" (not metal) produced would be an exponential inverse function of grade mined, the handling of lower grades would be compensated for by the availability of larger quantities of elements sought, and reserve estimates would depend only on the accuracy with which average crustal abundances were known.

Problems in extractive metallurgy, of course, are not considered in such an outlook.

This delightfully simple picture would supplant all other theories of ore deposits, invalidate the foundations of geochemistry, divest geology of much of its social relevance, and place the fate of the mineral industry squarely in the hands of economists and nuclear engineers.

Unfortunately this postulate is simply untrue in a practical sense for many critical minerals and is only crudely true, leaving out metallurgical problems, for particular metals, like iron and aluminum, whose patterns approach the predicted form. Sharp discontinuities exist in the abundances of mercury, tin, nickel, molybdenum, tungsten, manganese, cobalt, diamond, the precious metals, and even such staples as lead and zinc, for example. But how many prophets of the future are concerned about where all the lead or cadmium will come from for all those electric automobiles that are supposed to solve the smog problem?

Helium is a good example of a critical substance in short supply. Although a gas which has surely at some places diffused in a continuous spectrum of concentrations, particular concentrations of interest as a source of supply appear from published information to vary in a stepwise manner. Here I draw on data summarized by H. W. Lipper in the 1965 edition of the U.S. Bureau of Mines pulication *Mineral Facts and Problems.* Although an uncommon substance, helium serves a variety of seemingly indispensable uses. A bit less than half of the helium now consumed in the U.S. is used in pressurizing liquid fueled missiles and space ships. Shielded-arc welding is the next largest use, followed closely by its use in producing controlled atmospheres for growing crystals for transistors, processing fuels for nuclear energy, and cooling vacuum pumps. Only about 5.5 per cent of the helium consumed in the United States is now used as a lifting gas. It plays an increasingly important role, however, as a coolant for nuclear reactors and a seemingly indispensable one in cryogenics and superconductivity. In the latter role, it could control the feasibility of massive long-distance transport of nuclear-generated electricity. High-helium low-oxygen breathing mixtures may well be critical to man's long-range success in attempting to operate at great depths in the exploration and exploitation of the sea. Other uses are in research, purging, leak detection, chromatography, etc.

Helium thus appears to be a very critical element, as the Department of the Interior has recognized in establishing its helium-conservation program. What are the prospects that there will be enough helium in 2038?

The only presently utilized source of helium is in natural gas, where it occurs at a range of concentrations from as high as 8.2 per cent by volume to zero. The range, however, in particular gas fields of significant volume, is apparently not continuous. Dropping below the one field (Pinta Dome) that shows an 8.2 per cent concentration, we find a few small isolated fields (Mesa and Hogback, New Mexico) that contain about 5.5 per cent helium, and then several large fields (e.g., Hugoton and Texas Panhandle) with a range of 0.3 to 1.0 per cent helium. Other large natural gas fields contain either no helium or show it only in quantities of less than 5 parts per 10,000. From the latter there is a long jump down to the atmosphere with a concentration of only 1 part per 200,000.

Present annual demand for helium is about 700 million cubic feet, with a projected increase in demand to about 2 billion cubic feet annually by about 1985. It will be possible to meet such an accelerated demand for a limited time only as a result of Interior's current purchase and storage program, which will augment recovery from natural gas then being produced. As now foreseen, if increases in use do not outrun estimates, conservation and continued recovery of helium from natural gas reserves will meet needs to slightly beyond the turn of the century. When known and expected discoveries of reserves of natural gas are exhausted shortly thereafter, the only potential sources of new supply will be from the atmosphere, as small quantities of He³ from nuclear reactor technology, or by synthesis from hydrogen—a process whose practical feasibility and adequacy remain to be established.

Spending even a lot more money to produce more helium from such sources under existing technology just may not be the best or even a very feasible way to deal with the problem. Interior's conservation program must be enlarged and extended, under compulsory legislation if necessary. New sources must be sought. Research into possible substitutions, recovery and re-use, synthesis, and extraction from the atmosphere must be accelerated—*now* while there is still time. And we must be prepared to curtail, if necessary, activities which waste the limited helium reserves. Natural resources are the priceless heritage of all the people; their waste cannot be tolerated.[1]

Problems of the adequacy of reserves obtain for many other substances, especially under the escalating demands of rising populations and expectations, and it is becoming obvious to many geologists that time is running out. Dispersal of metals which could be recycled should be controlled. Unless industry and the public undertake to do this voluntarily, legislation should be generated to define permissible mixes of material and disposal of "junk" metal. Above all the wastefulness of war and preparation for it must be terminated if reasonable options for posterity are to be preserved.

The reality is that a healthy mineral resource industry, and therefore a healthy industrial economy, can be maintained only on a firm base of geologic knowledge, and geochemical and metallurgical understanding of the distribution and limits of metals, mineral fuels, and chemicals in the earth's crust and hydrosphere.

Premise IV

The assumption that world populations will soon attain and remain in a state of balance is central to all other premises. Without this the rising expectations of the poor are doomed to failure, and the affluent can remain affluent only by maintaining existing shameful discrepancies. Taking present age structures and life expectancies of world populations into account, it seems certain that, barring other forms of catastrophe, world population will reach six or seven billion by about the turn of the century, regardless of how rapidly family planning is accepted and practiced.

On the most optimistic assumptions, this is probably close to the maximum number of people the world can support on a reasonably sustained basis, even under strictly regularized conditions, at a general level of living roughly comparable to that now enjoyed in Western Europe. It would, of course, be far better to stabilize at a much smaller world population. In any case, much greater progress than is as yet visible must take place over much larger parts of the world before optimism on the prospects of voluntary global population control at any level can be justified. And even if world population did level off and remain balanced at about seven billion, it would probably take close to one hundred years of intensive, enlightened, peaceful effort to lift all mankind to anywhere near the current level of Western Europe or even much above the level of chronic malnutrition and deprivation.

This is not to say that we must therefore be discouraged and withdraw to ineffectual diversions. Rather it is a challenge to focus with energy and realism on seeking a truly better life for all men living and yet unborn and on keeping the latter to the minimum. On the other hand, an uncritical optimism, just for the sake of that good feeling it creates, is a luxury the world cannot, at this juncture, afford.

A variation of outlook on the population problem which, surprisingly enough, exists among a few nonbiological scholars is that quantity of people is of itself a good thing. The misconception here seems to be that frequency of effective genius will increase, even exponentially, with increasing numbers of people and that there is some risk of breeding out of a merely high level of mediocrity in a stabilized population. The extremes of genius and idiocy, however, appear in about the same frequency at birth from truly heterogeneous gene pools regardless of size (the data from Montgomery County, Maryland, are really no exception to this). What is unfortunate, among other things, about overly dense concentrations of people is that this leads not only to reduced likelihood of the identification of mature genius, but to drastic reductions in the development of potential genius, owing to malnutrition in the weaning years and early youth, accompanied by retardation of both physical and mental growth. If we are determined to turn our problems over to an elite corps of mental prodigies a more surefire method is at hand. Nuclear transplant from various adult tissue cells into fertilized ova whose own nuclei have been removed has already produced identical copies of amphibian nucleus-donors and can probably do the same in man (Joshua Lederberg, 1966, *Bull. Atomic Scientists,* v. 22, no. 8, p. 9). Thus we appear to be on the verge of being able to make as many "xerox" copies as we want or need of any particular genius as long as we can get a piece of his or her nucleated tissue and find eggs and incubators for the genome aliquots to develop in. Female geniuses would be the best because (with a little help) they could copy themselves!

The reality is that without real population control and limitation of demand all else is drastically curtailed, not to say lost. And there is as yet not the faintest glimmer of hope that such limitation may take place voluntarily. Even were all unwanted births to be eliminated, populations would still be increasing at runaway rates in the absence of legal limitation of family size, as Dr. Erlich has so passionately argued. The most fundamental freedom should be the right not to be born into a world of want and smothering restriction. I am convinced that we must give up (or have taken away from us) the right to have as many children as we want or see all other freedoms lost for them. Nature, to be sure, will restore a dynamic balance between our species and the world ecosystem if we fail to do so ourselves—by famine, pestilence, plague, or war. It seems, but is not, unthinkable that this should happen. If it does, of course, mineral resources may then be or appear to be relatively unlimited in relation to demand for them.

Premise V

The notion of the "technological fix" expresses a view that is at once full of hope and full of risk. It is a gripping thought to contemplate a world set free by nuclear energy. Imagine soaring cities, of aluminum, plastic, and thermopane where all live in peace and plenty at unvarying temperature and without effort, drink distilled water, feed on produce grown from more distilled water in coastal deserts, and flit from heliport to heliport in capsules of uncontaminated air. Imagine having as many children as you want, who, of course, will grow up seven stories above the ground and under such germ-free conditions that they will need to wear breathing masks if they ever do set foot in a park or a forest. Imagine a world in which there is no balance of payments problem, no banks, or money, and such mundane affairs as acquiring a shirt or a wife are handled for us by central computer systems. Imagine, if you like, a world in which the only problem is boredom, all others being solved by the state-maintained system of genius-technologists produced by transfer of nuclei from the skin cells of certified gene donors to the previously fertilized ova of final contestants in the annual ideal-pelvis contest. Imagine the problem of getting out of this disease-free world gracefully at the age of 110 when you just can't stand it any longer!

Of course this extreme view may not appeal to people not conditioned to think in those terms, and my guess is that it doesn't appeal to Dr. Weinberg either. But the risk of slipping bit by bit into such a smothering condition as one of the better possible outcomes is inherent in any proposition that encourages or permits people or industries to believe that they can leave their problems to the invention of technological fixes by someone else.

Although the world ecosystem has been in a constant state of flux throughout geologic time, in the short and middle term it is essentially homeostatic. That is to say, it tends to obey Le Chatelier's general principle—when a stress is applied to a system such as to perturb a state of near equilibrium, the system tends to react in such a way as to restore the equilibrium. But large parts of the world ecosystem have probably already undergone or are in danger of undergoing irreversible changes. We cannot continue to plunder and pollute it without serious or even deadly consequences.

Consider what would be needed in terms of conventional mineral raw materials merely to raise the level of all 3.3 billion people now living in the world to the average of the 200 million now living in the United States. In terms of present staple commodities, it can be estimated (revised from Harrison Brown, James Bonner, and John Weir, 1947, *The Next Hundred Years*, Viking Press, p.33) that this would require a "standing crop" of about 30 billion tons of iron, 500 million tons of lead, 330 million tons of zinc, and 50 million tons of tin. This is about 100 to 200 times the present annual production of these commodities. Annual power demands would be the equivalent of about 3 billion tons of coal and lignite, or about ten times present production. To support the doubled populations expected by the year 2000 at the same level would require, of course, a doubling of all the above numbers or substitute measures. The iron needed could probably be produced over a long period of time, perhaps even by the year 2000, given a sufficiently large effort. But, once in circulation, merely to replace losses due to oxidation, friction, and dispersal, not counting production of new iron for larger populations, would take around 200,000 tons of new iron every year (somewhat more than the current annual production of the United States), or a drastic curtailment of losses below the present rate of 1 per cent every two or three years. And the molybdenum needed to convert the iron to steel could become a serious limiting factor. The quantities of lead, zinc, and tin also called for far exceed all measured, indicated, and inferred world reserves of these metals.

This exercise gives a crude measure of the pressures that mineral resources will be under. It seems likely, to be sure, that substitutions, metallurgical research, and other technological advances will come to our aid, and that not all peoples of the world will find a superfluity of obsolescing gadgets necessary for the good life. But this is balanced by the equal likelihood that world population will not really level off at 6.6 or 7 billion and that there will be growing unrest to share the material resources that might lead at least to an improved standard of living. The situation is also aggravated by the attendant problems of disposal of mine wastes and chemically and thermally polluted waters on a vast scale.

The "technological fix," as Dr. Weinberg well understands, is not a panacea but an anesthetic. It may keep the patient quiet long enough to decide what the best long-range course of treatment may be, or even solve *some* of his problems permanently, but it would be tragic to forget that a broader program of treatment and recuperation is necessary. The flow of science and technology has always been fitful, and population control is a central limiting factor in what can be achieved. It will require much creative insight, hard work, public enlightenment, and good fortune to bring about the advances in discovery and analysis, recovery and fabrication, wise use and conservation of materials, management and recovery of wastes, and substitution and synthesis that will be needed to keep the affluent comfortable and bring the deprived to tolerable levels. It will probably also take some revision of criteria for self-esteem, achievement, and pleasure if the gap between affluent and deprived is to be narrowed and demand for raw materials kept within bounds that will permit man to enjoy a future as long as his past, and under conditions that would be widely accepted as agreeable.

The reality is that the promise of the "technological fix" is a meretricious premise, full of glittering appeal but devoid of heart and comprehension of the environmental and social problems. Technology and "hard" science we must have, in sustained and increasing quality, and in quantities relevant to the needs of man—material, intellectual, and spiritual. But in dealing with the problems of resources in relation to man, let us not lose sight of the fact that this is the province of the environmental and social sciences. A vigorous and perceptive technology will be an essential handmaiden in the process, but it is a risky business to put the potential despoilers of the environment in charge of it.

The Nub of the Matter

The realities of mineral distribution, in a nutshell, are that it is neither inconsiderable nor limitless, and that we just don't know yet in the detail required for considered weighing of comprehensive and national long-range alternatives where or how the critical lithophilic elements are concentrated. Stratigraphically controlled substances such as the fossil fuels, and, to a degree, iron and alumina, we can comprehend and estimate within reasonable limits. Reserves, grades, locations, and recoverability of many critical metals, on the other hand, are affected by a much larger number of variables. We in North America began to develop our rich natural endowment of mineral resources at an accelerated pace before the rest of the world. Thus it stands to reason that, to the extent we are unable to meet needs by imports, we will feel the pinch sooner than countries like the U.S.S.R. with a larger component of virgin mineral lands.

In some instances nuclear energy or other technological fixes may buy time to seek better solutions or will even solve a problem permanently. But sooner or later man must come to terms with his environment and its limitations. The sooner the better. The year 2038, by which time even current rates of consumption will have exhausted presently known recoverable reserves of perhaps half the world's now useful metals (more will be found but consumption will increase also), is only as far from the present as the invention of the airplane and the discovery of radioactivity. In the absence of real population control or catastrophe there could be fifteen billion people on earth by then! Much that is difficult to anticipate can happen in the meanwhile, to be sure, and to place faith in a profit-motivated technology and refuse to look beyond a brief "foreseeable future" is a choice widely made. Against this we must weigh the consequences of error or thoughtless inaction and the prospects of identifying constructive alternatives for deliberate courses of long-term action, or inaction, that will affect favorably the long-range future. It is well to remember that to do nothing is equally to make a choice.

Geologists and other environmental scientists now living, therefore, face a great and growing challenge to intensify the research needed to ascertain and evaluate the facts governing availability of raw material resources, to integrate their results, to formulate better predictive models, and to inform the public. For only a cognizant public can generate the actions and exercise the restraints that will assure a tolerable life and a flexibility of options for posterity. The situation calls neither for gloomy foreboding nor facile optimism, but for positive and imaginative realism. That involves informed foresight, comprehensive and long-range outlooks, unremitting effort, inspired research, and a political and social climate conducive to such things.

Conclusions and Proposed Actions

Every promising avenue must be explored. The most imperative objective, after peace and population control, is certainly a workable breeder reactor—with all it promises in reduced energy costs, outlook for desalting saline waters and recovering mineral products from the effluent wastes, availability of now uselessly remote mineral deposits, decrease of cutoff grades, conservation of the so-called fossil "fuels" for more important uses, and the reduction of contaminants resulting from the burning of fossil fuels in urban regions.

But, against the chance that this may not come through on schedule, we should be vigorously seeking additional geological sources of U^{235} and continuing research on controlled nuclear fusion.

A really comprehensive geochemical census of the earth's crustal materials should be accelerated and carried on concurrently, and as far into the future as may be necessary to delineate within reasonable limits the metallogenic provinces of our planet's surface, including those yet poorly-known portions beneath the sea. Such a census will be necessary not only in seeking to discover the causes and locations of new metalliferous deposits, but also in allowing resource data to be considered at the design stage, and in deciding which "common rocks" to mine first, should we ever be reduced to that extreme. Of course, this can be done meaningfully only in context with a good comprehension of sequence and environment based on careful geologic mapping, valid geochronology, perceptive biogeology, and other facets of interpretive earth science.

Programs of geophysical, geochemical, and geological prospecting should meanwhile be expanded to seek more intensively for subglacial, subsoil, submarine, and other concealed mineral resources in already defined favorable target areas—coupled with engineering, metallurgical, and economic innovation and evaluations of deposits found.

Only as we come to know better what we have, where it is, and what the problems of bringing it to the market place are likely to be will it be feasible to formulate the best and most comprehensive long range plans for resource use and conservation. Meanwhile, however, a permanent, high-level, and adequately funded monitoring system should be established under federal auspices to identify stress points in the mineral economy, or likely future demands, well in advance of rupture. Thus the essential lead time could be allowed in initiating search for new sources or substitutes, or in defining necessary conservation programs.

Practices in mixing materials during fabrication and in disposal of scrap metal should be examined with a view to formulating workable legislation that will extend resource lifetimes through more effective re-use.

Management of the nation's resources and of wastes from their extraction, beneficiation, and use should be regarded in the true democratic tradition as national problems and not left entirely to the conscience and discretion of the individual or private firm. Where practices followed are not conducive to the national, regional, or local welfare, informed legal inducement should make them so.

Research into all phases of resource problems and related subjects should be maintained at some effective level not dependent on political whimsey. It would be a far-sighted and eminently fair and logical procedure to set apart some specific fraction of taxes derived from various natural resources to be ploughed back into research designed to assure the integrity of the environment and the sufficiency of resources over the long term.

Much of the work suggested would naturally be centered in the U.S. Department of the Interior and in various state agencies, whose traditionally effective cooperative arrangements with the nation's universities should be enlarged.

[Universities] . . . are also central to the problem of sustaining a healthy industrial society. For they are the source of that most indispensable of all resources—the trained minds that will discern the facts and evolve the principles via which such a society comes to understand its resources and to use them wisely. The essential supplements are adequate support and a vision of the problem that sweeps and probes all aspects of the environmental sciences the world over. The times cry for the establishment of schools and institutes of environmental science in which geologists, ecologists, meteorologists, oceanographers, geophysicists, geographers, and others will interact and work closely together.

I can think of no more fitting way to close these reflections than to quote the recent words of Sir Macfarlane Burnet (p. 29, *in* "Biology and the appreciation of life," The Boyer Lectures, 1966, ABC, 45p.)—*"There are three imperatives: to reduce war to a minimum; to stabilize human populations; and to prevent the progressive destruction of the earth's irreplaceable resources."* If the primary sciences and technology are to be our salvation it will necessarily be in an economic framework that evaluates success by some measure other than rate of turnover, and in the closest possible working liaison with the environmental and social sciences.

Note

[1] In January 1974 the federal government terminated its helium storage and conservation program. [Ed.]

25   Global Energy Resources

The Ford Foundation

The 1973 oil embargo and the events that it set in motion caused those concerned with matters of energy supply to reevaluate the state of knowledge of energy resources both in the United States and in the rest of the world. The major question to be answered was whether there was a sufficient resource base available to allow an orderly transition from present patterns of energy resource use to some new pattern that would be required in the future.

No attempt has been made in this study to develop a new resource assessment, nor has it been necessary, in the context of a twenty-year horizon, to even engage in a careful critique of the resource data base. Work in progress at Resources for the Future and estimates from a large number of published sources were relied on for information on specific sources.

The rapidly increasing prices for conventional fuels and the recurring

Reprinted from *Energy: The Next Twenty Years*, Copyright 1979, the Ford Foundation. Reprinted with updated statistics by permission of Ballinger Publishing Company.

concern about the adequacy of supplies of oil and gas—the two fuels used most widely at present—have shifted attention to the availability of nonconventional fossil fuel resources. As a result, more emphasis than usual has been placed on evaluating the probable size of these resources, because as real prices of the remaining natural gas and oil rise, these resources should become economically recoverable and be extremely useful in making the inevitable transition to renewable resources that will occur sometime in the next century.

There is no universally accepted set of definitions for making resource assessments. The term "reserves" in nearly all classifications includes that part of the resource base (all of the resources that exist in the crust of the earth) that has been identified with what is considered a high degree of accuracy and can be recovered economically using existing technology.

The reserve base increases as real prices rise, with the development of new and lower cost extraction technology, and as new information about resources is developed. "Recoverable resources" is that part of the resource base that it is believed will be discovered and eventually recovered using technology that will be developed and at prices that will prevail in the future. A diagram widely used to classify resource information is shown in Fig. 1, which illustrates the meaning of many of the commonly used resource terms.

Conventional Resources

Coal

World resources are shown in Table 1, which gives the most recent estimates of "geological resources" and reserves of bituminous coal and anthracite by continent and by countries with major coal resources. The term "geological

FIG. 1. Classification of mineral resources.

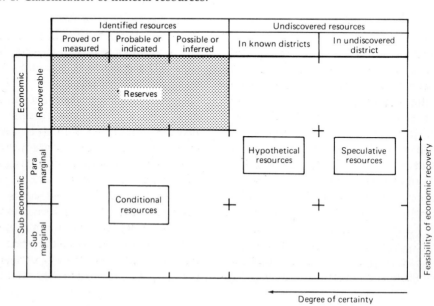

Table 1 World resources of bituminous coal and anthracite (10^6 tons of coal equivalent).[a]

Continent (countries)	Geological resources	Reserves
Africa	172,714	34,033
Botswana	(100,000)	(3,500)
South Africa	(57,366)	(26,903)
America	1,308,541	126,839
United States	(1,190,000)	(113,230)[b]
Canada	(96,225)	(8,706)
Asia	5,494,025	219,226
China	(1,424,680)	(98,883)
USSR	(3,993,000)	(82,900)
Australia and South Seas	213,890	18,164
Australia	(213,760)	(18,128)
Europe	535,664	94,210
Federal Republic of Germany	(230,300)	(23,919)
Poland	(121,000)	(20,000)
United Kingdom	(163,576)	(45,000)
Total	7,724,834	492,472

Source: World Energy Conference, *World Energy Resources 1985–2020* (New York: IPC Science and Technology Press, 1978), pp. 66–67.

Note: Numbers shown in parentheses reflect only the resources of the country with the largest reserves within that continent and are not addable to the totals.

[a]"Equivalent coal" has a heating value of 25.2 million (10^6) Btus per short ton.
[b]These values are lower than those in Table 2 as they do not include sub-bituminous coal and lignite and represent recoverable coal, not coal in place, as does Table 2.

resources" as defined by the World Energy Conference includes only that part of the resource base that may be expected to become of economic value in the future. In addition there are estimated to be an additional 2.4 trillion (10^{12}) tons of coal equivalent of sub-bituminous coal and lignite, for a world coal resource total of 10.1 trillion tons of coal equivalent.

From Table 1 it is clear that three countries dominate the bituminous coal and anthracite geological resources and reserves of the world. The U.S.S.R. has 51.7 percent of all the geological resources, China has 18.4 percent, and the United States has 15.4 percent; these countries together have more than 85 percent.

The world resources of all ranks of coal are so large that the known reserves could supply the total energy demand of the world for 180 years at current rates of consumption, while the geologic resources would last more than a thousand years.

The data on U.S. resources and reserves of coal are more complete than for most other countries. The coal resources of the United States are generally divided into three regional groupings. The eastern region includes the deposits found in the Appalachian area; the interior region includes the deposits from Indiana to west Oklahoma and Texas; and the western region includes all the coal west of the interior region deposits.[1]

The most useful resource classification for understanding the potential role for coal in the United States over the next 50 years is the "demonstrated reserve base." This is the quantity of coal in place under certain specified depth and seam thickness criteria and for which there is a high degree of geologic information and engineering evaluation available. All the coal in the demonstrated reserve base can be recovered economically and legally. The proportion that will actually be recovered depends on the method of mining and other factors and can range from 40 to 90 percent of coal in place.

Table 2 shows the demonstrated reserve base by area, method of mining, and rank of coal. Approximately 68 percent of all reserves are suitable for underground mining. The strippable reserves are concentrated in the West, with 60 percent of the total in the states of Montana, Wyoming, and North Dakota. These reserves consist entirely of sub-bituminous coal and lignite. Low-sulfur coals occur mainly in Montana and Wyoming in the West and in West Virginia in the East. The low-sulfur West Virginia coals are very high quality bituminous coking coals—that is, coals that produce a coke when heated in the absence of air.

Total U.S. resources of coal are very large. Even if just the demonstrated

Table 2   Demonstrated reserve base[a] of coal in the United States on January 1, 1980 (million short tons)

Area and method of mining	Anthracite 1980	Bituminous 1980	Subbituminous 1980	Lignite 1980	Total[b] 1980
East of the Mississippi River:					
Underground minable	7,087.9	162,997.3			170,085.2
Surface minable	129.6	40,428.6		1,083.0	41,641.2
Total[b]	7,217.5	203,425.9		1,083.0	211,726.4
West of the Mississippi River:					
Underground minable	116.4	26,845.9	118,815.0		145,777.3
Surface minable	7.8	9,001.1	63,220.1	42,980.9	115,209.9
Total[b]	124.2	35,847.1	182,035.0	42,980.9	260,987.2
Total east and west of the Mississippi River:					
Underground minable	7,204.3	189,843.2	118,815.0		315,862.5
Surface minable	137.4	49,429.7	63,220.1	44,063.9	156,851.1
Total[b]	7,341.7	239,272.9	182,035.0	44,063.9	472,713.6

Source: Department of Energy, Energy Information Administration, *Demonstrated Reserve Base of Coal in the United States on January 1, 1980.*
[a]Includes those parts of the measured and indicated resource categories as defined by the EIA and represents 100 percent of the coal in place. ("Glossary")
[b]Data may not add to totals shown due to rounding.

Table 3 World proved reserves of petroleum as of January 1, 1982

Geographic region	10^9 Barrels
Asia-Pacific	19
Western Europe	25
Middle East	363
Africa	56
Western Hemisphere	122
Subtotal	585
Communist areas	86
Total	671

Source: Oil and Gas Journal 79, no. 52 (December 28, 1982): 86–87.

Table 4 Estimates of total world ultimately recoverable reserves of crude oil for conventional sources

Year	Source	10^9 Barrels
1942	Pratt, Weeks, and Stebinger	600
1946	Duca	400
1946	Pogue	555
1948	Weeks	610
1949	Levorsen	1500
1949	Weeks	1010
1953	MacNaughton	1000
1956	Hubbert	1250
1958	Weeks	1500
1959	Weeks	2000
1965	Hendricks (USGS)	2480
1967	Ryman (Esso)	2090
1968	Shell	1800
1968	Weeks	2200
1969	Hubbert	1350–2100
1970	Moody (Mobil)	1800
1971	Warman (BP)	1200–2000
1971	Weeks	2290
1975	Moody and Geiger	2000

Source: Workshop on Alternative Energy Strategies, *Energy: Global Prospects 1985–2000* (New York: McGraw-Hill, 1977), p. 115. Reproduced with permission.

reserves are considered, and using an average recovery rate of only 50 percent, the resource base would be able to supply coal at current U.S. coal consumption rates for about 350 years, while it would last 6600 years if all the coal resources were included.

Petroleum

Table 3 gives the most recent estimates of the proved world petroleum reserves by geographic region. Nearly 67 percent of the non-Communist proved reserves are in the Middle East and about 14 percent are in the Western Hemisphere. The largest reserves are in Saudi Arabia (27 percent), but Kuwait (12.4 percent), Iran (11.3 percent), Iraq (6.2 percent), and Abu Dhabi (5.7 percent) all have reserves as large as those of the United States.

Many estimates of the world's total ultimately recoverable reserves of crude oil have been made. Some of those made between 1942 and 1975 are shown in Table 4.

Except for the four earliest estimates—made between 1942 and 1948, before the discovery of the large Middle East oil deposits—the esti-

mates fall in the range of 1000 to 2500 billion (10^9) barrels. Estimates made since 1960 tend to be at the upper range, with most of them at approximately 2000 billion barrels. These include reserves in Communist countries. If one assumes that ultimate reserves in the Communist countries are about 15 percent of the total (the same percentage as proved reserves), then the non-Communist ultimate recoverable reserves will be about 1700 billion barrels.

Starting in the mid-1960s the rate of addition to world oil reserves slowed down appreciably, and most geologists expect this trend to continue. The continuing increase in world oil demand combined with this decline in reserve additions has been projected

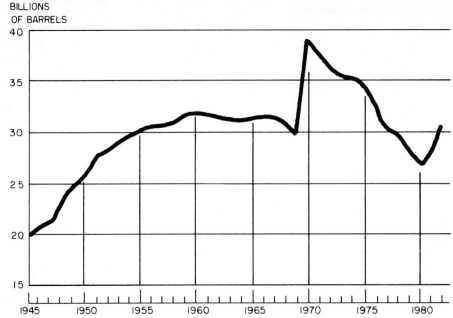

BILLIONS
OF BARRELS

FIG. 2. Proven Reserves of Crude Oil in the United States, 1945–82. Data from American Petroleum Institute (1945–75) and *Oil and Gas Journal* (1976–82).

to result in a peak in world oil production sometime in the period 1990 to 2010.

Figure 2 shows the estimated proved reserves of crude oil in the United States during the period 1945 to 1981. Between 1955 and 1969, proved reserves remained relatively constant at about 30 billion barrels. The Alaskan discovery at Prudhoe Bay increased reserves to nearly 40 billion barrels in 1970, but a sharp decline in the reserve values has occurred since then.

A number of estimates of U.S. undiscovered renewable petroleum resources starting in 1959 are shown in Table 5. Most of the earlier estimates were much higher than those that have been published since 1974. The five most recent ones range between 50 and 127 billion barrels (or from seven to nineteen years of supply at current U.S. rates of consumption), where the two extremes reflect differences in the probability of the stated amount being in fact discoverable and recoverable.

Natural Gas

Historical data on world natural gas resources are very limited. The large resources of natural gas are, except for the United States and, more recently, the United Kingdom, concentrated in regions where energy markets are small. Unless natural gas supplies could be delivered to large markets that could be reached by natural gas pipelines, they had limited commercial value, and as a result, they were poorly developed.

Proved reserves of natural gas for

Table 5 Estimates of U.S. undiscovered recoverable petroleum resources (adjusted to 1974)

Date	Source	10^9 Barrels
1959	USGS—Bulletin 1136[a]	387
1960	Weeks[b]	74
1962	National Academy of Sciences[c]	91
1962	National Fuels and Energy Study[d]	289–399
1962	Energy Policy Staff—DOI[e]	648
1965	USGS Circular 522[f]	264
1972	USGS Circular 650[g]	450
1974	Mobil[h]	88
1974	Hubbert[i]	67
1975	USGS—Resource Appraisal Group[j]	98
1975	National Academy of Sciences[k]	113
1975	USGS Circular 725[l]	50–127

[a]Paul Averitt, *Coal Reserves of the United States—A Progress Report,* January 1, 1960, U.S. Geological Survey Bulletin No. 1136 (Washington, D.C., 1969).
[b]L.G. Weeks, "The Next One Hundred Years Energy Demand and Sources of Supply," *Geotimes* 5, no. 1 (July–August 1960): 18.
[c]Marion King Hubbert, *Energy Resources* a report to the Commission on Natural Resources, National Academy of Sciences—National Research Council (Washington, D.C., 1962).
[d]Committee on Interior and Insular Affairs, *Report of the National Fuel and Energy Study Group on an Assessment of Available Information on Energy in the United States,* 83 Cong., 1 sess., September 31, 1962.
[e]Department of the Interior Energy Policy, *Supplies, Costs, and Uses of the Fossil Fuels,* prepared for the Atomic Energy Commission (Washington, D.C., June 1962).
[f]T.A. Hendricks, *Resources of Oil, Gas, and Natural Gas Liquids in the United States and the World,* U.S. Geological Survey Circular 522 (Reston, Va., 1965).
[g]Paul K. Theobald and others, *Energy Resources of the United States,* U.S. Geological Survey Circular 650 (Reston, Va., 1972).
[h]Robert Gillette, "Oil and Gas Resources: Did USGS Gush Too High?" *Science* 185, no. 4146 (July 12, 1974): 127.
[i]Marion King Hubbert, *U.S. Energy Resources: A Review as of 1972,* Committee on Interior and Insular Affairs, 93 Cong., 2 sess. (Washington, D.C., 1974).
[j]The Oil and Gas Branch Resource Appraisal Group, *USGS Mean* (Reston, Va.: U.S. Geological Survey, 1975).
[k]National Academy of Sciences—National Research Council, *Mineral Resources and the Environment* (Washington, D.C., 1975).
[l]Betty M. Miller, *Geological Estimates of Undiscovered Recoverable Oil and Gas Reserves in the United States,* U.S. Geological Survey Circular 725 (Reston, Va., 1975).

the world have only been published since 1967. In 1967, proved world reserves were 1042 trillion cubic feet (about 30 percent of the size of the world oil reserves in that year), and this value increased steadily to 2546 trillion cubic feet by 1975. In 1976 and 1977, estimated proved reserves declined, but by 1981 they were reported as 2911 trillion cubic feet.

Estimates of the world's ultimately recoverable natural gas reserves made between 1956 and 1975 are shown in Table 6. The more recent estimates range from 6000 to 12,000 trillion cubic feet (including reserves in the Communist area). Reserves of undiscovered recoverable natural gas in the Communist areas were estimated in 1975 to be 30 percent of the world total, compared to 15 percent for petroleum. The major large sources of potential supply are the United States and the Middle East (approximately 20 percent each) with the balance of the non-Communist world having about 30 percent, collectively.

Table 7 shows estimates of U.S. undiscovered recoverable resources made between 1956 and 1975. As with petroleum, the most recent estimates are much lower than the earlier ones. The two extreme values for the U.S.

Geological Survey estimates reflect differences in the probability that these resources will actually be found. The lower value would supply the current U.S. demand level for 17 years; the upper value, for 34 years.

Uranium

Data for uranium resources outside the United States are reported in two classes: "reasonably assured," which is similar to the U.S. "reserves" classification; and "estimated additional," which corresponds to the U.S. "probable" classification. Table 8 summarizes the world uranium resource data. For "reasonably assured" resources at forward costs up to $30/lb. of uranium oxide (U_3O_8), the major deposits are in the United States (32 percent), South Africa (18 percent), and Australia (17 percent). For "estimated additional" resources at that same cost cutoff, the major deposits are in the United States (53 percent) and Canada (27 percent).

Table 6 Estimates of world ultimately recoverable reserves of natural gas

Year made	Source	Reserves in 10^{12} cu. ft.	Reserves in 10^9 barrels oil equivalent
1956	U.S. Department of the Interior	5,000	850
1958	Weeks	5,000–6,000	860–1,035
1959	Weeks	6,000	1,035
1965	Weeks	7,200	1,240
1965	Hendricks (USGS)	15,300	2,640
1967	Ryman (ESSO)	12,000	2,070
1967	Shell	10,200	1,760
1968	Weeks	6,900	1,200
1969	Hubbert	8,000–12,000	1,380–2,070
1971	Weeks	7,200	1,240
1973	Coppack	7,500	1,300
1973	Hubbert	12,000	2,070
1973	Linden	10,400	1,800
1975	Kirby and Adams	6,000	1,030
1975	Moody and Geiger	8,150	1,400

Source: Workshop on Alternative Energy Strategies, *Energy: Global Prospects 1985–2000* (New York: McGraw-Hill, 1977), p. 150. Reproduced with permission.

Table 7 Estimates of U.S. undiscovered recoverable natural gas resources

Date	Source	10^{12} cu. ft.
1959	USGS—Bulletin 1136[a]	1,804
1962	National Fuels and Energy Study[b]	1,250
1962	Energy Resources Report[c]	800
1962	Department of Interior to AEC[d]	2,318
1965	USGS[e]	1,080
1972	USGS[f]	2,100
1973	Potential Gas Committee[g]	880
1974	USGS[h]	1,000–2,000
1974	Mobil[i]	443
1974	Hubbert[j]	361
1975	NAS[k]	530
1975	USGS[l]	332–655

[a]Paul Averitt, *Coal Reserves of the United States—A Progress Report*, January 1, 1960, U.S. Geological Survey Bulletin No. 1136 (Washington, D.C., 1961).
[b]Committee on Interior and Insular Affairs, *Report of the National Fuel and Energy Study Group on an Assessment of Available Information on Energy in the United States*, 83 Cong., 1 sess., September 21, 1962.
[c]Marion King Hubbert, *Energy Resources*, a report to the Commission on Natural Resource, National Academy of Sciences—National Research Council (Washington, D.C., 1962).
[d]Department of the Interior, *Energy Policy, Supplies, Costs, and Uses of the Fossil Fuels*, prepared for the Atomic Energy Commission (Washington, D.C., June 1962).
[e]T.A. Hendricks, *Resources of Oil, Gas, and Natural Gas Liquids in the United States and the World*, U.S. Geological Circular 522 (Reston, Va., 1965) adjusted to 1970.
[f]Paul K. Theobald and others, *Energy Resources of the United States*, U.S. Geological Survey Circular 650 (Reston, Va., 1972).
[g]Potential Gas Committee, "Potential Supply of Natural Gas in the United States" (1973).
[h]USGS news release, March 26, 1974.
[i]Mobil Oil Corporation, "Expected Value," *Science*, July 12, 1974.
[j]Marion King Hubbert, *U.S. Energy Resources: A Review as of 1972*, Committee on Interior and Insular Affairs, 93 Cong., 2 sess. (Washington, D.C., 1974).
[k]National Academy of Sciences—National Research Council, *Mineral Resources and the Environment* (Washington, D.C., 1975).
[l]The Oil and Gas Branch Resource Appraisal Group, *USGS Mean*, U.S. Geological Survey (Reston, Va., 1975).

At costs of up to $50/lb. of U_3O_8, the total amount increases from 2.2 to 2.9 million (10^6) tons in the "reasonably assured" category and from 1.9 to 2.9 million tons in the "estimated additional" category.

The widespread concern about the adequacy of uranium reserves and resources that was prevalent only two years ago has now abated. Projections of installed nuclear generating capacity, both for the United States and worldwide, have been revised downward sharply, so that even known

Table 8 World uranium resources (not including Communist countries) (thousand tons U_3O_8)

Country	Reasonably assured	Estimated additional
	Up to $30/lb. U_3O_8	
South Africa	398	44
Australia	376	57
Canada	217	510
Niger	208	69
France	48	31
India	39	31
Algeria	36	65
Gabon	26	6
Brazil	24	11
Argentina	23	—
Other	67	50
Foreign Total	1,462	874
United States	690	1,005
World Total	2,152	1,879
	Up to $50/lb. U_3O_8	
South Africa	452	94
Sweden	391	4
Australia	385	64
Canada	237	853
Niger	208	69
France	67	57
Argentina	54	—
India	39	31
Algeria	36	65
Other	142	119
Foreign Total	2,012	1,355
United States	920	1,505
World Total	2,932	2,860

Sources: DOE News, No. 79-47, April 25, 1979; and Organization for Economic Cooperation and Development, *Uranium: Resources, Production and Demand,* A Joint Report by the OECD Nuclear Energy Agency and the International Atomic Energy Agency (Paris, December 1977). The OECD-IAEA statistics employ a cost concept that is closer to full costs than that of "forward cost," used by DOE, except for Canadian resources, which are based on prices. The resulting lack of comparability, according to OECD-IAEA, is "not significant" within the general range of uncertainty in these estimates.

uranium resources will be adequate for twenty-five years or more.

Table 9 gives additional information on U.S. uranium resources and reserves and indicates that if potential resources are included, up to 4.1 million tons of U_3O_8 might be available at costs up to $50/lb. Additional very low-grade uranium resources are known to exist in very large deposits in the Chattanooga shales and in New Hampshire granite.

Unconventional Oil and Gas Resources

Petroleum

Enhanced Oil Recovery (EOR). The amount of unrecovered oil in place in the United States in reservoirs in which primary and secondary production has already taken place is estimated to be as much as 300 billion barrels, or nearly ten times the proved petroleum reserves of the United States. The location of these resources is known, but what is unknown is how much of this oil can be recovered economically.

Estimates of the amount of this oil that could be recovered at $15 per barrel generally range from 15 to 30 billion barrels. Table 10 shows the estimates by the Office of Technology Assessment of the amount of oil that could be recovered by EOR at various prices and for two different types of performance from new technologies under development. For oil at about $13.75 per barrel (1977 dollars), the ultimate recovery from EOR is estimated to range from 11 to 29 billion barrels (depending on process performance) or from approximately one-third to about the same amount as the proved U.S. petroleum reserves.

Recovering this known oil in the ground has been an intriguing problem for a long time, but achievements have been disappointing, partly because there is no technology uniformly applicable—apparently each field requires special treatment or nearly so. Yet the amounts involved are very large, and the absence of chance, as occurs in drilling new fields, continues to make this source highly attractive.

Tar Sands and Heavy Oils. Total known tar sand resources are estimated to be 3065 (10^9) barrels of oil-in-place. More than 81 percent of the total is in the Western Hemisphere; estimates for Canada are 2433 (10^9) barrels, for the United States 36 (10^9) barrels. Venezuela has a very large but as yet undefined resource. Other major tar sand resources are believed to exist in the U.S.S.R., Africa (Madagascar and Zaire), and Western Europe (Italy). These resources are expected to increase as a result of current and future exploration efforts. World commercial production of oil from tar sands in 1981 is estimated at 66.4 million barrels. In Alberta, Canada,

Table 9 U.S. uranium resources as of January 1, 1979 (10^3 tons U_3O_8)

Cutoff cost[a] ($/lb.)	Reserves	Potential			Total
		Probable	Possible	Speculative	
15	290	415	210	75	990
30	690	1,005	675	300	2,670
50	920	1,505	1,170	550	4,145

Source: DOE News, No. 79-47, April 25, 1979.
[a]Each cost category includes material in the lower category or categories. Not shown are an estimated 140,000 tons of U_3O_8 that could be produced as a by-product of phosphate and copper production. The cost concept underlying these estimates is that of "forward cost," which, roughly speaking, refers to capital and operating costs not yet incurred at the time an estimate is made and specifically excludes past expenditures for acquisition of land, exploration, and mine development. Thus forward costs are lower than production costs.

Table 10 Estimates of ultimate recoverable oil and daily production rates from enhanced oil recovery (EOR): advancing technology case with 10 percent minimum acceptable rate of return

		Price per barrel	Ultimate recovery[a] (10⁹ barrels)	Production rates 10⁶ barrels/day)		
				1985	1990	2000
High Process Performance	Upper tier	$11.62	21.2	0.4	1.1	2.9
	World oil	$13.75[b]	29.4	1.0	1.7	5.2
	Alternate fuels	$22.00[c]	41.6	1.3	2.8	8.2
		$30.00	49.2	[d]	—	—
	More than	$30.00	51.1	—	—	—
Low Process Performance	Upper tier	$11.62	8.0	0.4	0.5	1.1
	World oil	$13.75	11.1	0.5	0.7	1.7
	Alternate fuels	$22.00	25.3	0.9	1.8	5.1

Source: U.S. Congress, Office of Technology Assessment, *Enhanced Oil Recovery Potential in the United States* (Washington, D.C.: OTA, 1978), p. 7.
[a]These figures include 2.7 billion barrels from enhanced recovery processes that are included in the API estimates of proved and indicated reserves.
[b]13.75 is the January 1977 average price ($14.32 per barrel) of foreign oil delivered to the East coast, deflated to July 1, 1976.
[c]22.00 per barrel is the price at which the Synfuels Interagency Task Force estimated that petroleum liquids could become available from coal.
[d]Production rates were not calculated for oil at prices of $30 per barrel or higher.

about 53 million barrels were produced at two large mining and extraction plant projects; in California, production was more than 6 million barrels from several in situ projects, including commercial pilots. Italy and Venezuela each produced more than 3 million barrels.*

World deposits of heavy oils are believed to be very large, but very few countries have attempted to estimate the quantity, since under current conditions most of them cannot (except in a few special instances) compete with conventional sources of liquid fuels. Very large deposits of heavy oils are known to exist world-wide; those in Venezuela alone are reported at 890 billion barrels.

*The data on tar sands revised from the original by L.C. Marchant, U.S. Department of Energy, 1982.

Some heavy oils are currently being produced in the United States, particularly in California and Texas, by thermal methods; these reservoirs are included in conventional oil reserve estimates. Heavy oils will have to be produced by some form of enhanced oil recovery technique, perhaps by extensions of currently successful techniques of in situ combustion or steam injection. However, reservoirs where these techniques have been successful have a combination of oil viscosity and reservoir permeability that make this feasible. Many other reservoirs are not porous enough, and thermal reduction of viscosity does not lead to sufficient mobility of the liquid in place to permit pumping. Perhaps some combination of hydrofracturing and thermal recovery will be required.

Oil Shale. The world oil shale resources of all grades are extremely large, as shown in Table 11. Approximately one-third of the high grade deposits are found in Asia and about one-fourth in Africa, with the others approximately evenly distributed among the remaining continents. The total amount of energy in even the high-grade world oil shale resources is also very large—about 350 times that of the estimated ultimately recoverable resources of petroleum.

Although shale oil has been produced commercially for more than one hundred years and is still being produced in a few countries, costs have generally been higher than for conventional petroleum production. As petroleum prices rise, shale oil production will become economic. In 1978, best estimates are that production might become possible only at a large premium over conventional oil, and some severe environmental problems need to be solved. Shale oil has, of course, been "in the wings" for many decades, but has yet to appear on the stage with any impact. Whether its time has come should become apparent in the balance of the century. If so, the large resource base could provide liquid fuel for decades, at higher prices.

Natural Gas

Very little information exists with respect to world unconventional natural gas resources or reserves. In the United States, large resources of methane are known to exist in Devonian shales, in coal seams, in tight gas formations in various parts of the country (particularly in the West), in geopressured brines along the Gulf of Mexico coast, and perhaps in other places. There is currently some amount of commercial production of gas from Devonian shales and tight sand formations. Recently, considerable attention has been paid to potentially large gas reserves in deep sedimentary basins.

There is a wide range of uncertainty, both with respect to the size of the resource in place and with respect to what the supply-price curve might be for methane from each type of deposit. However, it is now believed that considerable unconventional methane could be obtained at prices below that required to produce methane from coal or to import liquefied natural gas.

Devonian Shales. The Devonian shales in the United States are concentrated in the eastern portion of the country. The estimated amount is 285 trillion cubic feet in the Appalachian Basin that extends from Michigan and lower New York State to the northern part of Alabama. This resource is about equal in size to the proved resources of conventional natural gas.

Coal Seams. Many coal seams contain significant amounts of natural gas that are released during mining. This gas constitutes a major safety problem, so that its removal and use prior to mining would increase safety and increase the size of the energy resource base. Some commercial use of the methane contained in coal has occurred in Europe, and several projects are now under test in the United States.

The total resources of methane in coal beds in the United States, varying with stipulated price and degree of optimism regarding this resource, are estimated to be in the range of 300 to 700 trillion cubic feet or from about the same size to more than twice as large as estimated resources of conventional natural gas.

Table 11 Order of magnitude of total stored energy in organic rich shale of the United States and principal land areas of the world

Continent or Country	Approximate area underlain by sedimentary rocks (10^6 square miles)	Shale containing 10–65 percent organic matter			Shale containing 5–10 percent organic matter		
		Shale in deposits (10^12 short tons)	Minimum organic content 10^12 short tons	Combustion energy content Q (10^18 Btu)	Shale in deposits (10^12 short tons)	Minimum organic content (10^12 short tons)	Combustion energy content Q (10^18 Btu)
United States	1.6	120	12	310	1,200	60	1,600
Africa	5.0	370	37	960	3,700	190	4,900
Asia	7.0	500	50	1,300	5,000	250	6,500
Australia	1.2	90	9	230	900	45	1,200
Europe	1.6	120	12	310	1,200	60	1,600
North America (including United States)	3.0	220	22	570	2,200	110	2,900
South America	2.4	180	18	470	1,800	90	2,300
World total	20	1,500	150	4,000±	15,000	750	20,000±

Source: D.C. Duncan and V.E. Swanson, "Organic-Rich Shale of the United States and World Land Areas," U.S. Geological Survey Circular 523 (1965), table 1.

Note: Estimates and totals are rounded.

Table 12 Estimates of gas resources in tight formations

Area	10^{12} cu. ft.
Green River Basin, Wyoming	240
Piceance Basin, Colorado	210
Uinta Basin, Utah	150
San Juan Basin, New Mexico	63
Northern Great PLains, Montana	130
	793

Source: U.S. Department of Energy, *Non-Conventional Natural Gas Resources,* National Gas Survey, U.S. DOE/FERC-0010, June 1978.

Tight Gas Formations. Most of the tight gas formations in the United States are in the western part of the country. Table 12 shows the estimated quantities of gas in each of the major tight gas formations. Total resources in these formations are about twice those of conventional natural gas.

Geopressured Brines. Most of the extensive geopressured zone formations that have been identified with any degree of reliability in the United States have been in the Gulf Coast Basin along the coast of Texas and Louisiana, both onshore and offshore. One of the largest deposits is in a zone 200 to 300 miles long in sedimentary deposits up to 50,000 feet thick. The gas is now believed to be dissolved in the hot water that is found in these deposits and in the shales within the geopressured zones.

A variety of estimates exist for the size of the natural gas in geopressured brines. The estimates range from 3000 to 49,000 trillion cubic feet. There is great uncertainty in the estimates of the size of the resource base. Even greater uncertainties are associated with the estimates of the amount of the resource base that can be economically recovered. One estimate puts the recoverable portion at only 5 percent of the total resource in place, which would give a range of recoverable resources of from 150 to 2450 trillion cubic feet. In another study, recoverable reserves were estimated at 260 trillion cubic feet, but the upper limit was given as 1150.

Geopressured zones are also found in South American (Columbia, Venezuela, Brazil, and others), Europe (France, Holland, Germany, United Kingdom, Norway, and others), Africa (Nigeria, Egypt, Algeria), Asia (Pakistan, India, Russia, Iran, Iraq, and others), Australia and Oceania, and in Taiwan and Japan. For these countries even less is known than for the United States about the size of the gas resources in place and the quantities recoverable at different costs, but with the large amount of gas potential, some production from these types of reservoirs should be developed over the next twenty to thirty years, at least from some of these deposits.

Renewable Resources

A wide variety of renewable resources has been used extensively in the past, and many continue to be used in various parts of the world. These include noncommercial fuels—animal dung, agricultural wastes, and wood—as well as water power, wind power, geothermal, and solar energy. Interest in the use of renewable resources has increased sharply in step with growing concern about the depletion of world oil and gas resources and about the environmental, health, and other risks associated with the use of coal and nuclear fuels.

The Adequacy of World and U.S. Energy Resources

From the preceding review of energy resources in the United States and the world the following conclusions can be reached:

1. Resource estimates are affected by a large margin of uncertainty, as the history of such estimates indicates. Because liquid and gaseous resources are hidden to a much greater degree than is true for coal, the uncertainties are correspondingly greater in the case of oil and gas. The discovery of Alaskan oil and gas in the 1960s and the Mexican discoveries of the recent past make the point forcefully. It is doubtful that surprises of this magnitude are ahead for coal. Given the large areas of the world that have not so far been adequately explored, one cannot say with certainty that we are approaching the end of the discovery of major oil and gas resources—except, in all likelihood, in the more thoroughly explored oil and gas regions of the world such as the United States, the countries bordering on the Persian Gulf, and parts of the Soviet Union.

2. Geographic concentration is especially strong in oil, less so in gas and coal. Table 13 shows world reserves and resources of conventional nonrenewable energy sources. In oil, the Middle East accounts for more than half of the reserves and between one-third and one-half of the estimated remaining recoverable resources. The North American continent and the Soviet Union run a relatively poor second and third, though the recent discoveries of oil in Mexico may push this particular resource there substantially closer to the Middle East. In coal, the United States, the Soviet Union, and China account for about two-thirds of reserves and for practically all the remaining recoverable resources. However, because these quantities are extremely large, even in countries with only a small share of reserves, the amount of energy is very large. For example, the remaining recoverable resources in Australia are estimated at over 100 billion (10^9) tons, which is almost as large as world petroleum reserves. For gas, the distribution is substantially more even across the globe. In the reserves category, the Soviet Union has by far the largest amount, about 35 percent of the world total, with the Middle East close behind with another 30 percent. The two together account for about 65 percent of the world total. North America and Western Europe are next largest, though the Mexican situation is likely to change these relationships somewhat. Distribution of remaining recoverable resources is even less concentrated.

3. Reduced to the common denominator of Btus (see Table 13), the world's fossil energy sources consist of three-fourths of coal in the case of reserves and about 90 percent in the case of remaining recoverable resources. The apparent degree of accuracy is somewhat deceptive. There is much disagreement, especially when estimates are made on a world scale, and the "true" ranges are probably much wider. Nonetheless, the dominance of coal is an inescapable fact, both for reserves and, even more so, for the less certain categories of remaining recoverable resources. It changes only, and radically at that, when uranium use in breeders is included.

Table 13 World reserves and resources of coal, oil, gas, and uranium (10^{15} Btu)

	Coal		Crude Oil		Natural Gas		Uranium[b]	
	Technically and economically recoverable reserves	Remaining recoverable resources,[a] excluding reserves	Technically and economically recoverable reserves	Remaining recoverable resources, excluding reserves	Technically economically recoverable reserves	Remaining recoverable resources, excluding reserves	Reasonably assured resources at $50/lb. or less cost	Estimated additional resources at $50/lb. or less cost
United States	4,945	30,780	168	1,856–2,230	210	1,672	276	452
Canada	250	1,333	35		60		71	256
Mexico	28	56	93		33		2	1
Other Western Hemisphere	278	83	145	302–534	83	788	24	5
Western Europe	2,528	3,445	139	145–261	147	358	153	38
Middle East	—	—	2,146	2,030–3,654	749	1,038	—	—
Africa	945	1,445	336	261–545	191	833	223	78
Asia-Pacific	1,861	2,917	116	313–603	123		131	28
USSR	3,056	64,450	412[c]		933[c]	2,332	N.A.	N.A.
China	2,750	17,224	116	365–713	26		N.A.	N.A.
Other Communist	1,028	1,195	17		10		N.A.	N.A.
Total	17,668	122,927	3,723	5,272–8,630	2,656	7,021	880	858

Sources: Coal reserves and resources: World Energy Conference, Conservation Commission, Coal Resources (Guildford, England: IPC Science and Technology Press, 1978). Oil resources: Richard Nehring, Giant Oil Fields and World Oil Resources, Report R-2284-CIA (Santa Monica, Calif.: Rand Corporation, 1978). The estimate shown for North America includes revision made by the author since the publication of this report. Gas resources: Joseph D. Parent and Henry R. Linden, A Survey of U.S. and Total World Production, Proved Reserves and Remaining Recoverable Resources of Fossil Fuels and Uranium as of December 30, 1975 (Chicago: Institute of Gas Technology, 1977). Uranium: DOE News, No. 79-47, April 25, 1979; and Organization for Economic Cooperation and Development, Uranium: Resources, Production, and Demand, A Joint Report by the OECD Nuclear Energy Agency and the International Atomic Energy Agency (Paris, December 1977).

Note: The values shown here differ somewhat from some of those exhibited elsewhere in this chapter, owing to reliance on different sources of information. These differences reflect the persistent uncertainty of energy resource estimates, but are not large enough to alter the conclusions drawn in the chapter. In all categories of the table, past production is excluded, as are the so-called unconventional sources of petroleum and natural gas. N.A. = information not available.

[a] Assumes 50 percent recoverability. As surface mining increases, the estimates become conservative.

[b] The values shown are quads (quadrillion Btus) of thermal energy generated in a light water reactor, assuming a burnup of 30,000 MWd/metric ton or uranium, 0.2 percent tails, 3 percent enrichment of U-235, 15 percent losses during conversion and fabrication. If converted into electricity, the available energy would be only about one-third as much. If used in a fast breeder reactor, the same amount of uranium would yield between sixty and one hundred times as much heat or electricity as it would if fissioned in a light water reactor.

[c] USSR figures are "explored reserves," which include proved, probable, and some possible.

Conclusions

It is in the light of these facts that one must evaluate the warning that "the world is about to run out of" any specific fuel. For one thing, the warning is hardly novel; it was first heard at least two centuries ago. The fact that it has so far proved wrong makes it legitimate to suspect it, but not to discount it entirely.

In the context of the twenty-year horizon of this study, we surely need not be concerned over physical inadequacy of coal, be this on the domestic U.S. scale or for the world as a whole. We cannot make the same assertion for petroleum or natural gas. Although surprises cannot be excluded and may even be expected, the present scale of consumption is such that even if we successfully throttle demand for these two fuels, the world will consume them at such a rate that conventional deposits will be depleted at least sometime in the next generation, if not sooner, and rising cost will reflect this prospect. Thus, the unconventional resources of petroleum and natural gas will loom increasingly large as an important part of the resource base.

For uranium, if breeder reactors are developed, the known uranium resource base would be extended by a factor of sixty to one hundred, so that the resource base would be comparable to remaining recoverable coal resources. Fuel other than uranium (i.e., thorium) would further extend the time horizon.

It is difficult to comprehend the size of the energy reserves that the world already has in hand. If only the reserves of fossil fuels were used and the growth rate in energy demand were 2 percent per year, then ten years of supply would still remain in the second decade of this century; if the remaining recoverable resources of fossil fuels are included, ten years of supply would still be available near the end of the next century. These calculations do not include the nuclear fuels; the large quantities of unconventional oil and gas resources, peat, or shale oil; or the large amount of renewable resources.

The United States is well enough endowed with fossil fuel resources and uranium to enable it to move without panic to a different pattern of energy sources (which is not to say that it will do so). What might that pattern be like? Barring absolute limits to the burning of fossil fuels due to the carbon dioxide problem, coal is likely to persist longest, among conventional fuels; fission-based electricity would endure if acceptable breeders can be designed; nonconventional sources of oil and gas would gradually take the place and extend the lifetime of conventional ones; renewable sources, especially solar energy, would find widening application; and possibly at a date substantially into the next century, fusion might offer a source of energy with a long horizon. A time profile of this kind would also hold for a number of other countries, but cannot be generalized for the world as a whole.

Barring misguided public policies, nations will presumably proceed from conventional to unconventional resources along a path that minimizes the rate of increase of cost. This means, for example, that nearly conventional oil and gas are likely to be exploited before there is large-scale production of synthetic fuels from coal, while timing of the entry of shale remains as uncertain as it has been for a long time. The picture is complicated not only because we need the wisdom to recognize the optimal path and the time to proceed on it, but also because so much is yet to be learned about the production costs of unconventional supplies and even more about the social, environmental, and health costs associated with their exploitation. Yet it is our judgment that we must obtain a much better hold on the nature, costs, and problems of this promising array of sources if we hope to make orderly transitions. Government leadership is an important ingredient in this endeavor.

Note

[1]The states included in each region are: *the East*—Alabama, Georgia, eastern Kentucky, Maryland, North Carolina, Ohio, Pennsylvania, Tennessee, Virginia, and West Virginia; *the Interior*—Arkansas, Illinois, Indiana, Iowa, Kansas, western Kentucky, Michigan, Missouri, Oklahoma, and Texas; and *the West*—Alaska, Arizona, Colorado, Montana, New Mexico, North Dakota, Oregon, South Dakota, Utah, Washington, and Wyoming.

26   Is Nuclear Energy Necessary?

Alvin M. Weinberg

Two questions dominate the nuclear issue: Is nuclear energy necessary? Can nuclear energy be made acceptable? Unless the answer to the first is affirmative, there is little incentive to devise the improvements in nuclear energy necessary to rescue it from its present malaise. In my view, nuclear energy is highly desirable, if not absolutely necessary. It can be made acceptable. And rather than continue the bitter confrontations between proponents of renewable sources and proponents of nuclear sources, the energy community ought to put its efforts into achieving an acceptable nuclear sys-

Alvin M. Weinberg is director of the Institute for Energy Analysis in Oak Ridge, Tennessee. Reprinted by permission of *The Bulletin of the Atomic Scientists,* a magazine of science and public affairs. Copyright © 1980 by the Educational Foundation for Nuclear Science, Chicago.

tem. We shall need all sources of energy; we cannot afford to reject nuclear energy because its current embodiments are faulted.

The fission chain reaction is a bit of a scientific fluke: for example, there is no theoretical reason why the number of neutrons emitted per fission must be sufficient to maintain a chain reaction. One cannot therefore argue that mankind would perish had fission not been discovered. Indeed, before 1939 energy futurologists, recognizing that fossil fuels were finite, speculated on the possibilities of drawing energy in the long run from the various solar sources—including wind, waves, ocean thermal gradients, and biomass—as well as geothermal sources. Nevertheless, the outlook at the time was rather pessimistic. Most pessimistic was the assessment of Charles G.

Darwin, descendant of the biologist. In his book, *The Next Million Years,* published in 1953, he predicted a brutish, Malthusian future for man unless he developed an inexhaustible energy source other than the sun. Darwin's candidate for such a source was controlled fusion.[1]

Many of us in the fission community have recognized that in the fission breeder man had another path to salvation from Darwin's ultimate Malthusian disequilibrium. It is therefore understandable that we addressed the development of the breeder with such enthusiasm: here was an embodiment of Darwin's inexhaustible energy source that would, to use H. G. Wells' words of 1914, "Set Man Free."[2]

Given this noble, perhaps noblest, of all technological dreams, it is almost incomprehensible to us why the world is now asking: Is fission necessary? Is not the relevant question: What can be done to eliminate the deficiencies of fission, rather than eliminating fission itself?

The Short Range: to 2000

What can we say about the necessity of fission in the near term, roughly to the year 2000? Obviously the need for fission depends upon the availability of alternative fuels or sources of energy, and upon the future demand for energy. These are not matters that can readily be settled for the whole world. In the United States, with its abundant coal, for example, a moratorium on fission would be less serious than it would be in Japan or France. The Institute for Energy Analysis (IEA), in its study "Economic and Environmental Implications of a U.S. Nuclear

Moratorium, 1985–2010," concluded that the United States could weather a limited moratorium with a loss of 0.5 percent in GNP. The moratorium would allow completion of all reactors under construction by 1985; no new reactors would be built after 1985.

The moratorium would place great pressure on coal or imported oil or both. It was estimated that between 18 and 27 billion tons of additional coal would be needed by 2010 to fuel those stations that would serve in place of nuclear plants not built; alternatively, the additional imported oil would amount to from 6 to 9 billion tons. Re-examining these estimates three years after they were first made, I would say the impact might be overstated, particularly because we expected electricity to capture 46 percent of the energy market by 2000, compared to its present market share of 30 percent. While electricity's market share will probably increase, it seems unlikely that it will increase this much. Nevertheless, in view of the great environmental problems associated with coal mining, let alone the political tensions created by expanded import of oil, I conclude that even this limited nuclear moratorium is very undesirable.

Most other countries do not have coal. For them, rejection of nuclear energy would certainly entail costly importation of coal and oil. A one-gigawatt (electric) oil-fired power plant uses 1.4 million tons of oil per year. Throughout the world 800 million tons of oil are burned annually in central electric power stations; this represents 25 percent of the world's consumption of oil in the year 1978. Should all the oil-fired plants be replaced over the next decade with nuclear plants, the pressure on oil would be reduced significantly.

In recent years it has become fashionable to fault this line of argument as simplistic. Rather than replace oil in electric power stations with coal or uranium, we are asked to believe we can so reduce our energy demand as to make many existing, let alone future, electric power plants superfluous. In any case, the use of electricity for such purposes as heating of houses or water is deemed inelegant and wasteful and ought to be discouraged.

It goes without saying that conservation must be central to any energy policy; indeed, much has already been accomplished. For example, D. Reister of the Institute for Energy Analysis points out that in the United States, the ratio of energy to gross national product has decreased by 10 percent between 1970 and 1978.[3] But as Stogaugh and Yergin point out in the Harvard Energy Futures study, conservation requires decisions by innumerable consumers; by contrast, increased supply requires far fewer decisions. Thus the prediction of how much conservation is actually achieved—as contrasted with how much is theoretically achievable—is intrinsically less certain than is the prediction of how much supply can be increased. (To be more accurate, since energy supply and demand must balance, at issue is the relative freedom of choice afforded by policies that depend on conservation rather than on increased supply.) The difficulty has recently been analyzed by P. C. Roberts of the United Kingdom.[4] Roberts gave evidence that, in the United King-

dom, the amount a family is likely to spend to retrofit its house with energy-conserving devices goes as the square of the family income. Roberts thus estimates that, even if fuel costs double, the amount of conservation induced by market forces in the next 25 years is about 25 percent of the theoretical. Conservation in houses, at least in the United Kingdom, does not happen automatically; subsidies of various sorts seem to be necessary. This is not to say that conservation is unimportant; it is simply that conservation in the residential sector is probably more difficult to achieve than in the industrial sector.

The current mood of rejection of electricity seems irrational. If oil is scarce and coal and uranium are abundant, it makes sense to replace oil with coal and uranium, even if in so doing one must resort to inefficient resistive heating or other devices that use electricity. After all, we are not driven by thermodynamic imperatives: economic or political considerations, such as reducing our dependence on foreign oil, certainly take precedence over the much discussed, but often irrelevant, stricture to improve the second law efficiency.

Two technical developments, the electric car and the heat pump, could swing the balance toward an electrical future dominated by large central stations. The recent announcement by General Motors of a car powered by a zinc battery whose lifetime is 50,000 kilometers, with a cruising speed of 80 kilometers per hour and a range between charges of 165 kilometers could, if realized, alter our attitude toward the electric future. In addition, the electrically driven heat pump is a proven device. In the United States 560,000 heat pumps were installed last year, and this number has been increasing by 40 percent each year.

If one concedes that:

- a predominantly electric future is at least as plausible as a nonelectric one,
- oil will continue to be scarce,
- the solar technologies will not penetrate on a large scale (cost, intermittency, storage?),
- coal is not generally available except in countries that possess indigenous deposits (would the United States be prepared, in 50 years, to mine an additional billion tons of coal, to be shipped overseas?),

then it seems inescapable that fission is necessary, at least in some parts of the world. Beyond this, if the cost per joule of coal reaches that of oil—that is $80 per ton of coal at present, possibly $140 per ton with oil at $40 per barrel— then even though the cost of nuclear reactors is high (say, $1,500 per kilowatt), electricity from current reactors probably will still be cheaper than electricity from coal-fired stations.

There are uncertainties in these assumptions. But it is because we cannot know with certainty even the next 20 years that we ought to preserve all our energy options, including nuclear. In this sense, I would judge nuclear energy in the short term to be a necessity.

The Long Term—Beyond 2030

We must look at the time, perhaps 50 to 100 years from now, when oil and gas have become scarce. Fission, if it survives, eventually would be based on breeders, though not necessarily fast breeders. Three issues dominate the long-term outlook for fission: the availability and acceptability of alternatives, the long-term energy demand, and carbon dioxide.

On the scale under discussion, only fusion or the various solar sources are large enough to compete with fission. Here I must admit to being agnostic. Fusion *may* turn out to be technically and economically feasible, but there is no way of knowing. And in good measure this is true of solar energy. Despite the many claims of technological and economic breakthroughs, the uncertainties remain. Such figures as 50 cents per peak watt, the aim of the Department of Energy for solar cells, are still no more than a hope.[5] Given these uncertainties about alternatives to the fission breeder, prudence requires us to develop the breeder, and to deploy it if the alternatives prove too costly or turn out to be unfeasible. In short, it is far too early to reject any of the long-term energy options, particularly the breeder.

Long-Term Scenarios

Here I can do no better than to borrow extensively from recent studies completed at the International Institute for Applied Systems Analysis (IIASA).[6] These scenarios, which are close to those developed by R. Rötty, at the IEA, contemplate a world population of 8 billion in 2030. Because China, Latin America and possibly India are likely to increase both their per capita energy expenditure and their populations faster than the developed world, it is plausible to expect these countries to use a much larger fraction of the world's energy than they do now. This, of course, could change somewhat if China's announced program of one child per family succeeds.

At present, the primary per capita energy demand averaged over the entire world comes to about 2.1 kilowatts per year. The total primary energy demand comes to 8.2 terrawatts (1 terrawatt = 1 trillion watts). If the average per capita demand grows to about 4.3 kilowatts per capita, the total primary energy demand might reach around 35 terrawatts; should it grow to only about 2.8 kilowatts per capita, then the total might be 22 terrawatts. These projections are designated by IIASA as "high" and "low"; the IEA scenario falls between these IIASA projections. (Table 1 summarizes the IIASA estimates of how the high and low energy demands might be met.)

Nuclear energy provides about one-fourth the world's energy in each of the scenarios, whereas hydro, solar, and others provide about 8 percent in the high scenario, 10 percent in the low. Can the renewables replace the 25 percent provided by nuclear energy?

Table 1 Projected supply patterns in 2030 (in terrawatts)

	1975	2030 high	2030 low
Oil	3.6	6.8	5.0
Natural gas	1.5	6.0	3.5
Coal	2.2	11.4	6.5
Light water reactor	0.12	3.2	1.9
Fast breeder reactor	—	4.9	3.3
Hydro	0.5	1.5	1.5
Solar	—	0.5	0.3
Other	0.2	0.8	0.5
Total	8.2	35.1	22.5

Source: International Institute for Applied Systems Analysis.

Here there is a sharp divergence, with nuclear opponents, such as A. Lovins, insisting that nuclear is unnecessary, that the various solar sources will suffice. To be sure, the total primary energy demand contemplated by these authors is around 17 terrawatts—an amount available from fossil sources according to IIASA, and a little more than twice the world's present energy budget. Whether a 17-terrawatt world, especially one that is increasingly urbanized, can be socially stable, no one can say. My own view is that in planning the future we do best to prepare for a higher rather than a lower energy demand. It almost goes without saying that the less energy shortages become the focus for social tensions, the better for all.

One must recognize that even the 5.2 terrawatts supplied by fission in the IIASA low scenario is formidable: some 1,500 reactors, each producing about 1,000 megawatts (electric) or 3,300 megawatts of heat. And if, in the very long run, fission takes over the role now played by oil and natural gas, the number of required reactors, even in IIASA's low scenario, would rise to over 4,000! In the high scenario these numbers would be roughly doubled.

Carbon Dioxide—A Sword of Damocles?

The possible constraint on fossil fuel, imposed by accumulation of carbon dioxide in the atmosphere, may be demonstrated by a few numbers. The world's total fossil fuel reserve is estimated to contain about 10 trillion tons of carbon. Should all of this be burned, and if 50 percent of the resulting carbon dioxide remains airborne, the total carbon dioxide in the atmosphere might increase some sevenfold. The temperature of the lower atmosphere is estimated by Manabe and Weatherald to increase by from 2.5 to 3 degrees centigrade for every doubling of carbon dioxide in the atmosphere.[7] Thus a sevenfold increase could raise the temperature of the lower atmosphere by as much as 8 degrees centigrade.

H. Flohn has estimated the climatic regimes that might prevail if various amounts of carbon dioxide were added to the atmosphere. He characterizes each regime by comparing it with similar regimes in the geologic past (Table 2).[8] Flohn's estimates of the temperature rise caused by a given addition of carbon dioxide are somewhat higher than those of other authors, largely because he includes the effect of other "greenhouse" gases.

Perhaps the most alarming aspect of Flohn's estimates is that the burning of only 20 percent of the world's fossil fuel would lead to an ice-free Arctic; and he argues that an ice-free Arctic would profoundly change the entire world's climate—to a regime that last occurred two million years ago. Flohn also offers evidence, from the very rapid disappearance of forests at the end of previous interglacials, that once a climate change begins, its ecological consequences could be manifested over a few decades.

Whether the climatic changes induced by carbon dioxide accumulation produced by fossil fuel would constitute the enormous catastrophe envisaged by Flohn is controversial. Nevertheless, we must ask: What is a prudent course in the face of such predictions, even granting their inherent uncertainty? The answer seems

obvious. Reduce consumption—a feat far easier to achieve in the energy-rich parts of the world than in the energy-poor parts; and be prepared to shift to energy sources that produce no carbon dioxide. In the latter category are the solar sources, the nuclear sources (fission and fusion), and geothermal.

If the geothermal source is as small as it now appears, and if fusion remains far from technical realization, the sun and uranium are the only remaining alternatives. And if one further concedes that the solar sources may remain expensive, especially if one tries to provide reliable power from inherently stochastic sources like the wind and the sun, then caution dictates that we be prepared to use the only other technically feasible, very large energy source that does not add carbon dioxide to the atmosphere—nuclear fission. To be sure, in making this judgment I am implicitly calculating the risk of nuclear energy to be less than the risk of a carbon dioxide catastrophe. Moreover, I am assuming that if and when (perhaps in a decade or two) the reality of carbon dioxide emerges as a clearly defined political issue, then the move to nuclear and solar, coupled with conservation, can be made fast enough to forestall the catastrophe envisaged by Flohn.

We cannot as yet invoke the carbon dioxide catastrophe as justification for a nuclear future because there is still so much scientific controversy surrounding it. But if we are, as Palmer Putnam put it in 1953, "prudent custodians of man's future," then we would be acting irresponsibly if we reject any energy source, including fission, that produces no carbon dioxide.

There is a likelihood that the world

Table 2 Climatic regimes[a]

Period	Degrees centigrade	Carbon dioxide[b] (parts per million)	Total carbon injected (billions of tons)	Fossil fuel reserve burned (present)	Time[c] (years, high)	Time[c] (years, low)
Present	0	330	180	1.8	1980	1980
1000 A.D.	1	405 ± 25	530	5.3	2015	2034
6000 before the present	1.5	455 ± 35	760	7.6	2037	2053
Interglacial 120,000 before the present	2	510 ± 50	1,000	10.0	2050	2073
Ice-free Arctic	4	755 ± 125	2,100	21.0	2110	2165

[a]After H. Flohn, "Possible Climatic Consequences of a Man-Made Global Warming."
[b]Hypothetical carbon dioxide concentration necessary to cause the temperature rise given in preceding column.
[c]Time to reach carbon dioxide concentrations given in Column 3, according to IIASA high and low scenarios.

may see many thousands of nuclear reactors within, say, the next 100 years. Is such a world feasible? In short, can man live with fission?

I set aside the issues of proliferation, of waste disposal, and of low-level radiation: the first because it is a political, not a technical question; the second because, as the recent Interagency Review Group established by President Carter asserted, satisfactory confinement of wastes for periods of several thousand years is technically feasible (after about 1,000 years the ingestion toxicity is comparable to that of the original uranium from which the wastes were derived); the third, because the hazard of low-level radiation has been grossly overplayed. But the matter of reactor accidents is not so easily disposed of. As the incident at Three Mile Island demonstrated, an accident in a nuclear plant is a real possibility; nuclear energy may not survive many incidents like Three Mile Island, even though no one was harmed there.

The *a priori* probability of an accident that releases sizable amounts of radioactivity was estimated by Rasmussen to be 50 millionths per reactor year. This number has since been criticized as being too low. In any event, the Three Mile Island accident—which released about a dozen curies of iodine-131 and according to the Kemeny commission, probably caused no bodily harm—occurred after a few hundred reactor-years of operation. Its *a priori* probability has been estimated at around one in 400 reactor-years.

For any given reactor, a satisfactory accident probability would appear to be 50 millionths per reactor-year. A particular reactor, during its 40 years of operation, would have a likelihood of one in 500 of suffering an accident. But if the world energy system involved as many as 5,000 reactors—that is, 10 times as many as are now either in operation or under construction—one might expect an accident that released sizable amounts of radioactivity every four years. Considering that a nuclear accident anywhere is a nuclear accident everywhere, I believe this accident probability is unacceptable. If man is to live with fission over the long term, he must reduce the *a priori* probability of accident by a large factor—say 100.

A relevant comparison is today's volume of air travel, which would have been impossible had not the accident probability per passenger mile been reduced drastically as the number of passenger miles increased. This was accomplished in air transport by a combination of technical and organizational improvements, and the same kinds of improvements will be needed in nuclear energy. Can we visualize the needed changes?

Perhaps the easiest are the technical improvements. Certainly the deficiencies that led to Three Mile Island will be corrected: German light water reactors, for example, have two relief valves in series so that if one fails to close, the other is available. Control panels will be provided with positive indication of valve position, something that was lacking at Three Mile Island. And other technical improvements will be forthcoming.

But beyond the technical fixes, various institutional changes are needed. Among these I mention first the establishment by the utilities of the Institute for Nuclear Power Operations and the Nuclear Safety Analysis Center. These are very significant developments. But I believe, in the long run, even more important is adherence to the principle of confined, permanent siting in relatively remote areas, which is already characteristic of most of the sites in the United States. We have estimated that a nuclear system of 615 gigawatts

(electric) could be accommodated in the United States on 80 of the present 100 sites, augmented by 20 new sites to replace those not suited for expansion. My proposal, at least for countries well embarked on nuclear energy, is essentially a moratorium on new sites, not on new reactors. The ultimate nuclear system would consist of large centers, located at those existing sites that are adequately remote, plus a few new sites that are also remote.

The advantages of clustered, permanent locations seem compelling to me. They include a larger cadre of experts available on site; better organizational memory, and therefore better operation; more effective security and easier control of fissile material; handling of low-level wastes and spent fuel elements on site; easier surveillance of decommissioned reactors; and, as at existing nuclear sites like Oak Ridge, a surrounding population that understands radiation and is prepared to respond in case of accident. Beyond this, the entities operating large clustered sites are likely to be stronger than are entities operating, say, a single reactor. This certainly appears to be the case in Canada, where Ontario Hydro operates Pickering, Bruce and Darlington, each with four or more reactors; and the Tennessee Valley Authority in the United States, which intends to confine all its reactors to the seven sites now under construction. Soviet nuclear specialists N. Dollezhal and Y. I. Koryakin (*Bulletin*, Jan. 1980) have proposed a similar confined siting policy for the Soviet Union.

The measures I suggest may not be sufficient to rescue nuclear energy from its present disaffection. But in view of the strong incentive to maintain the nuclear option, it seems important to devise the fixes that will make nuclear energy acceptable. To do less might impose on those who follow us a future much bleaker than one that uses all energy sources—including an acceptable nuclear energy.

Notes

[1]Charles G. Darwin, *The Next Million Years* (Garden City, N.Y.: Doubleday, 1953).

[2]H. G. Wells, *The World Set Free: A Story of Mankind.* (New York: Dutton and Company, (1914).

[3]D. Reister to Alvin Weinberg, Nov. 7, 1979.

[4]P. C. Roberts, "Energy and Society," presented at the Commission of the European Communities Energy Systems Analysis International Conference, Dublin, Ireland, Oct. 11, 1979.

[5]"Study on Solar Photovoltaic Energy Conversion," prepared for the Office of Technology Policy and the U.S. Department of Energy by the American Physical Society, New York, 1979.

[6]Wolf Häfele, "World-Regional Energy Modelling," presented at the Commission of the European Communities Energy Systems Analysis International Conference, Dublin, Ireland, Oct. 9, 1979.

[7]S. Manabe and R. T. Wetherald, "The Effects of Doubling the CO_2 Concentration on the Climate of a General Circulation Model," *Journal of Atmospheric Science* 32, 1 (1975), 3–15.

[8]H. Flohn, "Possible Climatic Consequences of a Man-Made Global Warming," presented at the Commission of the European Communities Energy Systems Analysis International Conference, Dublin, Ireland, Oct. 11, 1979.

Mineral Exploitation and Environmental Impact

Stripping shovel at the Sinclair Mine, Kentucky. The dipper of the big shovel holds 95.5 m^3 (125 yds^3) of earth and rock. (Photo courtesy National Coal Association.)

> *"The American Colossus was fiercely intent on appropriating and exploiting the riches of the richest of all continents—grasping with both hands, reaping where he had not sown, wasting what he thought would last forever."*
>
> Gifford Pinchot
> *Breaking New Ground*

During the past decade, there has been a more vocal expression of an increasing concern for the quality of our environment. Scientists, government officials, and public interest groups have called to our attention the hazards associated with the indiscriminate use of herbicides, pesticides, and fertilizers. The revelation of the impact of the careless disposal of toxic waste materials at Love Canal in Niagara, New York, has shocked the nation.

Almost every phase of the mineral industry, from exploration to production, transportation, and processing, has also been severely criticized. Minerals are an integral part of the environment, and production of minerals without modification of the environment is entirely unrealistic. The problem confronting us today is to find ways to ameliorate the impact of mineral production while still providing the nation with the minerals it requires at an acceptable cost. In this section we will describe some basic mining techniques, their impact on the environment, and legislation that has been enacted to minimize the environmental impact.

Surface Mining

Techniques

Solid ore deposits are exploited through either surface or subsurface mining techniques. The major factor determining the technique to be used is the thickness and nature of the material which overlies the ore deposit. If the overburden is relatively thin and easily worked, surface mining might be employed. If an ore deposit is deeply buried, it may be economically unfeasible or physically impossible to remove the overburden, and subsurface techniques must be used.

The advantages of surface mining when compared with underground mining are several: working conditions are safer, ore recovery is usually more complete (close to 100 percent in some instances), and the cost per unit of production is considerably lower.

There are five basic steps in most surface mining operations: (1) site preparation, including clearing vegetation and constructing access roads and ancillary installations; (2) removal and disposal or storage of overburden; (3) excavation and loading of ore; (4) transportation of the ore directly to market or to a processing plant or storage area; and (5) reclamation of the site, which may be integrated with the actual mining operation.

The principal surface methods employed in modern mining operations are generally classified as (1) open pit mining (quarry), (2) strip mining (area strip mining, contour strip mining, and mountain-top removal), (3) auger mining, (4) dredging, and (5) hydraulic mining.

In *open pit mining*, the amount of overburden removed is proportionately small compared with the quantity of ore recovered. Common examples of open pit mines are limestone and granite quarries and sand and gravel pits.

Strip mining has variations that depend upon the type of terrain. *Area strip mining* is practiced on relatively flat terrain. A trench, or "box cut," is made through the overburden to expose the ore, which is then removed. A second cut is made parallel to the first cut and the overburden is deposited in the cut previously excavated. The final cut leaves an open trench bounded on one side by a deposit of overburden (spoil bank) and on the other side by an undisturbed "highwall" (Figs. 1 and 2). *Contour strip mining* is practiced where deposits occur in rolling or mountainous terrain. If the ore body is flat-lying, it is mined by following the contour of the land. The overburden is removed and typically cast down the hillside. Additional cuts are then made to extract the ore. The process creates a shelf, or "bench," on the hillside, which is bordered by a highwall on one side and a steep slope on the other side (Fig. 3). Additional ore may be recovered through either mountain-top removal or auger mining. In *mountain-top* removal, overburden is removed and cast down to the bench left by contour strip mining. The ore is then removed, leaving a flat-topped hill.

Auger mining is a method of recovering ore by boring horizontally into the ore body. Cutting heads may be as large as 3 m (9.8 ft) in diameter, and holes may be drilled in excess of 61 m (200 ft) (Fig. 4). Augering is used to recover additional ore after contour strip mining has been completed.

Dredging operations utilize a suction apparatus or such mechanical devices as draglines or chain buckets. Dredging is used to recover sand, gravel and placer deposits from stream beds or offshore areas.

In *hydraulic mining*, powerful jets of water are used to wash down

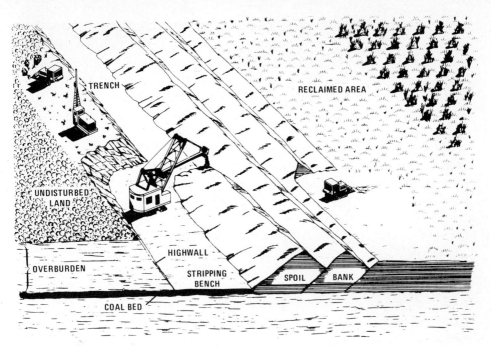

FIG. 1. Area strip mining with concurrent reclamation. (From Grim, E. and Hill, R., 1974, *Environmental Protection in Surface Mining of Coal*, EPA-670/2-74-093.)

the ore-bearing material, which is then fed into sluices. The ore separates out because of its higher specific gravity.

Environmental Impact. An estimated 23,430 km^2 (5.7 million acres) of land were disturbed by surface mining in the United States between 1930 and 1980. This amounts to 0.25 percent of the total land area in the country. But the disturbances are evident in every state and affect not only the land but the air and water. More than 9600 km (5965 mi) of streams and more than 12,141 ha (30,000 acres) of impoundments and reservoirs are seriously affected by coal mine operations alone. The geographic distribution of land utilized for mining is shown in Fig. 5. Seven commodities accounted for 93 percent of the acreage disturbed: coal (48 percent), sand and gravel (17 percent), building stone (13 percent), phosphate rock (5 percent), clay (4 percent), copper (3 percent), iron ore (3 percent). It is anticipated that surface mining will increase greatly in future years as demand for minerals and solid fuels increases and as the grade of mineral deposits available for exploitation diminishes further. Coal mining in the Western states will account for a disproportionate share of the increases in future years.

FIG. 2. Earth-moving machine with 60-cubic-yd dipper used in strip mining to expose coal deposit by digging away "overburden" (layers of earth and rock above the coal) and depositing it in furrow-like rows. The method shown is "area stripping," used for surface mining of coal in flat or fairly level country. (Bureau of Mines, U.S. Dept. of the Interior.)

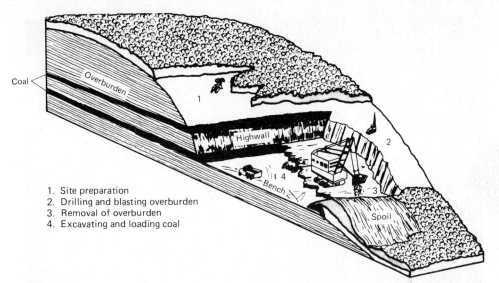

1. Site preparation
2. Drilling and blasting overburden
3. Removal of overburden
4. Excavating and loading coal

FIG. 3. Contour strip mining. (From Grim, E. and Hill, R., 1974, *Environmental Protection in Surface Mining of Coal*, EPA-670/2-74-093.)

Many factors influence the environmental impact of surface mining. The most significant are (1) mining technique, (2) chemistry of the ore, (3) nature and thickness of overburden and topsoil, (4) topography, (5) surface and subsurface hydrology, (6) climate, and (7) reclamation practices.

Some problems are shared by almost every operation. All mining operations create noise and dust and disturb or destroy scenic areas, croplands, and wildlife. Many mining operations pose a threat to water resources. The potential for pollution of surface water by siltation is greatest in areas of steep topography and in hydraulic mining and dredging operations. The potential for ground-water pollution is greater where area strip mining is practiced because of greater areal extent and because precipitation, instead of running off rapidly, infiltrates and migrates slowly through spoil piles, which may be high in sulfides and contain traces of lead, arsenic, or copper. Aquifers may be disturbed where large thicknesses of overburden are removed. In some operations ore is processed at the mine site, and the waste material (*tailings*) may represent a hazard. As examples, taconite (an iron ore mined in the Lake Superior area) tailings have been known to release asbestoslike materials; uranium tailings are radioactive; the leachate from sulfide ores is acidic and toxic to plants and animals.

Reclamation. The potential for reclamation is also dependent upon mining technique, climate, topography, thickness of topsoil, and nature

FIG. 4. Cutting heads of auger-mining machine boring into coal seam that has been exposed by contour strip mining. Coal is drawn out as the auger revolves and is loaded onto trucks by the conveyor belt in the background. (National Coal Association.)

of the overburden. In some instances it is not economically or even physically possible to restore the land to its original contour and vegetative cover. This is particularly true where there is little topsoil and overburden, the ore seam is very thick, and the climate is arid or semiarid. In other areas, however, reclamation is not only physically feasible but financially attractive. For example, sand and gravel pits on the fringe of urban areas are usually quickly reclaimed and converted to higher uses such as residential and commerical development.

The history of reclamation in the United States for the period 1930–70 was poor. Only 40 percent of the disturbed acreage was reclaimed—and most of that without the assistance of humans. Since then, however, there has been a marked improvement in the record. For example, approximately 80 percent of the land disturbed during

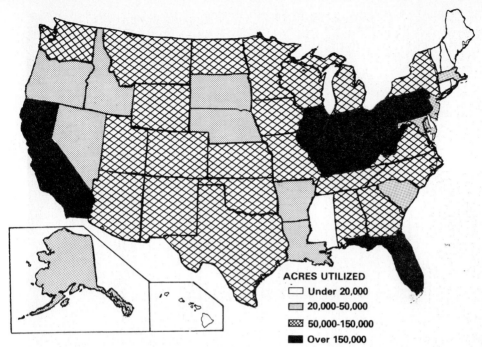

ACRES UTILIZED
- Under 20,000
- 20,000-50,000
- 50,000-150,000
- Over 150,000

FIG. 5. Geographic distribution of land utilized for mining, 1930 to 1980. (After W. Johnston and J. Paone, 1982, *U.S. Bureau of Mines Information Circular* 8862.)

1971 was reclaimed in that year alone; by 1980 a total of 47 percent of the acreage disturbed in the 50-year period since 1930 had been reclaimed (Fig. 6). The improved performance can be attributed to stricter state regulations. The geographic distribution of reclaimed acreage is shown in Figure 7.

Subsurface Mining

Solid Ores

Underground mining methods involve either sinking a vertical or inclined shaft or tunneling to the ore body. Peele (1941) indicates that the technique used to exploit an ore body depends on the following factors: (1) shape, size, regularity, and orientation of the body; (2) mineralogy and value of the ore; (3) strength and physical character of the ore and of the wall rock or overlying material; (4) relation of the deposit to the surface and to other ore bodies; (5) availability and cost of supporting materials; and (6) other economic, legal, and safety considerations.

Methods which ensure maximum extraction are generally employed

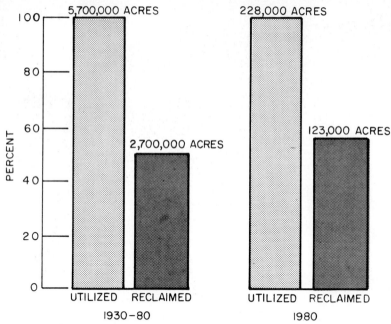

FIG. 6. Relationship of land utilized for mining to land subsequently reclaimed in the United States from 1930 to 1980 compared to mining and reclamation activity in 1980. (After W. Johnston and J. Paone, 1982, *U.S. Bureau of Mines Information Circular* 8862.)

when high-grade ore bodies are mined, whereas low-cost techniques are usually employed in mining low-grade ore bodies. For example, if the body is of low grade, the *room-and-pillar method* might be used. Surface and overlying rocks may be supported by permanent pillars left in the ore body; rooms, or stopes, are excavated between the pillars and the removal of the ore is incomplete and mining costs are minimal. Subsequently, additional ore may be recovered by simultaneously filling in the rooms with waste rock for support of the overlying material and removing ("robbing") the pillars. Or timbers may be used for permanent support while the pillars are removed. Either way, one must compare the additional mining costs of the fill or timber with the value of the additional ore extracted from the pillars.

Another approach involves the mining of a small section of an ore body and allowing the overlying material to cave in. This process, called *caving*, or *longwall mining*, is repeated in adjacent sections. No attempt is made to support the overlying material. The goal is to completely remove the entire seam or ore body in one operation. Working space is provided by timbers or other supports. Variations

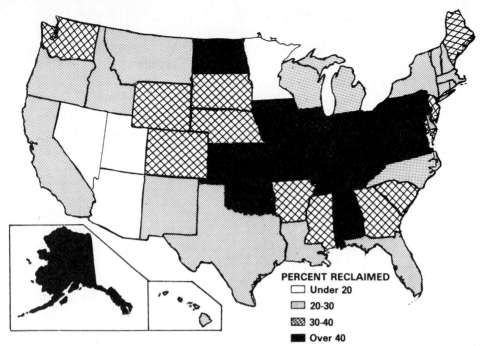

FIG. 7. Geographic distribution of reclaimed land utilized for mining from 1930 to 1982. (After W. Johnston and J. Paone, 1982, *U.S. Bureau of Mines Information Circular* 8862.)

and combinations of these basic techniques are numerous; the method employed depends upon the factors noted above.

Minor amounts of ores have been exploited through solution mining and in situ mining. In *solution mining*, water-soluble minerals such as halite (rock salt) are dissolved by hot water delivered through injection wells. Steam injection is also used to melt such minerals as sulfur, which is recovered as part of the steam injection program. *In situ mining* refers to the utilization or processing of a mineral resource in its natural place of occurrence. "In place" burning of coal or oil shale to recover heat, gas, or oil has been proposed as an alternative to the physical removal of coal or oil shale. In situ leaching of halite, potash, uranium, sulfur, and copper may be more economically feasible or practical, under some conditions, than other mining techniques.

The environmental impact of subsurface mining is not so visible as it is in a surface operation, but it is nonetheless significant. Subsurface mining does not involve the removal of topsoil or overburden, but it does generate spoil banks and tailings which may release toxic leachates. In coal mining areas, there is the possibility of underground fires which may ignite and burn slowly for decades. Wherever longwall mining is practiced there is considerable potential for surface subsidence—the sinking or gradual subsidence of the earth's surface. Subsurface operations may also encounter freshwater aquifers or rock units containing saline or acidic waters. Inasmuch as mines must be kept dry, all waters must be pumped to the surface and somehow disposed of. Finally, mine workers are subject to a variety of health and safety hazards, such as black lung disease, explosions, and cave-ins. (Fig. 8).

Fluid Resources

Underground water, oil, and natural gas are recovered by a variety of drilling techniques. In the petroleum industry, *rotary drilling* is the standard technique. A bit attached to the end of a "string" of drillpipe is rotated in the hole, and as the bit cuts into the rock, the entire assembly is lowered by attaching additional "joints" of drillpipe. Drilling mud is pumped into the pipe and circulated back to the surface around the outside of the pipe. Drilling mud serves to (1) cool and lubricate the bit, (2) bring drill cuttings to the surface, (3) prevent cave-ins, and (4) prevent blowouts associated with high-pressure oil and gas zones.

If drilling results in a commercial discovery, a string of casing is set. *Casing* is simply a large pipe that is run into the hole, cemented in place, and perforated opposite the producing horizons. Additional tubing, valves, pumps, and other equipment is then installed in order to recover the oil and gas. If drilling results in a dryhole, it will be necessary to case off freshwater aquifers before the well is plugged and abandoned.

If the natural or primary flow of oil or gas is not sufficient to sustain a commercial operation, several *enhanced recovery* techniques might be applied. Flow may be significantly increased by acidizing acid-soluble reservoir rocks, or by fracturing "tight" formations. If the oil is highly viscous, in situ combustion or steam injection will increase production. Injecting air, natural gas, or carbon dioxide into special wells to maintain high reservoir pressures also increases productivity. Water and certain solvents or detergents are sometimes used literally to clean the oil from the rocks, which is then recovered through a separation process and subsequent refining. Petroleum engineers must carefully weigh the cost of enhanced recovery techniques against the value of increased production.

The environmental impact of drilling for oil can be significant.

FIG. 8. Cave-in at abandoned mine shaft in Scranton, Pennsylvania. A 24-m (80 ft) crane disappeared when a concrete seal over the 88-m (288-ft) deep shaft gave way on May 22, 1982. (Michael Mullen, *Scranton Times.*)

Roads must be built to service the drilling site, which itself may have to be graded, and in the event of a discovery, storage and transportation facilities must be established. Care must be taken to avoid contamination of freshwater aquifers by setting surface or intermediate casing if borings come into contact with an aquifer. *Oil field brines* (saltwater that is almost invariably brought to the surface along with the oil) must be disposed of; for the most part this material is either evaporated in large ponds or reinjected through special injection wells.

An ever-present potential hazard is a *blowout*—an uncontrolled escape of oil or gas from a well. Blowouts are particularly difficult to contain when they occur offshore. Experience with large blowouts, such as the one at Santa Barbara, California, in 1969 suggests that the degree of sophistication that characterizes petroleum exploration and drilling does not characterize the industry's efforts to control oil spills at sea.

A significant problem associated with the withdrawal of large volumes of oil or water is that of subsidence. Specific examples and case histories of subsidence are presented in Reading 27.

Federal Legislation

Federal and state governments have enacted countless "environmental laws" which significantly affect exploration and mining activities; the most significant are considered here.

The National Environmental Policy
Act (NEPA) of 1969

The National Environmental Policy Act of 1969 (83 Stat. 852) requires a detailed environmental analysis (environmental impact statement) for all major actions on Federal land, with public review and response to the proposed action. An environmental impact statement (EIS) includes the following:

1. Description of the proposed action and environment affected
2. Relationship to land-use plans
3. Probable positive and negative impacts of the proposed action
4. Alternatives to the proposed action
5. Identification of unavoiable adverse environmental effects of the proposed action
6. The relationship between local short-term uses of the environment and long-term productivity in the area affected
7. Irreversible and irretrievable commitments
8. Other aspects of Federal policy thought to offset the adverse effects of the proposed action

The EIS thus represents a "full disclosure" document. The process encourages but does not mandate minimization of environmental harm.

Surface Mining Control and
Reclamation Act (SMCRA) of 1977

The primary purpose of the Surface Mining Control and Reclamation Act, which applies to both the surface and underground mining of coal, is to reduce the adverse environmentl impact of coal mining and to ensure that land disturbed or otherwise affected is reclaimed or reinstated to its premining appearance and uses. It also provides for the acquisition and reclamation of lands that have been previously mined and abandoned. SMCRA was the first comprehensive Federal statute to regulate the impact of coal mining on a national scale. The primary responsibility for implementing the provisions of the act rests with the individual states, which may assume "exclusive jurisdiction" if certain conditions are met.

Case Histories and Readings

Land Subsidence in Western United States

It has already been noted that both the underground mining of solid resources and the withdrawal of underground fluids—water, oil, and gas—can cause the subsidence or sinking of land. Specific examples and case histories of subsidence are presented in Reading 27, which documents the magnitude of the problem in the western United States and the economic and political impact of this phenomenon. In an oil field in the Los Angeles–Long Beach Harbor area, for example, subsidence had reached a maximum of 8.7 m (29 ft) by 1967 and $100 million in damage. Since then, water injection has not only halted further subsidence but has in fact led to a slight increase in surface elevations in the area.* The draining of peat bogs, the irrigation of certain soils, and the application of surface loads can also cause subsidence. Subsidence can also lead to catastrophic failures of dams or levees, and public agencies and private developers should be aware of the problem and the methods for stopping or reducing this hazard.

The Legacy of Uranium Tailings

In Reading 28, David Comey explores the health hazards associated with uranium tailings. He estimates that the total number of deaths resulting from projected production of uranium ore will be 5,741,500 over the next 80,000 years. Few of these deaths will occur during our

---

*X. Colazas, Department of Oil Properties, City of Long Beach. Personal communication, 1982.

lifetimes and close to half will occur outside the United States, but two provocative moral issues are posed: (1) Do we have the right to consume electricity from nuclear fission plants for the next few decades, forcing thousands of future generations to suffer lethal consequences? and (2) Do we have the right to consume electricity produced from nuclear fission in the United States, forcing millions in other countries who do not benefit from this electricity to suffer lethal consequences?

For years building contractors in Grand Junction, Colorado, used uranium tailings as a convenient and cheap source for landfill and concrete mix, used in the construction of homes, schools, and commercial buildings. Studies in 1970–71 in the Grand Junction area indicated a high incidence of cleft lip and cleft palate, a significantly lower birth rate, and a significantly higher death rate from congenital anomalies. It is clear that something must be done to eliminate both the immediate and the long-term hazards associated with uranium tailings.

Environmental Impact Analysis:
The Example of the Proposed
Trans-Alaska Pipeline

The social, economic, political, and environmental impact of resource development must be anticipated if we are to achieve a better balance between the losses and gains associated with development activities. Environmental impact analysis is the process by which we can predict the far-reaching effects of our attempts to exploit natural resources. Perhaps the most ambitious efforts to develop an environmental impact statement were those directed at the trans-Alaska pipeline project. In Reading 29, David Brew reviews these efforts and demonstrates how environmental considerations enter into the decision-making process. It is obvious that there are many interacting components and that special analytical techniques must be developed for each project.

Surface Mining Reclamation in
Appalachia: Impact of the 1977
Surface Mining Control and
Reclamation Act

In Reading 30, Larry Sweeney reviews reclamation techniques that have been successfully applied in Appalachia and comments on some of the controversial aspects of SMCRA.

27 Land Subsidence in Western United States*

Joseph F. Poland

Introduction

Volcanic eruptions, earthquakes, tsunamis, and landslides are instantaneous events that often have disastrous consequences. On the other hand, land subsidence due to man's activities, which I will be discussing in this paper, ordinarily is a relatively slow process that may continue for several decades. Subsidence may produce conditions or stresses that trigger some instantaneous event such as the failure of a dam or a levee, and public agencies should be aware of such potential hazards; but in many areas that have experienced appreciable subsidence, the problems created, although of considerable economic significance, are not hazards to human life.

Subsidence may occur from one or more of several causes, including withdrawal of fluids (oil, gas, or water), application of water to moisture-deficient deposits above the water table, drainage of peat lands, extrac-

*Publication authorized by the Director, U.S. Geological Survey.

Poland, J. F., 1969, "Land Subsidence in Western United States," in *Geologic Hazards and Public Problems*, May 27–28, 1969, Conference Proceedings, Olson, R. A. and Wallace, M. W., eds. U.S. Govt. Printing Office, pp. 77–96. Lightly edited by permission of the author. The paper was originally read before the conference and illustrated with slides.

Mr. Poland is a research hydrologist with the Ground Water Branch of the U.S. Geological Survey, Sacramento, California.

tion of solids in mining operations, removal of solids by solution, application of surface loads, and tectonic movements (including earthquakes).

In western United States, the subsidences of appreciable magnitude and area have been caused chiefly by the withdrawal of fluids, but also by application of water to moisture-deficient deposits and drainage of peat lands. This paper, therefore, will be limited to a brief description of these three types of subsidence, the problems created, and remedial measures, actual or potential.

Subsidence of Organic Deposits Due to Drainage

The peat lands which underlie roughly 450 square miles of the Sacramento-San Joaquin Delta constitute one of the largest areas of organic deposits in western United States. These peat deposits are as much as 40 feet thick. Drainage of the Delta islands for agricultural use began shortly after 1850. The land surface of many of the islands, initially about at sea level, is now 10 to 15 feet below sea level. Protective levees have been raised as the island surfaces have subsided. Leveling by Weir shows that the surface of Mildred Island subsided 9.3 feet from 1922 to 1955 at an average rate of 0.28 foot per year. Weir (1950) concluded that the causes of the subsidence were (1) oxidation of the deposits dewatered by lowering the water table to permit cultivation (by aerobic bacteria primarily, and probably the major cause), (2) compaction by tillage machinery, (3) shrinkage by drying, (4) burning, and (5) wind erosion.

The lower the island surfaces sink below the water surface in the Delta channels, the greater the stress on the protecting levees. This past winter, a levee reach on Sherman Island failed and the island was flooded. Although the water is being pumped out of Sherman Island, the levee-maintenance problem will increase in the Delta as continued drainage for cultivation lowers the island surfaces farther.

Subsidence Due to Application of Water (Hydrocompaction)

Locally, along the west and south borders of the San Joaquin Valley, moisture-deficient alluvial-fan deposits above the water table have subsided 5 to 15 feet after the application of water (Lofgren, 1960). These deposits are composed chiefly of mudflows and water-laid deposits and have higher clay content than the non-subsiding deposits, according to Bull (1964), who concluded that the compaction by the overburden load occurred as the clay bond supporting the voids was weakened by wetting.

This near-surface subsidence, or hydrocompaction, has been a serious problem, resulting in sunken irrigation ditches and undulating fields and has damaged canals, roads, pipelines, and transmission towers. It is particularly serious in construction and maintenance of large canals. As a preventive measure, deposits of this type along about 20 miles of the San Luis section of the California Aqueduct and along about 50 miles of the Aqueduct in Kern

County were precompacted by prolonged wetting, prior to canal construction. The estimated cost of this operation was $25 million.

According to Lofgren (1969), moisture deficient alluvial deposits that compact on wetting also have been reported in Wyoming, Montana, Washington, and Arizona, where subsidence of as much as 6 feet after wetting has created problems in engineering structures. Also, moisture-deficient loessial depostis as much as 100 feet thick covering extensive areas in the Missouri River basin have caused problems in the construction of dams, canals, and irrigation structures. Precompaction by wetting has been the usual solution, once this property of the sediments is recognized.

Subsidence Due to Withdrawal of Fluids

Subsidence due to withdrawal of fluids is by far the most common type of man-made regional subsidence. It may occur over oil and gas fields or over intensively exploited ground-water reservoirs. In either case, the cause is the same. The withdrawal of water reduces the fluid pressure in the aquifers and increases the effective stress (grain-to-grain load) borne by the aquifer skeleton. In ground-water reservoirs, the increase in effective stress in the permeable aquifers is immediate and is equal to the decrease in fluid pressure. The aquifers respond chiefly as elastic bodies. The compaction is immediate, but usually is small and mostly recoverable if fluid pressures are restored.

On the other hand, in the confining clays or the clayey interbeds, which have low hydraulic conductivity and high compressibility, the vertical escape of water and adjustment of pore pressure is slow and time-dependent. In these fine-grained beds, the stress applied by the head decline becomes effective only as rapidly as pore pressures decay toward equilibrium with pressures in adjacent aquifers. It is the time-dependent nature of the pore-pressure decay in these fine-grained beds that complicates the problem of predicting compaction or subsidence.

Intensive ground-water withdrawal and decline of head in heterogeneous confined aquifer systems in unconsolidated to semiconsolidated deposits of late Cenozoic age have produced the major areas of subsidence in western United States. Therefore, I will first review the dimensions and problems of subsidence due to ground-water withdrawal, and then comment briefly on subsidence of oil fields.

Subsidence Due to Ground-Water Withdrawal

In the Houston-Galveston area of the Texas Gulf Coast, 1 to 6 feet of subsidence has occurred over an area of about 1,500 square miles. This is due almost wholly to lowering of artesian head in the ground-water reservoir, although there are subsidiary depressions due to oil-field subsidence.

In south-central Arizona, subsidence of 1 to 3 feet has been defined by leveling in several areas where water levels have been lowered 150 to 250 feet. The maximum known subsidence in southern Arizona is in the Eloy-Casa Grande area where subsidence of 7.5 feet occurred between 1949 and 1967. The extent of the area is not defined.

At Las Vegas, Nevada, subsidence of 3 feet was indicated by leveling in 1963.

Figure 1 shows the principal areas of subsidence in California. In the Sacramento-San Joaquin Delta, we have the organic deposits that I described earlier. The areas of subsidence due to ground-water withdrawal include the Santa Clara Valley, which is at the south end of San Francisco Bay, and where about 250 square miles have been affected and maximum subsidence by 1967 was 13 feet in San Jose. Then we have the large area in the San Joaquin Valley extending from about Los Banos on the west side south to Wasco on the east side, and the area at the south end which is referred to here as the Arvin-Maricopa area. The maximum subsidence in the San Joaquin Valley is on the west side and was about 26 feet in 1966. To the south, in southern California, the location of the Wilmington oil field is shown, and about 20 miles to the northwest of the Wilmington oil field is the Inglewood oil field in the Baldwin Hills (not shown).

There is one other area of subsidence that I might mention. That is Antelope Valley (Lancaster area in Fig. 1). It is just north of the San Gabriel Mountains and at least 160 square miles have been affected; subsidence in Lancaster is at least 3 feet.

Figure 2 shows the magnitude and extent of subsidence in the San Joaquin Valley but not for the same period of time in all areas, because the year span is determined by the available leveling control. The maximum subsidence is on the west side of the valley southeast of Los Banos.

As of 1963, subsidence exceeded 20 feet west of Fresno, and extensive

FIG. 1. Areas of land subsidence in California. Major subsidence due to fluid withdrawal shown in black; subsidence in Delta caused by oxidation of peat.

areas had subsided 12 to 20 feet. On the east side, in the area between Tulare and Wasco, maximum subsidence was 12 feet by 1962; and that is the latest leveling control in that area. South of Bakersfield the maximum subsidence was 8 feet in 1965, which is the latest complete leveling in that area. The total area that is affected by more than one foot of subsidence exceeds 3,500 square miles or almost a third of the San Joaquin Valley. Each of the subsiding areas is underlain by a confined aquifer system in which the water level has been drawn down 200 feet or more—on the west side as much as 450 feet—by the intensive withdrawal. The dotted line is the position of the California Aqueduct

which passes through the western and southern areas of subsidence.

The land subsidence in the Santa Clara Valley from 1960 to 1967 is shown in Fig. 3; it was nearly 4 feet in San Jose in the 7 years, and this happened to be the period of most rapid land subsidence in the Santa Clara Valley. The total subsidence has been about 13 feet in downtown San Jose. You will note that from 1960 to 1967 there was about 2 feet of subsidence at the south end of San Francisco Bay.

Figure 4 illustrates the change in altitude (subsidence) of a bench mark in downtown San Jose, where the elevation changed from 98 feet above sea level to about 85 feet above sea

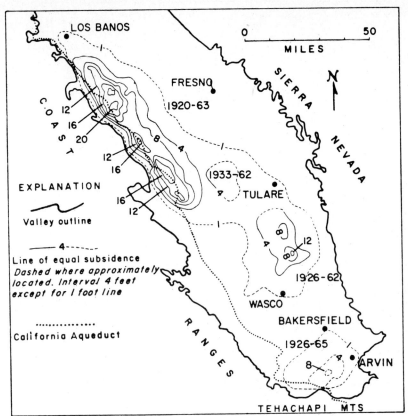

FIG. 2. Map showing the magnitude and areal extent of subsidence in southern San Joaquin Valley.

level from 1912 to 1967, representing a subsidence of 13 feet. The hydrograph shows the fluctuation of water level in a nearby well. I call your attention to the fact that during a period of artesian-head recovery from 1936 to 1943, the subsidence stopped; it presumably began again about 1947 and reached its steepest rate of about 0.7 foot a year in the early 60's, due to the rapid decline in head from 1959 to 1963. The rate of subsidence in San Jose has decreased substantially in the past 2 years because there has been a winter water-level recovery of about 30 feet above levels of the middle 1960's.

Figure 5 shows the relation of the subsidence of 25 feet occurring in western Fresno County in the San Joaquin Valley between 1943 and 1966 and a decline of approximately 400 feet in the water level in nearby wells. This bench mark is at the locus of maximum subsidence in the San Joaquin Valley.

Subsidence can be measured by two methods: by repeated leveling of bench marks at the land surface, which is the common way of measuring it, and also is the only way to get full areal coverage, or, it can be measured at one site, by measuring compaction directly. Releveling is the basis

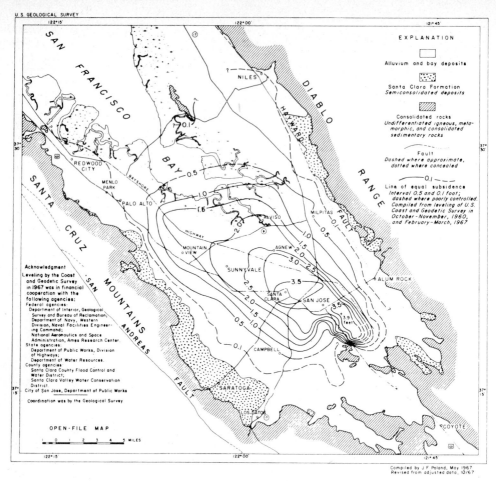

FIG. 3. Land subsidence from 1960 to 1967, Santa Clara Valley, California.

for knowing how much subsidence has occurred in the San Joaquin Valley. The level net in the subsidence areas plus the ties to bedrock total about 1,500 miles of leveling. Considering that first-order leveling costs about $200 a mile and second-order leveling about $100 a mile, releveling that entire area requires a substantial amount of funds.

Subsidence can be measured at a single point by using what can be elegantly called a bore-hole extensometer or can be referred to simply as a compaction recorder. Figure 6 is a simplified diagram of the type of compaction recorder operated in the San Joaquin and Santa Clara Valleys—consisting of an anchor connected to a stainless steel cable that passes over sheaves at the land surface and is kept taut by a counterweight. The cable is connected to a recorder. If there is compaction between the land surface and the anchor, the cable moves up with respect to the land surface and the magnitude can be recorded. The measured compaction equals land sub-

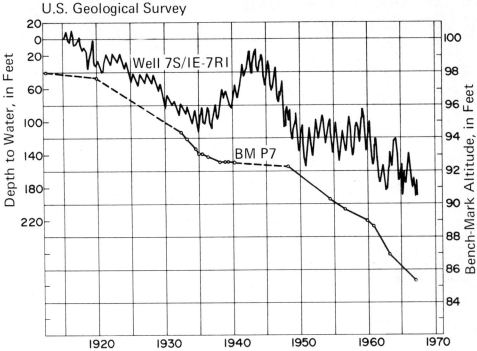

FIG. 4. Change in altitude at bench mark P7 in San Jose and change in artesian head in nearby well.

sidence if the anchor is below the compacting interval.

Figure 7 illustrates a record obtained on the west-central side of Fresno County at Cantua Creek from 1958 to 1966 of compaction in three wells—one about 500 feet deep, another 700, and a third, 2,000 feet deep. At the time the recorders were installed, the 2,000-foot well was almost as deep as any of the water wells nearby. Measured compaction 1958 to 1966 in the 500-foot well was small, in the 700-foot well, about 2 feet, and in the 2,000-foot well, more than 8 feet. The straight line connecting the three dots shows subsidence of a bench mark on the land surface as determined by leveling from distant stable bench marks and illustrates

that the compaction measured in the 2,000-foot well was almost equal to the subsidence of the land surface during this period. This type of multiple-depth installation can be utilized to find the magnitude and rate of compaction between the depth intervals; it can be used also to determine at what depth compaction is occurring.

Problems Caused by Subsidence

If the subsiding area borders the ocean or a bay as in the Santa Clara Valley, levees have to be built and maintained to restrain flooding of lowlands. If yearly high tides happen to coincide with times of excessive stream runoff, levees may be overtopped. The Christmas floods of 1955 caused inundation

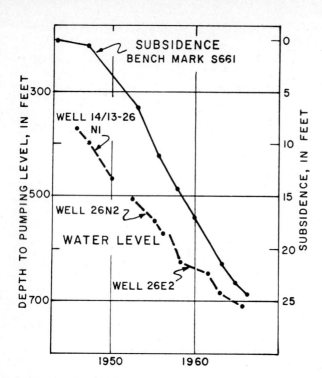

FIG. 5. Subsidence and artesian-head decline 10 miles southwest of Mendota.

of the town of Alviso at the south end of San Francisco Bay due to a combination of high tides and heavy stream runoff.

The differential change in elevation of the land surface in subsiding areas creates problems in construction and maintenance of water-transport structures, such as canals, irrigation systems, and drainage systems, and affects stream-channel grade. For example, the California Aqueduct passes through the subsiding areas on the west side of and at the south end of the San Joaquin Valley; also, a peripheral canal, when built, will be about at the east edge of the subsidence in the Delta.

The construction of the San Luis canal section of the aqueduct began in 1963. The map (Fig. 8) illustrates a problem that was faced in planning and construction of the canal in this subsidence area. The map shows subsidence in the 3 years ending in 1963. Obviously, then, planning involves the prediction of future subsidence, but beyond the physical characteristics that have to be considered, the economic and political aspects of the problem are very substantial. For example, how soon will Congress appropriate funds for the construction of distribution systems, and how soon will people who are going to utilize the canal stop pumping ground water and thereby decrease or stop subsidence? These kinds of questions complicate the problem of subsidence prediction.

Another problem that is common to the subsidence areas in the Santa Clara and San Joaquin Valleys, and also in south-central Arizona, results from the fact that the compaction of the depos-

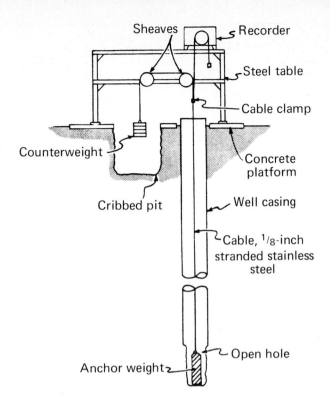

Sheaves

Recorder

Steel table

Cable clamp

Counterweight

Concrete platform

Cribbed pit

Well casing

Cable, 1/8-inch stranded stainless steel

Open hole

Anchor weight

FIG. 6. Compaction-recorder installation.

its develops compressive stresses on well casings. Many of the failures have been repaired by swaging out the ruptured casings and inserting liners. The costs of repair or replacement in central California probably have exceeded $10 million.

Subsidence of Oil Fields

Subsidence of a few feet has been noted in many oil fields and probably has occurred unnoticed in many more. The two oil fields in which subsidence has caused or contributed to major problems are the Wilmington and Inglewood oil fields in the Los Angeles coastal plain.

My comments on the Inglewood oil field in the Baldwin Hills area are summarized from a paper by Jansen,

Dukleth, Gordon, James, and Shields (1967). These men were all members of the State Engineering Board of Inquiry that investigated the Baldwin Hills Dam failure. In December 1963, the Baldwin Hills Reservoir was destroyed by failure of its foundation. Extensive damage was done to neighboring communities. This reservoir is on the northeast flank of the Inglewood oil field. Subsidence had been observed in the vicinity for many years and was estimated by Jansen and others (1967) to have been 9.7 feet between 1917 and 1963 at a point approximately one-half mile westerly of the reservoir. They concluded that "the earth movement which triggered the reservoir failure evidently was caused primarily by subsidence."

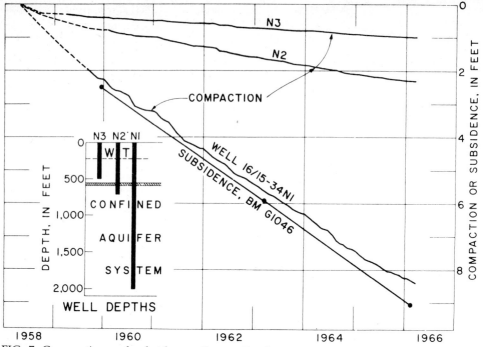

FIG. 7. Compaction and subsidence, Cantua Creek site.

In the Los Angeles-Long Beach Harbor area, the Wilmington oil field experienced a costly and spectacular land subsidence beginning in 1937, which had reached 29 feet at the center by 1967. In Fig. 9, the subsidence contours as of 1968 have been superimposed on an aerial photograph of the area. The small dimple to the northeast is the subsidence of the Long Beach oil field. The main part of downtown Long Beach, Terminal Island and Pier A are shown. The area is intensively industrialized, and this is one reason why the remedial costs have been so great. Extensive remedial measures have been necessary to keep the sea from invading the subsiding lands and structures, because much of the subsiding area initially was only 5 to 10 feet above sea level. The remedial

measures for restraining the sea have been chiefly massive levees, retaining walls, fill, and raising of structures. Horizontal as well as vertical movement developed stresses that ruptured pipelines, oil well casings, and utility lines, and damaged buildings.

The cost of this remedial work to maintain structures and equipment in operating condition had exceeded $100 million by 1962. The Wilmington oil-field subsidence has been described in many papers, so I won't discuss it further except under methods for stopping or reducing subsidence.

Methods for Stopping or Reducing Subsidence

Decreasing fluid pressure in a confined system increases grain-to-grain load and causes compaction; also, in-

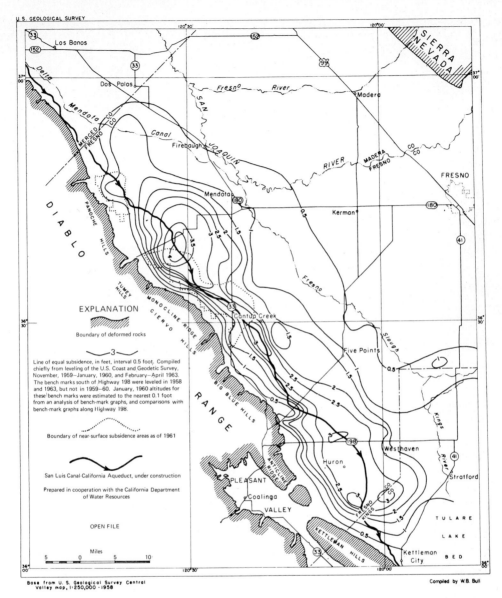

FIG. 8. Map showing land subsidence in the Los Banos-Kettleman City area, Calif., 1959–63.

creasing fluid pressure decreases grain-to-grain load and decreases or stops subsidence. In 1958, repressuring of the oil zones at Wilmington began, based on this premise and on the expectancy of increased oil recov- ery. The bench mark shown in Fig. 10 is on pier A, and you will note that it stopped subsiding in 1958 immediately after injection began nearby. It was essentially stable into 1961 and then rebounded about 0.5 foot by

FIG. 9. Aerial view of the city of Long Beach, California, showing total subsidence contours (1928–1968). (Photo courtesy of M.N. Mayuga, Dept. Oil Properties, City of Long Beach.)

FIG. 10. Graph of vertical movement of Pier A, Port of Long Beach, 1952–1965.

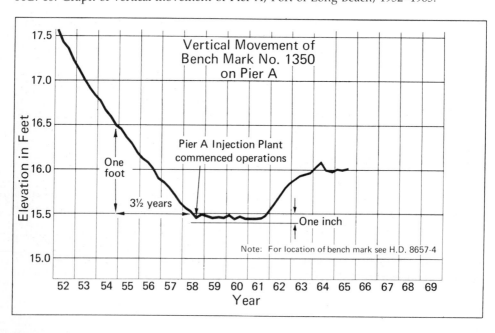

Geologic Hazards and Hostile Environments

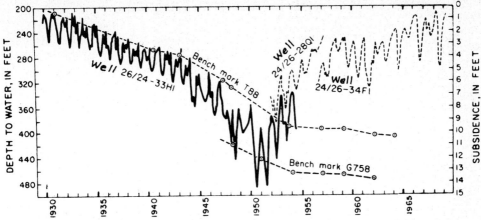

FIG. 11. Correlation of water-level fluctuations and subsidence near center of subsidence south of Tulare.

1964. The maximum rebound of bench marks due to the repressuring as of the present time is on the order of 1.1 foot. Maps prepared by the Department of Oil Properties, City of Long Beach, indicate that by 1968 the subsidence had been stopped in most of the Wilmington oil field by the repressuring operation. I think this is an outstanding demonstration of subsidence control. And I think the people responsible deserve a great deal of credit for overcoming many problems—legal and otherwise—in controlling this subsidence.

Although no subsiding groundwater basins have been extensively repressured through wells, water imported to the Tulare-Wasco area in the southeastern part of the San Joaquin Valley through the Friant-Kern Canal since 1951 has reduced the pumping draft and caused substantial recovery of water level. Figure 11 shows that subsidence of bench marks near Delano, south of Tulare, decreased greatly after 1954, due to recovery of artesian head. Importation of water to the Santa Clara Valley is helping to raise the arte-

sian head and decrease subsidence there, also. As a matter of record, water is or soon will be imported to all the major areas of subsidence due to ground-water withdrawal in California, so that ten years from now subsidence rates in many of these areas probably will be reduced greatly.

References

Bull, W. B., 1964, Alluvial fans and near-surface subsidence in western Fresno County, Calif.: U.S. Geol. Survey Prof. Paper 437-A, p. A1–A71.

Jansen, R. B., Dukleth, G. W., Gordon, B. B., James, L. B., and Shields, C. E., 1967, Earth movement at Baldwin Hills Reservoir: Am. Soc. Civil Engineers Proc., Jour. Soil Mech. Found. Div., SM4, no. 5330, p. 551–575.

Lofgren, B. E., 1960, Near-surface land subsidence in western San Joaquin Valley, Calif.: Jour. Geophys. Research, v. 65, no. 3, p. 1053–1062.

———, 1969, Land subsidence due to the application of water: Geol. Soc. America, Rev. Eng. Geology 2, p. 271–303.

Weir, W. W., 1950, Subsidence of peat lands of the Sacramento-San Joaquin Delta, Calif.: Calif. Univ. Agr. Expt. Sta., Hilgardia, v. 20, no. 3, p. 37–56.

28   The Legacy of Uranium Tailings

David Dinsmore Comey

A typical coal-fired power plant probably kills about 50 people per year with its air pollution whereas a typical nuclear power plant of the same capacity probably kills an average of 0.01 people per year. A coal-fired power plant kills more people every few days than a nuclear power plant will in its 30 or so years of existence.—Bernard L. Cohen

Last year an article in the *Bulletin* by Bernard L. Cohen opened with the above provocative statements.[1] The health impact of 50 deaths per gigawatt-year from coal fired plants was almost entirely due to sulfur oxides released from the plant stack.* The health impact of 0.01 deaths per gigawatt-year from a nuclear plant came from radioactive effluents released during normal plant operation, and assumed there would be no deaths from nuclear plant accidents. Other comparisons by nuclear proponents have reached similar conclusions.[2]

Evidence is now emerging, however, that the health impact from the nuclear fuel cycle has been grossly underestimated. Robert O. Pohl, professor of

physics at Cornell University, has made a major contribution to our knowledge of these effects in an important but as yet unpublished paper. Pohl concluded that due to thorium-230 decay alone, the health impact of nuclear-generated electricity is at least 394 deaths per gigawatt-year.[3] Thus, if the number of U.S. nuclear power plants projected for the year 2000 by the former Atomic Energy Commission are actually built, the result will be at least 5,741,500 future deaths from lung cancer over the next 80,000 years. A little over 3 million of these deaths will be among the population of the eastern United States, the remainder in the rest of the Northern Hemisphere.[4]

How could such enormous health effects have been overlooked? Probably because almost everyone has focused on emissions from the nuclear power plants and the fuel reprocessing plants, and virtually ignored the other end of the uranium fuel cycle.

Presently mined uranium ore contains 0.1 to 0.2 percent uranium and only 0.71 percent of *that* uranium is fissionable uranium-235.† Thus more than 100,000 tons of uranium ore must be mined to produce one gigawatt-year of electricity.

At the uranium mills, located in the

Reprinted by permission of *The Bulletin of the Atomic Scientists,* a magazine of science and public affairs, v. 31, n. 7, pp. 43–45. Copyright © 1975 by the Educational Foundation for Nuclear Science, Chicago. The late Mr. Comey was director of environmental research for Business and Professional People for the Public Interest when this article was written.

*One gigawatt-year equals 8,760,000 megawatt-hours (the amount of electricity generated in a year by continuous operation of a million-kilowatt power plant).

†It is the fissile uranium-235 which is fabricated into fuel elements for the nuclear reactors used commercially in the United States to generate electricity. After the uranium is mined, milled and enriched, it is fabricated into these fuel elements or rods used in the reactor's core. The many stages of the nuclear fuel cycle are discussed by John Holdren in "Hazards of the Nuclear Fuel Cycle," *Bulletin* (Oct. 1974).

western United States, the uranium ore is crushed and ground, then the uranium is chemically separated and the residue discarded on a tailings pile. When the pile dries out, radioactive radon-222 gas routinely escapes into the atmosphere, where it travels long distances.[5] Radioactive isotopes formed by the decay of radon-222 include the "radon daughters" polonium-218 and polonium-214. These isotopes of radon have chemical and physical properties that cause them to be deposited in the bronchial epithelium or tissue of the human lung, leading to high radiation dose rates to a specific region of the lung, the bronchus, where lung cancers are most likely to arise.[6]

Thorium-230

The isotope in the uranium mill tailings which is the source of the radon and its lethal daughters is thorium-230, whose radioactive half-life is 80,000 years.[7] This assures that radon-222 will continue to be produced from a tailings pile for more than a million years, with consequent health effects on human beings.

The U.S. Environmental Protection Agency, in its 1973 report on the uranium fuel cycle, estimated that approximately 5 percent of the radon generated in a tailings pile 16 feet thick would escape from the pile as a gas. Based on this release rate, the EPA estimated that in the first 100 years, a typical uranium mine tailings pile of 250 acres would result in 200 health effects.[8] A recent modification by EPA of their air dispersion model has reduced that to approximately 60 health effects during the first 100 years.[9]

Since those 'health effects' are lung cancers, for which even the most medically advanced survival rate is less than 5 percent, Pohl has assumed that 57 lung cancer deaths would result in this 100-year period from each such pile. Because only 0.00091 of the thorium-230 in the pile will decay during this 100 year period, however,[10] the total number of deaths over time will be enormous; and based on the EPA assumption that 159 gigawatt-years of electricity[11] will be generated from the uranium-235 that produced this tailings pile, Pohl calculates the fatality rate to be 394 deaths per gigawatt-year.*

If this estimate of 57 deaths per uranium tailings pile per 100 years and the EPA's estimate of 37,620 tons total production of uranium yellowcake per mill are correct,[12] then given the 6,897,000 tons of yellowcake that will be required for the 1,090 gigawatts of reactors the AEC expects will be operating in the United States by the year 2000, I estimate that the total number of deaths resulting from this uranium production will be 5,741,500 over the next 80,000 years.†

*$(57/159)/0.00091 = 394$ deaths per gigawatt-year.

†$(57/0.00091)/37,620 = 1.665$ deaths per ton; $6,897,000$ tons $\times$ 1.665 deaths per ton = 11,483,000 deaths; half of these deaths (5,741,500) will occur in the next 80,000 years during the first half-life of the thorium-230.

The AEC estimated a cumulative requirement of 2,187,000 tons of yellowcake by the year 2000. At their estimated annual consumption of 157,000 tons per year, 30 years operation would result in a total of 6,897,000 tons: $(157,000 \times 30) + 2,187,000 = 6,897,000$. This is for 1,090 gigawatts of nuclear plants on line by the year 2000. See *Nuclear Power Growth 1974–2000*, WASH-1139(74) (Washington, D.C.: Atomic Energy Commission, Feb. 1974), p. 3, Table 4, Case D. Hereinafter referred to as WASH-1139(74).

Pohl has probably erred on the conservative side in his calculations. For one thing, he uses the 1973 EPA estimate that one such model tailings pile equals 159 gigawatt-years of electricity.[13] The AEC's 1974 report, WASH-1139(74), indicates that 142.5 gigawatt-years per pile is a more realistic figure.[14] This increases the fatality rate to 440 deaths per gigawatt year.*

Moreover, Pohl's calculations assume no growth in population either in the United States or in the world over the 1970 population. While such a calculation is very valuable in terms of forming a constant baseline from which the health impact from various growth models can then be calculated, in itself it probably is not realistic.

The EPA report assumed that the U.S. population reached 300 million in the year 2020 and then stabilized. The world's population was assumed to increase at a 1.9 percent annual growth rate from a 1970 base of 3,560,000,000.[15] When this EPA growth model is taken into account, the fatality rate rises to 982 deaths per gigawatt-year.† This corresponds to

12,825,000 deaths over the next 80,000 years—if AEC predictions of 1,090 gigawatts of nuclear power plant capacity in the United States by the year 2000 are met, and if these plants operate for 30 years.*

Based on the foregoing, it would seem to be a myth that the lethal health effects from coal-generated electricity are 5,000 times greater than the lethal health effects of nuclear-generated electricity as estimated by Cohen and others.[16] The deaths induced by the decay of thorium-230 in uranium mill tailings alone seem to swing the statistics in the reverse direction, and further analysis of other parts of the nuclear fuel cycle may identify additional health effects that have been overlooked.

The Atomic Industrial Forum, the American Nuclear Society and others may argue that very few of the thorium-induced deaths will occur during our lifetimes, and that it is unfair to make such a comparison of *current* deaths from coal-generated electricity with *future* deaths from nuclear-generated electricity. But that makes the disparity a moral issue: Do we have the right to consume electricity from nuclear fission plants for the next few

*(57/142.5)/0.00091 = 440 deaths per gigawatt-year.

†This assumes that the non-U.S. health effects (lung cancers) after the first 100 years result from a stable world population corresponding to the average during the first 100 years. This means 90 health effects outside the United States during the first 100 years, 44 health effects inside, for a resulting figure of 982 deaths per gigawatt-year: ([(44 + 90)/142.5]/0.00091) × 0.95 = 982 deaths per gigawatt-year.

This calculation assumes that all the uranium-235 made into reactor fuel is actually fissioned to make electricity. In reality, however, about one-quarter of it is not fissioned and remains trapped in the spent fuel elements. The AEC seems to assume that all but 1.3 percent of it will be recovered by reprocess-

ing the spent fuel and by shipping the recovered uranium to an enrichment plant where the uranium fuel cycle will start once again (WASH-1139(74), p. 26). So far this has not been the case, however, and the economics of fuel reprocessing make it unlikely in the future. Correcting the above calculations for this factor would raise the fatality rate well over the 1,000 deaths per gigawatt-year level.

*See first note, col. 1. (134 × 0.95)/0.00091/37,620 = 3.719 deaths per ton. Thus 6,897,000 tons = 25,650,000 deaths; half of these deaths (12,825,000) will occur during the first 80,000 year half-life of thorium-230.

decades, forcing thousands in future generations to suffer the lethal consequences?

Theoretically we could have a national referendum and decide that 50 deaths per gigawatt-year from coal-generated electricity was a debt we are willing to pay ourselves. But how can we presume to make such a decision involving the lives of our descendents?

Even if no future generations were involved, there would still be a moral issue: Close to half the thorium-induced deaths will occur outside the United States, but the electricity will be consumed by the U.S. economy. That is likely to be regarded as a paradigm example of imperialism.

What is to be done?

In the case of coal-fired plants, the lethal health effects of the sulfur oxides can be considerably reduced either by removing them from the stack gases after the coal is burned, or through desulfurization.[17]

Oak Ridge National Laboratory estimates that 90 percent of the health effects of radon-222 from tailings piles can be eliminated by using a 20-foot earth cover over the pile;[18] but past experience with erosion by wind and flooding makes it clear that this would be an intermediate solution for only a few decades, and even this would cost about a billion dollars.

Removal of long-lived radionuclides such as thorium-230 from the tailings prior to their disposal could be accomplished at considerable economic cost. However, inasmuch as the disposal of high-level radioactive wastes from reactor spent fuel remains unresolved, this additional step would hardly assist in solving that problem.

Mixing thorium-contaminated tailings with cement and pumping the slurry down into abandoned uranium mines may present other long-term problems, not to mention the fact that the volume of the tailings is greater than the volume of the uranium ore removed from the mine, so not all of it would fit.[19]

Any solution is probably going to raise the price of uranium substantially, and may result in making nuclear fission plants so uneconomical in comparison with alternate sources of electricity that no further construction of nuclear plants will occur.

This eventuality, however, should not be allowed to result in the neglect of the more than hundred million tons of uranium tailings already accumulated in the western United States.

Notes

[1]Bernard L. Cohen, "Perspectives on the Nuclear Debate," *Bulletin*, 30 (Oct. 1974), p. 35.
[2]Leonard A. Sagan, "Human Costs of Nuclear Power," *Science*, 177 (1972), p. 487; Richard Wilson, "Kilowatt Deaths," *Physics Today*, 25 (1972), p. 73. See also Lester B. Lave and Linnea C. Freeburg, "Health Effects of Electricity Generation from Coal, Oil and Nuclear Fuel," *Nuclear Safety*, 14 (Sept.–Oct. 1973), 423.
[3]Robert O. Pohl, "Nuclear Energy: Health Effects of Thorium-230" (Ithaca, N.Y.: Department of Physics, Cornell University, May 1975). Pohl's study was done with the support of a fellowship by John Simon Guggenheim Memorial Foundation.
[4]Based on air distribution model used in *Environmental Analysis of the Uranium Fuel Cycle, Part I—Fuel Supply* (Washington, D.C.: U.S. Environmental Protection Agency, Oct. 1973), p. 73, as modified by letter of July 2, 1975 from William H. Ellett, EPA's Bioeffects and Analysis Branch to Bernard L. Cohen, University of Pittsburgh. The model and the report will be hereinafter referred to as the '1973 EPA Report.'
[5]Ellett to Richard Wilson, Harvard University, July 2, 1975. Ellett said: "We do not agree that

the 3.64 day half-life of radon confines the impact of tailings piles to remote areas. Both our studies and [Lester] Machta's indicate about 50 percent decay between the average tailings piles sites and the eastern U.S. seaboard."

[6]EPA Report, Part I, p. 64. Polonium-218 and polonium-214 emit alpha particles with very high energies (6.0 and 7.7 million electron volts respectively). See Merril Eisenbud, *Environmental Radioactivity* (New York: Academic Press, 1973), p. 208.

[7]Pohl, "Nuclear Energy."

[8]EPA Report, Part I, p. 73.

[9]EPA Report and Ellett to Cohen, July 2, 1975.

[10]Pohl, "Nuclear Energy," Fig. 1.

[11]EPA Report, Part I, p. 25.

[12]EPA Report, Part I, pp. 24–25. 'Yellowcake' is U_3O_8, or uranium oxide, which is ultimately converted into reactor fuel.

[13]EPA Report, Part I, p. 25.

[14]The average light water reactor requires 264 tons per year of uranium yellowcake to produce 1 gigawatt-year of electricity. See WASH-1139(74), p. 24, Table 12. The EPA's model mill is assumed to produce 37,260 tons over its lifetime: 37,620/264 = 142.5 gigawatt-years.

[15]EPA Report, Part 1, p. A-26ff; Part III, p. D-15.

[16]An article by Lave and Freeburg concluded that a pressurized water reactor "appears to offer 18,000 times less health risk than a coal-burning power plant" (*Nuclear Safety*, 14 (Sept.– Oct. 1973), 423). The authors also concluded that "Radiation exposure to the public from the current effluents of uranium mines and mills and plants involved in feed-materials production, isotopic enrichment, and fuel fabrication is not considered significant compared with doses from power-plant or reprocessing-plant effluents. For example, it has been estimated that the total population dose from current uranium-mill effluents per annual fuel requirement produced for a 1,000-megawatt electrical power plant is no more than 0.06 man-rem, primarily from airborne thorium-230."

[17]Significant breakthroughs in both technologies have been recently announced by the Illinois Institute of Technology and the Battelle Memorial Institute.

[18]EPA Report, Part I, p. 46.

[19]Pohl notes the interesting fact that the volume of uranium tailings involved in generating one gigawatt-year of electricity is of the same order of magnitude as the volume of fly ash from a gigawatt-year's worth of coal-generated electricity. This seems to have been overlooked by those environmentalists who argue that the use of uranium to generate electricity has far less impact on the environment than use of coal. At least fly ash can be mixed with concrete for construction purposes: such uses of uranium tailings (for example, Grand Junction, Colorado) are no longer permitted.

29 Environmental Impact Analysis: The Example of the Proposed Trans-Alaska Pipeline

David A. Brew

Introduction

The precedents that have been and will be set by the proposed oil-pipeline system and soon-to-be proposed gas-pipeline system in Alaska will have far-reaching implications for petroleum development in the arctic parts of the Western and possibly the Eastern Hemisphere. Some of the most important precedents will concern the acquisition, analysis, and use of environmental data. The Alaskan example is of interest to all groups involved in arctic resource development because it provides information on predicted environmental impacts and on the methods used in arriving at the predictions.

This paper represents an attempt on the part of the author to summarize pertinent elements of the experience derived from the preparation of a complex environmental analysis for the benefit of others concerned in similar endeavors.

The purposes of this paper are (1) to describe the reasons for analyzing environmental impact and discuss (a) the implications of the National Environmental Policy Act (NEPA) of the

Brew, David A., 1974, "Environmental Impact Analysis: The Example of the Proposed Trans-Alaska Pipeline, *U.S. Geol. Survey Circ. 695,* 16 pp.

The author is on the staff of the U. S. Geological Survey.

United States and of similar laws in other countries to governmental and industrial decision-making processes, (b) the economic and public interest factors in the industrial decision-making process, and (c) the basic need to develop ways of minimizing the environmental costs that mankind must pay now and in the future; (2) to describe the general methodology needed to analyze environmental impact rigorously and objectively; (3) to describe in some detail how this methodology was applied to the proposed trans-Alaska pipeline and related systems; (4) to describe the main types of impact predicted from that analysis; (5) to examine the alternatives to the proposed pipeline; and (6) to analyze briefly from the author's viewpoint the approval of the proposed trans-Alaska oil-pipeline system as an example of the degree to which environmental considerations influenced the decision-making process.

It is difficult to discuss these points disinterestedly, without advocating one view or another, because many of the issues and factors are politically sensitive and subject to opposing interpretations when differing value frameworks are used. Nevertheless, because the lessons to be learned from the Alaskan pipeline example are important, the author has attempted to examine the ramifications of the impact analysis and of the decision deliberately and objectively.

This paper is modified from a paper prepared for presentation to the Fifth International Congress of the Foundation Francaise d'Etudes Nordiques (Brew and Gryc, 1974). The interested reader is referred to that paper for a more complete discussion of the analysis of the government's decision (point 6, above).

Proposed Trans-Alaska Pipeline

The Secretary of the Interior of the United States has granted a permit to the Alyeska Pipeline Service Company for a 48-inch oil-pipeline right-of-way across Federal land in Alaska between a point south of Prudhoe Bay on the North Slope and Port Valdez, an arm of Prince William Sound, on the south coast (Fig. 1). The company will design, construct, operate, and maintain the pipeline system.

The pipeline will be about 789 miles (1,270 km) long, some 641 miles (1,030 km) of which will be across Federal land. The pipeline system will also include pump stations, campsites for use during construction, airfields for use during both construction and operation of the pipeline, a communication system, lateral access roads, and pits or quarries for construction materials. The marine terminal site on Port Valdez will consist of a tank farm, dock, and related facilities. Prior to construction of the pipeline north of the Yukon River, a road will be built for access and the movement of equipment, materials, and personnel during construction. This road, which is proposed to become part of the State of Alaska highway system, will be about 361 miles (580 km) long.

Construction of the proposed pipeline system will result in three additional significant developments not directly included in the pipeline application: (1) an oil field complex at Prudhoe Bay on the North Slope, (2) a probable gas transportation system, and (3) a marine tanker system operating between Port Valdez, Alaska, and various destination ports.

The pipeline and its related developments will constitute a complex engineering system that will result in changes in the existing abiotic, biotic, and social and economic systems of Alaska and adjacent areas. In addition, the pipeline system will affect the economics of energy use and the strategy of energy supply in the United States. The phrase "environmental impact" has gained general use in denoting changes that would occur in existing systems if a proposed course of action were to be adopted.

Reasons for Analyzing Environmental Impact in the Arctic and Other Regions

There are philosophical, economic, social, and legal reasons for attempting to analyze environmental impact in the Arctic and elsewhere. These different reasons are linked together in a complicated way, but the social reasons (those pertaining to the physical well-being of humans and their surroundings) have been dominant and have in some countries led to legal requirements.

People have only recently realized that some of the effects of the industrial revolution are potentially severely

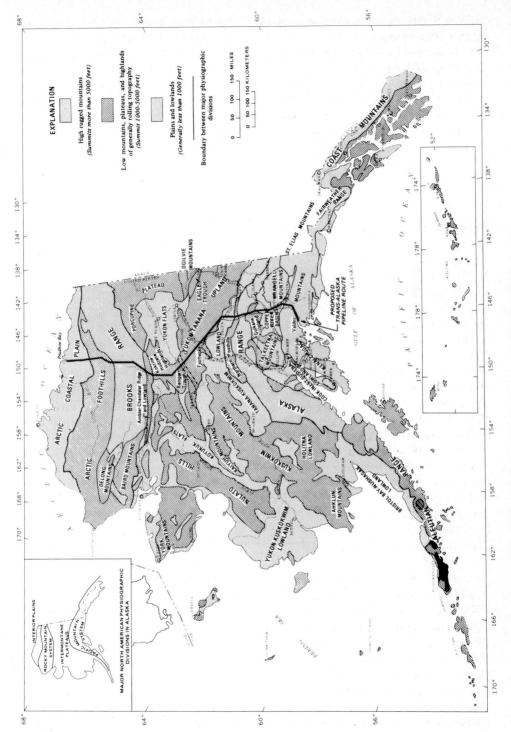

FIG. 1. Route of proposed oil-pipeline system. Modified from Wahrhaftig (1964, pl. 1).

detrimental to the life support system that must sustain present and future generations. The natural environment, as contrasted with the social and economic environment that man creates, is particularly susceptible to damaging stresses.

Many now believe that the greatest long-term benefits of health and enjoyment are possible only if the natural environment is maintained in a condition as close as possible to that existent before the world population explosion and industrial revolution. If people are to work toward this goal, then it is necessary to strive systematically to repair the damage already done to the natural environment and to avoid or minimize damage from current and future human development.

The Arctic is as yet practically untouched by modern industrial society, and detrimental effects can and should be avoided. If humans are to develop the resources of the Arctic, then they must choose from all the methods of exploration, extraction, and transportation available those that will cause the least environmental damage. The choice must therefore be based on predictions of the consequences to the environment of the various alternative methods. Environmental impact analysis is the process by which these predictions are made.

In addition to these social and philosophical reasons, industry has imposed on itself reasons for predictive analysis relating to the economic advantages of safe operation and of minimization of capital construction and of operating and maintenance costs. These analyses have for many years been an element in decision making by the pipeline and other industries, but it is now becoming clear that there are economic and social advantages in demonstrating that industry has a proper and positive concern for the environment and in particular for the effects that petroleum development, petroleum transportation, and their complex interactions have on the many facets of the environment. As the people of the United States continue to become more environmentally conscious, it will be advantageous for the oil industry to establish and maintain a position of positive environmental consciousness and action.

In the United States the legal requirements for analyzing environmental impact are contained in the National Environmental Policy Act (NEPA) of 1969.[1] The primary purpose of the United States Congress for that enactment was to establish a Federal policy in favor of protecting and restoring the environment. The wording of the act is such that all aspects of man's surroundings are the subject of Federal concern, and the intent is to make environmental considerations a real part of the governmental decision-making process.

The United States' NEPA contains strong directives to Federal agencies to follow this new policy. One section "authorizes and directs that, to the fullest extent possible *** the policies, regulations, and public laws of the United States shall be interpreted and administered in accordance with" the policy of the act. Another section of the act directs agencies to give "appropriate consideration" to environmental values in all decisions. Other sections relate to existing Federal agency policies and other aspects of environmental impact analysis and considera-tion in Federal authorizations of differ-ent types. Yet another section of the act establishes that Federal agencies must predict the environmental effects of proposed actions and of their alter-natives and describe them in an "en-vironmental impact statement" at an appropriate time in the decision-making process so that environmental considerations can be an actual part of that decision-making process. The en-vironmental impact statement (U.S. Federal Task Force on Alaskan Oil Development, 1972) on the proposed trans-Alaska pipeline which provides the background for this paper was prepared in compliance with that sec-tion of NEPA.

Although Canada does not have an act comparable to the NEPA of the United States, it is clear that the intentions of the Canadian Govern-ment are similar to those of the United States Government in requiring that environmental considerations shall be a part of resource development in arctic regions and that legal require-ments will be imposed on any appli-cants who propose pipeline construc-tion and operation in the Canadian Arctic. Those requirements will in-clude specific points considering the preservation and protection of the environment; therefore, the analysis of the environmental impact of any pro-posed pipeline system in northern Canada will be required.

There are of course both in the United States and in Canada many other laws and regulations which per-tain to the construction and operation of pipeline systems. They are only indirectly related to the analysis of environmental impact and are incor-porated in all planning and design of pipeline projects for the Arctic.

European countries do not yet have the legal requirements for pipeline systems based on exclusively environ-mental factors like those just discussed for the United States and Canada. Nevertheless, there are governmental regulations regarding the design, con-struction, and operation of pipelines, and those regulations are indirectly related to environmental considera-tions. The different governmental procedures, regulations, and national codes now existing in Western Europe have been examined in a paper by Watkins (1971). It is impossible from the author's vantage point in the United States to comment on whether environmental impact analysis re-quirements are likely to become a part of pipeline regulations in Western Eu-rope in the near future. Also, it is not known to the author whether environ-mental impact analysis of pipeline systems has been practiced or is being practiced in the Soviet Union (Pryde, 1972).

Although increased attention is be-ing given environmental questions in Europe (Verguèse, 1972), the impres-sion is that environmental impact analysis as discussed in this paper has not been practiced in other parts of the world.

A broader and more important rea-son for analyzing environmental im-pact transcends specific legal require-ments. People appreciate now as never before that they exist on an earth that has finite limits and tolerances. In the present century laymen and scientists alike have recognized many symptoms of environmental perturbation that cause concern. These symptoms, and the technical prediction abilities now available, can be used to demonstrate that people can inadvertently and adversely affect their total environment. If people are to continue to enjoy a healthy existence on earth, those effects must be minimized. The costs of minimizing must themselves also be minimized and must be assigned economically as well as socially.

Environmental impact analysis is a process that uses existing information, existing symptoms, and prediction techniques to forecast what environmental impact effects will be. The control of adverse environmental impact effects must be based on the best information available, and the best information available is obtained through environmental impact analysis.

Environmental Impact Analysis of the Trans-Alaska and Trans-Alaska-Canada Pipeline Systems

The components that are essential in an environmental impact analysis (and the interactions between them) are well illustrated in the example of the proposed trans-Alaska pipeline system and its related developments. Environmental impact analysis requires interrelating several components: analytical method, baseline environmental data, impact linkage data, impacting project data, analysts, and coordination (Fig. 2). To be applicable to geographically large, technologically complex, and environmentally sensitive projects, the analytical method should be formulated for the specific environmental situation and proposed project at hand. The general case methodology requires (1) system-

atic description of the environment including identification and classification of its sensitive elements, (2) systematic description of the project that would be doing the impacting including identificaiton and classification of the impacting factors inherent in it, (3) systematic accumulation of information related to linkages between impacting factors and the environment, (4) analysis of the interrelationships between the sensitive environmental elements and the impacting factors (including indirect and secondary feedback-type relations), (5) prediction of the net effects of those relations, and (6) preparation of an environmental impact report describing the results of the impact analysis. All these components and requirements were successfully included in the analysis of the proposed trans-Alaska pipeline. The actual analysis was made by a task force of resource scientists who were assigned the roles of impact analysts for specific resource topics or disciplines.

The environmental component and the impacting effects information just referred to could be combined to form an information matrix of specific design for the analysis of the impact of the proposed trans-Alaska pipeline and appurtenant systems. Needless to say, the matrix would be complicated and cumbersome, but it would synoptically depict in simplified fashion which impacting effects would impact on which environmental systems or components of those systems. In this regard, it is pertinent to comment on the approach to environmental impact analysis that is contained in U.S. Geological Survey Circular 645 (Leopold and others, 1971). As discussed later

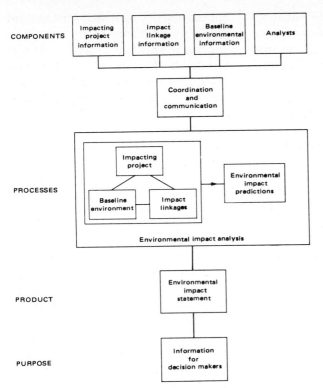

FIG. 2. Components and development of environmental impact analysis.

COMPONENTS

- Impacting project information
- Impact linkage information
- Baseline environmental information
- Analysts

Coordination and communication

PROCESSES

Impacting project

Baseline environment — Impact linkages → Environmental impact predictions

Environmental impact analysis

PRODUCT

Environmental impact statement

PURPOSE

Information for decision makers

under "Guidelines," the methodology used in the trans-Alaska pipeline analysis rigorously excluded value judgments throughout the process with such judgments were unavoidable; then alternative value framework judgments were presented. The approach described in Circular 645 is quite different in that it presents a nonspecific design approach to impact analysis and in that it admittedly "portrays many value judgments" (p. 1). It was released as a "preliminary effort to fill an interim need" (p. III). The nonspecific design described in that circular may be applicable to many environmental impact situations in which the magnitude and complexity of both the impacting project and of the environmental framework are rela-

tively limited, but it is not well suited to a project with the geographic, ecologic, and engineering complexities of the proposed trans-Alaska pipeline system.

Environment

For the purpose of a comprehensive impact analysis, the environment must be defined and environmental baseline data must be gathered for the total human environment. The total human environment consists of both the biotic and abiotic natural physical systems and the various superimposed socio-economic systems that are related to people and to their use of the natural physical systems.

Systematic description of the existing environment for impact analysis purposes should accomplish several related purposes: (1) It should inform the reader of the larger environmental framework and ecosystems within which the impact would occur; (2) it should afford the preparer the opportunity to look at a particular topic with the impact potential in mind and to identify sensitive components; (3) it should establish the limitations of the information framework and the degree to which the environmental factors can be quantified; and (4) it should provide reference to more detailed information if it is available.

In accomplishing these purposes for the analysis of the proposed trans-Alaska pipeline, a task force of experts on different environmental topics developed and compiled descriptive baseline information for the proposed pipeline route from Prudehoe Bay to Port Valdez, the marine tanker route from Port Valdez to west-coast ports, and the hypothetical pipeline routes from Prudhoe Bay across Alaska into Canada (Fig. 3). These experts were drawn from several Federal agencies including the U.S. Coast Guard, the Environmental Protection Agency, the National Oceanic Survey, the Environmental Data Service, the U.S. Corps of Engineers, the U.S. Geological Survey, the Bureau of Land Management, the Bureau of Out-door Recreation, the Bureau of Indian Affairs, the Bureau of Sport Fisheries and Wildlife, the National Park Service, and the National Marine Fisheries Service. Other baseline data were obtained from the Institute of Social, Economic, and Government Research at the University of Alaska, the Education Systems Resources Corporation of Arlington, Va.,

and several departments of the State of Alaska government.

Topics considered under natural physical systems for the environment of the proposed pipeline were physiography and geology, climate, air quality, water resources, vegetation, insects, fish and wildlife, and wilderness; topics considered under superposed socioeconomic systems were land use, population and labor force, the Alaskan Native community, composition of income and employment in Alaska, prices and costs, the oil and gas industry in Alaska, mining, fisheries, agriculture, forestry, electrical power systems, and transportation. For the proposed marine tanker transportation route, topics considered under natural physical systems were coastline and marine geology, climate and weather, physical oceanography, chemical oceanography, marine vegetation, biological oceanography, marine mammals, terrestrial mammals, and birds; under superposed socioeconomic systems the topics were fisheries, recreation, and marine transportation.

Impacting Project

To evaluate the impact on the environment, the analysts must know what will cause it, how it will occur, and what it will be in both its primary and secondary manifestations. To facilitate analysis the project that will cause the impact should be presented systematically in a project description which has already been reviewed for technical adequacy, agreement with specifications, and conformance with good environmental practice.

For the proposed trans-Alaska pipeline project, the Alyeska Pipeline Service Company prepared two project

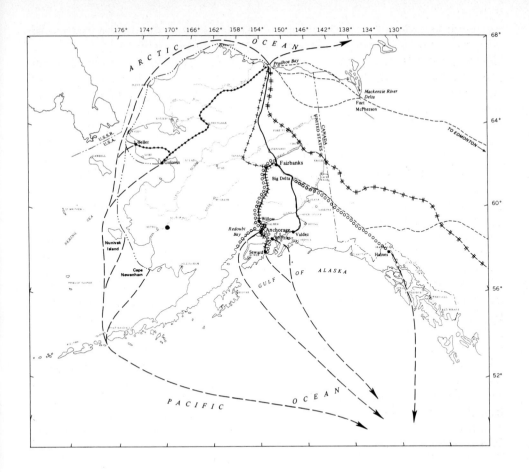

**ALTERNATE ROUTES FOR TRANSPORTING
NORTH SLOPE OIL**

——————————— Proposed trans-Alaska pipeline
+++++++++++++ The Alaska Railroad
+ + + + + + + Alaska Railroad extension
+ ++ ++ ++ ++ Trans-Canada resource railroad route
- - - - - - - - - Trans-Canada corridor
— — — — Marine transportation route
ooooooooooo Pipeline route to southern Alaska ports
••••••••••••• Overland pipeline route to western Alaska ports
—— ... —— Offshore pipeline route to western Alaska ports

FIG. 3. Alaska and northwestern Canada showing alternate routes for transporting North Slope oil.

descriptions, one covering the proposed pipeline system and related structures and one covering the proposed tanker transport system that would connect the terminus of the pipeline with destination ports. The descriptions were reviewed by ad hoc review groups, and the results of those reviews transmitted to the impact analysts and to the company. Following receipt of those reviews, Alyeska provided supplementary project description material which was also used by the impact analysts. The State of Alaska in cooperation with the Alyeska Pipeline Service Company prepared a description of the probable physical developments that would occur in the oil-field area.

Inasmuch as no proposal describing a specific gas transportation system had been received by the Department of the Interior, it was not certain what route or type of system would eventually be proposed. Accordingly, it was necessary to evaluate the impact of four hypothetical gas-pipeline routes: (1) Prudhoe Bay along the Arctic Slope to near the Mackenzie River Delta and on to Fort McPherson and south to Edmonton, (2) Prudhoe Bay across the Brooks Range to Fort McPherson and south to Edmonton, (3) Prudhoe Bay to Port Valdez along the route proposed for the Alyeska oil-pipeline system, and (4) Prudhoe Bay to Big Delta and east to Edmonton.

Identification of Impacting Effects

Just as systematic examination and description of the environment is needed to identify sensitive environmental components, systematic examination of the impacting project is needed to identify, classify, and quantify as much as possible its impacting effects. The review and evaluation of the project description should attempt to classify those effects in terms of their predictability: some effects will be unavoidable and therefore predictable; others will be probabilistic and therefore predictable only statistically, perhaps by comparison with performance records from similar projects.

The evaluation should also classify the impacting effects by their time of occurrence, by whether they are direct or indirect, and by whether they are primary, secondary, or tertiary.

The project descriptions of the proposed trans-Alaska pipeline system and associated developments were analyzed to identify and classify the impacting effects that would modify the existing environment. In all cases the unavoidable impact effects were differentiated from the threatened environmental impact effects (defined as effects with a probability of occurrence of less than 1). The effects during the construction, operation, and postoperation phases were differentiated from each other, as were direct and indirect effects. This was done first for primary impacting effects, and a similar process was carried out for secondary impacting effects; where applicable, tertiary impacting effects were also considered.

Within this complicated framework of distribution of effects were considered such primary and secondary effects as those listed subsequently.

Primary and secondary impacting effects associated with proposed trans-Alaska hot-oil pipeline, Arctic gas pipelines, and proposed tanker system

A. Primary effects associated with arctic pipelines:
   1. Disturbance of ground
   2. Disturbance of water (including treated effluent discharged into water)
   3. Disturbance of air (including waste discharged to air and noise)
   4. Disturbance of vegetation

5. Solid waste accumulation
6. Commitment of physical space to pipeline system and construction activities
7. Increased employment
8. Increased utilization of invested capital
9. Disturbance of fish and wildlife
10. Barrier effects on fish and wildlife
11. Scenery modification (including erosional effects)
12. Wilderness intrusion
13. Heat transmitted to or from the ground
14. Heat transmitted to or from water
15. Heat transmitted to or from air
16. Heat to or from vegetation
17. Moisture to air
18. Moisture to vegetation
19. Extraction of oil and gas
20. Bypassed sewage to water
21. Man-caused fires
22. Accidents that would amplify unavoidable impact effects
23. Small oil losses to the ground, water, and vegetation
24. Oil spills affecting marine waters
25. Oil spills affecting freshwater lakes and drainages
26. Oil spills affecting ground and vegetation
27. Oil spills affecting any combination of the foregoing
B. Secondary effects associated with arctic pipelines:
   1. Thermokarst development
   2. Physical habitat loss for wildlife
   3. Restriction of wildlife movements
   4. Effects on sports, subsistence, and commercial fisheries
   5. Effects on recreational resources
   6. Changes in population, economy, and demands on public services in various communities, including Native communities, and in Native populations and economies
   7. Development of ice fog and its effect on transportation
   8. Effects on mineral resource exploration
C. Primary effects associated with tanker system:
   1. Treated ballast water into Port Valdez
   2. Vessel frequency in Port Valdez, Prince William Sound, open ocean, Puget Sound, San Francisco Bay, southern California waters, and other ports
   3. Oil spills in any of those places
D. Secondary effects associated with tanker system:
   1. Effects on sports and commercial fisheries
   2. Effects on recreational resources
   3. Effects on population in Valdez and other communities

Impact Linkage Information

In addition to the baseline environmental and project impact effect data, it is necessary to compile data that pertain to the linkages or paths between and within various kinds of impacting effects and the various environmental topics. The linkages are of many kinds and types. They include all the information needed to actually predict an environmental impact other than the baseline environmental information and project information.

The following example drawn from the Alaska pipeline impact analysis illustrates what is meant by linkage data: Baseline information is available on the distribution of salmon and herring (at various life stages) in time and space in the Port Valdez-Valdez Arm area adjacent to Prince William Sound. Project information provided by Alyeska Pipeline Service Company indicates an estimated 2.4 to 26 barrels of oil per day would be introduced from a ballast-treatment facility into the waters of Port Valdez at a point 100 feet beneath the surface in a concentration intended not to exceed 10 parts per million. The linkage data are those needed to define the spatial and temporal paths that the effluent will take in the dynamic hydroenvironment, the changes it will cause en route in the water's chemical and physical properties, and the effects that the changed water will have on any given salmon or herring resource population at a specific time and place. Linkage data therefore in this case depend on hydrographic information that is properly part of the baseline environment description and also include knowledge of how different concentrations of different hydrocarbons in the water affect the fish population at different life stages.

In the analytical process related to the trans-Alaska pipeline proposal, the compilation of linkage data accompanied compilation of the baseline environmental data and was included both with the baseline data and with the impact analysis results. Where there were conflicting data, the analysts considered all in making the analysis and preparing the results of their analysis.

The limitations on the impact linkage data were the greatest problem encountered in the impact analysis. In many cases the complicated pathways and linkages that would exist between the project and the environment have not yet been studied to the point that predictions can be made with confidence. In such cases the analysts simply stated that rigorous prediction was not possible.

Analysts and the Analysis

Proper execution of an environmental impact analysis depends in a major way on the impartiality of the analysts. Regardless of the actual process used, including the use of predictive mathematical or simulation models, the analytical process is sensitive to the abilities and interests of those responsible for evaluating the interaction of the impacting project and the environment. At the present time the state of the art is such that the actual prediction process is likely to involve subjective steps requiring the judgment of the analyst.

Unfortunately, the educational traditions and occupational roles of scientists and technicians do not prepare them for dispassionate analysis of the type required. Resource scientists typically acquire and maintain strong conservation or development biases that reflect the customary work of their disciplines. Engineers and technicians normally adhere to strong developmental biases. The environmental impact analyst, to arrive at a prediction of impact, must discard prior conviction in favor of careful, thorough, systematic, and objective evaluation.

An example illustrates the problem. A government ornithologist responsible for research on endangered bird species and for enforcement of laws and regulations designed to protect them is given the task of objectively evaluating what will happen to those species if certain endangering transportation developments occur. He immediately faces conflict because the evaluation involves acceptance (for analytical purposes) of events which he has opposed for his entire career. An equally valid hypothetical example can be constructed using a development-oriented mining engineer faced with demonstrating sincere concern for the environment in evaluating the effect of an open-pit mine on the nesting area of an endangered species.

The impact analysis of the proposed trans-Alaska pipeline was made by resource scientists who had been given guidance regarding the analysts' proper function. As they proceeded to determine the type and extent of environmental impact that would occur in their respective fields, they were urged to be as rigorous and objective as possible and to utilize fully all available information. There undoubtedly were minor problems in the impact analysis of the proposed pipeline that arose from conflicts the analysts felt between the bias of traditional roles and the impartiality required in impact analysis.

Coordination and Communication

Coordination and communication compose the last essential component in environmental impact analysis. Coordination is needed to insure that all topics are handled in the same systematic way by all the analysts. Because of the intricate way in which an impacting project can produce reaction chains of impact effects, it is extremely important that preliminary results of analyses be communicated rapidly and fully from one analyst to another. This communication is facilitated by the awareness of all analysts that the results of others may have a direct effect on the analysis that they themselves are making. Careful attention is required to devise a system that provides this communication and that also generates the kind of information needed at the proper time for use of others involved in the analytical process.

Coordination within the impact analysis task force for the proposed trans-Alaska pipeline was accomplished by a core group of five persons including the chairman. Members of the core group worked closely in providing guidance and information to the individual analysts.

A principal effort was to insure that developments from one analysis that might bear on another were communicated rapidly; the draft material was circulated completely, insofar as possible, to all analysts, and all suggestions made by other analysts, the chairman, and the core group were considered before preparing the final draft of the impact statement. This procedure of replicate internal review results in a final report whose parts are the work of individual analysts and not the work of interpretation of the task force chairman or the core group.

Guidelines

Environmental impact analysis requires both philosophical and technical/analytical guidelines. If an analysis and the report of the analysis are to be scientifically, technically, and legally defensible, they must be prepared to the highest standards of objective scientific inquiry. Any attempts to bias the results of an environmental impact analysis cannot escape the notice of careful evaluators of the results.

Inclusion of objective environmental impact analysis information in any complex decision framework adds one more element that possibly conflicts with other elements in the framework. The decision makers of today are (or should be) prepared to evaluate such conflicting elements by applying their values in such a way as to produce a politically, socially, economically, and environmentally just decision. If an impact analysis is to provide this objective information to the decision makers, then the analysts and the analytical process should function in the traditional scientific way.

The technical and analytical guidelines established for any impact analysis should reflect these philosophical points. They should place a high degree of responsibility on the individual analyst for the assembly and compilation of pertinent environmental material and for the understanding of pertinent impacting effects. Similar responsibility exists for preparation of an objective report concerning the results of the impact analysis. Specific guidelines given to the analysts should facilitate preparation of their material in format compatible with that required by any existing laws and suitable for use in communicating with other analysts during the environmental impact analysis process.

A specific guideline worth emphasizing is that, insofar as is possible, value-judgment factors should be omitted from environmental impact analysis; when omission is impossible the specific value framework used should be specified. This is the only way of assuring that the values exercised in the decision are the values of the decision makers rather than the values of those who prepared the environmental impact statement. A related guideline of utmost importance to the impact analysis process is for the analysts to recognize fully and completely that *they* are not to decide the issue.

This discussion of environmental impact analysis would be incomplete without mention of the extremely critical relation between the impact analysis and the "decision point." Stated otherwise, what type and scope of environmental analysis can and should be made at the different "decision points" in the overall decision-making process that accompanies evolution of a project? The information available about a pipeline or any other project varies with time, from a relatively low level in the conceptual stage to a high level in the final construction and operation stages. Depending on the project, its location, the types of impacts possible, and other factors, the pertinent environmental information may or may not follow a similar path. The determination that the available environmental information is adequate in scope and quantity to constitute an element in the decision process depends mainly on the value framework of those responsible for the decision. As environmental awareness and conscience develop, certain critical elements of environmental information should become acknowledged as requirements for just decisions, in the same way that predicted cost and profit data are now universally accepted as critical factors in the analysis of economic feasibility.

In the case of the proposed trans-Alaska pipeline, policymakers in the Department of the Interior decided that the available environmental information was adequate for impact analysis. This was determined before the environmental data had been compiled, and it is therefore questionable to what extent the amount of environmental data actually available influenced the decision to proceed. In retrospect, however, the baseline environmental information available at the start of the impact analysis was approximately comparable in quantity and quality to data available on the proposed pipeline project.

In several ways the completion of the impact analysis of the proposed trans-Alaska pipeline represents a successful endeavor. Because there was no precendent for the analysis, it was necessary to design the procedures to be used, find and assemble the people to work on it, and establish the philosophical and technical guidelines that would result in a scientifically sound product. The methods used in the analysis and in the preparation of the report were rigorous and objective, the analytical group was independent of exterior influence, and the results are a milestone in the developing science of environmental impact analysis.

Main Types of Impact Predicted by the Analysis

The analysis indicated that environmental impact would result from (1) the construction, operation, and maintenance of the proposed oil-pipeline system, including the accompanying highway north of the Yukon River, and of a gas transportation system of some kind, (2) from oil field development, and (3) from the operation of the proposed marine tanker system. Because of the scale and nature of the project, the impact would occur on abiotic, biotic, and socioeconomic components of the human environment far beyond the relatively small area (940 sq. mi. out of 572,000 sq. mi. of land area) of Alaska that would be occupied by the oil-pipeline system and the oil field. The impact paths between the project itself and the affected parts of the environment would be of varying complexity and length and would involve linkage factors that are not all well known.

Of the impact effects that would occur, some, like those associated with wilderness intrusion and public access north of the Yukon River, could be considered either beneficial or adverse depending on the value framework used. Some of the effects on socioeconomic parts of the environment would be classified as beneficial by most persons. Most of the other impact effects would in some way alter the existing environment in a way that was not demonstrably beneficial and would in that sense be adverse. Such effects would occur both on natural physical systems and on the superposed socioeconomic systems.

Some impact effects are unavoidable

and can be evaluated with a degree of certainty. Others could result from the occurrence of a threatened event of some kind which would impact the oil or gas transportation systems. These threatened impact effects cannot be evaluated with comparable certainty.

The principal unavoidable effects would be (1) disturbances of terrain, fish and wildlife habitat, and human environs during construction, operation, and maintenance of the oil pipeline, the highway north of the Yukon River, the oil field, and the gas pipeline that would probably follow, (2) the effects of the discharge of effluent from the tanker-ballast-treatment facility into Port Valdez and of some indeterminate amount of oil released into the ocean from tank-cleaning operations at sea, and (3) effects associated with increased human pressures of all kinds on the environment. Other unavoidable effects would be those related to increased State and Native corporation revenues, accelerated cultural change of the Native population, and extraction of the oil and gas resource.

Changes in stable terrain caused by construction and maintenance procedures could produce rapid and unexpected effects, including slope failure, modification of surface drainage, accelerated erosion and deposition, and other disturbances as a result of the permafrost thawing that would follow destruction of the natural insulating properties of the tundra. Placement of gravel pads and berms would especially affect surface drainage. The excavation of borrow materials and placement of the pipeline ditch in and near flood plains and streambeds would also cause changes in stream erosion

and deposition. About 83 million cubic yards of construction material, mostly gravel, would be required for the oil pipeline. The general noise, commotion, and destruction of local habitat could cause many species of wildlife to leave an area amounting to about 60 square miles.

Socioeconomic effects during construction would include accelerated inflation, increased pressures on existing communities for accommodations and public services, and job opportunities for perhaps 25,000 persons at peak times (including multiplier effects); unemployment in Alaska, however, would continue to be relatively high.

The main disturbances during operation would be (1) thawing in permafrost leading to possible foundation instability and differential settlement, (2) some barrier effects of aboveground oil-pipeline sections on large mammal (especially caribou) migrations in the Brooks Range, Arctic Coastal Plain, and Copper River Basin areas, and similar effects of any aboveground sections of gas pipeline that would eventually be built, and (3) adverse but unquantifiable effects on the marine ecosystem of Port Valdez and perhaps Valdez Arm and Prince William Sound proper from the discharge of an estiamted 2.4 to 26 barrels of oil per day from the ballast-treatment facility and on the marine ecosystem in general from discharge of an indeterminate amount of oil from tank-cleaning operations at sea. These last effects would in turn affect the fishing industry to some unquantifiable extent.

Other main operational effects would include (1) the gradual conversion of about 880 square miles of the North Slope wildlife habitat to an area with widely spaced drilling pads, roads, pipelines, and other structures, with the accompanying adverse effects on the tundra ecosystem, (2) the many diverse effects on wilderness, recreational resources (including hunting and fishing), and general land-use patterns that would result from increased public access to the now relatively inaccessible region north of the Yukon River, (3) acceleration of the cultural change process that is already underway among Alaskan Natives and some adverse modification of local Native subsistence-resource base as a result of secondary effects, and (4) additional state revenues of about $300 million per year and subsequent expenditure of those revenues for public works and activities throughout Alaska. Immediately after the end of construction, unemployment would probably increase.

The main threatened environmental effects would all be related to unintentional oil loss from the pipeline, from tankers, or in the oil field. Oil losses from the pipeline could be caused by direct or indirect effects of earthquakes, destructive sea waves, slope failure caused by natural or artificial processes, thaw-plug instability (in permafrost), differential settlement of permafrost terrain, and bed scour and bank erosion at steam crossings. Any of these processes could occur at some place along the route of the proposed pipeline. Oil loss from tankers could be caused by accidents during transfer operations at Valdez and at destination ports like Puget Sound, San Francisco Bay, and Los Angeles, and by tanker or ship casualties resulting from collision, grounding, ramming, or other causes along the tanker routes.

The potential oil loss from pipeline failure cannot be evaluated because of the many variables involved, but perfect no-spill performance would be unlikely during the lifetime of the pipeline. Various models of oil loss from the tanker system indicate that an average of 1.6 to 6.0 barrels per day could be lost from the whole system during transfer operations and an average of 384 barrels per day or about 140,000 barrels per "average" year could be lost from tanker casualties. This modeled loss would occur in incidents of undetermined size at unknown intervals and at unknown locations. This is considered to be a maximum or "worst case" casualty discharge volume.

Oil spilled from the pipeline as a consequence of one of the threats mentioned could, depending on location, volume, time of year, and other factors, result in adverse effects on all the biota involved. Not all the linkages and impact paths are known, but vegetation, waterfowl, and freshwater fisheries could all be affected and then affect Native subsistence use to an unquantifiable extent.

Oil spilled in tanker, casualties or transfer operations would affect the marine ecosystem to an extent that would be determined by many variable factors. The salmon and other fishery resources of Prince William Sound would be especially vulnerable to such spills. Over the long term, however, persistent low-level discharge from the ballast-treatment facility and tank-cleaning operations at sea could have a greater adverse effect than short-lived larger spills.

The probable eventual construction and maintenance of a gas pipeline would, if it were not in the oil-pipeline corridor, result in a separate corridor with many of the same effects described for the proposed oil-pipeline corridor. Those effects and those impacts on the environment would be in addition to those predicted for the proposed oil-pipeline system.

Analysis and Comparison of Alternatives

The environmental impact analysis also included consideration of various alternatives to the oil transportation system proposed. The three main types of alternatives examined were those available to the Secretary of the Interior, those concerning alternative routes and transportation systems, and those concerning energy and policy alternatives. The information regarding energy and policy alternatives was compiled and prepared by a special task force made up of representatives of various Federal agencies.

The alternatives considered available to the Secretary of the Interior were granting the permits that had been applied for, denying the permits, or deferring any action. The environmental impact implications of those different actions were examined.

Alternative routes and systems for the transportation of oil included (1) pipelines from Prudhoe Bay to other ice-free ports in southern Alaska such as Redoubt Bay, Whittier, Seward, and Haines, (2) marine transportation systems including ice-breaking and subsurface tankers, (3) both offshore and overland pipelines to terminal ports on the Bering Sea, (4) trans-Alaska-Canada pipelines to Edmonton includ-

ing coastal offshore and onshore routes to the Mackenzie River Delta, routes inland across the eastern Brooks Range to Fort McPherson, and routes across the central Brooks Range (along the proposed oil-pipeline route) to Fairbanks, Big Delta, and east along Alaska Highway, (5) railroad and highway transportation modes including an Alaska railroad extension from Prudhoe Bay to a southern Alaska port and a new trans-Alaska-Canada railroad route and highway system, and (6) other oil transportation schemes including land, sea, air, and in other energy forms (fig. 3). Some of the alternate oil-pipeline routes that were considered and analyzed are, as noted previously, the same as those routes considered and analyzed for gas pipelines.

The energy and policy alternatives examined were (1) reduction in demand, (2) increased oil imports to the United States, (3) additional production from outer continental shelf and onshore areas, (4) modification of natural gas pricing, (5) nuclear stimulation of natural gas reservoirs, (6) increased use of coal as solid fuel and as source for synthetic fuels, (7) nuclear fuel, (8) synthetic sources, oil shale, tar sands, coal, (9) geothermal power, (10) hydroelectric power, and (11) exotic energy sources and improved efficiency systems.

The analysis of the environmental impact of the alternative trans-Alaska-Canada route and other routes provided information that was used for a comparison between those routes and the proposed trans-Alaska route. This comparison resulted in relative ranking of important impact effects for all the routes. The comparison process included (1) identification of combinations of routes and modes of transportation, (2) identification of specific unavoidable impact effects taking into account the abundance and vulnerability of the resources involved, the length of the route along which they would be affected, and the impact factors that would probably be involved, (3) identification of specific major threatened impact factors, and (4) ranking the different routes against each other on an arbitrary relative-magnitude scale. In the comparison no attempt was made to imply absolute magnitude or to weight any impacts or impact factors in relation to each other.

For the terrestrial environment the unavoidable environmental impacts that were compared included terrain disruption related to oil pipeline, terrain disruption related to terminal, construction material requirements, induced terrain disruption, physical space commitment, surface- and ground-water effects, air-quality effects, vegetation and habitat disruption, and effects on fisheries, on wildlife including birds, on recreation and esthetics, on wilderness, on communities, and on Native culture and subsistence. For the marine environment the unavoidable impacts considered included effects on Alaskan terminal port waters, destination port waters, fisheries, and wildlife including birds. The threatened environmental impact factors that were compared for the terrestrial environment included seismic risk to the pipeline, seismic risk to the terminal, permafrost degradation, slope failure, flooding risk, and, for the marine environment, tanker casualties and oil-transfer operations.

Synthesis of the material included in the comparison of the different routes and transportation modes resulted in several conclusions which were reported in the final environmental impact statement: (1) No single generalized oil-pipeline route appeared to be superior in all respects to any other; (2) in comparing the unavoidable impacts upon the terrestrial abiotic systems, it appeared that all the trans-Alaska routes would have less impact than the trans-Alaska-Canada routes; (3) in comparing the unavoidable impacts upon various terrestrial biotic systems, it appeared that the trans-Alaska route to a Bering Sea port would probably have the least impact and the trans-Alaska-Canada coastal route the next lowest impact; (4) in comparing the unavoidable impacts upon socioeconomic systems, it emerged that the trans-Alaska-Bering Sea port route would probably have the least impact and the trans-Alaska-Canada coastal route would be next; (5) in comparing unavoidable impacts on the marine environment, all the trans-Alaska-Canada routes would have less impact than the trans-Alaska routes; (6) in comparing the threatened environmental impact factors for the terrestrial environment, the trans-Alaska-Canada coastal and inland routes were found to pose the least threat; and (7) comparing the threatened environmental impact factors for the marine environment, the trans-Alaska-Canada routes would be lowest because no direct marine transportation of oil would be involved.

It should be kept in mind that different levels of information were available for the proposed Alyeska Pipeline Service Company oil-pipeline system, for the alternate oil transportation systems and routes, and for gas transportation systems and routes. The difference affected the analysis and therefore all the comparisons.

The information in the environmental impact statement and in related documents released as a result of legal action after the final statement established that one of the important environmental questions involved comparison of an overland one-corridor oil- plus gas-pipeline transportation system through Alaska and Canada with a two-corridor system involving an oil pipeline through Alaska connecting to a tanker route and a gas pipeline through Alaska and Canada. Although this question was not considered in detail in the final environmental impact statement, the author believes it is important to reiterate here the pertinent facts available during the final decision process and the conclusions he drew from that information. It is emphasized that different conclusions could be (and were) drawn from the same information by other parties and persons.

Any combination of overland pipeline plus tanker systems or of tanker systems alone would impose the threat of oil pollution on the marine environment. The most important causes would be contamination resulting from intentional oil discharge from a ballast-treatment facility and from possible tank-cleaning operations at sea and from unintentional oil loss during transfer operations and from oil-tanker casualties. LNG (Liquefied Natural Gas) tanker systems would impose some threat of unintentional gas loss resulting from ship casualties. If LNG tankers were operating from the same ports as the oil tankers, they would contribute to increased vessel density and thereby indirectly to oil-tanker casualty frequency.

Overland gas- and oil-pipeline systems would impose the threat of environmental impact from rupture and unintentional loss of oil or gas. The most likely cause of rupture would be earthquakes and their attendant ground effects. The most likely impact from gas-pipeline rupture would be fire that could spread into areas adjacent to the pipeline. The most likely effect of oil-pipeline rupture would be oil lost onto the land and into lakes and streams and the various secondary effects that such loss would cause.

Any overland oil- and gas-pipeline systems would intrude the wilderness and would utilize physical space for the pipeline alinement as well as for camps, pump stations, airfields, and so forth; the accompanying roads would provide access along the pipeline corridors. Access would bring with it increased recreational opportunity and increased human pressures on the wilderness resources.

A combination of oil and gas pipelines in one corridor (not necessarily on a single or contiguous rights-of-way) would localize and restrict these effects and thus require less space, cause less wilderness intrusion, provide less access, have less effect on fish and wildlife habitat, and probably have less overall effect on the migration of large animals.

Based on these considerations, and without specifying one or another corridor or transportation mode, it is the author's opinion that one corridor containing both oil and gas pipelines would have less environmental impact and thus incur less environmental cost than would separate corridors.

Considering (1) the threat to the marine environment that any tanker system would impose, (2) the threat that zones of high earthquake frequency and magnitude would impose on pipelines, and (3) the apparent lesser environmental impact of a single corridor as compared with two corridors, it is the author's opinion that environmental impact and cost would be least for a gas- and oil-transportation system that (1) avoided the marine environment, (2) avoided earthquake zones, and (3) placed both oil and gas pipelines in one corridor. The onland trans-Alaska-Canada routes to Fort McPherson and through the Mackenzie Valley to Edmonton would meet these criteria for minimizing environmental impact and would, from that point of view, be preferable. Of the possible onland routes, the inland route across the Brooks Range between Prudhoe Bay and Fort McPherson appears on some grounds to cause the least overall adverse environmental impact.

These conclusions of the author are subject to one additional qualification. To reach market areas, any solely overland oil and gas transportation system ending near Edmonton would have to be extended beyond the geographic limits that were set for the environmental impact analysis and would thus entail additional environmental impact. The extended construction and operation would, however, be entirely in areas now traversed by oil and gas pipelines and no unusual problems would be encountered nor would any new transportation corridors be created.

As noted earlier, other conclusions are possible, and indeed the official conclusion of the U.S. Department of the Interior and of the U.S. Congress was that the environmental costs of a trans-Alaska-Canada hot-oil pipeline would be approximately the same as those of the proposed trans-Alaska hot-oil pipeline.

Conclusion

To this point this paper has been concerned mainly with the scientific and analytical aspects of environmental impact analysis and with the impact analysis of the proposed trans-Alaska oil pipeline and its alternatives. This final section examines some human and political aspects of this previously discussed material.

The scientist involved in environmental impact analysis should use the same standards and practices that he uses in scientific work. He is not only responsible for using the best available information and using it in the best possible way, but also for making sure that no personal biases enter the analytical procedure. By keeping impact analysis entirely scientific, it is possible to produce information which provides an objective input to the decision process.

Nevertheless, it is likely that almost all environmental impact analysis will be conducted in situations which are influenced by external pressures. These pressures lead to the imposition of time constraints and to requirements that the analysis be made without significant amounts of additional information that would normally be acquired through extensive research. Although the involved scientist will be affected by these pressures, he will not be absolved of the responsibility of conducting the analysis in a rigorous scientific fashion.

The coordination of an environmental impact analysis and the communication of the results to the decision makers will necessarily involve scientists who are in a supervisory role. The external political and economic factors which often are a major part of the decision makers' value framework are likely to be brought to bear directly on those scientists. They therefore have the responsibilities of (1) maintaining the scientific standards of the analytical group against any external pressures and (2) communicating the results of the environmental impact analysis effectively to the decision makers. More and more scientists will have these responsibilities in the future as the world's decision makers require more and better scientific input on questions of critical environmental significance.

The way in which the United States Government received and used the environmental information of the trans-Alaska pipeline impact analysis is a most important part of this story. The results of the analysis were provided to the decision makers in the written environmental impact statement and in discussions between the policy makers and the core group of the task force that made the impact analysis. Clear documentation of how the results of the environmental impact analysis and other pertinent information were used in the decision is contained in a document[2] released on May 11, 1972, by the Department of the Interior. This document notes that the major considerations involved in the decision on the proposal were (1) United States energy and crude oil posture, (2) national security aspects, (3) choice of market for North Slope oil, (4) the proposal for the trans-Alaska pipeline, (5) alternative methods of transporting North Slope oil, and (6) further deferral of action. Brew and Gryc (1974) analyzed the document in relation to the information contained in the final environmental impact statement.

The decision document concludes that the environmental consequences of either a trans-Alaska or trans-Alaska-Canada oil-pipeline route are acceptable when weighed against the advantages to be derived from the construction. The Department of the Interior concluded that the Alyeska proposal was acceptable. Similar conclusions are contained in the testimony of Secretary of Interior Morton before the Joint Economic Committee of Congress (U.S. Congress, 1972).

The conflict between environmental values and resource development values that the trans-Alaska pipeline exemplifies demonstrates the continuing need for research on impact analysis and for environmental impact analysis as an essential early component in industrial and governmental decision making and also demonstrates that the scientist can and must interact with the engineer and with those in decision-making roles if the critical human goal of compatibility of environmental and resource-developmental factors is to be achieved.

Notes

[1] 42 United States Code 4332.

[2] "Applications for pipeline right-of-way and ancillary land uses, Prudhoe Bay to Valdez, Alaska," and "Application by State of Alaska for

right-of-way for highway—Statement of reasons for approval." U.S. Dept. Interior, Office of Communications, Washington, D.C., May 11, 1972.

References Cited

Brew, D. A., and Gryc, George, 1974, The analysis of impact of oil- and gas-pipeline systems on the Alaskan Arctic environment: Internat. Cong. Foundation Francaise d'Etudes Nordiques "Arctic oil and gas—problems and possibilities," 5th, Le Havre, May 2–5, 1973, Proc. (in press).

Leopold, L. B., Clarke, F. E., Hanshaw, B. B., and Balsley, J. R., 1971, A procedure for evaluating environmental impact: U.S. Geol. Survey Circ. 645, 13 p.

Pryde, P. R., 1972, The quest for environmental quality in the USSR: Am. Scientist, v. 60, p. 739–745.

U.S. Congress, 1972, Natural gas regulation and the trans-Alaska pipeline—Hearings before the Joint Economic Committee, June 7, 8, 9, and 22, 1972: Washington, U.S. Govt. Printing Office.

U.S. Federal Task Force on Alaska Oil Development, 1972, Final environmental impact statement, proposed trans-Alaska pipeline (6 volumes): U.S. Dept. Interior interagency rept.; available from the Natl. Tech. Inf. Service, U.S. Dept. Commerce, Springfield, Va., 22151, NTIS PB-206921.

Verguèse, Dominique, 1972, Europe and the environment—Cooperation a distant prospect: Science, v. 178, p. 381–383.

Wahrhaftig, Clyde, 1964, Physiographic divisions of Alaska: U.S. Geol. Survey Prof. Paper 482, 52 p.

Watkins, R. E., 1971, The influence of governmental regulations on the design, construction, and operation of pipelines in western Europe: Am. Petroleum Inst., 22d Ann. Pipeline Conf., Dallas, Tex., Apr. 26–28, 1971.

Editor's note: The Trans-Alaska oil pipeline was completed on June 20, 1977. The pipeline has a capacity of 1.6 million barrels per day, but the "throughput" has been less than capacity (1.4 to 1.5 million barrels per day) because of West Coast market conditions.

30   Surface Mining Reclamation in Appalachia: Impact of the 1977 Surface Mining Control and Reclamation Act

Larry R. Sweeney

Coal is one fossil fuel in the United States whose extraction rate can be increased. The United States possesses about 31 percent of the world's known coal reserves. A third of these deposits can be extracted by surface mining,

Reprinted from *Journal of Soil and Water Conservation*, v. 34, pp. 199–203, with permission of the author and the Soil Conservation Society of America. Copyright © 1979 by scsa.

assuming current technologic and economic conditions. Regardless of environmental and sociological constraints on its use, coal represents the only abundant energy resource capable of meeting the nation's projected needs (5).

Unregulated mining results in substantial environmental damage during and after extraction. In addition to a general disturbance of the ecosystem, surface mining leaves a site in a condition that discourages productive post-mining land use.

State governments reacted to this situation by enacting surface mining and reclamation statutes appropriate for their particular geographic and climatic conditions. Most of these laws have been amended and strengthened several times.

West Virginia enacted the first surface mining law in the United States in 1939 (19). It is the only state that lies entirely within the Appalachian Province, the most productive coal region in the world (20). As such, it has become a center of controversy in recent years over improved surface mining methods and stronger regulatory legislation. However, West Virginia, along with other states, has repeatedly demonstrated that environmentally damaging aspects of surface mining can be reduced through the use of proper mining techniques and reclamation practices.

A review of laws regulating surface mining and reclamation in 15 eastern states as of April 1, 1975, indicated a variety of regulations pertaining to application procedures, fees, penalties, and other requirements (14). Most of these regulations came about in response to public demand, as did the Surface Mining Control and Reclamation Act of 1977, which established nationwide standards for surface mining activities. Progressive improvement of state regulations, enactment of federal legislation, and intelligent enforcement of regulations have resulted in marked improvement in extraction and reclamation by the surface mining industry.

Surface Mining Techniques

Improved mining and reclamation technologies, along with increased awareness of the potentially damaging effects of surface mining, have resulted in several relatively new mining and reclamation methods. These methods often result in landforms and soils with qualities superior to those occurring naturally in coal mining areas, but not without increased production costs, which are passed from the mine operator to the consumer.

In the past, indiscriminate dumping of overburden materials on the downslope (the old "shoot and shove" method) constituted the largest single source of sediment from surface mining in Appalachia (4). New mining methods have been devised to improve mining efficiency and effectively control pollution. However, the mining technique itself cannot be considered a complete reclamation plan, but rather a method that must be supplemented by reclamation procedures to be effective (13).

The most productive method of surface mining is mountaintop removal. With complete removal of the overburden on a ridge or mountain, an entire coal seam can be recovered. This method has proven especially effective for recovery of coal left by contour mining methods. Gargantuan machinery permits more efficient handling of overburden today than was previously possible, making the reworking of old contour mined sites economically attractive for mountaintop removal. Mountaintop removal methods result in areas of flat to gently rolling reclaimed land, a valuable commodity in mountainous Appalachia.

A recent study (9) in eastern Kentucky reported the following average values for land in the mining regions: unmined land, $300 per hectare; mined but unreclaimed land, $150 per hectare; contour-mined and reclaimed land, $500 per hectare; and land mined by the mountaintop-removal technique and reclaimed, $850 per hectare. These gains in property value are due primarily to creation of level or gently sloping land. This is true even for reclaimed contour-mined areas, which feature benches of nearly level land. The recent federal legislation prohibits the formation of such benches by requiring that overburden be restored to the original contour.

Much of the land overlying coal seams in Appalachia consists of unmanaged forests. This land is too steep for use even as pasture or meadow. The level or gently sloping lands produced by mining and reclamation have the potential for a variety of uses, including wildlife areas, pasture and meadow, cultivated crops, orchards, vegetable production, and residential, commercial and industrial development. However, the Eastern Kentucky study (9) found that more than 90 percent of the surface-mined land in the region surveyed was unmanaged forest prior to mining and remained in that use after reclamation. Thus, the potential improvement in land use following proper mining and reclamation may be unrealized in some areas.

When large volumes of surplus overburden result from mountaintop removal, an overburden storage method may be employed to dispose of excess spoil. The most popular method is the head-of-hollow fill.

Vegetation is first removed from a hollow, which is then filled with the spoil material, compacted in relatively shallow layers. The outslope is restricted to 50 percent or less, and terraces are installed at specific intervals to reduce runoff velocity. This fill is constructed with an internal rock drainageway to permit groundwater and naturally infiltrating water to move through the fill without saturating it (13). When completed, the ravine is transformed into a plateau-like area with carefully engineered slope and drainage control, which, like mountaintop-removal areas, offers excellent developmental potential. Overburden is handled once rather than twice as in convential contour mining (4). The cost of a head-of-hollow fill depends on the method of mining it supplements, the distance the spoil must be hauled, the equipment used, and other site preparation factors.

Even with these two progressive techniques, mountaintop removal and head-of-hollow fill, revegetation and subsequent reclamation depend upon surface conditions of the regraded mine spoil. Many state and federal laws have been enacted to insure that resulting conditions will be conducive to plant growth.

Segregation of overburden to remove and bury toxic materials was practiced by many operators before the law required them to do so. The new federal legislation mandates more complete overburden segregation, including removal and stockpiling of topsoil to be replaced on the reshaped surface after backfilling and before revegetation. On prime farmland a major soil horizon must be removed, stockpiled, and replaced in its original sequence (12).

Prior to today's stringent reclamation laws, topsoiling was rarely practiced. It was much less expensive to pile all materials together and thereby reduce handling costs. But operators soon realized the value of the original topsoil as a plant growth medium once more stringent revegetation requirements necessitated its use. However, topsoiling is often unattractive economically in Appalachia because of the terrain and the topsoil's shallow depth. Actual practice has demonstrated that a soil can be constructed from available material (soils or suitable geologic strata) that will have a depth of at least 5 feet and provide a favorable medium for plant growth, often superior to the natural soils of the area (6).

At most mining sites the original soil material is considered the best medium for plant growth. Compared to spoil, the original soil material usually is less stony, retains water better, contains fewer toxic materials, and lacks the dark-colored components that may absorb enough solar energy to restrict vegetative growth (13).

Research in North Dakota (8) showed that as little as two inches of soil material placed back on mine spoils benefited plant growth and production and reduced surface crusting and runoff by increasing infiltration. This effect, like that of a surface mulch, was lost when the soil was mixed into the upper few inches of the spoil.

In the North Dakota study about 30 inches of topsoil produced maximum yields, regardless of topsoil quality, although maximum yields increased as topsoil quality improved.

In comparing topsoiled and nontop-soiled sites in southern West Virginia, J. N. Jones, Jr., Dan Amos, and I (unpublished data) found that two-year-old topsoiled sites were similar to two-year-old nontopsoiled sites in most respects except solum depth and coarse fragment content. The topsoiled sites were deeper and generally contained slightly fewer coarse fragments. Apparently topsoiling produced no other immediate advantages, particularly with regard to soil physical properties. In examining nontopsoiled mine soils up to 10 years old, we also found the mine soils superior to the natural soils in overall fertility, and the coarse fragment content of the A horizons was less in the mine soils.

Aggregation (soil structure) is reduced in the removal and regrading process. One study (10) concluded that infiltrating water moved more rapidly and carried much less sediment with it on nontopsoiled sites. On both topsoiled and nontopsoiled sites, however, infiltrating water reached the water table faster than might be expected from movement of the wetting front. This implies considerable channel flow. In the same study, oxygen concentrations following wetting recovered more rapidly on nontopsoiled sites than on topsoiled sites.

The cost of handling topsoil depends on the distance moved, the thickness of the layer to be stockpiled, and the terrain (13). Expenses can be minimized if the handling process is well planned before mining.

Preservation of Aesthetic Values

The aesthetic value of a particular landscape is probably the most difficult parameter to measure in economic terms. Particularly in the Appalachian region, surface mining has been long criticized for its destruction of the landscape's aesthetic characteristics.

Most surface mining in Appalachia has involved contour mining techniques. Contour mining removes overburden from above the coal seam, following the outcrop of the coal seam around the contour of the mountain. A level, frequently broad bench is created at the coal seam, together with a vertical rock face (highwall), the height of which is determined by the thickness of the overburden that can be removed economically.

To minimize aesthetic degradation, federal regulations (12) now require all contour-mined lands to be returned to their approximate original contour. Pennsylvania was the first state to address the approximate original contour question. It did so in legislation enacted in the late 1960s. West Virginia followed suit in mid-1977.

Highwalls create negative aesthetic values but apparently have little or no degrading effect on the environment. In contour restoration, spoil material is placed back onto the mined area and regraded against the highwall. This usually results in loose, unconsolidated spoil material being left at or near the angle of repose, a condition highly susceptible to erosion.

This "haulback" method of contour mining, initiated in the early 1970s in West Virginia (4), causes less disturbance by keeping spoil off the downslope area and facilitates the deep burial of toxic materials. However, it not only results in steep slopes, which erode easily, but also eliminates the potentially useful level benches created by normal contour mining.

Haulback is an expensive operation because of the large volume of spoil that must be handled. Costs range from $1,240 to $6,180 per hectare (13). It is difficult to return land to its original contour after mining, particularly if the ratio of coal to overburden is large.

These costs, which are passed on to consumers, may be one reason for the decline of coal production in some eastern states. Statistics gathered by the West Virginia Coal Association show that West Virginia's coal production in 1978 dropped to that in 1922. This continues a steady, downward trend in coal production that began in 1971, when the West Virginia Surface Mining and Reclamation Amendments were enacted. Their implementation in subsequent years, coupled with more stringent clean air standards and labor disputes, contributed to the continuing decline (17).

West Virginia's laws governing surface mining and reclamation are considered the toughest in the nation, especially since 1971. While coal production has declined, West Virginia has led the nation in the number of acres reclaimed for nine consecutive years (18). This has been due mainly to the use of a coal tax to reclaim previously abandoned surface mines. Originally established in 1963 at $30 per acre, this tax rose to $60 per acre in 1971.

Development of National Standards

Largely because of the diversity among state regulations, a movement began in Congress in 1971 to establish national standards for mining and reclamation.

The 90th Congress held hearings on implementing such standards, but no bills resulted from these hearings. Nor did the 91st Congress produce any proposals. In 1973 the House of Representatives passed a bill to regulate the surface mining industry, but Congress adjourned before the Senate could complete its consideration of this bill. A Senate vote to reconsider this bill in 1974 failed by one vote. The 93rd Congress was then left with the task of drafting a new bill, which it accomplished after many hearings and amendments. When the 93rd Congress adjourned in December 1974, this bill was vetoed by President Ford because of what he felt would be the bill's adverse economic impact. President Ford vetoed a similar bill passed by the 94th Congress in 1975, and the Congressional vote to override the veto fell short of the required two-thirds majority by three votes. The legislation finally enacted by Congress and signed by President Carter in 1977 was originally introduced by the 95th Congress in early 1976 (7).

In general, the federal legislation is designed to ensure that surface-mined land is restored to a condition that can support its original use or an approved higher or better use. The law includes special standards for mining on steep slopes, mountaintop removal, and the protection and restoration of prime farmlands. It also regulates the surface effects of underground coal mining (1). Permits to engage in surface mining may be issued for a period of five years, and they are renewable. They can be transferred under certain conditions also. Areas may be deemed unsuitable for mining and a permit application denied. Reclamation must

be achieved in a manner consistent with the local environment and climate, and it must produce landforms and conditions compatible with state and local land use plans (9).

A comparison of the Surface Mining Control and Reclamation Act of 1977 with the West Virginia law shows them to be similar. They are nearly identical in haulroad requirements, sediment control, extraction methodology, and regrading. The federal law calls for higher fees and performance bonds, more engineering and preplanning detail, and a longer bond period (five years versus two growing seasons), but far less frequent inspections (six months versus 15 days) (18).

The federal act is administered by the U.S. Department of the Interior's new Office of Surface Mining (OSM). However, the act states, "Because of the diversity in terrain, climate, biological, chemical, and other physical conditions in areas subject to mining operations, the primary governmental responsibility for developing, authorizing, issuing and enforcing regulations for surface mining and reclamation operations subject to this act should rest with the states." Therefore, state governments are the primary administrators, but Interior will oversee operations within a state until the state has submitted a regulatory scheme for approval by the secretary of the interior. States had until March 3, 1980, to file their state programs. The original filing date, February 3, 1979, was extended six months because of OSM's failure to meet the act's deadlines in issuing final regulations, then extended again. Allowing time for OSM to review proposed state programs, operators are not likely to face the permanent program rules until at least mid-1980 (9).

The act allows a state, as the regulatory agency with local jurisdiction over mining operations, to prescribe any other requirements it deems appropriate for conditions within its boundaries. The act also requires that notice be given and a hearing provided for all holders of interest in the lands to be affected before a permit is issued. In addition, citizens are authorized to bring suit against either the secretary of the interior or the state regulatory agency to compel enforcement of the act (9).

As mentioned, many federal reclamation requirements are similar to West Virginia's laws. According to Ben Greene, president of the West Virginia Surface Mining and Reclamation Association, the new federal legislation "will, in no way, alter or improve the quality of reclamation in our state" (20). Nevertheless, the act has faced heavy criticism, even in West Virginia. In December 1977, 40 West Virginia mining companies were the first of several plaintiffs to file suit to test the act and its accompanying regulations (20).

Perhaps the lawsuit with the greatest potential implications is one currently under litigation in Virginia. Early this year a federal judge in Big Stone Gap, Virginia, suspended enforcement of the act until its constitutionality could be determined. This temporary injunction was overturned by an appeals court, but a permanent injunction has not yet been ruled on.

The temporary injunction suspending enforcement of the act challenged the act's constitutionality on the basis that it infringes on powers reserved to states by the Tenth Amendment, permits taking of private property without compensation in violation of the Fifth Amendment, authorizes shutdown orders without hearings in violation of the due process clause of the Fifth Amendment, and violates the equal protection clause of the Fifth Amendment. The surface mining companies and landowners who filed this suit claim that the regulations will unfairly damage surface mining interests in Virginia. The companies state that 95 percent of Virginia's strippable coal reserves lie on slopes exceeding 20 degrees and that mining on such slopes will be economically impossible with the requirement in the federal law that the land be returned to its original contour (3).

According to the Mining and Reclamation Council of America, about 1,000 small surface mine operators plan to go out of business before the end of this year because of the anticipated impact of the new controls. Southwestern Virginia already is feeling this impact. Surface mining activity is down 40 percent. Only 24 permits were issued under Virginia's laws in 1978; this number compares with the 375 permits issued in 1974 (15).

Some of the federal act's requirements are more controversial than others. One particularly controversial item involves the performance bond.

A performance bond must be posted before issuance of a permit to insure that money will be available if the operator is negligent, or if for some other reason reclamation would need to be completed by a third party. The new federal act requires that performance bonds continue in effect for at least 5 years (10 years in areas with limited rainfall) after revegetation has been accomplished. This means that the bond must continue in effect at least 8 to 10 years.

Insurance companies are becoming more reluctant to continue guaranteeing performance bonds because of the increased longevity of bonding regulations. As one West Virginia insurance agent observed, "The only companies that are writing bonds are those that don't understand the federal law" (2). Because so little of the major insurance companies' business involves surface mine performance bonding (1% general performance bonds; 0.6% reclamation bonds), the companies are unwilling to risk so much for such a small amount of business.

This situation penalizes small, independent operators more than large corporations. The only operators who can afford to post a bond for themselves may well be those who have cash collateral for the bond. One projection estimates that the bond requirement for the average small mine will range from $1.5 million to $2 million. At an annual premium rate of $20 per $1,000, the annual bond cost would range from $20,000 to $40,000, an amount that will be difficult for many small operators to absorb (2), assuming they can be bonded in the first place.

While the effects of a decline in surface mining are obvious on those directly employed by the industry, such a decline will also have its impact on underground mining in Appalachia. A study by the Stanford Research Institute (16) concerning the impact of surface mining on the economy of West Virginia revealed that surface mining has significant implications for employment in the deep mining industry. Many deep mines are only marginally profitable, and a company may supplement its production from such a mine with surface mine operations. This subsidizing of deep mine production with surface mining methods was found to be the principal means of maintaining production levels. The Stanford study concluded that if surface mining operations are related to deep-mine employment in the same proportion as is production, then 6,000 to 8,000 deep miners are affected in some way by surface mining in West Virginia.

The only true test of the impact of the new federal legislation will be its implementation. That final effect remains to be determined, but it is being monitored closely by all concerned. Mining and reclamation methods have improved greatly in the past decade, but not without certain costs. Advanced technological methods, many of which have been dictated by state and federal laws, no doubt have contributed to ever-increasing energy costs. But these improved methods have, in turn, reduced the degrading effects of surface mining. In the end, attempting to increase coal production while reducing adverse environmental effects raises production costs, which always are borne by consumers. How much consumers are willing to pay and how highly they value their environment and the existence of any realistic alternatives are questions yet to be resolved. As one surface mine operator commented with regard to the approximate original contour requirement, "As long as consumers are willing to pay for it, we can do anything. We can line it with gold if you tell us to" (11).

References Cited

[1]Carroll, James. 1978. *New federal agency tackles surface mining problems.* J. Soil and Water Cons. 33(2): 77–79.

[2]Coal Outlook. 1979. *Coal buying panic foreseen.* Pasha Publications, Washington, D.C. January 8: 1.

[3]Dellinger, Paul. 1979. *Enforcement of mine act suspended.* Roanoke Times and World News (February 15): A1, A8.

[4]Greene, Benjamin C., and William B. Raney. 1974. *West Virginia's controlled placement.* In Proc., Second Res. and Applied Tech. Symp. on Mined-Land Reclamation Nat. Coal Assoc., Louisville, Ky.

[5]Hayes, Earl T. 1979. *Energy resources available to the United States, 1985 to 2000.* Science 203 (4377): 233–239.

[6]McCormack, D. E. 1974. *Soil reconstruction: For the best soil after mining.* In Proc., Second Res. and Applied Tech. Symp. on Mined-Land Reclamation Nat. Coal Assoc., Louisville, Ky.

[7]Mink, D. T. 1976. *Reclamation and rollcalls: The political struggle over surface mining.* Environmental Policy and Law 2(4): 176–180.

[8]Power, J. F., R. E. Ries, and F. M. Sandoval. 1976. *Use of soil material on spoils—effect of thickness and quality.* N. Dak. Farm Res. Bimonthly 34(1): 23–24.

[9]Randall, Alan, Orlen Grunewald, Angelos Pagoutalos, Richard Ausness, and Sue Johnson. 1978. *Estimating environmental damages from surface mining of coal in Appalachia: A case study.* EPA-600/2-78-003. U.S. Environmental Protection Agency, Washington, D.C. 131 pp.

[10]Rogowski, A. S., and E. L. Jacoby, Jr. 1977. *Water movement through Kylertown strip mine spoil.* Paper 77-2057. Am. Soc. Agr. Eng., St. Joseph, Mich.

[11]Spencer, Charles. 1977. *Mine cutbacks.* Raleigh Register (July 27): 1.

[12]United States Congress. 1977. *Surface mining control and reclamation act of 1977.* 95th Congress. Washington, D.C.

[13]U.S. Environmental Protection Agency. 1973. *Processes, procedures and methods to control pollution from mining activities.* EPS-430/9-73-011. Washington, D.C. 390 pp.

[14]U.S. Environmental Protection Agency. 1976. *Erosion and sediment control. Surface mining in the eastern United States.* EPA-625/3-76-006. Washington, D.C.

[15]Warren, Lucian. 1979. *Publication of mine regulations eagerly awaited.* Roanoke Times and World News (March 11): B9.

[16]West Virginia Surface Mining and Reclamation Association. 1973. *The surface mining issue: A reasoned response.* Charleston.

[17]West Virginia Surface Mining and Reclamation Association. 1976. *West Virginia surface mine production declines for 5th straight year.* Green Lands 6(1): 2–4.

[18]West Virginia Surface Mining and Reclamation Association. 1977. *Our law—everybody's law.* Green Lands 7(3): 5, 17.

[19]West Virginia Surface Mining and Reclamation Association. 1977. *The issue and the industry.* Green Lands 7(4): 5, 17.

[20]West Virginia Surface Mining and Reclamation Association. 1978. *A year of federalism.* Green Lands 8(1): 9–11.

Editor's note: In *Hodel et al.* v. *Indiana et al.* the State of Indiana, the Indiana Coal Association, several coal mine operators, and others challenged the constitutionality of the following provisions of SMCRA: that applicants for mining permits be required to submit reclamation plans; that states establish an administrative procedure for determining the suitability of particular lands for surface mining; that mined land be restored to its approximate original contour; that "prime farmland" be protected; and that topsoil be segregated and preserved for use during reclamation. Some of the same issues arose in *Hodel* v. *Virginia Surface Mining and Reclamation Association, Inc. et al.* along with a challenge to prescribed performance standards for mining on "steep slopes." In both of these cases the, U.S. Supreme Court held that, in the context of a facial challenge, the Act is constitutional and that many of the challenges were premature inasmuch as they did not involve a concrete controversy concerning a particular surface mining operation or a specific parcel of land.

Supplementary Readings

(Anonymous), *Electric Power and the Environment*—A Report Sponsored by the Energy Policy Staff, Office of Science and Technology, U.S. Gov. Printing Office, 1970, 71 pp.

(Anonymous), University of California, Santa Barbara, Calif., 1970, *Santa Barbara Oil Pollution, 1969*, Dept. of Interior, Federal Water Pollution Control Admin.

Averitt, P., 1970, Coal Resources of the United States, January 1, 1970, *U.S. Geological Survey Bull. 1322*, 24 pp.

Barnea, J., 1972, Geothermal Power, *Sci. Amer.*, v. 225, n. 1, pp. 70–77.

Barnett, H., 1967, The Myth of Our Vanishing Resources, *Trans-Action*, v. 4, pp. 6–10.

Berryhill, H. L., Jr., 1974, The Worldwide Search for Petroleum Offshore—A Status Report for the Quarter Century, 1947–72, *U.S. Geological Survey Circular 694*, 27 pp.

Blumer, M., 1971, Scientific Aspects of the Oil Spill Problem, *Environmental Affairs*, v. 1, pp. 54–73.

Brobst, D., and Pratt, W., eds., 1973, United States Mineral Resources, *U.S. Geological Survey Prof. Paper 820.*

Brooks, D. B., 1966, Strip Mine Reclamation and Economic Analysis, *Natural Resources Jour.*, v. 6, pp. 13–44.

Brooks, J. W., et al., 1971, Environmental Influences of Oil and Gas Development in the Arctic Slope and Beaufort Sea, *U.S. Dept. Interior Resource Publication 96*, 24 pp.

Brown, T. L., 1971, *Energy and the Environment*, C. E. Merrill, Columbus, Ohio, 141 pp.

Cameron, E. N., ed., 1973, *Mineral Position of the United States—1975–2000*, Univ. of Wisconsin Press, Madison, 159 pp.

Cheney, E. S., 1974, U.S. Energy Resources: Limits and Future Outlook, *Amer. Scientist*, v. 62, n. 1, pp. 14–22.

Cloud, P. E., Chr., Committee on Resources and Man, 1969, *Resources and Man: A Study and Recommendations*, W. H. Freeman and Co., San Francisco, 259 pp.

Committee on Geological Sciences, National Research Council, 1972, *The Earth and Human Affairs*, Canfield Press, San Francisco.

Dole, H., Chr., 1974, Development of Oil Shale in the Green River Formation, Report of the Committee on Environment and Public Planning, *The Geologist*, supplement to v. 9, n. 4, 8 pp.

Duncan, D. C., and Swanson, V. E., 1965, Organic-Rich Shale of the United States and World Land Areas, *U.S. Geological Survey Circular 523*, 30 pp.

Flawn, P. T., 1966, *Mineral Resources—Geology, Engineering, Economics, Politics, Law*, Rand-McNally & Co., New York, 406 pp.

Fowler, J., 1975, *Energy and the Environment*, McGraw-Hill, New York.

Freeman, S. D., et al., 1974, *A Time to Choose: America's Energy Future*, Ballinger Publishing Co., Cambridge, Mass.

Gough, W. C., and Eastland, B. J., 1971, The Prospects of Fusion Power, *Sci. Amer.*, v. 224, n. 2, pp. 56–64.

Gregory, D. P., 1973, The Hydrogen Economy, *Sci. Amer.*, v. 228, n. 1, pp. 13–21.

Hammond, A., et al., 1973, *Energy and the Future*, Amer. Assoc. Advancement Sci., Washington, D.C., 184 pp.

Hill, A., and McCloskey, M., 1971, Mineral King: Wilderness versus Mass Recreation in the Sierra, in *Patient Earth*, Harte, J., and Socolow, R. H., eds., Holt, Rinehart, and Winston, New York, pp. 165–80.

Hubbert, M. King, 1971, The Energy Resources of the Earth, *Sci. Amer.*, v. 224, n. 3, pp. 61–70.

Kesler, S., 1976, *Our Finite Mineral Resources*, McGraw-Hill, New York.

Landsberg, H., 1964, *Natural Resources for U.S. Growth: A Look Ahead to the Year 2000*, Johns Hopkins Univ. Press, Baltimore.

Landsberg, H., Chr., 1979, *Energy: The Next Twenty Years*, Ballinger Publishing Co., Cambridge, Mass.

Laporte, L., 1975, *Encounter with the Earth: Resources*, Canfield Press, San Francisco.

Leopold, L. B., Clarke, F. E., Hanshaw, B. B., and Balsley, J. R., 1971, A Procedure for Evaluating Environmental Impact, *U.S. Geological Survey Circular 645*, 13 pp.

Lewis, R. S., and Spinrad, B., eds., 1972, *The Energy Crises*, A Science and Public Affairs Book. Educational Foundation for Nuclear Science, Chicago, 148 pp.

Lovering, T. S., 1968, Non-Fuel Mineral Resources in the Next Century, *Texas Quarterly*, v. 11, n. 2, pp. 127–47.

Lovins, A., 1971, *World Energy Strategies: Facts, Issues and Options*, Friends of the Earth International, San Francisco.

McDivitt, J. F., and Manners, G., 1974, *Minerals and Men*, Johns Hopkins Univ. Press, Baltimore, 192 pp.

McKelvey, V. E., Tracey, J. I., Jr., Stoertz, G. E., and Vedder, J. F., 1969, Subsea Mineral Resources and Problems Related to Their Development, *U.S. Geological Survey Circular 619*, 26 pp.

Murdoch, W. W., ed., 1971, *Environmental-Resources, Pollution and Society*, Sinauer Assoc., Inc., Stamford, Conn., 440 pp.

National Academy of Sciences, 1975, *Mineral Resources and the Environment*, Natl., Academy Sci., Washington, D.C.

National Academy of Sciences, 1980, *Energy in Transition 1985–2010*, W. H. Freeman and Co., San Francisco.

Park, C. F., and Freeman, M. C., 1968, *Affluence in Jeopardy: Minerals & the Political Economy*, Freeman, Cooper and Co., San Francisco, 468 pp.

Park, C. F., and Freeman, M. C., 1975, *Earthbound, Minerals, Energy and Man's Future*, Freeman, Cooper and Co., San Francisco, 279 pp.

Park, C., and MacDiarmid, R., 1970, *Ore Deposits*, 2d ed., W. H. Freeman and Co., San Francisco.

Pecora, W. T., 1968, Searching Out Resource Limits, *Texas Quarterly*, v. 11, n. 2, pp. 148–54.

Perry, H., 1974, The Gasification of Coal, *Sci. Amer.*, v. 230, n. 3, pp. 19–25.

Potter, J., 1973, *Disaster by Oil, Oil Spills: Why They Happen, What They Do, How We Can End Them*, MacMillan, New York, 301 pp.

Rickert, D., Ulman, W., and Hampton, E., eds., 1979, Synthetic Fuels Development—Earth Science Considerations, *U.S. Geological Survey Prof. Paper 1240-A*, 45 pp.

Risser, H. E., 1973, Energy Supply Problems for the 1970's and Beyond, *Environmental Geology Notes No. 62*, Illinois State Geol. Survey, Urbana, 12 pp.

Risser, H. E., 1973, The U.S. Energy Dilemma: The Gap between Today's Requirements and Tomorrow's Potential, *Environmental Geology Notes No. 64*, Illinois State Geol. Survey, Urbana, 64 pp.

Risser, H. E., and Major, R. L., 1967, Urban Expansion—An Opportunity and a Challenge to Industrial Mineral Producers, *Environmental Geology Notes No. 15*, Illinois State Geol. Survey, Urbana, 19 pp.

Rose, D. J., 1974, Energy Policy in the U.S., *Sci. Amer.*, v. 230, n. 1, pp. 20–29.

Rowe, J., 1979, *Coal Surface Mining: Impacts of Reclamation*, Westview Press, Boulder, Colo.

Ruedisili, L., and Firebaugh, M., 1982, *Perspectives on Energy*, 3d ed., Oxford Univ. Press, New York, 475 pp.

Schlee, J., 1968, Sand and Gravel on the Continental Shelf off the Northeastern United States, *U.S. Geological Survey Circular 602*, 9 pp.

Skinner, B., 1976, *Earth Resources*, Prentice-Hall, Englewood Cliffs, N.J.

Starr, Chauncey, 1971, Energy and Power, *Sci. Amer.*, v. 255, n. 3, pp. 36–49.

Steinhart, C. E., and Steinhart, J. S., 1972, *Blowout: A Case Study of the Santa Barbara Oil Spill*, Duxbury Press, Boston, Mass., 138 pp.

Swanson, V. E., Chr., 1974, Environmental Impact of Conversion from Gas or Oil to Coal for Fuel, Report of the Committee on Environment and Public Planning, *The Geologist*, supplement to v. 9, n. 4, 8 pp.

Theobold, P. K. et al., 1972, Energy Resources of the United States, *U.S. Geological Survey Circular 650*, 27 pp.

U.S. Bureau of Mines, (published annually), *Minerals Yearbook*, Metals, minerals and fuels., vols. I and II.

U.S. Bureau of Mines, 1976, Mineral Facts and Problems, *U.S. Bureau of Mines Bull. 630*.

Walsh, J., 1965. Strip Mining: Kentucky Begins to Close the Reclamation Gap, *Science*, v. 150, pp. 36–39.

Warren, K., 1973, *Mineral Resources*, John Wiley and Sons, New York, 272 pp.

Wenk, Edward, Jr., 1969, The Physical Resources of the Ocean, *Sci. Amer.*, v. 221, n. 3, pp. 166–67.

White, Donald, 1965, Geothermal Energy, *U.S. Geological Survey Circular 519*, 17 pp.

Yerkes, R. F., Wagner, H. C., and Yenne, K. A., 1969, Petroleum Development in the Region of the Santa Barbara Channel Region, California, *U.S. Geological Survey Prof. Paper 679-B*, pp. 13–27.

FILMS

A Gift from the Earth (Argonne Films: 26 min.)

The Atom Underground (U.S. Atomic Energy Commission, 1969: 20 min.)

The Bitter and the Sweet (U.S. Atomic Energy Commission, 1971: 30 min.)

The Bottom of the Oil Barrel (Time-Life Films and BBC-TV: 40 min.)

Coal—Bridge to the Future (Modern Talking Picture Service, Inc.: 28 min.)

Energy for the Future (Encyclopedia Britannica: 17 min.) 38 min.)

Energy—The Nuclear Alternative (Churchill Films, Inc., 2d ed. 1980: 22 min.)

Geothermal: Nature's Boiler (Owen Murphy Productions, Inc., 1977: 8 min.)

How Safe Are America's Atomic Reactors? (Impact Films)

The Minerals Challenge (U.S. Bureau of Mines, 1972: 27 min.)

Moving Earth—The Story of Mined Land Subsidence Control (U.S. Department of the Interior, 1975: 30 min.)

No Act of God (National Film Board of Canada: 28 min.)

Nuclear Energy: Power for Today and Tomorrow (Modern Talking Picture Service, Inc.: 28 min.)

Nuclear Power and the Environment (U.S. Atomic Energy Commission, 1969: 14 min.)

Nuclear Power in the United States (U.S. Atomic Energy Commission, 1971: 28 min.)

Oil (Shell Oil Co., 1973: 18 min.)

Oil Spill!! Patterns in Pollution (Association-Sterling Films: 17 min.)

Oil Well (Shell Oil Co., 1976: 21 min.)

The Ravaged Land (John Wiley and Sons, 1971: 15 min.)

The Role of Coal (Indiana University, 1979: 17 min.)

Santa Barbara—Everybody's Mistake (National Educational Television, 1971: 30 min.)

Sunbeam Solutions (Time-Life Films and BBC-TV: 38 min.)

Torrey Canyon (Time-Life Films, 1970: 26 min.)

Unconventional Gas Resources (Morgantown Energy Technology Center, 1980: 29 min.)

III
WATER RESOURCES
AND THE
ENVIRONMENT

This irrigation well in Arizona pumps at a rate of 1500 gallons per minute. A considerable amount of ground water is being withdrawn for irrigation with almost no recharge taking place. As a result, the nation's water supply is being depleted at the rate of 21 billion gallons per day. (Photo by U.S. Dept. of Agriculture.)

Hydrologic Cycle

Water is our most vital resource. Its availability, quality, and use as a facility for recreation or as a diluent for wastes represent critical problems in many places.

Approximately 71 percent of the earth's surface is covered by water, but more than 97 percent of all this water is in the ocean basins and is not generally available for direct use. Solar radiation causes the evaporation of a small part of this water, which then becomes involved in the hydrologic cycle (Fig. 1) as precipitation, or *meteoric water*. Precipitation includes all forms of water that falls from the atmosphere to the earth's surface. There it may evaporate or be absorbed by the soil (*infiltration*), temporarily accumulate on the surface as snow and glacial ice, or run off in well-defined courses or as diffused surface water. The areal distribution of average annual precipitation in the United States is shown in Fig. 2.

Soil is a temporary storage bank for moisture in the hydrologic cycle. It may support plant life, which then returns moisture to the atmosphere through transpiration. If soil is saturated, the excess water will percolate downward through underlying unsaturated materials, finally reaching the zone of saturation, where all the pores are full of water. Water in this zone is called *ground water*.

Approximately 29 percent of the annual precipitation in the United States reaches streams and rivers or is stored as ground water. The remaining 71 percent is lost through evaporation or evapotranspiration. We can do little to diminish the losses associated with tran-

FIG. 1. The hydrologic cycle. (From *U.S. Geological Survey Professional Paper 942, 1977.*)

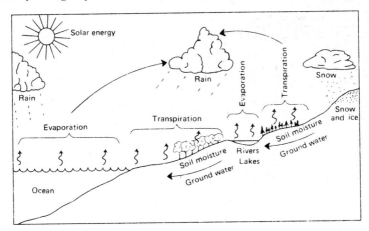

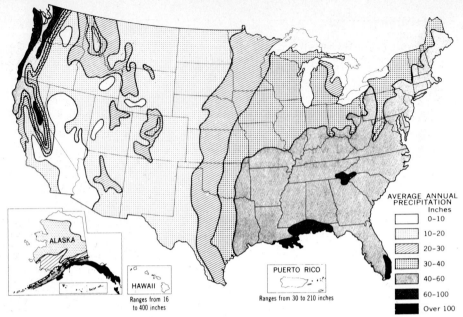

FIG. 2. Average annual precipitation in the United States. (From U.S. Dept. of the Interior.)

spiration and evaporation. We have, in fact, increased evaporation losses by creating reservoir lakes and large-scale irrigation projects.

Surface Water

Occurrence

The availability of surface water is influenced in part by the balance between average precipitation and potential evapotranspiration. Humid regions, where average precipitation exceeds potential evapotranspiration, generally have abundant water. The climate of the eastern United States is classified as humid, with sufficient water for most vegetation and for a perennial surplus that appears as runoff in streams and in lakes. The plains states are generally considered subhumid to semiarid, with average precipitation and potential evapotranspiration nearly the same. The western states are characteristically semiarid or arid. Except in some humid mountain ranges, the moisture requirement of most vegetation is not satisfied in full, and water is generally not available for overland flow to streams or for permanent lakes. Figure 3 shows the areal distribution of areas of water surplus and water deficiency in the United States.

One of the most direct indicators of the availability of surface water is average annual runoff (Fig. 4). *Runoff* includes that part of precipita-

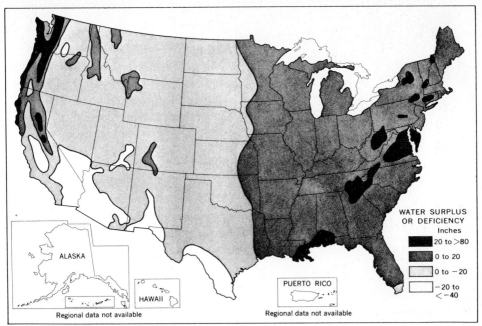

FIG. 3. Areas of water surplus and water deficiency in the United States. (From U.S. Dept. of Interior.)

tion which flows over the surface of the land as *sheet wash* and appears in surface streams. The aggregate flow of streams in the United States averages about 21.6 cm (8.5 in.) a year, or 4560 billion liters (1200 billion gallons) per day. This flow is about five times greater than present withdrawals by humans and twenty times greater than the actual consumption of water. However, the variability of streamflow and the variability of the geographic distribution of streams are basic obstacles to the full use of streams. Planned use of surface water is, for the most part, at a level near minimum natural flow. Flows above minimum levels are generally unused, although a small part of these flows may be stored in reservoirs. Where stream yield does not match present patterns of stream use, surface water is either wasted or in short supply. The geologist A. M. Piper (1965) has observed that "water demand is commonly large when yield of the water sources is small, and large demands commonly arise in areas remote from large water sources." This mismatch is a crucial problem for water supply management.

Water Quality

The pollution of surface water may generally be attributed to municipal, industrial, and agricultural wastes. The range of pollutants is

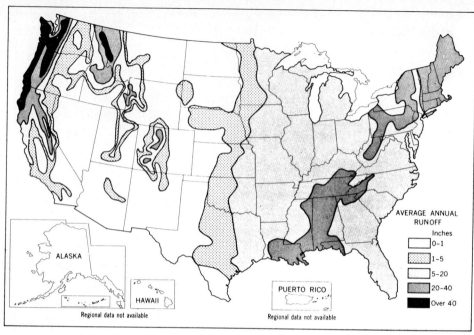

FIG. 4. Average annual runoff in the United States. (From U.S. Dept. of the Interior.)

wide and includes heavy metals such as mercury and lead, radioactive isotopes, pathogenic bacteria and viruses, animal wastes, fertilizers, pesticides, eroded soils, sediments (Reading 19), and a variety of exotic chemical compounds. These pollutants can create serious health problems, destroy aquatic organisms and their habitats, and degrade water supplies. The most troublesome pollutants to control are those from agricultural sources.

Legislation addressing the problem of water pollution was enacted more than 30 years ago with passage of the Federal Water Pollution Control Act (FWPCA) of 1948. Since then, plans for implementing and enforcing water quality standards in all 50 states have been established. The act has been amended on numerous occasions, updating methods of water pollution control dating from 1899, requiring permits for pollutants discharged into navigable waters, prohibiting dumping of radioactive waste, authorizing the Environmental Protection Agency (EPA) to issue construction grants for municipal wastewater treatment plants, requiring the EPA to conduct extensive research on water quality problems, and providing for public participation in regional planning enforcement. Many rivers and "dead" lakes which only a few decades ago were so highly polluted that fish and other aquatic life could no longer exist are again usable as a result of rigorous enforcement of FWPCA and the cooperation of industry and municipalities.

Ground Water

Ground water is one of the earth's most important resources. Many major cities throughout the world are almost totally dependent on deep aquifers. In the United States, the volume of ground water is almost seven times that of all surface water, and currently supplies about 50 percent of the country's drinking water. Not only is ground water important in arid regions where surface flows may be low or nonexistent, it is also important in humid regions where surface flow is either inadequate or unusable.

Aquifers

Ground water is stored in rock formations known as *aquifers*—in pore spaces, fractures, or solution cavities. The parameter which measures the void space in an aquifer is referred to as its *porosity*—the percentage of the total volume of rock that is occupied by open spaces. The capacity of a rock to transmit water is called its *permeability*, or *effective porosity*.

Porosity may be developed as a primary feature of an aquifer or it may be the product of secondary processes. For example, as sediments are deposited, voids between the grains may remain unfilled, and the aquifer is said to have *primary porosity*. Compaction and cementation will reduce but rarely eliminate the volume of void space. Organic reefs also contain open spaces which are seldom completely filled after a structure has been buried by sediments. When such secondary processes as solution, recrystallization, or the development of fractures and joints increase the porosity of sedimentary rocks, they are said to have *secondary porosity*. Figure 5 illustrates several types of porous rocks.

A ground-water reservoir, or aquifer, is made up of rock strata or sediments sufficiently porous and permeable to yield water to wells or springs. If the aquifer intersects a sloping land surface, ground water will flow onto the land surface as a seep or a spring. If the aquifer intersects a stream channel, water will discharge into the stream. Relatively shallow aquifers that are not confined by overlying impermeable beds are known as *unconfined,* or *water-table,* aquifers (Fig. 6). The "water level" (water table) in these aquifers is the top of the saturated zone and changes with changing seasonal precipitation.

Other aquifers are deeper and isolated from the surface by an im-

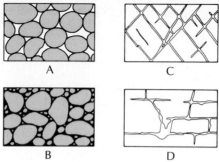

FIG. 5. Types of porous rocks. (A) Well-sorted sedimentary deposit having high porosity. (B) Poorly sorted sedimentary deposit having low porosity. (C) Rock rendered porous by fracturing. (D) Rock rendered porous by solution. (From *U.S. Geological Survey Water Supply Paper 489*, 1923.)

permeable rock unit called an *aquiclude* (Fig. 7). A *confined,* or *artesian,* aquifer is not as sensitive to seasonal variations in precipitation since annual recharge is often small compared to the amount of water that has been stored in the aquifer over the centuries. If the water in a confined aquifer is under pressure from the weight of water at higher levels (*hydrostatic head*), the pressure will force the water to an elevation above the top of the aquifer. This hydrostatic pressure level is known as the *potentiometric surface* (Fig. 7).

Geologic Occurrence. The most common aquifers are associated with porous and permeable sedimentary rocks and surficial deposits. Most igneous and metamorphic rocks have a crystalline texture and lack sufficient primary porosity and permeability to be good aquifers. Of course, any rock that possesses joints, fractures, or fault zones may contain ground water. In the Thousand Springs area of Idaho, for example, water discharges directly from the fractured basaltic lava flows of the Snake River plain.

Unconsolidated surficial deposits are among the best aquifers (Appendix B). They are highly porous and permeable, and in humid climates their water tables lie close to the surface and are easily recharged. Surficial deposits are abundant along the Atlantic coastal plain and the Gulf coast, in glacial deposits of the United States Central Lowlands,* adjacent to many river channels, and along the flanks of steep mountain ranges.

*Ohio, Michigan, Indiana, Illinois, Wisconsin, Minnesota, Iowa, North Dakota, and South Dakota.

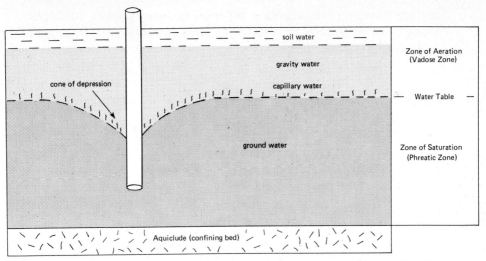

FIG. 6. Ground-water terminology for an unconfined, or water-table, aquifer. The water table is free to rise or fall in response to available water supply. Pumping produces a local cone of depression in the water table.

Movement of Ground Water

Quantitative studies of the flow of ground water through porous rocks are based on an empirical relationship known as *Darcy's law*, which may be expressed as

$$Q = KA \frac{dh}{dl}$$

where

Q is the volume of water flowing through a cross section of area A in a given time,

K is the hydraulic conductivity of the material,

dh/dl is the hydraulic gradient in which the quantity dh represents the change in head between two points, and dl is the distance between the two points.

Hydraulic conductivity (K) is a measure of the permeability of the porous material; that is, a measure of its ability to transmit water. In the United States, Q has been most commonly expressed in gallons per day, but metric units can also be used.

· The rate of movement of ground water is important, particularly in those problems related to pollution. For example, if a harmful substance is accidentally introduced into an aquifer it becomes a matter of great urgency to determine when the substance will reach wells or surface streams. The equation for ground-water velocity combines Darcy's law with the basic velocity equation of hydraulics ($v = Q/An$), and may be expressed as

$$V = \frac{K\,(dh/dl)}{7.48\ n}$$

where

V   is the velocity in feet per day
K   is the hydraulic conductivity of the material
dh/dl is the hydraulic gradient where dh/dl represents the change in head between two points, and dl the distance between the points, and n is the effective porosity.

If hydraulic conductivity is expressed in gallons per day per square foot, the equation must include the number of gallons in a cubic foot (7.48) in order to obtain velocity in feet per day. It is important to note that this equation gives only an approximate velocity and that both porosity and permeability vary widely along most flow paths.

Water Quality

Although ground water is free of sediments and naturally occurring pathogenic organisms, water quality may be affected by other natural contaminants. Chloride and sodium ions in sufficient amounts may be responsible for brackish or even saline water. Calcium, magnesium, and iron ions contribute to water hardness (these may be removed however, with home water softeners). Sulfur may occur either as a sulfate salt or as hydrogen sulfide. A high iron or sulfide content renders water unpalatable. Trace amounts of many toxic metals have also been reported, particularly where aquifers are in contact with intrusive igneous rocks.

Ground water is easily polluted when chemicals and micro-organisms are introduced into the flow system. This most commonly occurs when wastes are disposed near shallow aquifers (Part IV). Some of these wastes are removed naturally. Most soils and rocks can physically filter out some of the suspended solids, and chemical precipitation may remove other polluting substances. Some clay minerals have active surfaces that can react with an array of pollutants (a process known as *ion exchange* or *adsorption*). Many micro-organisms cannot adapt to the subsurface environment, but anaerobic organisms (those that do not require oxygen) may thrive there. The opportunity for dispersion and dilution of pollutants not removed by physical filter-

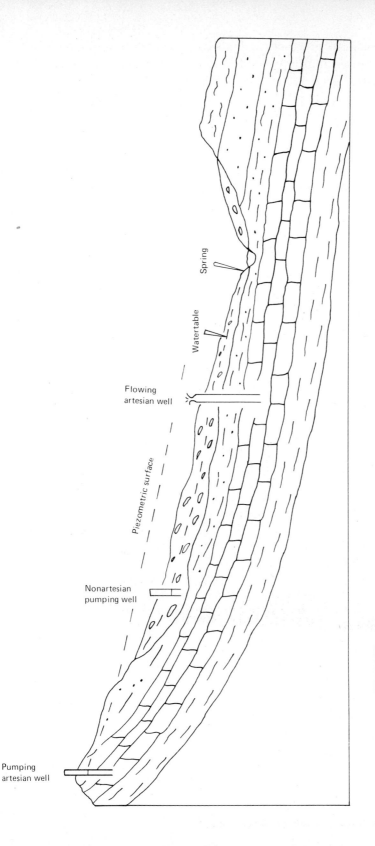

Spring

Watertable

Impermeable confining beds (aquiclude)

Impermeable confining shale (aquiclude)

Flowing artesian well

Piezometric surface

Nonartesian pumping well

Unconfined gravel aquifer

Artesian aquifer

Pumping artesian well

FIG. 7. Common occurrences of ground water. (Adapted from *U.S. Geological Survey Water Supply Paper* 639.)

ing and chemical processes is limited, but the lack of oxygen and sunlight reduces the opportunity for degradation of the pollutants.

The Safe Drinking Water Act of 1974, which authorizes the establishment of national standards for drinking water, provides specifically for the promulgation of Federal regulations to protect ground water. The Act established a permit system for injection wells, allows the EPA to initiate civil suits for violation of the ground-water quality provisions of the law and also requires EPA to supply Congress with detailed reports on ground-water problems resulting from, for example, surface spills, abandoned injection or extraction wells, the application of pesticides and fertilizers, and pollution from ponds, pools, lagoons, pits, or other surface disposal areas.

Case Histories and Readings

Ground Water—A Mixed Blessing

C. L. McGuinness explores the advantages and disadvantages of the development of ground water in Reading 31. The ubiquitous occurrence of ground water has enabled humans to occupy large areas that otherwise could not have been settled. But many problems are associated with the full development of ground water. McGuinness cites the High Plains of Texas as a dramatic example of the economic implications of ground-water withdrawal and replenishment.

The Changing Pattern of Ground-
Water Development on Long Island,
New York

In coastal aquifers a boundary exists between freshwater and saltwater. Overpumping of an aquifer may lead to saltwater incursion and pollution of freshwater wells. The authors of Reading 32 document the response of the hydrologic system on Long Island, New York, to population trends and attendant changes in the use and disposal of water.

Geological Influence on Regional
Health Problems

Bedrock geology determines the trace elements that enter groundwater flow systems and become available for the use of plants and animals. The variety of regional diseases which can be related to the geochemistry of trace elements available in local water supplies is explored by Jean Spencer in Reading 33. Spencer notes that the "normal municipal water treatment hardly alters the percentages of the various trace elements in the water distributed to consumers."

31 Ground Water—A Mixed Blessing*

C. L. McGuinness

Introduction

Ground water—the water of wells and springs—is one of the world's most valuable resources. Indeed, in the not distant future it will prove to be not only valuable but vital. Yet, many difficulties will be encountered in the full exploitation of this resource that will be essential in tomorrow's society. It is the purpose of this paper to develop this theme briefly, in terms of conditions in the United States.

First let us define this ground water that is going to occupy a good deal of our attention in the future. Ground water is the water of the zone of saturation as defined by O. E. Meinzer (1923, p. 21); it is the water under hydrostatic pressure in the pores and crevices of the rocks that is free to move under the influence of gravity from places where it enters the zone of saturation to places where it is discharged. It is a phase of the hydrologic cycle, and it is that fact which both makes it a valuable, renewable re-

*Publication authorized by the Director, U. S. Geological Survey.

McGuinness, C. L., 1960, Ground Water—A Mixed Blessing, *Proceedings of Section 20, Twenty-First Session International Geological Congress,* Copenhagen.

Reprinted from *Proceedings of Section 20, Twenty-First Session International Geological Congress,* Copenhagen 1960 by permission of the author and the Danish Geological Survey.

The late Mr. McGuinness was Chief of the Ground Water Branch of the U. S. Geological Survey.

source and creates many of the difficulties that attend its full utilization.

Ground water is one of the earth's most ubiquitous resources, and therein lies much of its value. It exists wherever three conditions are met: (1) water from precipitation or streamflow penetrates beneath the surface in quantities sufficient to exceed the field capacity of the soil—that is, moves downward through the zone of aeration under the influence of gravity; (2) the rocks beneath the soil are permeable enough to transmit this water; and (3) the rate of infiltration is sufficient that the zone of saturation formed in the lower part of the permeable rocks will be built up to a perceptible thickness by the time the lateral outflow increases to a rate equal to the rate of infiltration from the land surface. These conditions are met—that is, ground water exists at least intermittently—in a very large part of the world. However, it is only under three further conditions that ground water becomes a usable resource: (1) the rocks in the zone of saturation are permeable enough to yield useful supplies of water to wells, springs, or streams; (2) the zone of saturation is permanent, or at least persists long enough each season to allow practical exploitation; and (3) the mineral substances dissolved by the water as it travels through the ground do not reach such concentrations as to make the water unfit for use at the place where a supply is needed. These further conditions, stringent as they are,

are met commonly enough that at least the small quantities of potable water needed for domestic supply are available very widely. Only where impermeable rocks or permanently frozen ground extend to great depths, or where the climate is exceedingly dry, is ground water entirely or practically absent; even in large regions having these characteristics, ground water may be available locally.

It is this near-universality of occurrence that has made ground water so important in the past in enabling human occupation of many large regions that otherwise could not have been settled. As our areal reconnaisances of ground-water geology are extended and as methods of prospecting for and developing ground water are refined, this resource will continue to contribute to the settlement of primitive areas for some time to come. But it is not this aspect of ground-water development that will produce the most headaches; rather, it is in the optimum, or maximum, exploitation of water supplies in areas more abundantly endowed by nature that the principal difficulties have been encountered, and will continue to be.

Full Development of Ground Water— A Future Necessity

The time is coming—coming rapidly— when full exploitation of our water resources will be, not just a desirable objective, but an absolute necessity not only to social progress but even to human survival. Vast water supplies remain nearly or quite untouched, even in some highly developed countries; but, as the world's population continues to increase and the standard of living advances, the margin between supply and demand will become ever narrower. Then, full development will be the order of the day, with all its complexities, its difficulties, and its cost. True, the time may come when advances in nuclear science may provide us an energy source so cheap that sea water can be converted to fresh and piped at reasonable cost anywhere in the world, and when it comes we can forget our water-supply problems. But that happy day is decades off, if not centuries, and our water problems are with us *now*.

Problems and Solutions

It is to be hoped that the reader is now convinced, as is the writer, that the world's supply of usable water will have to be exploited thoroughly and efficiently, and that ground water will have to play its full part. Let us now see what we will have to do to get the ground water we will need.

In many areas, large and small, and for many uses, ground water is by far the most economical source. This situation applies especially where the demand is small or moderate—the few hundred gallons a day needed for a family, or the few tens or hundreds of thousands needed for a town or industry of modest size. Locally, in geologically favorable areas, it applies even to demands of millions or tens of millions of gallons a day. However, surface water commonly is the source of supply for the largest cities and the largest agricultural, industrial, and power-generation projects.

Ground water has many advantages over surface water—it is generally potable and palatable without treatment or can be made so with only simple treatment; its uniform chemical quality and general lack of organic matter simplify any treatment that may be necessary; its uniform temperature makes it attractive as a cooling agent in the summer; it is relatively immune from evaporation and from contamination; and, where the local supply is adequate to meet the demand, the overall cost of obtaining and treating it is generally less than that for surface water.

It is when the demand strains the limits of supply and when surface water is not available at reasonable cost that we begin to realize the mixed nature of the blessing that is ground water. It is then that we become aware that, in comparison to surface water, ground water is more difficult and expensive to locate, to evaluate quantitatively, and to manage. These difficulties are inherent in our lack of detailed knowledge of the places where ground water occurs, of the principles that govern its occurrence and movement in complex geologic situations, of methods for evaluating it quantitatively in these situations which defy practical mathematical analysis, and of methods for exploiting it efficiently under limitations imposed by economics and by law. To obtain the knowledge needed to manage this resource will be much more costly, gallon for gallon, than to learn how to manage surface water— yet, where we must have water and where other sources are not available, we are going to pay the cost or we are going to do without. Our job, then, is to recognize the high cost of optimum ground-water development, and to exert our best efforts to reduce that cost to the minimum.

The Ground-Water Situation in the United States

So far the discussion has been essentially philosophical rather than practical. To get our feet back on the ground, let us describe briefly certain aspects of the ground-water situation in the United States. Then we will have a background for delineating very briefly, and again philosophically, the problems associated with full development of ground water.

Taken as a whole, the United States is abundantly endowed with water. An average of 30 inches of rain and snow falls annually. Of this, about 8½ inches escapes the demands of evapotranspiration and flows back to the sea. This 8½ inches represents our recoverable water supply, except as it might be increased by modifications of vegetal growth to reduce evapotranspiration, by artificially induced increases in total precipitation, or by conversion of saline water.

As measured, the 8½ inches is surface water, but ground water plays no mean part in generating it, for it is the overflow of ground-water reservoirs that maintains most of the fair-weather flow of streams. It has been estimated [W. B. Langbein, U.S. Geol. Survey, oral communication, 1950] that perhaps one-third to two-fifths of the total streamflow in the United States represents ground water whose contribution can be identified in the streamflow graph. A much larger portion of the streamflow—under some conditions virtually all—represents water that has passed beneath the land surface and moved through a shallow, temporary zone of saturation in the soil before entering the streams, but this storm-flow component cannot be distinguished on the hydrograph and, moreover, it hardly constitutes ground water that can be utilized as such.

In the United States the total water supply, as measured by the precipitation, is large in the East and Southeast and decreases generally westward and northward, except in the high mountains of the West and especially of the Pacific Northwest. The geology and the air temperature modify this picture extensively so far as available water supply is concerned, however.

The accompanying map gives a good general idea of the availability of ground water in the United States. It was compiled by Dr. H. E. Thomas of the Geological Survey for his book written for the Conservation Foundation (1951). It is based largely on published reports on the Geological Survey and its cooperating State agencies and on unpublished information in field offices of the Geological Survey. It shows areas where supplies of 50 gallons per minute per well are, or are not, commonly available. It shows also whether the productive aquifers consist of unconsolidated or consolidated rocks, or both.

The map does not show specifically where supplies of hundreds or thousands of gallons per minute are available from single wells. It does so in a general way, however, because in the greatest part of the patterned areas wells may yield several hundred gallons per minute or more. This fact is brought out by a similar map, divided into 10 parts for the principal ground-water regions, prepared by Dr. Thomas for his report to a congressional committee in 1952. This later map shows slightly more detail than the earlier, as it distinguishes among the principal types of water-bearing rocks—limestone, sandstone, volcanics, and so on. However, the patterned areas, which on the later map show a capability for "moderate to large yields to wells," are very nearly coextensive with those on the accompanying map.

The map shows that almost the entire Atlantic and Gulf Coastal Plain is underlain by productive unconsolidated aquifers—strata of sand and gravel—which are of Mesozoic and Cenozoic age. In Florida and adjacent States large parts of the Coastal Plain are underlain also by consolidated-rock aquifers (largely Tertiary limestone beneath or interbedded with the sand and gravel). The Midwestern States have both unconsolidated (Pleistocene glacial drift) and consolidated (Paleozoic limestone and sandstone) aquifers. The Northeastern States have many small but productive valley deposits of glacial outwash sand and gravel. The entire Mississippi River basin has similar alluvial deposits, shown in solid black. The High Plains have a widespread Tertiary alluvial deposit which yields water to thousands of irrigation wells. The desert valleys of the Basin and Range province of the Southwest and such large valleys as the San Luis Valley of Colorado and the Central Valley of California contain productive alluvial sand and gravel of Tertiary and Quaternary age. The great Tertiary and Quaternary lava plateaus of Idaho and the Pacific Northwest are productive ground-water sources. And, some of the glacial-outwash deposits of the Northwest are among the most permeable aquifers known and are generously recharged.

In most of the areas shown unpatterned on the map, supplies of ground water adequate for at least domestic use are widely available. One of the most consistently dependable areas for small to moderate supplies is the Piedmont, extending from Alabama to its extension in the New England province, where fractured crystalline bedrocks of Precambrian and Paleozoic age are penetrated by many hundreds of thousands of domestic and small municipal and industrial wells. Locally, supplies of several hundred gallons per minute have been obtained. In the glaciated area which covers New England, New York, and the States to the west as far south as the Ohio and Missouri Rivers, glacial drift and the underlying bedrock yield small to moderate, and locally large, supplies to wells. In the unglaciated Middle West and Midcontinent areas, weathered and fresh sedimentary rocks, largely Paleozoic, generally yield small supplies, but there are some areas where water is difficult to get in more than meager quantities. This difficulty exists also in a considerable part of the glaciated area of western Minnesota and the Dakotas.

In the dry plateau country of the Southwest water is often difficult or expensive to obtain, although the risk of failure can be reduced nearly to zero if the stratigraphy is studied carefully before drilling is attempted. In the mountainous areas, drilled wells are not too successful because the rocks are hard and the fractures tend to be drained, but springs and dug wells supply the relatively sparse rural population, and municipalities use streams, or wells in the larger alluviated valleys, for their supply.

Now we have a general idea where the ground-water resources of the United States are. Where are these resources exploited? Obviously, the large-scale withdrawals are confined to the patterned areas on the map, but where in particular?

In 1955 the total estimated withdrawal of fresh ground water in the continental United States averaged about 45.7 billion U.S. gallons per day [MacKichan, 1957, p. 13]. This was a little more than a fifth of the total of 221 bdg of fresh water withdrawn from all sources for uses other than hydropower generation. We have no more recent figures for the whole country, but this year the 1955 data will be brought up to date, and it will be surprising indeed if they do not show an increase.[1] The 1955 figure for ground water represented better than a 50-percent increase over the comparable figure of 30 bgd for 1950 [MacKichan, 1951, p. 7], and though the increase from 1955 to 1960 may not be as large percentagewise it is virtually certain to be substantial.

The largest single use of ground water is that for irrigation—30 bgd in 1955. A third of the total was pumped in California; Texas was next with 6.5 bgd, and Arizona third with 4.7 bgd. Thus more than two-thirds of the total for irrigation was pumped in 3 of the 48 States.

Industrial use of fresh ground water, including fuel-electric power generation, averaged 9.2 bgd in 1955. The use was greater in the East than in the West but was still substantial in two of our previously named Western States—about half a billion gallons per day each in Texas and California. Industrial use is growing all over the

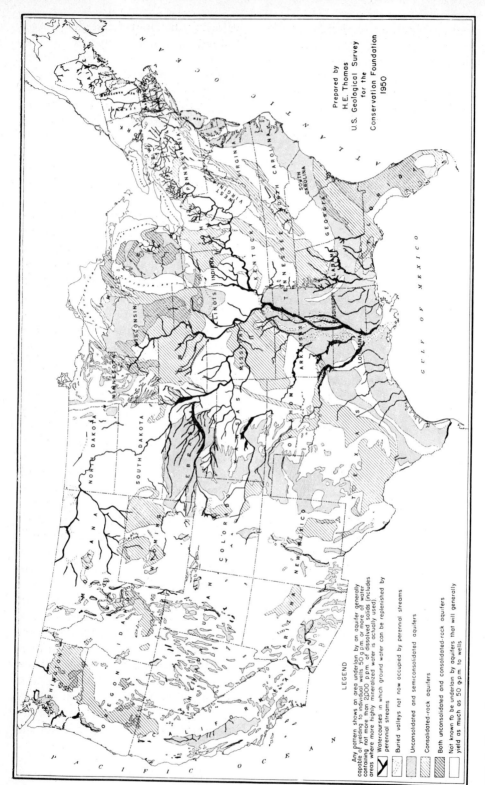

LEGEND

Any pattern shows an area underlain by an aquifer generally
capable of yielding to individual wells 50 g.p.m. or more of water
containing not more than 2,000 p.p.m. of dissolved solids (includes
areas where more highly mineralized water is actually used)

Watercourses in which ground water can be replenished by
perennial streams

Buried valleys not now occupied by perennial streams

Unconsolidated and semiconsolidated aquifers

Consolidated-rock aquifers

Both unconsolidated and consolidated-rock aquifers

Not known to be underlain by aquifers that will generally
yield as much as 50 g.p.m. to wells

Prepared by
H.E. Thomas
U.S. Geological Survey
for the
Conservation Foundation
1950

FIG. 1. Ground-water areas in the United States.

438   Water Resources and the Environment

country, but perhaps especially in the South and West.

Public-supply use of ground water was 4.7 bgd in 1955. It of course accorded with population distribution and was greatest in the Northeast and Middle West. Here again, however, Texas and California were leading States—first and second, respectively.

Rural use of ground water, the smallest of the four major categories, was 1.8 bgd in 1955. It was rather evenly distributed, being larger than average in the populous Middle Western and Northeastern States and—again—in Texas and California, and smallest in the small New England States and thinly populated Western States.

Thus, in 1955 we had 10 States each pumping more than a billion gallons per day, and 3 more very close to that total and no doubt over it by now. Of the 13 only 3—California, Texas, and Arizona—exceeded 1.5 bgd, but these did it handsomely, with totals of 11.1, 7.2, and 5.0 bgd.

Of our 13 leading States 8, including the top 6, are among the 17 Western States—the States where the precipitation is prevailingly less than the national average. This fact is a tribute to the productive aquifers that nature has placed in the West, in spite of the reduced precipitation, but it has a sinister meaning also. It means that a large part of the Nation's serious water problems are to be found in these water-rich yet water-poor States. Just a quick look at one of these problems will show how important and how serious they are.

An Example of a Water-Supply Problem

About as spectacular a case as any is that of the southern High Plains of Texas—the so-called "South Plains." The aquifer of the South Plains and the adjacent area in New Mexico is a remnant of a formerly continuous alluvial apron which in Pliocene time extended for hundreds of miles east of the Rocky Mountains but which since has been cut off from the Rockies by erosion and has itself been cut into segments. The aquifer in the South Plains of Texas contains an enormous amount of ground water in storage—about 200 million acre-feet, or about 65 million million U. S. gallons, as of 1957–58 [J. E. Cronin, U. S. Geol. Survey, written communication, 1959]. But this aquifer, in common with its related parts in much of the rest of the High Plains, has one distinguishing feature—an extremely low rate of replenishment. This rate is estimated to be something like 50,000 acre-feet per year in the Texas portion of the southern High Plains. Both storage and recharge in the New Mexico portion are perhaps a third as great as in the Texas counterpart. Thus we have an aquifer which contained 250 to 275 million acre-feet of water in storage as of 1957–58, and from which something like 40 million acre-feet of water had already been pumped, but the replenishment of which currently is at a rate equivalent to less than 1 percent of the withdrawal. Furthermore, because of the vastness of the aquifer the current pumping, heavy as it is, has not yet reduced the natural discharge at the edges of the Plains; thus the withdrawal to date has come wholly from storage.

The economy of the South Plains is based largely on irrigation with ground water, which in recent years has grown at a rate unparalleled elsewhere in the country. But the ground water is being "mined." Surface water is not the answer—there is very little of it in the region, and under present economic conditions the amount that could be developed would meet only a very small fraction of the total water demand. Artificial recharge through wells of a part of the rainwater that gathers in many ponds or "sinks" in wet weather is being investigated actively and will be helpful locally, but it is no more an answer to the total problem than is water from streams. In the same class so far as a permanent solution is concerned are conservation measures to reduce waste of the pumped water: these will help to prolong the life of the ground-water supply, but no measures that are practical under present conditions can prolong it indefinitely at the current rate of withdrawal. On the other hand, to leave the water in the ground means losing the wealth which it, like any other minable resource, is capable of creating. How the South Plains will adjust economically to the depletion of the ground-water resource has yet to be determined, but that a problem of the first magnitude is involved is obvious.[2]

The case of the South Plains is exceptional in the rate of increase in withdrawal and in the dramatic contrast between withdrawal and replenishment, but as a problem of shortage of water to meet future demands it is anything but unique. Many ground-water basins in the arid Southwest are currently being overdrawn, even though they may be replenished at rates that are equivalent to a very substantial part of the withdrawal. Similar problems of at least local shortage exist in many of the heavily pumped urban areas of the East. In both East and West, many problems of shortage in quantity are created or complicated by the threat of encroachment of saline water due to pumping—and this in such inland areas as the Tularosa Basin—Hueco. Bolson area of New Mexico and Texas just as truly as in the heavily pumped coastal areas of California, Texas, Florida, and Long Island, New York. [See Reading 32.]

The Basic Problem of the Future

Perhaps we have made our point that many serious problems of ground-water availability and quality exist in the United States. There are even more important problems, however—those involved in finding ways to make ground water meet its full share of the responsibility for satisfying our water demands of the future. These latter problems can be characterized briefly as follows:

1. Evaluating aquifer systems.
2. Devising methods of water management that are compatible with both hydrologic and social realities, or can be made so by means that can reasonably be achieved.

The first task is that of the geohydrologist—that geologic and hydrologic expert who represents a synthesis of the ground-water geologist, the chemist, the physicist, the mathematician, the meteorologist, and the engineering scientist. The task, in a word, is to describe specific aquifer systems in terms of their geologic and hydrologic boundaries and of their response to external forces, as a basis for predicting reliably their hydrologic behavior under any condition that may be assumed to be likely in the future.

Quantitative ground-water hydrology has come a long way in recent decades in Europe, in America, and elsewhere. Aquifer-test methods for non-steady-state conditions became a reality in the United States with the work of C. V. Theis (1935). His basic equation for an infinite and isotropic aquifer and a steady pumping rate has been modified by many later workers to provide useful means for handling aquifers, and problems, whose departures from the stringent assumptions of the original Theis equation can be assigned specific values.

But all this is not enough. As every geologist knows, variations in the lithology and dimensions of permeable rocks, as of other rocks, are infinitely complex. It is easy to show that problems can be set up in which difficulties introduced by even modest variations in permeability and storage coefficient, in dimensions, and in boundary conditions of an aquifer are beyond the ability of even the most refined mathematics to solve in a reasonable time. The answer to this seemingly unscalable obstacle is the analog model, of which the electrical model appears to be more promising than the hydraulic or heat-flow model. Set up on the basis of geologic and hydraulic information obtained by conventional means, the model will show the response of complex aquifer systems with a reliability that depends solely on the amount and accuracy of data fed into it. And herein is our problem as ground-water geologists and hydrologists. The more complex problems need a lot of data, and data cost money. Yet, decisions are going to have to be made, and they will be made either on the basis of scientific data or on the basis of social or political judgment which may or may not prove to be sound.

The problem of getting enough data for aquifer-system description is one that ground-water geologists and other water-oriented scientists can get their teeth into, but they will not find it easy to solve. That geologic complexities will make aquifer-system analysis difficult and costly is the most important of what might be called the natural reasons why ground water is a mixed blessing. Other problems are man-caused, or at least strongly man-influenced. These are economic, legal, psychological—what might be called social problems. Let us mention the most important ones briefly.

Perhaps one of the most urgent problems is the need for wide public recognition of the extent to which ground water and surface water are interconnected and interdependent, and the extent to which it will be necessary to develop them on a coordinated, integrated basis to meet the maximum water demands of the future. In such developments ground water will have a most vital part to play, because ground-water reservoirs comprise by far the greatest part of the total storage facilities available for evening out the effects of fluctuations in replenishment and in demand.

Evaluating their storage capacity and devising means for getting water into and out of them at the proper times are a part of the task of the geohydrologist and the water-management engineer.

This enormous storage capacity of the ground-water reservoirs, which is potentially so important to the water management of the future, is responsible for some of today's problems, however. Where unplanned, unrestricted development has taken place, it has tended to encourage withdrawal at a rate that cannot be sustained indefinitely except by means, such as artificial recharge, which cost money that the initial developers never contemplated spending. Still, this can be considered only one phase of the difficulties involved in achieving public understanding of the procedures necessary for optimum water management.

The physical problems of water management, even when all the necessary data are at hand, are staggering. In effect, they will require setting up a second analog model, into which can be fed the data on ground water and on the streams, which, so far as the aquifers are concerned, are a part of the boundary conditions determining aquifer-system response. To these geohydrologic data must be added information on expected places and amounts of water withdrawal and disposal. This information, like all the rest, obviously is only as dependable as the means by which it is generated, but nevertheless it must be given numbers.

But suppose we have all the hydrologic data we need, including reliable predictions of future needs, and the master model or computer by means of which the procedures necessary for efficient water management can be indicated readily and clearly. Let us make another very optimistic assumption and say that we have found economical means of performing each of the operational steps of management. Have we won the battle? Not yet. We still have the social problems resulting from the facts that aquifers and streams cross jurisdictional lines, and that water laws in adjacent jurisdictions may be based on conflicting legal principles and may incorporate incorrect hydrologic concepts. We might point out that aquifers and river basins often do not coincide, but this is a physical rather than a social problem and can be handled scientifically if enough data can be obtained.

So here is the crux of our problem: It is the need for vastly accelerated scientific studies of the geology of water, and for extensive interpenetration of the knowledge and the thinking of all concerned with water—geologists and other scientists, water-management authorities, lawyers and legislators, and the general public. The problems of ground water in particular and of water in general are many and serious and will grow more numerous and more serious before we begin to get them under control. But the very fact that we can recognize and analyze them signifies that we have the intellectual means for solving them, and we will have only ourselves to blame if we fail to do so.

Notes

[1]In 1970 the total estimated withdrawal of fresh ground water in the continental United States averaged about 67.8 billion U. S. gallons per day (C. R. Murray and E. B. Reeves, 1972, Estimated

Use of Ground Water in the United States, 1970: *U. S. Geol. Survey Circ. 676*). This was a little more than a fifth of the total of 320 bgd of fresh water withdrawn from all sources for uses other than hydropower generation. The 1970 figure for ground water represented better than a 67 per cent increase over the comparable figure of 45.7 bgd for 1955. The largest single use of ground water continued to be for irrigation—45 bgd in 1970. Industrial use of fresh ground water, including fuel-electric power generation averaged 9.4 bgd in 1970. Public-supply use of ground water was 9.4 bgd in 1970 while rural use of ground water was 3.6 bgd in 1970. These two uses record a doubling of the 1955 figures. [Ed.]

[2]Pumping, from 1953–1961, averaged 5 million acre-feet per year. Pumping in 1973 was 4.1 million acre-feet, and the estimated supply at that time was 176 million acre-feet. By the year 2015, pumping is predicted to decrease to 95,000 acre-feet per year, and less than 2½ per cent of the supply of 1973 will exist. Precipitous declines in agricultural production are forecast by 1990 and it is predicted that irrigated acreage will decline from 4 million acres at present to 125,000 acres by 2015. (*Water Policies for the Future*, Final Report to the President and to the Congress of the U. S. by the Nat'l Water Comm., U. S. Govt. Printing Office, Wash., D. C. June, 1973, 239 pp.) [Ed.]

References

MacKichan, K. A. (1951): Estimated use of water in the United States, 1950: *U. S. Geol. Survey Circ. 115*, 13 p.

MacKichan, K. A. (1957): Estimated use of water in the United States, 1955: *U. S. Geol. Survey Circ. 398*, 18 p.

Meinzer, O. E. (1923): Outline of ground-water hydrology, with definitions: *U. S. Geol. Survey Water-Supply Paper 494*, 71 p.

Theis, C. V. (1935): The relation between the lowering of the piezometric surface and the rate and duration of discharge of a well using ground-water storage: *Am. Geophys. Union Trans.*, p. 519–524.

Thomas, H. E. (1951): The conservation of ground water: New York, McGraw-Hill, 327 p.

Thomas, H. E. (1952): Ground water regions of the United States—their storage facilities; The Physical and Economic Foundation of Natural Resources, v. 3: *U. S. Cong., Interior and Insular Affairs Comm.*, Washington, U. S. Govt. Printing Office, 78 p.

32   The Changing Pattern of Ground-Water Development on Long Island, New York

R. C. Heath, B. L. Foxworthy, and Philip Cohen

Introduction

Even before the severe drought that is now (1965) affecting the Northeastern United States, Long Island was well known among water specialists for its

Heath, R. C., Foxworthy, B. L., and Cohen, Philip, 1966, "The Changing Pattern of Ground-Water Development on Long Island, New York," *U. S. Geol. Survey Circ. 524*, 10 pp.

The authors are on the staff of the United States Geological Survey.

underground-water resource, mainly as a result of both the magnitude of the ground-water resource and the unique aspects of man's utilization of that resource. The current drought has focused increased attention upon the vast amount of ground water in storage on Long Island and upon the large quantity of water being pumped from the system. In 1963, for example, an average of about 380 mgd (million gallons per day) was pumped from Long Island wells; these wells tap a fresh ground-water reservoir that has an estimated storage capacity of 10 to 20 trillion gallons. Nearly all the water pumped was for domestic and industrial use, and this pumpage probably represents one of the largest such uses of a single well-defined ground-water reservoir anywhere in the world.

The history of ground-water development on Long Island has been thoroughly documented, largely as a result of studies made by the U. S. Geological Survey in cooperation with the New York State Water Resources Commission and Nassau and Suffolk Counties. The water development has followed a general pattern which, although somewhat related to population density and local waste-disposal practices, has been controlled largely by the response of the hydrologic system to stresses that man has imposed upon the system. The purpose of this report is to summarize the highlights of the historical pattern of ground-water development on Long Island and to consider briefly the insight that the history of development affords regarding the future development and conservation of Long Island's most valuable natural resource.

Geologic Environment

Long Island (Fig. 1) has a land area of about 1,400 square miles and is geographically a large detached segment of the Atlantic Coastal Plain. The island is underlain by crystalline bedrock, the uppermost surface of which ranges in altitude from about sea level at the northwest corner of the island to about 2,000 feet below sea level in the southeastern part of Suffolk County (Fig. 2).

The bedrock is overlain by a wedge-shaped mass of unconsolidated sedimentary deposits that attain a maximum thickness of about 2,000 feet.

FIG. 1. Long Island and vicinity.

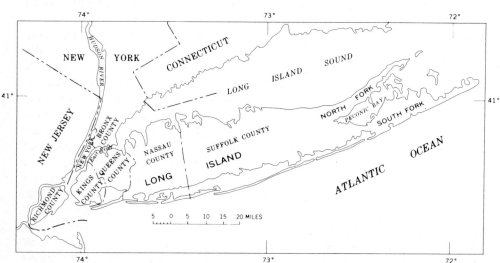

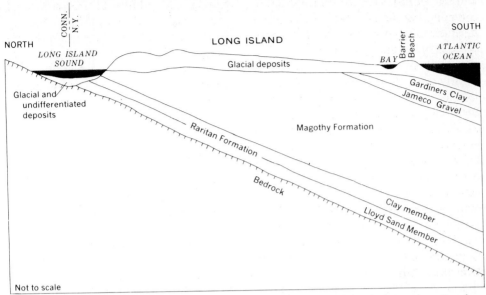

FIG. 2. Diagrammatic section showing general relationships of the major rock units of the ground-water reservoir in Nassau County.

These deposits constitute the ground-water reservoir of Long Island and can be divided into six major stratigraphic units, which differ in their geologic ages, mineral composition, and hydraulic properties. These units are, from oldest to youngest, (1) Lloyd Sand Member of the Raritan Formation, (2) clay member of the Raritan Formation, (3) Magothy Formation, (4) Jameco Gravel, (5) Gardiners Clay, and (6) glacial deposits. (Suter and others, 1949). The first three units listed are of Cretaceous age, and the last three are of Pleistocene age.

The Lloyd Sand Member of the Raritan Formation has a maximum thickness of about 300 feet and consists mainly of fine to coarse sand and some gravel and interbedded clay. It forms the basal water-bearing unit of the ground-water reservoir. The clay member of the Raritan Formation is

composed mainly of clay but locally contains considerable sand; it also has a maximum thickness of about 300 feet. Hydraulically, the clay member is a leaky confining layer for the Lloyd Sand Member—retarding, but not preventing, vertical leakage of water to and from the Lloyd.

The Magothy Formation on Long Island is partly correlative with the Magothy formation in New Jersey. It consists of complexly interbedded layers of sand, silt, and clay and some gravel in the lower part. The complexity of the interbedding and the character of fossils it contains suggest that the formation was mainly laid down under continental (flood-plain) conditions. The Magothy Formation is the thickest unit of the ground-water reservoir on Long Island, attaining a maximum thickness of about 1,000 feet. Its horizontal permeability differs widely from place to place and is considerably higher than its vertical permeability. It commonly yields more than 1,000 gpm (gallons per minute) per well. Water in the formation is largely under artesian conditions.

Near the north and south shores of the island, the Magothy Formation locally is overlain by the Jameco Gravel. The maximum thickness of the Jameco is about 200 feet. It consists mainly of medium to coarse sand, but locally contains abundant gravel and some silt and clay. The Jameco Gravel is moderately to highly permeable and yields as much as 1,500 gpm per well. Water in the formation occurs under artesian conditions.

The Gardiners Clay is mainly restricted in extent to two moderately narrow bands that parallel the north and south shores, and it is commonly underlain by either the Jameco Gravel or the Magothy Formation.

The surface of Long Island is composed mostly of material deposited either directly by Pleistocene continental ice sheets or by melt water derived from the ice sheets. These glacial deposits consist mainly of sand and gravel outwash in the central and southern parts of the island, and mixed till and outwash atop and between the hills in the northern part of the island. The glacial outwash deposits are highly permeable and therefore permit moderately rapid infiltration of precipitation.

Hydrologic System

The four major water-bearing units of the ground-water reservoir of Long Island are the glacial deposits, Jameco Gravel, Magothy Formation, and Lloyd Sand Member of the Raritan Formation (Fig. 2). These four units contain mostly fresh ground water; however, locally they contain salty ground water or they are hydraulically connected with salty water of the ocean, sound, or bays. Under natural conditions recharge to the ground-water reservoir resulted entirely from the infiltration of precipitation, which is estimated to have averaged roughly 1 mgd per square mile (Swarzenski, 1963, p. 35). Most of the ground water moved laterally through the glacial deposits and discharged into streams or into bodies of salt water bordering the island without first reaching deeper waterbearing zones. Most of the remainder of the ground water moved downward through the glacial deposits into the Jameco Gravel or Magothy Formation, and from there part flowed laterally to the ocean and the remainder flowed downward through the clay member of the Raritan Formation into the Lloyd Sand Member. (See Fig. 4).

Estimates of ground-water discharge under natural conditions can be developed by extrapolation of data listed by Pluhowski and Kantrowitz (1964, p. 38–55) for the Babylon-Islip area, a large and reasonably representative part of Long Island. Those data suggest that about 90 percent of the total recharge ultimately discharged from the glacial deposits (mainly by seepage to streams), and about 10 percent discharged by subsurface outflow from the Magothy Formation, the Jameco Gravel, and the Lloyd Sand.

The water table on Long Island (Fig. 3) and also the piezometric (pressure) surfaces of the underlying artesian

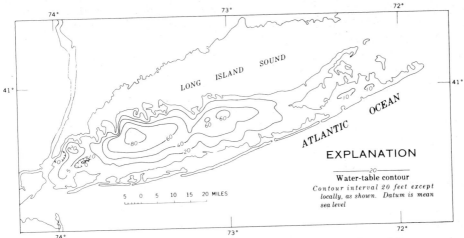

FIG. 3. Generalized contours on the water table (the upper surface of the ground-water reservoir) in 1961.

aquifers (which have about the same general shape as the water table) form elongate mounds following roughly the configuration of the land surface. Two prominent highs characterize the water table—one centered in Nassau County and one centered in Suffolk County. Northwestern Queens County also has a small high in the water table. Other notable features are the cones of depression that extend below sea level in Kings and Queens Counties; these cones are in areas of past or current local overdevelopment of ground water.

Changes in Ground-Water Development with Time

Phase 1—Predevelopment Conditions

Ground-water development on Long Island has progressed and is progressing through several distinct phases. Under natural or predevelopment conditions (Fig. 4), the hydrologic system was in overall equilibrium and long-term average ground-water recharge and discharge were equal. The general positions of the subsurface interfaces between fresh and salty water in each of the previously described geologic units were stable, reflecting the overall hydrologic balance. The interfaces were virtually at the coasts in the glacial deposits and were off-shore in the underlying units.

Phase 2

In the initial stage of development (Fig. 5), which began with the arrival of the first European settlers, virtually every house had a shallow well drawing water from the glacial deposits and a cesspool returning waste water to the same deposits. As the population increased, individual wells were abandoned and public-supply wells were installed in the glacial deposits. The individual cesspools, however, were retained and little water was lost from

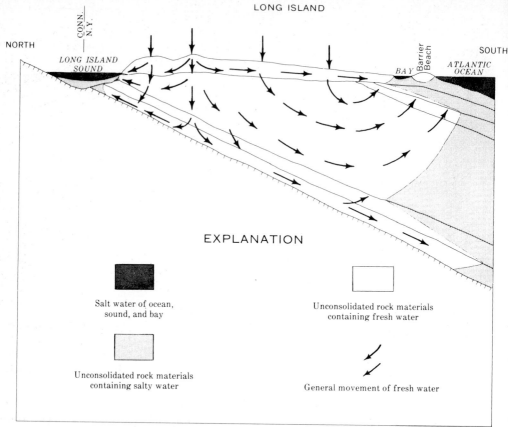

FIG. 4. Diagrammatic section showing predevelopment (phase 1) generalized ground-water conditions. Contacts between rock units are as shown in Fig. 2.

the system during use. Although a considerable amount of ground water was being withdrawn, practically all of it was returned to the same aquifer from which it was removed. In general, therefore, the system remained in balance, and the positions of the interfaces between fresh and salt water remained practically unchanged. However, this cycle of ground-water development and waste-water disposal resulted in the pollution of the shallow ground water in the vicinity of the cesspools.

Phase 3

In time, as the cesspool pollution spread, some shallow public-supply wells had to be abandoned and these were replaced with deeper public-supply wells, most of which tapped the Jameco Gravel and the Magothy Formation. Supply wells were also constructed in the deeper units at places where the glacial deposits contained water with objectionable amounts of dissolved iron or other troublesome natural constituents. Most of the water

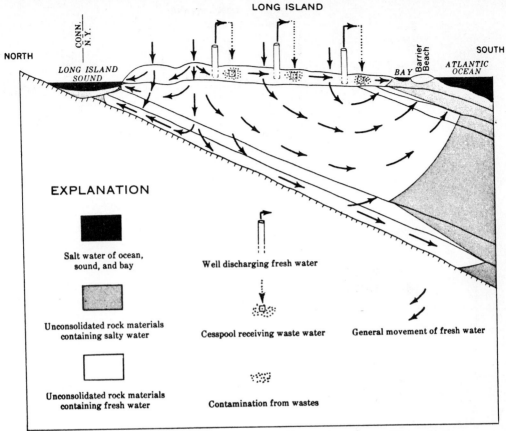

FIG. 5. Diagrammatic section showing generalized ground-water conditions during phase 2 of ground-water development (shallow supply wells and waste disposal through cesspools). Contacts between rock units are shown in Fig. 2.

withdrawn from the deeper units was returned to the shallower glacial deposits by means of cesspools, and subsequently discharged to the sea by subsurface outflow or by seepage to streams (Fig. 6).

As a result of the withdrawal of water from the Magothy Formation and the Jameco Gravel, and the concurrent decrease in hydraulic heads in these units, the downward movement of ground water from the overlying glacial deposits locally was increased. However, the increased downward

movement only partially compensated for the withdrawals of water from the Magothy and Jameco deposits. Locally, a hydraulic imbalance developed in the Magothy and Jameco deposits and caused a decrease in the amount of fresh ground water in storage and a landward movement of salty water.

Phase 4

The next major phase in the development of ground water on Long Island (Fig. 7) was the introduction of large-scale sewer systems—notably in that

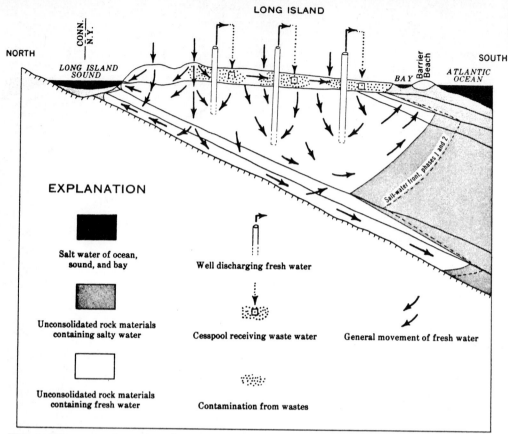

LONG ISLAND

NORTH

CONN.
N.Y.

LONG ISLAND
SOUND

SOUTH

BAY

Barrier
Beach

ATLANTIC
OCEAN

Salt-water front, phases 1 and 2

EXPLANATION

Salt water of ocean,
sound, and bay

Well discharging fresh water

Unconsolidated rock materials
containing salty water

Cesspool receiving waste water

General movement of fresh water

Unconsolidated rock materials
containing fresh water

Contamination from wastes

FIG. 6. Diagrammatic section showing generalized ground-water conditions during phase 3 of ground-water development (deep supply wells and waste disposal through cesspools). Contacts between rock units are as shown in Fig. 2.

portion of Long Island that is part of New York City (Kings and Queens Counties). Most of the pumped ground water that previously had been returned to the ground-water reservoir by means of cesspools was thereafter discharged to the sea through the sewers. Whereas the net draft on the ground-water system during the preceding phases of development was negligible, virtually all the ground water diverted to sewers during phase 4 represented a permanent loss from the system. The newly im-

posed stress on the ground-water system locally resulted in a rapid landward encroachment of salty water into the previously fresh ground-water reservoir. The most dramatic example occurred during the 1930's in Kings County (the Borough of Brooklyn), which by that time had been completely sewered for many years. In 1936, decreased natural recharge owing to urbanization and increased ground-water withdrawals, which during the previous few years averaged more than 75 mgd, caused

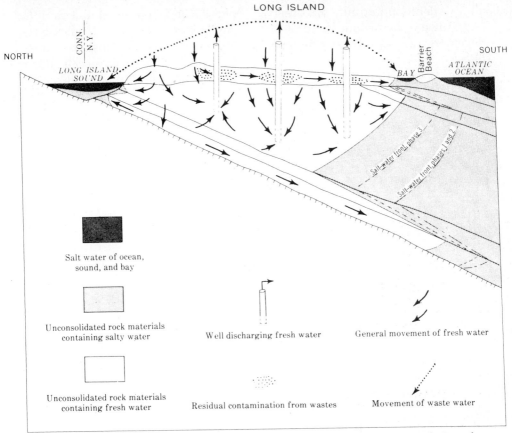

NORTH

CONN.
N.Y.

LONG ISLAND
SOUND

LONG ISLAND

SOUTH

Barrier
Beach

BAY

ATLANTIC
OCEAN

Salt-water front phase 3

Salt-water front phases 1 and 2

Salt water of ocean,
sound, and bay

Unconsolidated rock materials
containing salty water

Well discharging fresh water

General movement of fresh water

Unconsolidated rock materials
containing fresh water

Residual contamination from wastes

Movement of waste water

FIG. 7. Diagrammatic section showing generalized ground-water conditions during phase 4 of ground-water development (deep supply wells and waste disposal through sewers to adjacent salt-water bodies). Contacts between rock units are as shown in Fig. 2.

ground-water levels in Brooklyn locally to decline to as much as 35 feet below sea level (Lusczynski, 1952, pls. 1 and 2). This local overdevelopment caused contamination of large parts of the ground-water reservoir in that area from sea-water encroachment.

In 1947 virtually all pumping for public supply in Kings County was discontinued and the Borough was thereafter supplied with water from the New York City municipal-supply system, which utilizes surface-water reservoirs in upstate New York. A

notable exception was ground-water withdrawal for air-conditioning use. Such usage was permitted, however, only under the condition that the water was returned to the ground-water reservoir by means of injection wells (locally referred to as "diffusion" wells).

Present Areal Differences in Ground-Water Development

The present pattern of ground-water development on Long Island affords an excellent opportunity to observe and evaluate the historic trend of that development, because all the major phases of development described herein, except the predevelopment phase, can be observed now in different subareas of the island (Fig. 8). Moreover, once the transitory status of present development in each subarea is recognized in relation to the pattern

FIG. 8. Water-development subareas in 1965.

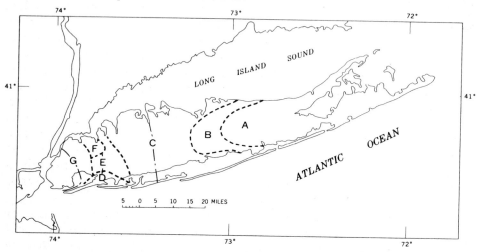

EXPLANATION

Subarea   *Characteristics*

A   Phase 2 of development. Pumpage mainly from shallow privately owned wells. Waste water returned to shallow glacial deposits through cesspools; local contamination of glacial deposits by cesspool effluent. System virtually in balance; positions of salt-water fronts unchanged.

B   Transition between phase 2 and 3. Pumpage from privately owned and public-supply wells. Waste water returned to shallow glacial deposits by way of cesspools; areas of cesspool-effluent contamination spreading. System virtually in balance.

C   Phase 3 of development. Pumpage mainly from deep public-supply wells; waste water returned to shallow glacial deposits by way of cesspools. System locally out of balance, causing local salt-water intrusion.

D   Phase 4 of development. Pumpage almost entirely from deep public-supply wells; waste water discharged to the sea by way of sewers. System out of balance; salty water actively moving landward.

E   Phase 4 of development. Pumpage almost entirely from deep public-supply wells; waste water discharged to the sea by way of sewers. System out of balance; may be subject to salt-water intrusion in the future.

F   Very little ground-water development. Water supply derived from New York City municipal-supply system; waste water discharged to the sea by way of sewers. System in balance.

G   Very little ground-water development. Water supply derived from New York City municipal-supply system; waste water discharged to the sea by way of sewers. Large areas contain salty ground water owing to former intensive ground-water development and related salt-water intrusion.

of historical trends, it becomes possible to predict and perhaps forestall some of the undesirable aspects of those trends.

Subarea A (Fig. 8) includes roughly the eastern two-thirds of Suffolk County. Except for several small communities, the subarea is largely rural and has the lowest population density on Long Island. On the whole, the subarea can be characterized as being in phase 2 of ground-water development (Fig. 5)—that is, most of the wells in the subarea tap the shallow glacial deposits and supply water to single-family dwellings. The bulk of this water is returned to the glacial deposits through individually owned cesspools, and in overall aspect the ground-water system is still in hydraulic balance.

Subarea B, in central Suffolk County, is experiencing the impact of the suburban expansion associated with the entire New York City metropolitan area. Farms and woodlands are giving way to housing developments, and most of the pumpage in the subarea is now from large-capacity public-supply wells that tap the glacial deposits. However, most of the sewage disposal is still through individually owned cesspools. Thus, the area is in a transition between phase 2 and phase 3 of development. Cesspool pollution still is not widespread, but is substantial enough to be of concern to local government agencies. Accordingly, plans are currently (1965) being made to construct sewers in the area and to gradually replace the wells that tap the glacial deposits with wells that will tap the Magothy Formation.

Subarea C includes the westernmost part of Suffolk County and the eastern two-thirds of Nassau County. Mainly because it is closer to New York City, this subarea was subjected to intensive suburban development earlier than was subarea B. Therefore, the population density and, accordingly, the water requirements in subarea C are substantially greater than in subarea B. Virtually the entire water supply for subarea C is obtained from large-capacity public-supply wells. The part of the subarea that is in western Suffolk County obtains most of its water supply from public-supply wells, of which about half tap the glacial deposits and most of the remainder tap the Magothy Formation. In the part of the subarea that is in Nassau County, most of the public-supply wells tap the Magothy Formation.

Except for a few communities along the coast, most of subarea C is not sewered; practically all the domestic sewage is disposed of through individually owned cesspools. Thus, on the whole the subarea is in phase 3 of development (Fig. 6). The system locally is out of balance owing to this development; however, substantial widespread salt-water encroachment has not yet occurred. Plans are being made to install sewers throughout the subarea.

Subareas D and E, which include parts of western Nassau and southeastern Queens Counties, are moderately to highly urbanized and are almost completely sewered. Practically the entire water supply for these subareas is derived from wells tapping the Magothy Formation, Jameco Gravel, and the Lloyd Sand Member of the Raritan Formation. Thus, these subareas are mainly in phase 4 of development and are characterized by a hydrologic imbalance (Fig. 7). The imbalance, which is accentuated because more than 70 mgd of water derived from the ground-water reservoir of these subareas currently is being discharged to the sea by way of sewage-treatment plants, is most clearly manifested in subarea D, where salty water is moving landward (Lusczynski and Swarzenski, 1960; Perlmutter and Geraghty, 1963). If the present trend continues, subarea D (the area of active salt-water encroachment) probably will expand at the expense of subarea E.

Subarea F, in northeastern Queens County, receives nearly its entire water supply from the New York City municipal-supply system. The subarea is sewered; however, because ground-water pumpage is negligible, the ground-water system is largely in balance.

Subarea G is the most highly urbanized and receives virtually all its water from the New York City municipal system. The entire subarea is sewered. As previously noted, large areas in Kings County were invaded by salty water because of substantial overdevelopment and the resulting decline in ground-water levels. Similarly, salty water had invaded the ground-water reservoir in parts of western Queens County. Water levels in Kings County have recovered appreciably since the mid 1940's, when the consumptive ground-water uses were drastically reduced. Presumably, the salty water is retreating seaward and is being diluted by recharge derived from precipitation, but precise data regarding these changes are lacking.

Conclusion

Ground water probably will continue to be the major source of water for most of Long Island (except for Kings and Queens Counties) for at least the next several decades. Moreover, if the present trends continue, the ground-water resources of the island probably will continue to be depleted—perhaps at an accelerated rate. The historic trends of ground-water development and the present status of development strongly suggest that such depletion will in time cause salt-water contamination of larger and larger parts of the ground-water reservoir. Moreover, the areas in which such contamination occurs, in addition to extending inward from the coasts, probably will also extend farther and farther eastward as the population continues to expand in that direction.

Several alternative methods of conserving and augmenting the ground-water resources of Long Island are currently being considered. These include, among others, desalting of sea water with the use of atomic energy, artificial recharge, and the reclamation of water from sewage. The consequences of such possible measures are highly significant inasmuch as the future well-being of several million people is at stake. However, even with the most promising of conservation methods, wise management will be required to gain the fullest use from the available fresh-water supply while also preventing undue hardships resulting from local overdevelopment of the ground-water reservoir. Fully effective management requires:

1. Recognition of the unity of the hydrologic system of Long Island.

2. The best obtainable scientific information about the system and how it functions.

3. Sound evaluation of the various alternative methods of water development and conservation, guided by available scientific information—including the hydrologic consequences of the historic and present-day changing pattern of ground-water development on Long Island.

References

Lusczynski, N. J., 1952. The recovery of ground-water levels in Brooklyn, New York from 1947 to 1950: U. S. Geol. Survey Circ. 167, 29 p.

Lusczynski, N. J, and Swarzenski, W. V., 1960, Position of the salt-water body in the Mago-

thy(?) Formation in the Cedarhurst-Woodmere area of southwestern Nassau County: N. Y. Econ. Geology, v. 55, no. 8, p. 1739–1750.

Perlmutter, N. M., and Geraghty, J. J., 1963, Geology and ground-water conditions in southern Nassau and southeastern Queens Counties, Long Island, New York: U. S. Geol. Survey Water-Supply Paper 1613-A, 205 p.

Pluhowski, E. J., and Kantrowitz, I. H., 1964, Hydrology of the Babylon-Islip area, Suffolk County, Long Island, New York: U. S. Geol. Survey Water-Supply Paper 1768, 119 p.

Suter, Russell, de Laguna, Wallace, and Perlmutter, N. M., 1949, Mapping of geologic formations and aquifers on Long Island, New York: New York State Power and Control Comm. Bull. GW-18, 212 p.

Swarzenski, W. V., 1963, Hydrology of northwestern Nassau and northeastern Queens Counties, Long Island, New York: U. S. Geol. Survey Water-Supply Paper 1657, 90 p.

33 Geological Influence on Regional Health Problems

Jean M. Spencer

Introduction

Medical geology, regional pathology, and geographical disease all describe the relationship between health and geology. In the past, certain localities were regarded with suspicion as areas with high percentages of particular types of disease. Current investigation into many of these suspicions demonstrates a decided link between various

Reprinted from *The Texas Journal of Science*, v. 21, p. 459–469 (1970) with permission of the author and the Texas Academy of Science.

Ms. Spencer is on the faculty at Baylor University.

chronic diseases and particular geological environments.

Iodine Deficiency

Perhaps the best known relationship between a disease and geology is that of thyroid disease caused by an environmental iodine deficiency. Iodine is essential to normal function of the thyroid gland which is located at the base of the neck. Iodine is also present in tissue and blood. Its deficiency causes goiter or enlargement of the thyroid in many persons living in iodine-deficient areas. The term "Derbyshire neck" reflects the frequency of goiter cases in Derbyshire, England.

If a pregnant woman has a severe iodine deficiency, her child may be born a cretin, as were common at one time in high goiter-rate areas of Switzerland and Mexico. A cretin is an idiotic dwarf, small in stature, who rarely reaches or exceeds a mental age of 10.

Much of the northern United States, especially the Great Lakes region and the northwestern states, has iodine-deficient soils and water, reflected in the high goiter rate before the use of iodized salt. Similar relationships exist throughout the world as in Central Mexico, Great Britain, Nepal, and Switzerland.

The National Nutrition Survey found goiter incidence increasing in the United States because iodized salt is not used as widely as before World War II. In some areas of the United States goiter is still endemic, particularly among the poor (Schaefer and Johnson, 1969).

Iodine also appears to play an important role in the incidence of female breast cancer. In the United States breast cancer is the leading cause of death for women between 40 and 44, and the leading cause of death for women dying from all types of cancer between the ages of 35 and 55. The death rate from female breast cancer is higher in iodine-deficient areas than in other areas (Fig. 1). This relationship is

FIG. 1. Distribution of death rate in the United States for female breast cancer. *After* Bogardus and Finley, 1961—Breast cancer and thyroid disease. *Surgery,* 49 (4): 464.

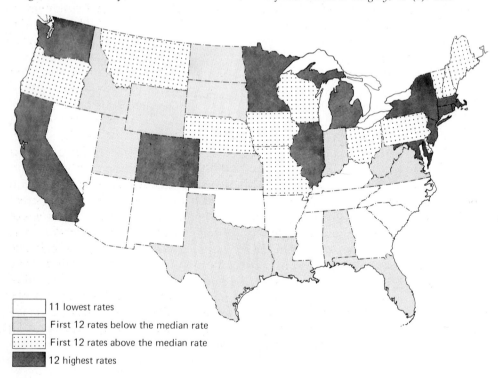

☐ 11 lowest rates

▨ First 12 rates below the median rate

⬚ First 12 rates above the median rate

■ 12 highest rates

observed in other countries as well as in the United States. Countries with high incidences of goiter and breast cancer include Thailand, Mexico, England, Wales, the Netherlands, and Switzerland. Countries with low incidences of goiter and breast cancer include Japan, Chile, and Iceland (Bosgardus and Finley, 1961).

Fluorine Deficiency

Fluorine, another environmental halogen, is the object of much investigation. This element is known to produce strong teeth mottled with brown in children raised where the natural fluorine concentration in the water supply is well above that recommended by the U. S. Public Health Service. This condition is common to some areas of west Texas. Where fluorine is deficient in the water supply (considerably below one part per million) children develop soft teeth that decay easily.

Not so well known perhaps, is the fact that the elderly also need fluorine to maintain good bone density. Bone is composed of the protein collagen, which acts as reinforcing material, and the mineral apatite, which gives strength to the bone. Calcium found in bone is present in apatite crystals.

Apatite, with a composition of $Ca_5(OH, F, Cl)(PO_4)_3$, accepts either fluorine or the hydroxyl radical into its crystal lattice in bone. When fluorine is present, apatite crystals are larger, more perfect, and the bone is more resistant to degeneration. More calcium remains in the bone, and density is maintained. When the hydroxyl radical substitutes for the fluoride ion, degeneration occurs more readily, and various types of bone fractures become more common.

Gradual loss of bone density in men and women over 40 can result in loss of as much as half the total skeletal mass by age 70. This loss of bone density leads to the condition known as osteoporosis. The bent-over little old ladies of the 6th decade and beyond, many with what is called a "dowager's hump," are suffering from collapse of the spinal vertebrae due to osteoporosis and subsequent breaking. The incidence of collapse and fracture doubles during each 5-year period after age 40 among women and after 50 among men.

Two regional studies comparing high and low fluoride areas show osteoporosis higher in the low fluoride areas and rare in high fluoride areas (Bernstein, et al., 1966). However, the amount of fluoride in drinking water needed to avoid osteoporosis in the elderly is high enough to cause some mottling of teeth in children.

While osteoporosis is more common among elderly women, the calcium released by loss of bone density frequently deposits in and lines the aortas of men, producing aortic calcification and subsequent narrowing of the aorta, the major artery leading from the heart. Adult men, in every age group, living in low fluoride areas have 2 to 3 times more aortic calcification than those living in high fluoride areas (Anon., 1967). Aortic calcification in turn, often produces aneurysms or dilation of the calcified blood vessels causing other cardiovascular problems.

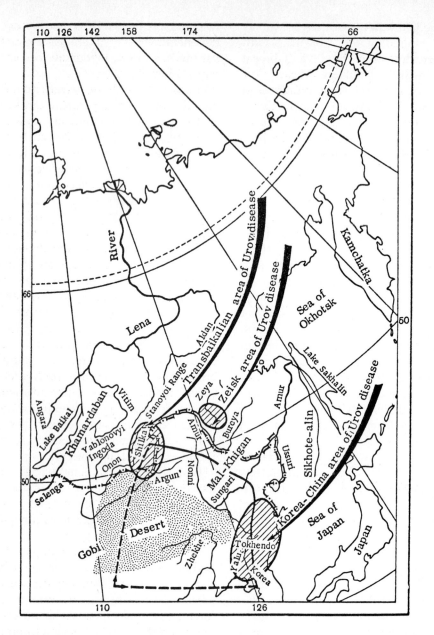

FIG. 2. Distribution of Urov endemic areas in the Far East. *From* Khobot'ev, 1960—
Bio-geochemical provinces with calcium deficiency. *Geochemistry*, 8: 831.

Calcium Deficiency

A disease produced by an environmental calcium deficiency is the Urov disease of the Eastern Transbaikal of the Soviet Union and northern China (Fig. 2). In these areas people and domestic animals gradually develop enlarged and stiffened joints early in life. Bone growth is arrested. In one village in the Amur province 44% of the population was afflicted. Most of these were children 11 to 15 years of age (Khobot'ev, 1960).

While various other elements (lead, cadmium, gold, iron, radium), organic matter, and iodine deficiency have been blamed for the disease, these causes have not been proved. The "sick" water, however, "is deficient in dissolved solids, especially calcium, while 'healthy' water contains more dissolved solids and is richer in calcium" (Khobot'ev, 1960). Where the disease is endemic, waters are not only calcium deficient but strontium and barium enriched.

The disease does not occur where there are limestone outcrops but is endemic to marine Jurassic deposits (sandstones and shales) and zones of igneous rocks including a granite unusually low in calcium (Khobot'ev, 1960).

Other Factors

In west Devonshire, England an uneven distribution of cancer deaths prompted a study in which cancer deaths for a 20-year period were plotted on a geologic map of the area. This study demonstrated an extraordinary distribution of cancer deaths for those persons living on certain geologic formations (notably Devonian sandstones). These death rates were not matched for persons living on adjacent areas (Allen-Price, 1960).

Many soils in the high cancer areas proved to have high concentrations of lead—100 ppm, with some samples containing more than 1000 ppm lead. Vegetables grown on these soils have high lead contents, none less than 10 ppm in the ash (Warren, 1963).

The development of highly sophisticated analytical techniques, such as the atomic absorption spectrophotometer, permits rapid microchemical studies of cells, soil, and water. This analytical progress facilitates investigation into the sources of trace elements and their actual role in physiological function.

To function properly, the body requires a number of trace elements in addition to the so-called bulk elements (those found in the body in high concentrations). Trace elements are usually, though not always, present in the body in parts per million. The amounts of trace elements required are small and pathological conditions can be produced either by a slight deficiency or a slight excess of a particular element, as in the case of selenium and fluorine.

Of the trace elements, manganese, iron, cobalt, copper, zinc, and molybdenum are most commonly associated with enzyme activity. Enzymes act as catalysts in biochemical reactions, and the poisoning effects of such elements as silver, mercury (+2), and lead may stem from the fact that they are enzyme-inhibitors when present in an enzyme in place of the desired element.

Diseases caused by trace elements were first known among miners.

Symptoms were caused by constant contact with large amounts of a particular element over a long period of time. Pitchblende miners in Austria acquired a condition called "Bergkrankheit"—mountain disease—from inhalation of the radio-active gas radon, which emits alpha particles. Mortality rate from lung cancer among these miners was 50 times that of the rest of the population (Alexander, 1959).

Elements can be inhaled, as in the case of the miners suffering from "Bergkrankheit," but normally they are ingested with food and water. The geology of an area in the form of its rocks and sediments, is the source of the elements that enter soil and water

and are made available to man and the plants and animals of his food supply.

However, the fact that an element is present in soil or rocks does not mean that it is necessarily available to plants or that it will dissolve in water. Its physical state and whether it is tightly bound in a mineral lattice or held to clay particles will determine its availability.

Deficiencies and poisonings in plants and domestic animals have been known and treated for many years. Many areas of the world with excesses or deficiencies of various trace elements are known. Figure 3 shows areas of animal diseases due to mineral deficiencies and poisonings in the United States.

FIG. 3. Mineral nutritional diseases in animals. *From* Beeson, 1957—Soil management and crop quality, in *Soil, The 1957 Yearbook of Agriculture.* U.S. Department of Agriculture, 259.

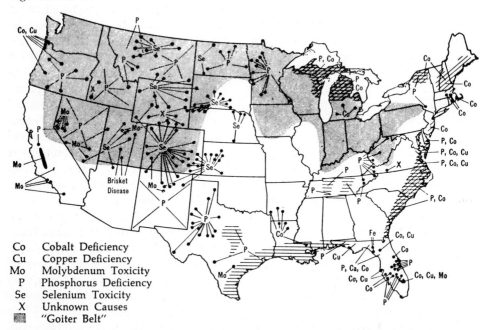

Co   Cobalt Deficiency
Cu   Copper Deficiency
Mo   Molybdenum Toxicity
P   Phosphorus Deficiency
Se   Selenium Toxicity
X   Unknown Causes
▨   "Goiter Belt"

I. Known areas in the United States where mineral-nutritional diseases of animals occur. The dots indicate approximate locations where troubles occur. The lines not terminating in dots indicate a generalized area or areas where specific locations have not been reported.

Because in the United States our food supplies come from many sources rather than local kitchen gardens, water is probably the major source of inorganic ions that reaches almost 100% of a resident population. It is therefore important to know what elements are present or lacking in a given water supply, the source of these elements, and their geochemistry and biochemistry. Normal municipal water treatment hardly alters the percentages of the various trace elements in the water distributed to consumers.

Studies into the physiological effects of hard and soft municipal water demonstrate a relationship between soft water and high cardiovascular death rate for those of middle and old age. None of the water supplies studied contained any elements considered toxic or in dangerous concentrations, and no element was found to be particularly significant as the cause of cardiovascular disease (Crawford, 1968). Crawford, working in England and Wales, found concentrations of manganese and aluminum higher in soft water-high cardiovascular areas and concentrations of boron, iodine, fluorine, and silica higher in hard water-low cardiovascular areas. In the United States similar studies also demonstrate that the softer the water the higher the death rate from cardiovascular disease (Schroeder, 1966).

In certain areas of Japan the high silica content of drinking water is thought to contribute to brain damage and arteriosclerosis in these areas (Henschen, 1966).

A chronic, slowly progressive kidney disease related to river areas is under investigation in Yugoslavia.

Chronic endemic nephropathy, or Balkan nephritis, gradually reduces kidney function of people living along the Sava, Drina, and Kalabara rivers of Yugoslavia and in very localized areas of Bulgaria and Romania (Fig. 4). It has not been found in any other areas of these countries. Hereditary and organic causes essentially have been ruled out. The disease is most common among those who live in the river valleys but not among those who live in nearby foothills. The farther a person lives from the river, the less severe the disease. Medical studies are attempting to implicate lead, uranium, cadmium, and nickel, but with no success (Griggs and Hall). The cause of this ultimately fatal disease is still unknown.

Biological data are being gathered on the effects of many other trace elements in the human body. Little, if any, is known about the sources of many of these elements and the regional aspects of disease with respect to their abundance or deficiency in the environment.

Zinc is an important constituent of many enzymes. It appears to concentrate in parts of the body where there is an increase in cell growth, such as in healing wounds and burns. Given in large amounts, it speeds wound healing (Pories, 1967).

Zinc-deficient diets in pregnant rats produce such birth defects as cleft lip, missing digits, and clubbed feet. These congenital malformations are also found in human infants, although no relationship to zinc has been established (Hurley, 1968).

Zinc-deprived humans have retarded longitudinal growth and physical maturing. In a group of zinc-

FIG. 4. Outline map of Yugoslavia and surrounding countries. Dotted lines indicate areas of chronic endemic nephropathy. *From* Griggs and Hall—Investigation of chronic endemic nephropathy in Yugoslavia, in *15th Annual Conference on the Kidney Proc.* National Kidney Foundation, 313.

deprived Egyptian boys who were also heavily infected with parasites and suffered from general malnutrition, only the institution of zinc therapy produced growth and physical maturing (Sandstead, *et al.*, 1968).

The presence of zinc in thermal springs in Germany has been implicated as an important factor in the sexual rejuvenation properties attributed to many of these springs.

Cadmium is apparently an antimetabolite for zinc. An elevated renal cadmium-to-zinc ratio often appears in persons dying of hypertensive disease. Cadmium is not present in the body at birth but increases in concentration with age. An increase in cadmium parallels an increase in hypertension (Schroeder, 1965). Coffee and tea are sources of cadmium, but what about its natural sources in soil and water?

These are only a few of the relationships, understood and unclear, between geology and health. Medical personnel are aware of many deviations from "normal" metabolism by excesses or deficiencies of trace elements. They are not, however, trained to recognize sources and geologic relationships of these elements. It is, therefore, the responsibility of those who are trained in the study of the earth to contribute their knowledge to this field that is important to all of us—health.

Literature Cited

Alexander, P., 1959—*Atomic Radiation and Life.* Penguin Books, Harmondsworth, Middlesex, England.

Allen-Price, E.D., 1960—Uneven distribution of cancer in west Devon. *Lancet,* June 1960: 1235–1238.

Anonymous, 1967—Flouride: Deficiency held a factor in osteoporosis. *Medical Tribune,* 8 (57): 12.

Beeson, K. C., 1957—Soil management and crop quality, in *Soil, The 1957 Yearbook of Agriculture.* U.S. Dept. of Agriculture, 258–267.

Bernstein, D. S., N. Sadowsky, D. M. Hegsted, C. D. Guri, and F. J. Stare, 1966—Prevalence of osteoporosis in high- and low-fluoride areas in North Dakota. *Jour. Amer. Med. Assoc.,* 198 (5): 499–504.

Bogardus, G. M., and J. W. Finley, 1961—Breast cancer and thyroid disease. *Surgery,* 49 (4): 461–-468.

Crawford, M. D., M. J. Gardner, and J. N. Morris, 1968—Mortality and hardness of local water-supplies. *Lancet,* April 1968: 827–831.

Griggs, R. C., and P. W. Hall, Investigation of chronic endemic nephropathy in Yugoslavia, in *15th Annual Conference on the Kidney Proc.* National Kidney Foundation: 312–328.

Henschen, F., 1966—*The History and Geography of Disease.* Delacorte Press, New York.

Hurley, L. S., 1968—The consequences of fetal impoverishment. *Nutrition Today,* 3 (4): 2–10.

Hurley, L. S., and H. Swenerton, 1966—Congenital malformations resulting from zinc deficiency in rats. *Soc. for Experimental Biology and Medicine Proc.,* 123: 692–696.

Khobot'ev, V. G., 1960—Biogeochemical provinces with calcium deficiency. *Geochemistry,* 8: 830–840.

Pories, W. J., 1967—Acceleration of healing with zinc sulfate. *Ann. Surg.,* 165: 432–436.

Sandstead, H. H., A. S. Prasad, A. R. Schulert, Z. Farid, A. Miale, Jr., S. Bassilly, and W. J. Darby, 1967—Human zinc deficiency, endrocrine manifestations and response to treatment. *Amer. Jour. Clinical Nutrition,* 20 (5): 422–442.

Schaefer, A. E., and O. C. Johnson, 1969—Are we well fed? The search for the answer. *Nutrition Today,* 4 (1): 2–11.

Schroeder, H. A., 1965—Cadmium as a factor in hypertension. *Jour. Chronic Dis.,* 18: 647–656.

———, 1966—Municipal drinking water and cardiovasular death rates. *Jour. Amer. Med. Assoc.,* 195 (2): 81–85.

Warren, H. V., 1963—Trace elements and epidemiology. *Jour. Coll. Gen. Practit.,* 6: 517–531.

Supplementary Readings*

Ackermann, W. C., 1971, The Oakley Project—A Controversy in Land Use, in *Environmental Geology Notes 46,* Bergstrom, R. E., ed., pp. 33–39, Illinois State Geol. Survey.

Baumann, D., and Dworkin, D., 1978, Water Resources for Our Cities, Resource Paper No. 78-2, *Assoc. of Amer. Geographers,* Washington, D.C., 35 pp.

Davis, G. H., and Wood, L. A., 1974, Water Demands for Expanding Energy Development, *U.S. Geological Survey Circular 703,* 14 pp.

Deming, H. G., Gilliam, W. S., and McCoy, W. H., 1975, *Water,* Oxford Univ. Press, New York, 350 pp.

Dunne, T., and Leopold, L., 1978, *Water in Environmental Planning,* W. H. Freeman and Co., San Francisco, 818 pp.

Durfor, C. N., and Becker, E., 1964, Public Water Supplies of the 100 Largest Cities in the United States, 1962, *U.S. Geological Survey Water-Supply Paper 1812,* 364 pp.

Freeze, R., and Cherry, J., 1979, *Groundwater,* Prentice-Hall, Englewood Cliffs, N.J., 604 pp.

*References to the hydrologic implications of waste disposal are included in the supplementary readings in Part IV.

Goldman, C. R., et al., eds., 1973, *Environmental Quality and Water Development*, W. H. Freeman and Co., San Francisco, 510 pp.

Hackett, O. M., 1966, Ground-Water Research in the United States, *U.S. Geological Survey Circular 527*, 8 pp.

Harte, J., and Socolow, R. H., 1971, The Everglades: Wilderness versus Rampant Land Development in South Florida, in *Patient Earth*, Harte, J., and Socolow, R. H., eds., Holt, Rinehart, and Winston, New York, pp. 181–202.

Hasler, A. D., 1947, Eutrophication of Lakes by Domestic Drainage, *Ecology*, v. 28, n. 4, pp. 383–95.

Jenkins, D. S., 1957, Fresh Water from Salt, *Sci. Amer.*, v. 196, n. 3, pp. 37–45.

Jones, D. E., 1967, Urban Hydrology—A redirection, *Civil Engineering*, v. 37, n. 8, pp. 58–67.

Leopold, L. B., 1968, Hydrology for Urban Land Planning—A Guidebook on the Hydrologic Effects of Urban Land Use, *U.S. Geological Survey Circular 554*, 18 pp.

Leopold, L., 1974, *Water—A Primer*, W. H. Freeman and Co., San Francisco, 172 pp.

Lohman, S. W., et al., 1972, Definitions of Selected Ground-Water Terms—Revisions and Conceptual Refinements, *U.S. Geological Survey Water-Supply Paper 1988*, 21 pp.

Maxwell, J. C., 1965, Will There Be Enough Water? *Amer. Scientist*, v. 53, pp. 97–103.

McGuinness, C. L., 1969, Scientific or Rule-of-Thumb Techniques of Ground-Water Management—Which Will Prevail? *U.S. Geol. Survey Circular, 608*, 8 pp.

Nace, R. L., 1967, Are We Running Out of Water? *U.S. Geological Survey Circular 536*, 7 pp.

Peixoto, J. P., and Kettani, M. A., 1973, The Control of the Water Cycle, *Sci. Amer.*, v. 228, n. 4, pp. 46–61.

Piper, A. M., 1965, Has the United States Enough Water? *U.S. Geological Survey Water Supply Paper 1797*, 27 pp.

Pluhowski, E. J., 1970, Urbanization and Its Effect on the Temperature of the Streams on Long Island, New York, *U.S. Geological Survey Prof. Paper 627-D*, 110 pp.

Rickert, D. A., and Spieker, A. M., 1971, Real-Estate Lakes, *U.S. Geological Survey Circular 601-G*, 19 pp.

Sayre, A. N., 1950, Ground Water, *Sci. Amer.*, v. 183, n. 5, pp. 14–19.

Schneider, W., and Spieker, A., 1969, Water for the Cities—The Outlook, *U.S. Geological Survey Circular 601-A*, 6 pp.

Schneider, W. J., Rickert, D. A., and Spieker, A. M., 1973, Role of Water in Urban Planning and Management, *U.S. Geological Survey Circular 601-H*, 10 pp.

Seaburn, G. E., 1969, Effects of Urban Development on Direct Runoff to East Meadow Brook, Nassau County, Long Island, New York, *U.S. Geological Survey Prof. Paper 627-B*, 14 pp.

Seaburn, G. E., 1970, Preliminary Results of Hydrologic Studies at Two Recharge Basins on Long Island, New York, *U.S. Geological Survey Prof. Paper 627-C*, 17 pp.

Smith, W. C., 1966, Geologic Factors in Dam and Reservoir Planning, *Illinois State Geol. Survey, Environmental Geology Note 13*, 10 pp.

Spencer, J. M., 1974, Geology, Health, and Phosphorus, *Jour. Geol. Education*, v. 22, n. 3, pp. 93–96.

Spieker, A. M., 1970, Water in Urban Planning, Salt Creek Basin, Illinois, *U.S. Geological Survey Water-Supply Paper 2002*, 147 pp.

Thomas, H. E., and Leopold, L. B., 1964, Ground Water in North America, *Science*, v. 143, n. 3610, pp. 1001–1006.

Thomas, H. E., and Schneider, W. J., 1970, Water as an Urban Resource and Nuisance, *U.S. Geological Survey Circular 601-D*, 9 pp.

Todd, D., 1959, *Ground Water Hydrology*, John Wiley and Sons, New York, 336 pp.

Walker, W., 1969, Illinois Ground Water Pollution, *Jour. Amer. Water Works Assoc.*, v. 61, pp. 31–40.

Weinberger, L. W., Stephan, D. G., and Middleton, F. M., 1966, Solving Our Water Problems—Water Renovation and Reuse, *Annals of the New York Academy Sciences*, v. 136, art. 5 pp. 131–154.

Films

The First Pollution (Stuart Finley, Inc.; 26 min.)

Groundwater—A Part of the Hydrologic Cycle (Cherry Film Productions Ltd., 1978: 29 min.)

Groundwater—The Hidden Reservoir (John Wiley and Sons, 1971: 19 min.)

How the Water Witch Drowns in the Dry Hole (National Water Well Association: 32 min.)

Inland Lake Demonstration Project (University of Wisconsin: 27 min.)

Little Plover River Project (University of Wisconsin: 27 min)

Secrets of Limestone Groundwater (Inside Earth Films: 14 min.)

The Subject Is Water (U.S. Geological Survey: 28 min.)

Water Circulation in Karst (Laboratoire Souterrain de Moulis, 1978: 26 min.)

The Water Cycle (Encyclopedia Britannica, 1980: 14 min.)

IV
WASTE DISPOSAL

*"We might someday be known as the generation that stood
knee deep in garbage while firing rockets at the moon."*
Photo by Ted Jones courtesy of Stuart Finley, Incorporated.

Nature of the Problem

Waste may be defined as the residue of human use of the earth's resources, including the by-products of production having no apparent economic value and, in time, the products themselves. Wastes may be solids, liquids, or gases and may be directly related to agricultural, industrial, or municipal activities. Part IV will be concerned primarily with the *geologic aspect* of dealing with solid and liquid wastes, hazardous wastes, and radioactive wastes.

Solid wastes are accumulating at the rate of more than 907 million kg (2000 million lbs) per day in the United States, and the rate of production is increasing at 4 percent a year. The per capita production of waste is highest in urbanized areas, where diverse industrial wastes are added to the garbage and rubbish generated by individuals (Table 1).

The disposal of solid wastes is one of the most serious problems confronting many municipalities. The current annual cost of waste collection and disposal is over $5 billion, an amount exceeded only by expenditures for schools and roads. Safe disposal sites are at a premium in many areas, and indiscriminate disposal can lead to serious health and environmental problems.

The principal methods of solid waste disposal are summarized in

Table 1 Classification of refuse materials [from W. Schneider, 1970]

Kind of refuse	Composition	Source
Garbage	Wastes from preparation, cooking, and serving of food; market wastes; wastes from handling, storage, and sale of produce.	Households, restaurants, institutions, stores, and markets.
Rubbish	Combustible: paper, cartons, boxes, barrels, wood, excelsior, tree branches, yard trimmings, wood furniture, bedding, and dunnage. Noncombustible; metals, tin cans, metal furniture, dirt, glass, crockery, and minerals.	Do.
Ashes	Residue from fires used for cooking and heating and from onsite incineration.	Do.
Trash from streets	Sweepings, dirt, leaves, catch-basin dirt, and contents of litter receptacles.	Streets, sidewalks, alleys, vacant lots.
Dead animals	Cats, dogs, horses, and cows	Do.
Abandoned vehicles	Unwanted cars and trucks left on public property	Do.
Demolition wastes	Lumber, pipes, brick, masonry, and other construction materials from razed buildings and other structures.	Demolition sites to be used for new buildings, renewal projects, and expressways.
Construction wastes	Scrap lumber, pipe, and other construction materials	New construction and remodeling.

Table 2 Methods of solid waste disposal [from W. Schneider, 1970]

Method	Description
1. Open dumps	Indiscriminate location and disposal practices with little or no effort to prevent nuisance, open burning, health hazards, or water pollution.
2. Sanitary landfill	Alternating layers of compacted refuse and soil. Soil cover reduces health hazards, odors, flying debris, etc. Generally located and operated to prevent pollution of ground water and surface water.
3. Incineration	Burning at high temperatures to reduce combustible wastes to inert residue.
4. Onsite disposal	Includes small-scale incinerators, home garbage disposal units, and septic tank systems.
5. Swine feeding	Involves the segregation and collection of raw garbage which is cooked and fed to swine.
6. Composting	Biochemical decomposition of organic materials to a humus-like end product.

Table 2. Special procedures are needed for liquid, hazardous, and radioactive wastes.

Sanitary Landfills

Sanitary landfills, which consist of alternating layers of compacted refuse and soil, have increasingly become a preferred method for the disposal of municipal solid wastes. Each day refuse is deposited, compacted, and covered with soil. Compaction and daily cover reduce the nuisance and health hazards associated with flies, rodents, wind-blown debris, and decaying garbage. Two types of sanitary landfills are common: *area landfill* on essentially flat land sites, and *depression landfill* in trenches or natural ravines, gulleys, or pits (Fig. 1).

Hydrogeologic Considerations

A major problem associated with landfills in humid regions is the generation of *leachate*. Leachate is a solution which forms when rain-

FIG. 1. Area and depression methods of constructing a sanitary landfill. (A) Area method. A bulldozer spreads and compacts the waste while an earth mover hauls and spreads cover material. (B) Depression method. A truck deposits its load into the depression, where the bulldozer spreads and compacts it. (Maryland Geological Survey.)

water or snowmelt percolates through the landfill and dissolves the soluble substances in it. The composition of leachate is highly variable, but it is generally high in coliform bacteria, carbonic acid, chlorides, nitrates, sulfates, and iron. If leachates reach ground-water flow systems, they can contaminate aquifers and surface waters (Figs. 2 and 3).

The major factors that influence the potential for water pollution are: (1) site topography, (2) thickness and type of surficial material, (3) nature of bedrock, (4) elevation of water table, and (5) surface drainage features. Flat upland areas, topographic rises, or the heads of gullies and ravines are safer sites for landfills than steep topography, depressions, or the lower reaches of gullies. Thick deposits (15 m; 49.2 ft) of impermeable clays or silt will contain leachate and permit deep trenching, whereas permeable deposits of sand and gravel allow

FIG. 2. Generalized movement of leachate through the land phase of the hydrologic cycle. (From W. Schneider, 1970.)

rapid migration of leachate. Impermeable bedrock, like shale, will also restrict the migration of leachate and protect deeper aquifers. Permeable or fissured and fractured rocks are unfavorable. The base of the landfill should be a safe distance above the highest seasonal elevation of the water table and a safe distance from surface water courses and floodplains.

The hydrologic effects of solid waste disposal in four geologic environments are shown in Fig. 3. In Fig. 3A, the potential for pollution is high because the base of the landfill site is in contact with the water table, and ground-water flow through the permeable sand and gravel will transport the leachate to shallow wells or surface water. In Fig. 3B, the leachate is confined to the vicinity of the site by the impermeable clay and silt. In Figs. 3C and 3D, the base of the landfill is above the water table, but because the surficial materials (sand and silt, and sand and gravel) permit migration of the leachate, both the shallow aquifers and the permeable bedrock aquifers may be polluted.

If a proposed landfill site fails to meet one or more of the criteria for a safe site, it will be necessary either to abandon the site or to employ methods for containment of leachate. For example, it may be necessary to divert drainage from the site or to seal the site with clay or plastic to contain the leachate, which can then be collected and treated (Fig. 4). It is also good practice to install wells to monitor the movement and composition of leachate.

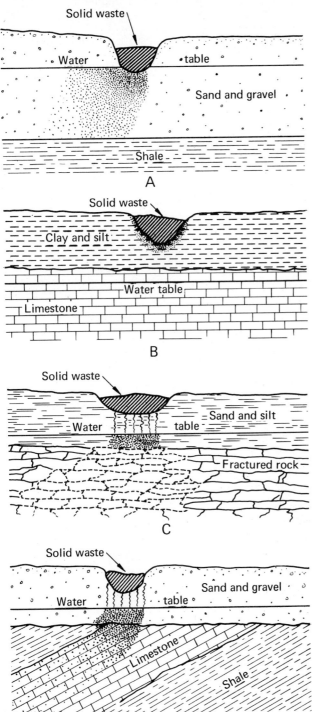

FIG. 3. Effects on ground-water resources of solid-waste disposal at at site (A) in a permeable environment, (B) in a relatively impermeable environment, (C) underlain by a fractured-rock aquifer, and (D) underlain by a dipping-rock aquifer. Leachate shown by dots. (From W. Schneider, 1970.)

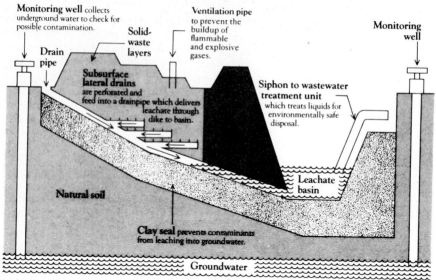

Monitoring well collects underground water to check for possible contamination.

Solid-waste layers

Drain pipe

Ventilation pipe to prevent the buildup of flammable and explosive gases.

Monitoring well

Subsurface lateral drains are perforated and feed into a drainpipe which delivers leachate through dike to basin.

Siphon to wastewater treatment unit which treats liquids for environmentally safe disposal.

Leachate basin

Natural soil

Clay seal prevents contaminants from leaching into groundwater.

Groundwater

FIG. 4. Diagram of secure landfill for hazardous chemical waste. The impervious clay seal and subsurface lateral drains prevent leachate from contaminating the groundwater. Decomposition of organic matter produces a variety of gases, including methane, which is flammable and explosive. It is therefore necessary to ventilate landfills properly to prevent the buildup of gases. (From Chemical Manufacturers Assoc.)

Septic Tank Sewage Disposal

The most satisfactory method for disposing of sewage is to pipe it through municipal sewers to sewage treatment facilities; unfortunately, millions of Americans live beyond the reach of the city sewer lines. Most of these people rely on an onsite disposal technique known as the *septic tank* system. It consists of a septic tank and *soil absorption field* (Fig. 5). Untreated sewage enters the tank, where larger solids settle to the bottom and greases and oils rise to the top. Anaerobic bacterial decomposition reduces the volume of the solids. Liquid effluent moves from the tank to the absorption field through perforated pipes. The effluent, which contains complex organic compounds, suspended matter, and bacteria, is a health hazard, and if not purified may cause typhoid fever, hepatitis, or dysentry. The system relies upon the percolation of effluent through the soil for purification. Aerobic bacteria in the soil should oxidize the organic matter and clay minerals should absorb some of the dissolved materials. Other materials are likely to be trapped in the pore spaces, but some dissolved

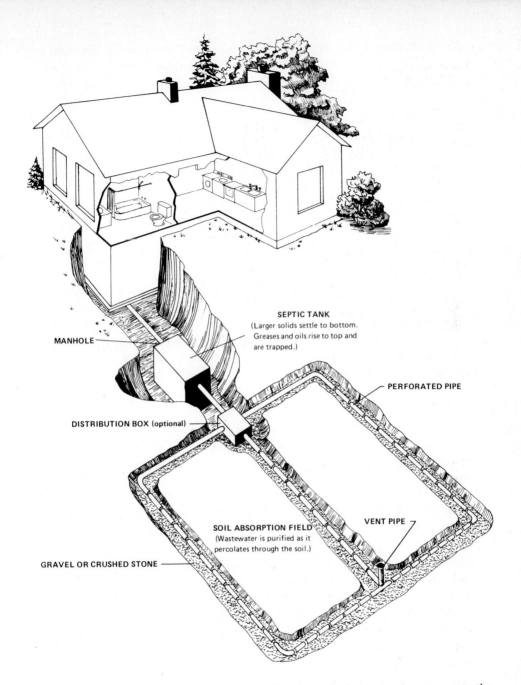

MANHOLE

SEPTIC TANK
(Larger solids settle to bottom.
Greases and oils rise to top and
are trapped.)

PERFORATED PIPE

DISTRIBUTION BOX (optional)

SOIL ABSORPTION FIELD
(Wastewater is purified as it
percolates through the soil.)

VENT PIPE

GRAVEL OR CRUSHED STONE

FIG. 5. Septic tank system. The main sewer line from the house carries sewage to the septic tank where solids undergo anaerobic decomposition. Liquid effluent moves from the tank to the soil absorption field by way of perforated pipe. The effluent is purified as it percolates through the soil. (From State of Wisconsin Department of Natural Resources.)

chemical compounds, such as phosphates, may move easily through most soil systems, contaminating ground and surface waters.

The effectiveness of a septic tank system depends upon the size and design of the tank, maintainence of the system, and the geology of the site. The major geologic factors to be considered are (1) soil suitability, (2) elevation of the water table, (3) depth to bedrock, (4) proximity to surface water courses, and (5) topographic slope.

Soil suitability is usually determined by a percolation test, which determines the infiltration capacity of the soil to accept the effluent. "Perc rates" less than 2.5 cm/hr (1 in./hr), for example, are too low for normal usage. Other soil properties such as texture, ion exchange capacity, and mineralogy influence the potential for purification and changes in percolation rates over time, but they are seldom measured. The maximum seasonal elevation of the ground-water table should be at least 1.5 m (4.9 ft) below the absorption field; bedrock should be at a depth greater than 1.5 m (4.9 ft). Bore holes and trenches can be dug to determine water-table elevation and depth to bedrock. The surface slope should generally be less than 15 percent, and the absorption field should be 15 m (49.2 ft) from a water course or water well.

Septic tank systems may fail for several reasons. Improper design, installation, and maintenance are common causes of failure. If the sludge in the bottom of the tank is not pumped out every two to four years, sewage in the tank may back up or overflow. If coarse-grained solids get into the absorption field, pore spaces in the soil will clog. Poorly designed tanks will also permit a high percentage of suspended solids to move directly from the septic tank to the absorption field.

Another reason for failure is that percolation rates may change over time (Fig. 6). A gradual reduction in percolation rates is common. This not only allows the effluent to rise to the surface in wet weather, forming a "septic bog," but it also creates anaerobic conditions in the soil, which inhibits the breakdown of soil-clogging organic products.

If the absorption field is located too close to the water table or permeable bedrock, the system is bound to fail. The effluent will not be purified before entering the ground-water flow system (Fig. 7). If the direction of ground-water flow is toward the well of a residence, the owners risk polluting their own water supply.

Management of Hazardous Wastes

Congress has defined hazardous wastes as discarded material that may pose a substantial threat or potential danger to human health or

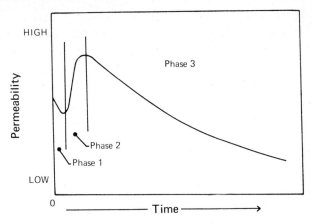

FIG. 6. The effect of prolonged submergence on soil permeability. The initial decrease in percolation rate in Phase 1 is probably due to swelling of soil particles and dispersion of soil aggregates. The increase in permeability in Phase 2 may be caused by the solution of entrapped air or soluble minerals. The gradual reduction in permeability in Phase 3 may be caused by mechanical and biological disintegration of soil aggregates and biological clogging of soil pores. (From J. Cain and M. Beatty, 1965, with permission of Soil Conservation Society of America, © 1965 by S.C.S.A.)

the environment when improperly handled. It includes material which is ignitable, corrosive, infectious, or toxic. Most of the hazardous wastes in the United States are generated by chemical, primary metals, electroplating, textile, petroleum refining, and rubber and plastics industries. The Department of Defense generates significant amounts of explosives, herbicides, and nerve gases, while the medical profession and mining industry contribute minor amounts of hazardous materials.

A small amount of hazardous wastes are recycled, but most are disposed of by either incineration or underground injection, or in sanitary landfills. Incineration can be used to decompose and detoxify some chemical wastes, and deep-sea incineration ships have been designed to take waste out in the ocean for burning.

Underground injection has been an important disposal technique for many years, and there are thousands of deep-well disposal units injecting oil-field brines and hazardous industrial wastes. The technique is not, however, without its problems. The subsurface disposal of hazardous liquid wastes at the Rocky Mountain Arsenal well near Denver, for example, led to the development of damaging earthquake swarms (Reading 7) and revealed our limited understanding of safe injection pressures. Subsurface injection has also polluted ground water and has proved to be a health hazard when this has occurred.

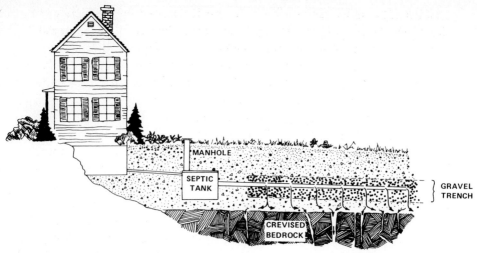

FIG. 7. Soil absorption field located too close to the underlying creviced bedrock. The effluent will not be purified before reaching the bedrock, and the ground water in the bedrock will be polluted. (From State of Wisconsin Department of Natural Resources.)

The geologic factors most important to take into account when evaluating the feasibility of underground injection are (1) ability of the host rock to accommodate the anticipated volume of the liquid waste, (2) ability of the host rock to accommodate any potential chemical reactions involving the host rock and the waste materials, (3) relationship of the host rock to freshwater aquifers, and (4) tectonic stability of the region. The storage capacity of the host rock will be influenced by porosity, permeability, thickness, areal extent, antecedent fluid saturation, and the chemical reactions that might take place between the hazardous waste and the rock. If the host rock can conveniently accept the total volume of waste, injection pressures will be minimal. The host rock must also be isolated from all freshwater aquifers by confining impermeable formations. If the rock is undergoing tectonic strain, additional pore pressures may lead to failure. Even under the best hydrogeologic conditions, it is important to monitor disposal wells when hazardous waste is involved.

Most hazardous wastes end up in sanitary landfills, although it is required that they be handled separately from the rest of the wastes. The Environmental Protection Agency has recommended phasing out land disposal of hazardous wastes, but in the interim, special cells with clay and plastic liners, leachate collection systems, and monitoring wells may be used.

A number of federal laws now address the problem of hazardous wastes. The most far reaching are the Resource Conservation and Recovery Act (RCRA) of 1976 and the Comprehensive Environmental Response, Compensation and Liability Act ("Superfund") of 1980. RCRA establishes procedures to improve safety measures, which include a stringent "cradle-to-grave" system for managing hazardous wastes. Final rules on the siting, design, operation, and monitoring of new land disposal facilities are scheduled to be issued in 1982. The Superfund is designed to deal with future catastrophies and also such past sins as New York's Love Canal and the "Valley of the Drums" near Louisville, Kentucky. Waste chemicals were stored at these sites in drums which are now corroding and releasing toxic materials into the soil and ground water. The lives of thousands of people have been affected. Cleanup costs at Love Canal alone may exceed $100 million, and annual costs for monitoring the area may be as much as $3 million. The Superfund includes provisions to clean up abandoned hazardous waste dumps that pose serious threats to human health or the environment.

Management of Radioactive Wastes

Radioactive wastes (*radwastes*) can be referred to as high level, transuranic, and low level. *High-level wastes* are the wastes of fission products with a high level of penetrating radiation, high rates of heat generation, and a long toxic life. *Transuranic wastes* generate little or no heat and have much lower radiation levels, but still contain significant concentrations of long-lived alpha radiation emitters. *Low-level wastes* include a variety of materials with low levels of radiation, or those that have been contaminated through use but do not fit into the other categories.

The amount of wastes generated since World War II is overwhelming. As of 1981, the United States' defense industry had generated about 287 million liters (76 million gallons) of liquid radioactive waste, commercial reactors were responsible for 8200 metric tons (9040 short tons) of spent fuel elements, and uranium mines produced 0.065 billion m³ (2.5 billion ft³) of radioactive mill tailings (Shapiro, 1981).

Radioactive waste problems are as complex and varied as the proposed solutions. Disposal has been under consideration since 1955, when the National Academy of Sciences met with a group of earth scientists to study the geologic, biologic, physical, and chemical ramifications of the problem. Some of the disposal methods that have been considered include deep bedrock disposal (depositing wastes in impermeable rocks underlying the continents), transmutation (the conversion of a radioactive nucleus to another isotope by bombarding it with radiation or nuclear particles), extraterrestrial disposal (placing wastes in space), ice-sheet disposal (placing wastes in polar regions), and sea-floor disposal (depositing wastes in submarine canyons where, it is hoped, sediments would bury them).

Deep continental bedrock formations appear to be favored by the U.S. Geological Survey, but no repositories for the disposal of high-level radioactive waste in deep bedrock exist today. Most of these wastes are now stored above ground or in near-surface underground tanks at three federal repositories or onsite at nuclear energy facilities. There are eleven areas in the United States now being used as burial sites for low-level waste materials.

Different rock types have been considered as host rocks for high-level repositories. Rock salt has received the most attention because of its low porosity and permeability, its plasticity, and its high thermal conductivity. Shale, limestone, granite, and basalt have also been considered because of their low permeability. Potential host rocks should also be (1) homogeneous, (2) tectonically and chemically stable, (3) of sufficient thickness and depth, and (4) free of water. The presence of aquifers, fault zones, fractures, abandoned mines or wells, hydrated minerals, or fluid inclusions pose a threat to the safety of a repository.

Because some radioactive substances remain toxic for hundreds of thousands of years (e.g., plutonium-239 for an estimated 250,000 to 500,000 years), the waste repositories must be totally isolated from the environment for a very long time. The impact of possible major climatic changes in the future (including continental glaciation) and fluctuations in sea level and erosion rates must be considered.

As long as we are without a safe system for the *disposal* of radioactive wastes we will have to continue to *store and manage* these wastes, and responsible management will require stable human institutions as well as stable geologic formations.

Readings

Hydrogeologic Considerations
in Liquid Waste Disposal

Industries and municipalities generate large volumes of liquid wastes which, in some instances, can be safely reclaimed as irrigation water while at the same time augmenting ground-water recharge and soil fertility. In Reading 34, S. Born and D. Stephenson review the advantages and disadvantages of disposing of industrial wastes through irrigation systems. They also note the hydrogeologic factors which must be considered if one is to avoid contaminating ground-water flow systems.

Nuclear Waste Disposal:
The Geological Aspects

The treatment and disposal of radioactive wastes are of international concern. In Reading 35, scientists at the Institute of Geological Sciences in London review the geologic aspects of high-level nuclear waste disposal. The geologic issues they explore are applicable to other countries that are studying the feasibility of deep bedrock disposal.

34 Hydrogeologic Considerations in Liquid Waste Disposal

S. M. Born and D. A. Stephenson

Municipal wastes have been disposed of by irrigation on sewage farms for almost a century, but the idea of disposing of industrial wastes by irrigation has evolved primarily since World War II.

The earliest irrigation disposal method was ridge-and-furrow irrigation, a process by which wastes are transported to furrowed plots of land and allowed to infiltrate the soil. But ridge-and-furrow irrigation has several serious shortcomings. A comparatively large amount of reasonably level land is required for the disposal site, and the site requires clearing and preparation prior to the application of waste water. The ridge-and-furrow method is also prone to flooding, which creates odors and may damage crops. In recent years, pressing demands for more intensive use of land have further diminished the popularity of the system for disposal of industrial wastes. It is still used locally, however, for disposal of stabilized municipal wastes.

An improved technique for waste-

Born, S. M. and Stephenson, D., 1969, "Hydrogeologic Considerations in Liquid Waste Disposal," *Jour. Soil and Water Conservation*, vol. 24, no. 2. Reprinted by permission of the authors and the Soil Conservation Society of America. Copyright 1969 by S.C.S.A.
Dr. Born is Director, State Planning Office, Madison, Wisconsin. Dr. Stephenson is Chairman of the Environmental Resources Unit, University of Wisconsin—Extension.

water irrigation was developed in 1947 and was first used by a canning company in Hanover, Pennsylvania.[3] Effluent water was applied to the land by sprinklers, giving rise to the name "spray irrigation." Waste water is thus disposed of by infiltration and evaporation.

Spray-irrigation disposal systems have numerous advantages: (1) The possibility of creating odor nuisances is minimized since the waste water is aerated during the application process, and oxygen-deficient conditions resulting from ponding and surface flooding at the irrigation site can be avoided due to the mobility of the sprinkler system. (2) No special land preparation may be necessary, and sloping areas and woodland can be irrigated. (3) Spray-irrigated land can be farmed. (4) The spray system is readily expanded to accommodate increased volumes of effluent, and the distribution system can be salvaged.

The method is not without its disadvantages. Unfavorable sites or poor management practices can lead to surface runoff from spray-irrigation areas. This runoff may be of sufficient magnitude to condemn the method. Wind can transport both sprayed effluent and odor to unwanted places. Some waste water cannot be sprayed without extensive pre-treatment, such as sedimentation with or without flocculants, cooling, and screening. Clogging of sprinkler nozzles by solids in the effluent or by precipitated chemicals can decrease disposal efficiency. Where practicable, however, spray irrigation is a satisfactory means of disposing of liquid wastes on land.

Spray-irrigation disposal systems vary widely in design, cost, and capacity. Choice of a system is largely controlled by the physical characteristics of the site. One efficient system, located at Seabrook Farms in New Jersey, disposes of 5 to 10 million gallons of process water daily on 84 wooded acres of loamy sand.[3] Total cost of the Seabrook installation was about $150,000.

Waste disposal by irrigation has benefits other than the immediate one of protecting surface-water quality. In some cases, nutrient-charged waste water can be used to fertilize cultivated lands. A team of scientists and engineers at Pennsylvania State University recently demonstrated that the application of treated sewage effluent to croplands increased hay yields 300 percent, corn yields 50 percent, corn silage yields 36 to 103 percent, and oat yields 17 to 51 percent.[4] Concentrations of various nutrients originally in the effluent were essentially removed from the plot. The crop functioned as a "living filter."

Augmentation of groundwater recharge by treated effluent water is another important potential benefit of irrigation with waste water. Regions long dependent on groundwater supplies are experiencing an overdraft on these reserves. The overdraft is expressed in terms of declining water table, increased water cost, local water quality deterioration, and even a water shortage. Recharge by irriga-

tion with waste water can provide a partial answer to this problem of depleted groundwater supplies. In the Pennsylvania State University experiment, 60 to 80 percent of the applied effluent entered the groundwater reservoir relatively free of nutrients. However, the potential danger of contaminating groundwater cannot be overemphasized.

The ideal approach to irrigation disposal of liquid wastes is to derive secondary benefits from the operation and simulataneously guard against endangering the quality of groundwater supplies. Careful selection of the disposal site and wise management of the irrigation installation maximizes the chance of achieving both of these goals. Fortunately, nature has provided a remarkable purification system of its own. Removal of solids and nutrients from effluent water is accomplished by the microbial population in the soil and by the soil and rock medium itself through adsorption by clays, ion exchange, precipitation, and filtration. Man can supplement the natural purification of irrigated waste water by renovation techniques such as the cropping practice mentioned previously.

Pollution from irrigated areas can be controlled by adequate preliminary evaluation of the proposed site with due consideration given to the geologic and hydrogeologic environments. These physical environments should be influential factors in the design of a waste-disposal system.

Determining Flow Systems

Definition of a groundwater flow system allows an operator of a plant producing liquid wastes to select the optimum disposal site available on his land and further permits him to know in which direction effluent waters will travel in the ground, at what rate they travel, and where they will surface. In general, the flow system of interest for irrigation purposes is a local (shallow) system; however, in some geologic environments, the local system is only a small element of a larger regional system where the flow path of the effluent is governed by a deeper flow pattern.

Until recently, surface-water bodies have been studied as entities separate from groundwater; now these systems are recognized as being interconnected and are studied as such. Groundwater is derived from surface water by infiltration through the soil and includes all water within the saturated zone below a water table. Thus, the upper limit of this zone of saturation defines the water table, which is a subdued replica of surface topography. A water table is high under uplands, and slopes toward lowlands. Where the land surface intersects the water table, a surface body of water is present.

Water in the ground ultimately returns to the surface and becomes run-off in the lowlands. Movement in a groundwater flow system is along flow paths from areas of high potential (upland or recharge zone) to areas of lower potential (lowland or discharge zone). Potential is water elevation expressed as feet above sea level. Lines connecting points of equal potential are called equipotential lines.

In a recharge zone, the groundwater gradient is downward from the water table; in a discharge zone, it is up toward the water table (Fig. 1). Evidence of a groundwater discharge zone is commonly a wetland area or marsh in humid regions and a playa or "alkali flat" in arid zones. Water may be present perennially in a discharge area because of the upward movement of groundwater. Irrigation, therefore, must be undertaken with caution in a discharge area because infiltration is at a minimum and the effluents remain on or near the surface.

The pattern of a groundwater flow from a recharge of a discharge area constitutes a dynamic flow system.[5] A flow system is controlled by topography but modified in flow direction and flow rate by the soil-rock conditions prevailing along individual flow paths. Groundwater flow systems can be defined in the field with empirical geologic and hydrogeologic techniques.

Each geographic area has unique rock-soil-water relationships since geology, topography, and climate vary regionally. On a gross pattern, however, it is likely that conclusions regarding flow systems for one area can be extrapolated to areas of similar or like environments. A mapped system has numerous applications with respect to water quality and quantity problems, including more efficient use of land for developmental, waste-disposal, and water supply purposes.

Effluent Infiltration and Movement

The ability of water to permeate the surface (infiltration capacity) is the critical element in efficient irrigation disposal operations. Successful disposal of waste water, especially by spray or ridge-and-furrow irrigation, allows the effluent to infiltrate the ground and undergo natural purifica-

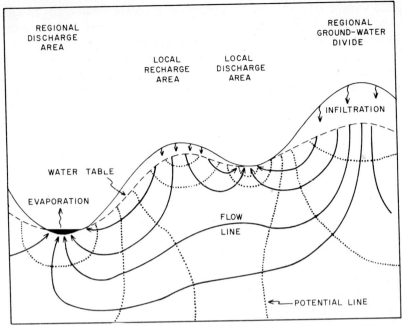

FIG. 1. Idealized groundwater flow system, homogenous soil conditions.

tion (particularly clarification) while percolating through the geologic medium prior to being discharged from the area. This minimizes malodorous effects and the possibility of surface water contamination at the discharge point.

Infiltration capacity is controlled by a number of variables.[7] Topographic relief controls the disposition of surface runoff and, therefore, to a large degree, the amount of water available for infiltration at a given location. As noted previously, topography further controls groundwater gradients and, hence, how the flow system operates at any specified point. Infiltration in discharge areas of humid and subhumid regions is sometimes limited by a prevailing upward gradient and saturation of the soil due to a frequent vertical proximity of the water table

and ground surface. In recharge portions of the flow system, the limiting factor controlling water intake is geological, that is the nature of the surface material (Fig. 2).

Infiltration is also influenced by rock type and the rock's weathering history which determine the nature and extent of pore space, the depth to impermeable zones, and the texture of the ground surface. Rates of infiltration are further reduced by mechanical compaction (by livestock, man, machines, and rain) and by pore clogging resulting from the erosion of natural materials and the in-wash of organic materials and solids in many industrial effluents. Surface runoff, which increases the chance of direct surface-water pollution, is one by-product of reduced infiltration.

Another influence on infiltration is

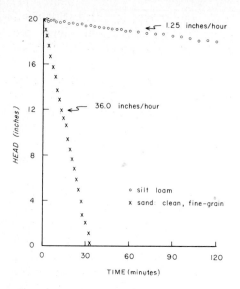

FIG. 2. Infiltration rate curves for two markedly different geologic materials (data from infiltration ring studies in northern Wisconsin by the authors).

the soil's state of saturation. Water is removed naturally from the soil profile by evaporation through the air-soil interface and by transpiration through vegetation. This removal produces a soil moisture deficiency which benefits the infiltration of effluent disposed of by irrigation. A dry soil will generally absorb at least twice as much water as a wet one. This fact explains the sometimes rapid decline in infiltration capacity that occurs under continued irrigation.

Having long-term records of rainfall and climate available is desirable so that pre-irrigation soil moisture conditions can be anticipated; such records are maintained by the U. S. Weather Bureau. A disposal site must be able to accommodate a volume of waste water over and above the natural "background" of precipitation. This consid-

eration is especially important in designing a spray-irrigation system for a humid region.

In many operations, generation of waste water is almost continuous. Variations in precipitation and soil moisture—and, therefore, infiltration capacity—must be provided for in the system, either by creating temporary holding capacity and thereby reducing discharge during periods of rainfall, or by utilizing sufficient acreage to cope with precipitation and liquid-waste disposal simultaneously.

Once effluent infiltrates the ground, it eventually enters a groundwater flow system. Velocity of water within the saturated zone is controlled by permeability and gradient (Darcy's Law). The permeability of a material is defined as "the ability of a soil or rock to transmit fluids." Permeability in rocks and soils is controlled by the nature of the interconnected open space, either as pores between the grains, or fractures, or solution cavities.

Permeability of soil or rock samples collected in the field can be routinely measured in a laboratory with an instrument known as a permeameter. However, laboratory testing may not be accurate because the natural packing of the sample material is difficult to duplicate in the testing apparatus, and a small sample may not be representative of field conditions where discontinuities in the soil or rock medium frequently occur. Where accurate data are mandatory, pumping tests of wells must be conducted. The groundwater gradient can be determined in the field by monitoring groundwater levels in a group of sand-point observation wells.

Calculating the velocity of effluent movement in the ground permits disposal-site operators to predict the length of time required for decay of ground-water mounds built up during irrigation. In a sprinkler rotation system, such predictions enable scheduling of "rest periods" for different parts of the site. Knowing the velocity and direction of groundwater flow also enables the operator to predict where and when effluent which has entered the ground will discharge. By monitoring ground-water quality along its flow path, the operator can ensure that waste water is renovated and that contaminated water is not entering surface water at the point of discharge. In view of the present outcry over polluted water and the strong legislative response by state and federal governments, such information may prove valuable should the concerned industry become involved in pollution litigation.

Evaluating a Disposal Site

A thorough knowledge of geologic conditions is necessary where waste water is to be disposed of by irrigation. Of particular importance is the geology of unconsolidated, surficial deposits. The thickness, nature and distribution of these deposits define the effectiveness of filtration of effluent water, the adsorptive capacity of soil types present for the removal of specific constituents in the effluent, and the location of impermeable horizons which impede downward movement of the waste water. The amount of pore space in the geologic medium above the water table permits the geologist to calculate the storage capacity of the unsaturated portion of the soil profile. This potential storage volume, together with the rates of infiltration and ground-water movement, limits the application of effluent on the disposal site.

Infiltration rates can be determined in situ with an infiltrometer.[2] Reasonable estimates of infiltration and permeability can also be determined by comparing textural properties of sediments at the site (grain size, shape, amount of clay and silt) with known hydraulic characteristics for given sediment types. Where relatively crude data are acceptable, such information can be derived by microscopic inspection of samples, thereby saving the cost of laboratory permeameter tests or expensive well pumping tests.

Occurrence of bedrock valleys of glacial or alluvial origin may permit the location of a satisfactory disposal site in otherwise unacceptable terrain. The existence of such valleys has been documented in geological studies of many areas. Relatively simple geophysical surveys can be used to define them where they are likely to be present.[1]

Zones of increased weathering, solution, and high permeability commonly are localized along fractures in geologic materials. Permeability along bedrock fractures is often many times greater than in adjacent, unfractured bedrock; velocity of groundwater movement is fast; and very little filtration may occur. Therefore, fractured bedrock, particularly limestone and dolomite, near the surface of a disposal site may provide "avenues of pollution" for downward-moving waste water. Fracture patterns oftentimes can be identified by field investigations and mapped on aerial photographs. The disposal site can then be positioned to minimize the danger of largely unfiltered, contaminated water being discharged from the area by these routes.

It was noted earlier that flow systems exist that are larger than local systems, in which case the discharge area is not necessarily the lowland adjacent to an upland recharge area. Instead, downward-moving water may follow a deeper flow system along fractures or solution channels and travel many miles along an unfiltered path to a regional discharge area[6] (Fig. 1). The source of pollutants at such a discharge point may be more difficult to locate, but the resulting pollution is of no less concern. Carbonate and volcanic terrains are especially susceptible to rapid travel of effluents. Regional geologic studies can be used to define such areas and prevent their use for disposal sites.

Preliminary Studies Imperative

National concern over the deteriorating quality of our surface water resources has resulted in federal and state laws for regulating the disposal of polluting effluents. Many small industries have been affected and, of necessity, have considered land disposal of process waters. A common method adopted by these industries has been disposal of effluent onto an irrigation plot. This practice can and does degrade both surface water and groundwater at irrigation sites where inadequate technical consideration has been given to location and maintenance.

Giving adequate consideration to geologic and hydrogeologic factors in selecting and operating disposal sites for liquid wastes minimizes the pollution potential. Costs associated with preliminary investigations can be high; but weighed against the consequence of inadequate planning, these costs may never be cheaper and should be an inherent part of any budget for an irrigation disposal facility.

Notes

[1]Drescher, William J. 1956. *Ground water in Wisconsin.* Inf. Circ. No. 3. Wisc. Geol. Surv., Madison, Wisc. 37 pp.

[2]Johnson, A. I. 1963. *A field method for measurement of infiltration.* Water-Supply Paper 1544-F. U. S. Geol. Surv., Washington, D. C. 27 pp.

[3]Lawton, G. W., L. E. Engelbart, G. A. Rohlich, and N. Porges. 1960. *Effectiveness of spray irrigation as a method for the disposal of dairy plant waste.* Res. Rept. No. 6. Wisc. Agr. Exp. Sta., Madison, Wisc. 59 pp.

[4]Parizek, R. R., L. T. Kardos, W. E. Sopper, E. A. Myers, D. E. Davis, M. A. Farrell, and J. B. Nesbitt. 1967. *Waste water renovation and conservation.* Study No. 23. Penn. State Univ., University Park, Penn. 71 pp.

[5]Toth, J. 1962. *A theory of ground water motion in small drainage basins in Central Alberta, Canada.* J. Geophys. Res. 67(11): 4375–4387.

[6]Winograd, I. J. 1962. *Interbasin movement of ground water at the Nevada Test Site.* Prof. Paper 450-C. U. S. Geol. Surv., Washington, D. C. pp. C108-C111.

[7]Wisler, C. O., and E. F. Brater. 1959. *Hydrology.* John Wiley and Sons, New York, N. Y.

35 Nuclear Waste Disposal: The Geological Aspects

Neil Chapman, David Gray, and John Mather

Radioactive wastes are one of the legacies of man's enterprises in harnessing the energy which results from nuclear fission and their treatment and disposal are the subject of considerable contemporary international concern. The two words "radioactive" and "waste" are in themselves highly emotive and, when linked, open dark areas of apprehension among the public such as have rarely been encountered before by any sector of national industry. Recent decisions by nuclear energy agencies throughout the world to find a solution to the ultimate disposal of these wastes have generated understandable feelings of fear and mistrust among those who feel they may be closely affected by such disposal. On the whole, their anxiety probably stems from the lack of easily accessible information on the problem itself and on the state of research into possible solutions. This concern has crystallised over the issue of "high-level" waste (HLW), the disposal of which by burial deep in rock formations has been mooted as the most feasible option. Such an option is currently the object of research programmes in several countries, including the UK. For example, the recent

The authors are affiliated with the Institute of Geological Sciences, London. This first appeared in *New Scientist*, London, the weekly review of science and technology, April 1978. Reprinted with permission of the authors and New Science Publications.

moratorium on nuclear power development in Sweden, until the major issue of waste disposal is accepted as resolvable, has led to the rapid instigation of both experimental and theoretical studies of geological disposal.

The object of disposal is to emplace the wastes in a repository such that their long-term containment is assured and their return to the biosphere in concentrations such as to create a biological hazard is eliminated. In this article we examine the rationale for deep disposal in rock in the United Kingdom and the investigative techniques which will be employed in determining the suitability or otherwise of potential repository sites. Our aim is to clarify the geological issues which will arise in investigations of such a site in order to ensure its long-term integrity, and the impact of development of a site on the local environment.

The wastes concerned are the by-products of reprocessed spent reactor fuel and contain fission products and actinides. Current policy is to incorporate these wastes into the structure of a borosilicate glass ("Researching radioactive waste disposal," by Drs. Frank Feates and Norman Keen, *New Scientist*, vol. 77, p. 426) producing blocks about 60 cm in width and 3 m long. The thermal capacity of the waste in these blocks necessitates the provision of some period of pre-disposal surface storage during which time the initially high heat output, caused by the decay of the short-lived nuclides, could be dissipated in an engineered facility. It is at that stage that the waste blocks would be transferred to a disposal site. It is estimated that the post-disposal problems associated with the heat output of the waste will last at most for 1000 years, as the thermal output will by then have declined to an insignificant value.

Disposal Options

Most countries, having defined the magnitude of the problem, have opted for a similar preparatory technique to that outlined above, although the recent American policy change to storage of spent fuel, rather than reprocessing, has led to some concomitant changes in disposal policies. Considerable thought has gone into possible subsequent methods of disposal, and into the question of the desirability or otherwise of maintaining retrievability (disposal versus controlled storage with the option of later exhumation should a future requirement arise). Many of the ideas put forward over the last decade have verged on science-fiction, yet all have been subject to feasibility studies. The nuclear industry is, after all, faced with isolating a hot, radioactive and toxic material for a length of time totally outside mankind's experience. In such an unprecedented situation every relevant idea must be examined. After discarding or relegating many suggestions (such as shooting the waste into the Sun or out of the Solar System by rocket or accelerator, burial under the Antarctic ice-cap, transfer to the Earth's mantle via an oceanic subduction zone) on the grounds of safety, cost or currently inadequate technology, the most feasible option is presently thought to be disposal to geological formations. This option has become the predominant topic of research in all countries concerned, as it is well within the bounds of present-day technological expertise and research capability.

Within the concept of disposal into continental rocks, several methods involving extremely deep (up to 10 km) burial or rock melting, have been discarded for the present on the basis of cost or unpredictable behaviour. This ability to predict the behaviour of the host-rock under known conditions is the crux of the whole problem. Any method which could invoke random effects which are difficult to model with certainty must be treated with circumspection. By applying this criterion an excavated underground facility in a suitable body of rock emerges as the most practical and easily modelled form of repository, and is now the focus of considerable international research effort.

The function of such an underground repository will be to accept vitrified blocks of waste over a period of decades by their emplacement in mined cavities in arrays designed so that the thermal profiles within the host-rock are predictable, and finally to cease operation and to be completely sealed from the surface. Current conceptual designs envisage emplacement of the blocks in boreholes spaced along the floors of a series of horizontal tunnels (*New Scientist* loc. cit.). The repository might be between 300 and 1000 m below the Earth's surface and would be serviced by inclined access tunnels or vertical shafts. This simplified design defines the basic spatial requirements which the geologist must provide; the nature of the waste and its requisite isolation period define the types of rock in which the repository could be situated.

It is not generally appreciated that a natural example of geological containment of high-level wastes for many millions of years exists at the site of a uranium mine in Gabon. The "Oklo phenomenon" as this has come to be termed, is an occurrence of natural chain fission reactions which took place in a uranium deposit in Precambrian times, and which were moderated by groundwaters. Many of the resultant actinides and fission-products are still fixed in the host rocks in the reactor zones despite 1800 million years of subsequent exposure to geological processes.

A Suitable Host Rock

The criteria which must be met by a suitable volume of rock can thus be nominated in an abstract sense (Table 1). It is the task of the geologist to attempt to match these demands to extant rock formations at sites which also satisfy other basic criteria related to non-geological factors such as access, availability and operability. The prime contenders for a host-rock are argillaceous (clay-rich) rocks, salt deposits (either bedded or in domed structures) and hard crystalline igneous or metamorphic rocks (granites, gabbros, gneisses, granulites, etc.). All are chosen for their relative impermeability, large-scale homogeneity, and massive and widespread nature. Salt, and some clay deposits, have the advantage of being self-sealing (open-

ings tend to close up by progressive creep); clays have a high ion-adsorption capacity which would impede any nuclide migration; and crystalline rocks can be identified which are generally dry, stable against geological movements, extensive both vertically and laterally, and inert to low quantities of heat.

The United Kingdom is fortunate in possessing an extremely diverse geology for such a small land area, and large volumes of suitable rock types occur throughout Britain. Most suitable hard crystalline host rocks are restricted to the upland regions of the north of England and Scotland; those in Devon and Cornwall generally carry much groundwater and are less acceptable. Argillaceous rocks are widespread spatially, but the more homogeneous, older and compacted units with the most mature thermal properties are commonest in Wales, northern England and southern Scotland, although other younger formations in England may warrant consideration. The last category of rocks, the salt deposits, as well as suitable interbedded salt and argillaceous formations are among the potential host rocks to be found in the fertile belt of the English lowlands. Here the bedded salts are of only limited thickness although mixed salt/clay sequences are not limited by this factor. Large, deep salt domes, such as those under study for disposal purposes in the United States, West Germany and Holland, are encountered at appropriate depths only under the North Sea. Salt was originally advanced as being perhaps the best host for a HLW repository, and research is now directed primarily at assessing the suitability of the me-

Table 1. Principal geological characteristics of a possible HLW disposal site

Spatial

1. A volume of effectively homogenous rock up to one cu. km in size with an upper surface at a depth of between 0.3 and 1 km depth

2. Ease of access to this rock volume by standard mining techniques together with a suitable area for surface operations

Physical

1. The rock should not be highly fractured or traversed at depth by open fissures, major discontinuities, fault or crush zones

2. In order to ensure containment of the waste there should be negligible through-flow of groundwater or potential access for surface waters. The rock would ideally be impermeable and devoid of mobile water

3. The thermal response of the rock should be such that containment of the waste is not jeopardised during the heating and cooling spectrum to which it will be subjected

Chemical

1. If any migration of actinides or fission-products from the waste blocks occurs, the rock should provide a geochemical barrier by means of ion-exchange or adsorption sufficient to impede egress to an uncontrolled zone or the surface

2. Chemical reaction with, or irradiation by, the waste likewise should not affect the integrity of the rock

3. The rock should be chemically stable against attack by any external agent over the requisite period of isolation

Latent and temporal

1. A region of low seismicity should be chosen with an extremely low probability of future significant earth movements or igneous activity within the necessary isolation period

2. The rock should be of extremely low conceivable commercial significance to lessen the danger of future mining or exploitation in the region should the site of the repository be forgotten

3. Climatic change is to be expected during the period of containment and the rock characteristics, taken in conjunction with the design of the repository, must be able to withstand events such as glaciations, interglacials, onset of pluvial or arid conditions, and sea-level changes

dium, with particular reference to the risk of flooding and the lack of a chemical barrier.

As part of a broadly based R&D programme, funded partly by the EEC and partly by the UKAEA, the Institute of Geological Sciences, a component body of the Natural Environment Research Council, is at present examining the suitability of the hard crystalline rocks, with a lower level of effort directed towards the argillaceous rocks and salts. The main part of the current research programme is directed at ob-

taining a clear understanding of their behaviour under the specific conditions of a HLW repository.

This research effort is split between laboratory and theoretical studies of rock and groundwater behaviour, and actual site investigations. Since the critical factor in containing the waste will be to inhibit access of groundwater to the glass and its subsequent leaching and migration in a fluid phase, the fluid content of the rock and its behaviour under the thermal and stress conditions of a repository are under investigation. This includes laboratory studies of fluid/rock and fluid/waste interactions, and modelling of thermal stressing, convection and permeability conditions. Combined with data on the thermal stability of various rock types and their thermomechanical properties these studies define the most suitable textural and mineralogical features for the host rock, and the nature of the waste which it can contain. Parallel studies are aimed at finding the best method of back-filling and sealing a completed repository, as the leakage routes provided by mined openings appear likely to be one of the principal problems in preventing access of water to the disposal zone. Various methods of filling and rock-grouting are being examined, as is the possibility of using plastic clays and other materials as both sealants and chemical barriers to nuclide migration.

Interim field studies are aimed at bridging the gap between laboratory work and actual site investigations by effectively scaling-up the existing theoretical models and developing new large-scale investigative techniques for use in the later site studies. These interim studies are to be carried out at sites with equivalent geological formations to those which may eventually prove suitable for a repository. They would involve groundwater research (relatively little is yet known about the behaviour of fluids at depth in UK crystalline rocks), the development of three-dimensional sub-surface borehole mapping techniques, and scaled heating experiments. The latter will use electrical heaters to simulate the thermal output of waste blocks and characterise the in-situ behavior of rock masses under such conditions. Peripheral experiments will investigate the possibility of using microseismic monitoring systems to evaluate any stress-induced cracking and develop techniques of remote monitoring of conditions deep within the rock.

Eventual site investigations will obtain a structural, textural, chemical and hydrogeological picture of the rock volume to determine overall suitability and optimum position within the zone for a repository. Such work will obviously necessitate a long-term research programme involving detailed surface and sub-surface geological surveying, the latter by drilling deep boreholes; geophysical surveying and seismic monitoring; groundwater flow analysis, fluid composition determinations, and water dating (to determine residence time in the proposed repository zone and hence the likely possibility of recharge of movement).

An Eventual Repository Site

It seems likely that these long-term site investigations will be performed in several candidate areas during the next 20 years, including some areas and host rocks other than those presently under consideration. These highly intensive studies will obtain major amounts of data on relatively small volumes of rock, thus providing an unprecedented opportunity to enlarge our understanding of their nature and behavior. Armed with this information it will be possible to proceed towards a final decision on the viability of disposal of high-level radioactive wastes in geological formations.

Supplementary Readings

Abbots, J., 1979, Radioactive Waste: A Technical Solution? *Bull. of the Atomic Scientists*, v. 35, n. 7, pp. 12–18.

Andersen, J., and Dornbush, J., 1967, Influence of Sanitary Landfill on Ground Water Quality, *Amer. Water Works Assoc. Jour.*, v. 59, n. 4, pp. 457–70.

Bascom, W., 1974, The Disposal of Waste in the Ocean, *Sci. Amer.*, v. 231, n. 2, pp. 16–25.

Bergstrom, R. E., 1968, Disposal of Waste: Scientific and Administrative Considerations, *Environmental Geology Notes No. 20*, Illinois State Geol. Survey, 12 pp.

Bredehoeft, J., England, A., Stewart, D., Trask, N., and Winograd, I., 1978, Geologic Disposal of High-Level Radioactive Wastes— Earth-Science Perspectives, *U.S. Geological Survey Circular 779*, 15 pp.

Cain, J., and Beatty, M., 1965, Disposal of Septic Tank Effluent in Soils, *Jour. Soil and Water Conservation*, v. 20, pp. 101–105.

Cartwright, K., and Sherman, F. B., 1969, Evaluating Sanitary Landfill Sites in Illinois, *Environmental Geology Notes No. 27*, Illinois State Geol. Survey, 15 pp.

Chapman, N., Gray, D., and Mather, J., 1978, Nuclear Waste Disposal: The Geologic Aspects, *New Scientist*, 27 April, pp. 225–27.

Council on Environmental Quality, 1970, *Ocean Dumping—A National Policy*, U.S. Govt. Printing Office, Washington, D.C.

Fox, C. H., 1969, *Radioactive Wastes*, U.S. Atomic Energy Comm., Washington D.C.

Franks, A. L., 1972, Geology for Individual Sewage Disposal Systems, *California Geology*, v. 25, n. 9, pp. 195–203.

Galley, J. E., ed., 1968, *Subsurface Disposal in Geologic Basins—A Study of Reservoir Strata*, Amer. Assoc. Petroleum Geologists Memoir 10, 253 pp.

Gross, M. G., 1970, New York Metropolitan Region—A Major Sediment Source, *Water Resources Research*, v. 6, pp. 927–31.

Hopkins, G., and Popalisky, J., 1970, Influence of an Industrial Waste Landfill Operation on a Public Water Supply, *Water Pollution Control Fed. Jour.*, v. 42, n. 3, pp. 431–36.

Landon, R. A., 1969, Application of Hydrology to the Selection of Refuse Disposal Sites, *Ground Water*, v. 7, n. 6, pp. 9–13.

McGauhey, P. H., 1968, Earth's Tolerance for Waste, *Texas Quarterly*, v. 11, n. 2, pp. 36–42.

Miller, D. W., ed., 1980, *Waste Disposal Effects on Ground Water*, Premier Press, Berkeley, Calif. 512 pp.

Otton, E., 1972, Solid-Waste Disposal in the Geohydrologic Environment of Maryland, *Maryland Geologic Survey Report of Investigation No. 18*, 59 pp.

Pierce, W. G., and Rich, E. I., 1962, Summary of Rock Salt Deposits in the United States as Possible Storage Sites for Radioactive Waste Materials, *U.S. Geological Survey Bull. 1148*.

Piper, A. M., 1969, Disposal of Liquid Wastes by Injection Underground—Neither Myth nor Millennium, *U.S. Geological Survey Circular 631*, 15 pp.

Schneider, W., 1970, Hydrologic Implications of Solid-Waste Disposal, *U.S. Geological Survey Circular 601-F*, 10 pp.

Shapiro, F. C., 1981, *Radwastes*, Random House, New York, 228 pp.

Summers, C. M., 1971, The Conversion of Energy, *Sci. Amer.*, v. 224, n. 3, pp. 149–60.

Zanoni, A., 1972, Ground-water Pollution and Sanitary Landfills—A Critical Review, *Ground Water*, v. 10, pp. 3–13.

Films

A Day at the Dump (Stuart Finley, Inc.: 15 min.)

The City's Waste: The Country's Wealth (Filmspace, 1980: 17 min.)

Clean Town, U.S.A. (Atlantic-Richfield Co.: 16 min.)

Cloud over the Coral Reef (Moonlight Productions, 1972: 27 min.)

Installing a Septic Tank—Soil Absorption System (University of Wisconsin, 1969: 22 min.)

The 3rd Pollution (Stuart Finley, Inc.: 23 min.)

Safety in Salt: The Transportation, Handling and Disposal of Radioactive Waste (U.S. Atomic Energy Commission, 1971: 29 min.)

Sanitary Landfill—One Part Earth to Four Parts Refuse (Bureau of Solid Waste Management, 1969: 24 min.)

Wealth Out of Waste (U.S. Bureau of Mines, 1974: 27 min.)

V
GEOLOGY AND URBANIZATION

For many people, today's world is an urban world. Approximately 70 percent of the United States' population lives in urban areas, and it is predicted that by the year 2000 this figure will rise to about 85 percent. The demands that population growth and expanding settlement patterns have placed on the physical environment can be overwhelming—especially in older, established cities where the less problematical areas have already been developed, and in areas where population explosion has created a demand for rapid, high intensity development. It has, therefore, become necessary to recognize and anticipate the problems associated with these demands and to plan intelligently to cope with them. Urban geology—a field of applied geology—can contribute significantly to the solution of many environmental problems associated with urbanization. In Reading 36 Donald R. Nichols reviews the role of the earth sciences in the urban environment.

Geologic and hydrologic hazards cause considerable economic loss and hardship to individuals and urban communities. In the United States alone, annual losses run into billions of dollars, hundreds of people are killed, and thousands of families are uprooted. What can communities do to avoid or to reduce these losses? What are the roles of the local, state, and federal government in decision-making? What type of earth science information is needed by urban and regional planners? These questions and others are the subject of Reading 37.

Human interaction with natural processes not only challenges our technological ingenuity but may also pose questions involving critical value judgments. Venice was founded 1400 years ago on low land in the middle of a shallow lagoon. Throughout its history, it has been particularly vulnerable to the destructive forces of erosion, flooding by storm waves, and the long-term effects of an eustatic rise in sea level. Yet the island setting afforded a measure of security against hostile neighbors. The city's artesian aquifers furnished abundant and high quality water, and Venetians could also rely on the waters of the lagoon and the tides to dilute and flush away their wastes. But during the twentieth century, nearby industrial construction and intensive pumping of water, oil, and gas combined to cause subsidence and increase the city's vulnerability to flooding. Industrial wastes have defiled the lagoon, and acid vapors foul the atmosphere. The unique landscape and artistic treasures of Venice are clearly threatened. There are those who argue for immediate action and preservation at all cost.

The ruins of St. Pierre, Martinique, after the eruption of Mont Pelée in 1902. A fiery cloud of dust and vapor (nuée ardente) destroyed this port of 30,000 inhabitants. (Library of Congress.)

On the other side, however, are those who argue that Venice cannot survive as a viable city without continued industrial development of the port and the nearby mainland. Even interim measures, as well as long-term solutions, appear to be incompatible with both the preservation of Venice and the continued development of the port and mainland. In this respect, the controversy is similar to those involving many preservationists and developers. But it is more than a local dispute over the solution to a complex environmental problem: "Venice is not only an Italian but a worldwide patrimony. . . ." In the closing reading of this book, Carlo Berghinz explores the physical problems facing Venice and the controversy over the proposed solutions.

36 Earth Sciences and the Urban Environment

Donald R. Nichols

Cities have been built and have either prospered or declined and died because of the natural attributes of their locations. Only recently has there been much awareness of the role that earth sciences can play in defining the urban environment and in offering alternatives for its development and enjoyment.

Commerce traditionally has been regarded as the basis of urbanization. Commerce itself, however, is governed by the availability of resources, such as construction materials, land, water, and minerals, and by the ease of transportation, which in turn is governed by energy, topography, or the presence of navigable waterways. Where these resources have combined favorably to support large populations, cities have prospered and grown; where they have not, cities have gradually declined or even been obliterated overnight by the forces of nature. For example, the depletion of mineral resources created many ghost towns in the West; many ancient and even some relatively modern cities have disappeared because of catastrophes such as drought, duststorms, earthquakes, and volcanic eruptions.

Modern technology can overcome some resource problems by transporting water, by transmitting energy and fuel, and by moving minerals to factories, but the direct and indirect costs to

Reprinted from U.S. Geological Survey Annual Report, Fiscal Year 1975. U.S. Government Printing Office.

the nation's economy and environment are skyrocketing; also, opposition to the export of critical resources is growing in many areas. In addition, byproducts of modern technology produce air and water pollution that make cities ugly, uncomfortable, and even hazardous to both the health and the life of their inhabitants.

Natural Resources

Natural resources are the basis for urbanization. Without construction materials, cities could not be built; without minerals, industry could not develop and expand and most urban jobs would not exist; without fuel, factories would close and houses would be cold and unlit; without a potable water supply, society could not function; without prime agricultural land, vast concentrations of people could not be fed. Because these resources are severely limited in many areas, it is important that they be identified and judiciously developed if urban amenities are to be preserved.

Construction materials are essential to urban growth. Sand, gravel, and crushed stone are rapidly becoming scarce commodities in some urban areas because nearby sources are exhausted or are being built over or because their development is opposed as a nuisance. Such materials have a low unit value; thus, transporting them is a major factor in their cost, which roughly doubles for every 32 kilometers (20 miles) of transport.

Although the development of most mineral commodities does not conflict with urbanization, some jurisdictions have recognized the need to divert urbanization from potential resource areas. For example, Pima County in Arizona has adopted an ordinance prohibiting surface development over potentially minable copper deposits. The ubiquitous oil-well pumps in commercial, industrial, and even residential districts of Los Angeles, California, and Oklahoma City, Oklahoma, are striking testimony to the priority that society has placed on fuel resources.

As oil and gas become more scarce, other sources of fossil fuels become more important. It remains to be seen, however, whether the development of coal and oil shale, where it conflicts with urbanization and the quest for environmental quality, will be pursued. In any event, the rapid depletion of fossil fuels requires development of alternative energy sources if our present form of society is to be maintained. Even more fundamental to an increasing global population is the expansion of food production, which requires preservation of good agricultural land and water supplies and the production of fertilizers. These prerequisites are being threatened, however. For example, prime agricultural land is being converted to housing tracts, and the mining of phosphate, a critical fertilizer ingredient, is in conflict with other land use values in some areas.

Although water is usually not considered a depletable commodity, as are minerals and fuels, adequate good-quality water supplies are of special concern in most urban areas. This concern arises from the increasing demands placed on locally limited water supplies and the effects of urbanization on water quality. Urban areas compete for existing surface-water supplies from streams and reservoirs; suitable reservoir sites are few and often are in conflict with other land uses. Only by regional or, in some cases, statewide or interstate water-resource planning can serious shortages be averted.

Continuous heavy withdrawal of ground water, coupled with urbanization that paves recharge areas and that greatly increases storm runoff, has severely depleted supplies in many areas. Additionally, industrial and residential wastes have polluted both surface- and ground-water sources. In the case of ground water, the effects of existing waste-disposal practices (for example, poorly designed landfills, leaky sewer systems and septic tank systems) may not become evident for decades; where serious ground-water pollution has occurred, as it has in some areas of acid mine-water drainage, water supplies may not become potable for tens of years, if ever. It is essential, therefore, that critical ground-water supplies be identified and wisely managed.

Just as conflicts over the use of the land surface grow, so has competition for the use of the subsurface. The shallow subsurface has long been used to store and filter effluents from septic tank systems and industrial and urban wastes. In recent decades, deeper zones have been used to store industrial wastes and toxic fluids. Existing underground openings in mines and new, specially designed excavations are used increasingly, particularly in urban areas; storm-water runoff is stored temporarily in tunnels in the

Chicago, Illinois, area; natural gas is stored for peak urban use in manmade caverns in many areas; factories and warehouses occupy excavations beneath Kansas City, Missouri; tunnels are being designed, built, or expanded for utilities and rapid transit systems in many larger cities. Development of all underground resources requires extensive and detailed subsurface geologic and hydrologic data to determine their capacity for development.

Hazards

Natural hazards are taking an increasing toll of life and property. Catastrophic landslides that have claimed large numbers of lives include earthquake-generated avalanches that killed 30,000 Peruvians in 1970 and 830,000 Chinese in 1556. Not all destructive landslides are triggered by earthquakes, however. Almost 3,000 people were killed in 1963 when a large earth mass, lubricated by a rising water table, slid into the Vaiont Reservoir in Italy. The resulting overflow destroyed several towns. [See Reading 13.] We court similar disasters in the United States by building reservoirs in valleys underlain by active faults or on the flanks of volcanoes that have produced debris avalanches.

Aside from such infrequent catastrophic events, thousands of landslides occur annually throughout the United States, especially in urban areas where residential development, seeking new ground for expansion, has moved into "view" sites on hillslopes, many of which are already unstable. Cutting into these slopes, adding water to lawns, concentrating runoff from roofs and streets, and other acts of man cause landslides estimated to cost over $200 million annually. Typical damage during a single winter has been documented for only a few areas; damages of more than $25 million occurred in the nine-county San Francisco Bay region (Taylor and Brabb, 1972), $6.25 million in the Los Angeles area (Yelverton, 1971), and $250,000 in the vicinity of Seattle, Washington (Tubbs, 1974).

Most losses from flooding and erosion are predictable and can be avoided. Despite that, flooding is the most widespread natural hazard in the United States and accounts for the largest average annual property losses (White and Haas, 1975, p. 255). Nearly all cities were founded beside rivers and along coasts to be close to both the water supply and a convenient means of transportation and to enjoy the scenic views. When man encroaches on river flood plains or coastal plains, he is subject to havoc from rapid runoff of melting snow, intense rainfall, hurricanes, tsunamis—tidal surges caused by submarine landslides—and accelerated erosion and sedimentation (Table 1). Erosion along shorelines of rivers, along Lakes Michigan and Erie, in coastal California, along the Gulf Coast, and along the Atlantic is both costly and hazardous to owners of waterfront property. Major engineering works to prevent flooding and erosion in many cases succeeded only in postponing the hazard or in diverting it and intensifying it in another area.

Earthquakes pose the greatest potential hazard to many urban areas of the United States. A great earthquake today in Los Angeles or San Francisco, California, or a repeat of the Charles-

Table 1 Examples of catastrophic losses to property and lives from single hydrologic events affecting low-lying land

Event	Property losses (millions of dollars)	Fatalities
Rapid runoff from melting snow (Vanport, Oreg., flood, 1948)	$103	51
Intense rainfall (Rapid City, S. Dak., 1972)	128	238
Hurricane Agnes (1972)	3,100	117
Hurricane Camille (1969)	1,400	258
Tsunami (Hawaii, 1946)	25	159
Tsunami (Crescent City, Calif., 1964)	11	10
Submarine landslides (Valdez and Seward, Alaska, 1964)	27	42

ton, South Carolina, earthquake of 1886 would probably result in the loss of tens of thousands of lives and many billions of dollars in property damage. Most of these losses would be caused by building collapse owing to vibration or to failures in the ground beneath them (landslides and liquefaction). Losses also would result from surface ruptures along active faults that underlie many urban towns along the Pacific coast and in other parts of the West.

Subsidence of the land surface owing to ground-water withdrawal occurs in many densely populated areas in Alabama, Florida, Missouri, Texas, Louisiana, Arizona, and California. Individual houses and even entire cities have suffered major property damage; losses are likely to continue for several years, even after the causes are identified and eliminated. Land subsidence of as much as 2.5 meters (8 feet) in the Houston-Galveston area of Texas has been estimated to cost $110 million. Even greater amounts of subsidence have occurred in San Jose (4 meters, 13 feet) and Long Beach (8 meters, 26 feet), California, the latter being due to oil withdrawal. [See Reading 27.]The Baldwin Hills Reservoir failure in Los Angeles that killed five people in 1954 is generally attributed to subsidence resulting from oil-well operations. Subsidence also occurs in other states over formerly mined areas, natural cavities and caverns, heavily pumped oilfields, and partly devastated ground-water reservoirs.

Volcanic activity during historic time in the United States fortunately has been infrequent and restricted to remote areas. Elsewhere, however, the complete burial of Pompeii, Italy, in 79 A.D. and of St. Pierre on the island of Martinique in 1902 and the rapid growth to a height of 366 meters (1,200 feet) of the Parícutin Volcano on farmland in Mexico in the mid-1940's serve as reminders that we live on a dynamic planet. Large areas in the Western United States, including many urban areas (Hilo, Hawaii; Portland, Oregon; and Seattle, Washington) are adjacent to major volcanic eruptive centers. A residential subdivision near Hilo is located on a series of lava flows, the most recent of which erupted from Kilauea in 1955. The renewal of volcanic activity in 1975 on Mount Baker,

The Winter Park, Florida, sinkhole of May 1981. The sinkhole grew rapidly over a period of several days and caused more than $2 million damage. This sinkhole, like many others, formed when ground-water levels were low. (Rich Deuerling, Bureau of Geology.)

east of Bellingham, Washington, is a cause for concern in the Puget Sound area. The location and frequency of future volcanic activity are unknown, but the consequences are predictable in some areas. [The 1980 eruption at Mount St. Helens is reviewed in Reading 3.]

The safe disposal of wastes is closely interrelated with geologic and hydrologic phenomena and conditions. Certainly, a better understanding of geologic processes and environments, combined with the application of established engineering practices, could have averted the Buffalo Creek, West Virginia, disaster, where failure of coal-mine waste piles claimed 118 lives in 1972. The potential hazards from nuclear waste disposal; many types of industrial and mining operations; overconcentrations of fertilizers, herbicides, and insecticides; livestock feedlots; and municipal sewage and trash disposal have long been recognized. For example, a recent newspaper article (*Washington Post*, September 29, 1975) cited authorities in

Tampa, Florida, as saying, "High levels of radiation, which could double the chances of lung cancer, have been found in houses built on reclaimed phosphate mining lands in southwest Florida." Similar accounts have come from Colorado, Missouri, and other States.

Constraints and Opportunities in Land Use Management Afforded by Earth Sciences

Man has lived with and surmounted natural hazard and finite resource problems without much serious consequence throughout his history. Why should there suddenly be such strong concern for earth-science problems? Some reasons are obvious. The total dependency of urbanization and technology on natural resources and on a stable and conducive urban environment was no problem as long as (1) resources were plentiful, (2) population growth did not exceed the capability of the land and water to absorb the resultant wastes, and (3) complex economic and industrial centers did not concentrate in hazard areas. Recognition of the constraints that the Earth places on its exploitation is a first step in developing opportunities for continued growth through effective land and water management.

Natural resources can be harvested in such a way that the land surface and subsurface can still be put to other vital uses. Planning and development to take advantage of these uses, however, rest upon (1) recognition of resource areas, (2) definition of geologic and hydrologic controls, (3) ultimate potential uses of the land, (4) environmental consequences of development,

(5) imaginative management, and (6) an enlightened public that has the will and patience to conserve resources and guide sequential use of the land.

After resource extraction, open-pit mines can be used for recreational lakes, for flood storage and desilting basins, for buried underground structures, and for disposal of solid wastes and subsequent reclamation and development of the surface. Ground-water recharge areas can be used for nonintensive agriculture and grazing or as parkland. Treated sewage can be injected into aquifers to form barriers against salt-water intrusion. Underground mines sometimes can be reclaimed for solid-waste disposal, storage, or even for industrial sites.

Future development can minimize natural hazards by avoiding high risk areas and, at the same time, provide for such environmental amenities as open-space and recreational areas. Applying earth-science information can avert most hazard losses in new construction and reduce them in existing structures. Where hazards to structures cannot be mitigated, alternative land uses are possible. Golf courses, nurseries, agriculture, parks, and many other land uses are compatible with active faulting, landsliding, flooding, and volcanic hazards. Existing mechanisms that can be used to encourage future development and redevelopment in low-risk areas include public education, land use planning, zoning, public acquisition, and tax incentives, when they are properly based on a knowledge of natural conditions and processes.

Less catastrophic but equally or more costly problems, such as subsidence, swelling soils, erosion, and

A plant nursery, undeveloped open space, a freeway, and a cemetery are land uses most compatible with the hazards posed by this active fault. Other uses might include a drive-in theater, golf course, a riding stable, and other recreational activities. (U.S. Geological Survey.)

weak foundation conditions, can be overcome by knowing where and why they occur, how severe they are, and how to engineer for them. For example, corrective construction design and practice, stimulated by building codes, allow development on swelling or compressible soils without serious consequences.

Status of Earth Sciences in the Urban Environment

The U.S. Geological Survey, its sister State surveys, and the earth-science professions have a long and proud history of service to society. Geologists, hydrologists, and topographers, through nationwide mapping programs that have identified mineral, water, and energy resources, have provided the basis for current urban and technological development.

Today, earth scientists are becoming more active in relating their work directly to urban problems. More scientists, at all levels of government and in the private sector, are working in urban areas, making their studies more easily understandable to nonscientists, and participating in public forums. Consequently, many people now have a much greater awareness of the role that the earth sciences play in governing the quality of life in the urban environment.

This awareness is reflected in legislation such as the National Environmental Policy Act (Public Law 90-190) of 1969, the Flood Insurance Program of 1968, which was amended in 1973 to include mudslide insurance, and the Federal Water Pollution Control Act (Public Law 92-500) of 1972, and the act establishing the Environmental Protection Agency. In addition, several Federal agencies are actively seeking a greater application of earth science in their programs. For example, the Department of Housing and Urban Development joined with the Geological Survey in sponsoring a program to develop and apply earth-science information in support of land use planning and decisionmaking in the San Francisco Bay region. Similarly, the Department of Transportation and the Appalachian Regional Commission have provided funds for the Geological Survey to conduct geologic, geographic, hydrologic, and topographic studies in urban areas and to relate them to urban and regional planning needs. Leading this effort was a new

program initiated by the Survey in 1971 to conduct a series of interdisciplinary earth-science studies in representative urban areas across the country. Many new and innovative products and techniques are outgrowths of the program, including regional topographic maps at scales of 1:100,000 and 1:250,000; orthophoto quadrangles at a variety of scales and with overprinted contours; slope maps; maps delineating potential hazards such as floods, active faults, landslides, and mine subsidence; maps delineating the availability of ground water, the potential for copper deposits, and other mineral and construction resource maps; and land capability studies.

At the same time, many States have undertaken earth-science studies in urban areas and have adopted legislation for their application to land use problems. Alabama, California, Florida, Illinois, and Texas are among the States that have provided leadership in collecting, interpreting, and applying geologic, hydrologic, and soils data to urban decisionmaking. Two noteworthy examples of a State's concern for geologic hazards are California Senate Bills 351 (1971) and 520 (1972). Senate Bill 351 requires that all community general plans include a seismic safety element which assesses hazards from earthquake faulting, ground shaking, ground failure, and seismic sea waves. Senate Bill 520 initially prohibited the construction of any structure intended for human occupancy on the trace of an active fault but was later amended to exclude single-family residences.

Many local governments have independently faced up to their responsibilities to minimize hazards and conserve resources in urban areas. As early as 1952, both the City and the County of Los Angeles, California, required geologic studies to be made before construction was allowed in hazard areas. Since 1972, four counties and a major city in the San Francisco area have employed geologists on their staffs to review development plans, assist in general planning, and guide public-works construction. Geologists are now similarly employed in Tucson, Arizona, Boulder and Lakewood, Colorado, King County in Washington, and other areas. Long Island, New York, Orange County in California, and Houston, Texas, are among the many urban areas that have adopted controls to limit the further withdrawal and degradation of ground-water resources.

Private industry, which had employed geologists, hydrologists, and geotechnical engineers largely to correct costly foundation failures, now is employing an increasing number of private consultants in advance of site selection and development. Earth scientists contribute not only to the siting and construction of major urban-related structures such as dams, powerplants, highways, and office and apartment buildings, but also to the planning and designing of new communities.

The Future

As people continue to concentrate in cities, earth science becomes increasingly important in maintaining an acceptable environment. Urbanization, already so dependent on limited mineral, water, energy, and agricultural resources, will stagnate and wither unless critical resources are conserved, new resources are found, and the effects of potential catastrophic natural disasters are minimized. The future also affords many new opportunities to correct past mistakes and avoid new ones. As cities age, redevelopment should not concentrate populations in previously developed areas of potential hazards and costly foundation conditions. In the past, development generally proceeded first in areas of flat ground having good foundation conditions, but often on flood plains, sand and gravel deposits, ground-water recharge areas, and prime agricultural land, and avoided land more costly to develop (often unstable hillslopes and other hazard areas). When land was plentiful and cheap, earth science was not critical to development. More recently, development has tended to "fill in" the still open hazard areas. With less land available, increasing development costs, and major potential hazards, earth science becomes more critical in guiding urban development.

As the use of earth science grows, so will the ability of the science to devise means of collecting and applying knowledge. Computer-assisted mapping and display techniques are evolving that combine a variety of earth-science and related data and will allow legislators to rapidly assess planning solutions and the consequences of alternative courses of actions (Van Driel and Stewart, 1976). Opportunities also are in sight to use Landsat and other Earth satellites more fully in monitoring environmental and land use changes. New geophysical techniques, remote sensing tools, and the data collection and relay capacities of satellites can greatly reduce the cost of data collection and the speed with which data can be both interpreted and synthesized.

Earth-science information is useless, however, without effective communication and application. Earth, natural, and social scientists must work diligently to communicate with one another, with decisionmakers, and with the public. Where earth-science expertise exists in support of the planning process, opportunities are enhanced for partnerships between local, State, and Federal agencies to effectively address urban problems. A start has been made under the Geological Survey's pilot Urban Area Studies program. However, a much greater effort is necessary if the earth sciences are to assume their full responsibilities in helping to guide future urbanization.

References

Kiersch, G. A., 1964, Vaiont Reservoir disaster: Civil Eng., vol. 34, no. 3, p. 32–39.

Taylor, F. A., and Brabb, E. E., 1972, Map showing distribution and cost by counties of structurally damaging landslides in the San Francisco Bay region, California, winter of 1968–69: U.S. Geol. Survey Misc. Field Studies Map MF-327, scale 1:500,000.

Tubbs, D. W., 1974, Landslides and associated damage during early 1972 in part of west-central King County, Wash.: U.S. Geol. Survey Misc. Inv. Series Map 1-852-B, scale 1:48,000.

Van Driel, J. N., and Stewart, J. C., 1976, The role of computer mapping in the decision-making process for Montgomery County, Maryland: Rev. Public Data Use, vol. 4, no. 4, p. 31–40.

White, G. G., and Haas, J. E., 1975, Assessment of research on natural hazards: Cambridge, Mass., MIT Press, 487 p.

Yelverton, C. A., 1971, The role of local governments in urban geology, in Nichols, D. R., and Campbell, C. C., eds., Environmental planning and geology: Washington, D.C., U.S. Geol. Survey and U.S. Dept. of Housing and Urban Devel., p. 76–81.

37 Suggestions for Improving Decisionmaking to Face Geologic and Hydrologic Hazards

U.S. Geological Survey

What Is the Cost of Not Attempting to Reduce Losses from Geologic and Hydrologic Hazards?

Decisionmakers must weigh the cost of potential losses from geologic and hydrologic hazards against the cost of loss-reduction actions. Economic loss

Reprinted from Hays, W., ed., 1981, Facing Geologic and Hydrologic Hazards—Earth Science Considerations, U.S. Geological Survey Professional Paper 1240-B, Chap. 6.

is one measure of the cost of geologic and hydrologic hazards to the Nation. Table 1 gives estimates of both the average annual loss and the potential for sudden loss as a result of earthquakes, floods, ground failures, and volcanic eruptions. Economic loss, however, is only a fraction of the true impact because these hazards cause considerable hardship to individuals and communities, including death and physical injury, psychological trauma,

Table 1 Estimates of average annual losses and the potential for sudden loss from geologic and hydrologic hazards in the United States [Some loss estimates may be too high or too low by a factor of 2]

Hazard	Annual loss (in billion dollars)	Sudden loss potential (in billion dollars)
Earthquakes ground shaking, surface faulting, earthquake-induced ground failures, tsunamis).	0.6	50
Floods flash floods, riverine floods, tidal floods	3	5
Ground failures landslides, expansive soils, subsidence	4	6
Volcanic eruptions tephra, lateral blasts, pyroclastic flows, mud flows, lava flows	2	3

[2]Past data are too limited to determine average annual losses for volcanic eruptions. The 1978 eruption of Mount St. Helens in the State of Washington will provide a reference for the future.

disruption of lives, and a reduction in the overall stability of the community. For example, even though the 1972 Hurricane Agnes is recognized in terms of economic impact as the greatest natural disaster in the United States, its total impact should also include more than 118 deaths, more than 250,000 families uprooted in Pennsylvania, upturned graves and markers, reduced salinity in the Chesapeake Bay which affected the shellfish industry, and interrupted power and transportation.

The history of the Nation's long-term response to natural disasters suggests that public policy and action are not always adequate to reduce the very large and increasing economic and social costs. Hence, public officials and the populace must be continuously reminded of the threat and the magnitude of the economic and social disruptions that will accompany future

earthquakes, floods, ground failures, and volcanic eruptions.

Recent information (Schiff, 1980; Moorhouse and others, 1980; Smith, 1980) suggests that significant loss reduction can be achieved for some hazards with the use of simple low-cost practices. Such practices must be identified and implemented for all geologic and hydrologic hazards.

What Is the Difference between Good and Bad Decisions?

Earth-science information is only one of several types of information needed to devise methodologies for reducing losses from geologic and hydrologic hazards. Decisionmaking on matters concerning a community's vulnerability to earthquakes, floods, ground failures, and volcanic eruptions typically involves choosing from a variety of different alternatives. Decisionmakers have been described "as socio-political men who must bargin with diverse clients, knowing that the "public good" is defined in many conflicting ways by intensely competitive and self-interested groups. Such decisionmakers know that goals are fluid, multiple, inconsistent, multidimensional, and incommensurable. They also know that no fixed solutions are possible, regardless of their technical or economic elegance" (Alston and Freeman, 1975).

In spite of the difficulties in reaching decisions which meet the various definitions of the public good, the process of decisionmaking to reduce losses from natural hazards must include information about the land—surface form and drainage pattern, soil and rock properties, and its historical record of responding to natural hazards and man's activities. To neglect this type of information has proved to be costly and unwise.

What Can Communities Do?

Decisionmaking to avoid or to reduce losses from geologic and hydrologic hazards is restricted by economic, social, and public policy factors. The principal restraint is stated by the question. "How much will it cost?" If a community decides to attempt to reduce losses from geologic and hydrologic hazards, its planners and decisionmakers must face the possibility of increased costs and decide what actions are conservative and prudent.

As communities accept the premise that costs associated with specific loss-reduction actions such as avoidance, land-use zoning, engineering design, and insurance are prudent, the ques-

tion that will be asked is, "How much are we willing to pay?" An initial requirement for answering this question is for the community to determine:

• The physical causes of each natural hazard and the probability of each hazard occurring locally.
• The current local annual loss and the potential for sudden loss from each hazard.
• The local distribution of levels of relative severity expected from each hazard.
• The potential loss as a function of time and loss-reduction actions.

What Is the Benefit-Cost Ratio of Reducing Losses from Geologic and Hydrologic Hazards?

No widely accepted method exists for determining benefit-cost or risk-benefit ratios for specific loss-reduction actions. However, the following excerpt from *The Nature, Magnitude, and Costs of Geologic Hazards in California and Recommendations for Their Mitigation* [Alfors et al., 1973] provides some insight into benefit-cost analysis of the ground-shaking hazard:

Given a continuation of present conditions, it is estimated that losses due to earthquake shaking will total $21 billion (in 1970 dollars) in California between 1970 and 2000. Most of the damage and loss of life will occur in zones of known high seismic activity; structures that do not comply with the Field and Riley Acts, passed in 1933, will be especially vulnerable. If the present-day techniques for reducing losses from earthquake shaking were applied to the fullest degree, life loss could be reduced up to 90 percent, and the total value of losses could be reduced by as much as 50 percent. Total

Table 2 Earth science information needed to reduce losses from geologic and hydrologic hazards

Reduction decision
Technical information needed about the hazards from earthquakes, floods, ground failures, and volcanic eruptions.

Avoidance
Where has the hazard occurred in the past? Where is it occurring now? Where is it predicted to occur in the future?
What is the frequency of occurrence?

Land-use zoning
Where has the hazard occurred in the past? Where is it occurring now? Where is it predicted to occur in the future?
What is the frequency of occurrence?
What is the physical cause?
What are the physical effects of the hazard?
How do the physical effects vary within an area?
What zoning within the area will lead to reduced losses to certain types of construction?

Engineering design
Where has the hazard occurred in the past? Where is it occurring now? Where is it predicted to occur in the future?
What is the frequency of occurrence?
What is the physical cause?
What are the physical effects of the hazard?
How do the physical effects vary within an area?
What engineering design methods and techniques will improve the capability of the site and the structure to withstand the physical effects of a hazard in accordance with the level of acceptable risk?

Distribution of losses
Where has the hazard occurred in the past? Where is it occurring now? Where is it predicted to occur in the future?
What is the frequency of occurrence?
What is the physical cause?
What are the physical effects of the hazard?
How do the physical effects vary within an area?
What zoning has been implemented in the area?
What engineering design methods and techniques have been adopted in the area to improve the capability of the structure to withstand the physical effects of a hazard in accordance with the level of acceptable risk?
What annual loss is expected in the area?
What is the maximum probable annual loss?

costs for performing the loss reduction work would be about 10 percent of the total project loss, which with 50 percent effectiveness provides a benefit to cost ratio of 5:1.

According to Terry Margerum (1980), "for most geologic hazards, the loss amount is generally reduced well over 90 percent when construction codes are applied."

What Are the Local-State-Federal Roles in Reducing Losses from Geologic and Hydrologic Hazards?

A program for reducing losses from geologic and hydrologic hazards is

Table 3 Suggested contributions for government

Federal

Provide national and regional earth-science information on each hazard.
Develop federal legislation and policy to support short- and long-term loss-reduction programs.
Provide technical assistance.
Provide advice on preparedness.
Encourage short- and long-term planning to reduce losses.
Provide support for research on scientific, engineering, and socioeconomic problems.
Conduct postdisaster surveys.

State

Formulate preparedness plans.
Adopt legislation.
Enforce State laws and regulations designed to reduce losses from specific hazards.
Create councils of interdisciplinary experts, such as the California Seismic Safety Commission and
   the Utah Seismic Safety Advisory Council, to recommend public policy and loss-reduction
   programs.
Aid in identifying risks to communities.
Aid in identifying sources of funding and technical expertise to use in loss reduction programs.
Provide support and, whenever possible, funding for research.

Local

Identify a community leader to rally local support for loss-reduction actions, favoring short-term
   solutions.
Collect, archive, and update earth-science information which affects loss-reduction actions.
Modify land-use and development ordinances to reflect the best available knowledge of geologic
   and hydrologic hazards.
Increase public awareness and encourage individual preparedness.

likely to be more effective if plans are formulated and conducted by local governments. Because geologic and hydrologic hazards may well extend beyond the jurisdiction of a single local government, neighboring governments must avoid conflicting plans and policies. This consideration is particularly important in flood-plain management; for example, one community's plan to zone flood-prone land for aesthetic and recreational use may be jeopardized by another community's plan to reserve the opposite stream bank for industrial uses. Alternatively, an upstream community could conceivably adopt a policy of urbanization and structural flood protection that would lead to an increased probability of floods for a downstream community.

R. A. Platt and others (1980) suggest that the substate regional level is the most appropriate one for coordinating local plans and policies. Regional coordination units are more effective if they are able to offer incentives to the communities that enter into agreements and coordination of loss-reduction plans.

A team of interdisciplinary experts from all levels of government and the private sector can provide guidance for planning and decisionmaking to reduce losses. Each expert and representative of government and the private sector contributes whatever they do best.

References

Alfors, J. T., Burnett, J. L., and Gay, T. E., Jr., 1973, The nature, magnitude, and costs of geologic hazards in California and recommendations for their mitigation: California Division of Mines and Geology Bulletin 198, 112 p.

Alston, R. M., and Freeman, D. M., 1975, The natural resources decision-maker as political and economic man—Toward a synthesis: Journal of Environmental Management, v. 3, p. 167–183.

Burton, I., Kates, R. W., and White, G. F., 1978. The environment as hazard: New York, Oxford University Press, 240 p.

Davenport, S. S., and Waterstone, Penny, 1979, Hazard awareness guidebook: Planning for what comes naturally: Austin, Texas Coastal and Marine Council, 41 p.

Margerum, Terry, 1980, The big quake—What local governments can do: Berkeley, Calif., Association of Bay Area Governments, 36 p.

Moorhouse, D. C., James, S. E., and Patwardhan, A. S., 1980, Loss reduction planning for earthquakes and earthquake prediction—A preliminary model for businesses, in Hays, W. W., ed., Earthquake prediction information: U.S. Geological Survey Open-File Report 80-843, p. 262–278.

Platt, R. A., McMullen, G. M., Paton, R., Patton, A., Grahek, M., English, M. R., and Kusler, J. A., 1980, Intergovernmental management of floodplains: Boulder, Colo., Institute of Behavioral Science, University of Colorado, Monograph 30, 317 p.

Schiff, A. J., 1980, Pictures of earthquake damage to power systems and cost-effective methods to reduce seismic failures of electric power equipment: West Lafayette, Ind., Center for Earthquake Engineering and Ground Motion Studies and School of Mechanical Engineering, Purdue University, 49 p.

Smith, S. M., 1980, Earthquake predictions and their effects on preparedness—A public education perspective, in Hays, W. W., ed., Earthquake prediction information: U.S. Geological Survey Open-File Report 80-843, p. 307–328.

Ward, D. B., 1978, Communicating seismic safety information for public policy development, in Hays, W. W., ed., Communicating earthquake hazard reduction information: U.S. Geological Survey Open-File Report 78-933, 426 p.

38   Venice Is Sinking into the Sea

Carlo Berghinz

The flood caused by a storm surge of extraordinary height early in November 1966, during which more than eighty percent of Venice was submerged, suddenly called the world's attention to the city. The problem of Venice's survival appeared dramati-

Reprinted from *Civil Engineering*, v. 41, pp. 67–71, 1971, by permission of the author and the American Society of Civil Engineers. Copyright 1971 by A.S.C.E.

Mr. Berghinz is Director, Electroconsult, Milan, Italy.

cally urgent. The inventory of the damages to the town and to its artistic patrimony, evaluated at more than $70 million, and the increasing frequency of such catastrophes, were too serious to allow further delays.

What is the real danger menacing Venice, and why is it becoming more and more serious?

Venice and its Lagoon

Venice was founded in the seventh century A.D. by the mainland population in the middle of a large lagoon in order to ensure safety.

The Venetian people prevented siltation by diverting the three rivers (Brenta, Sile and Piave) that flow in the lagoon, and built wall protections (the "murazzi") along the seaward side of the lagoon. Venice has always been "wed to the sea" and only in the past century has it been connected with the mainland (Poro Marghera and Mestre) with a 3-mile railroad bridge. In 1935 a highway bridge was built.

The lagoon extends along the Adriatic sea for 35 miles and is some 6 miles wide (see Fig. 1). Total area is about 210 square miles. Tidal surges flow in the lagoon through three inlets: Lido, Malamocco and Chioggia. The tide entering from each inlet commands a defined part of lagoon, which is therefore considered as divided in three basins.

One-third of the lagoon is occupied by large shallow permanent pools and by a system of canals, partly natural and partly artificial, ranging from 3 ft to more than 50 ft of depth. Some 20 percent of the area, located mainly north and southwest, has been gradually enclosed by dikes and transformed into fishery ponds. Toward the mainland the lagoon bottom rises and lies within the range of normal tide fluctuations, giving birth to mud and sand flats, called "barene." The wet barene are submerged by each tide, the dry barene only during spring tides. The barene cover more than 40 percent of the lagoon; a part of them has been reclaimed to expand the Marghera's industrial zone. Some 5 percent of the lagoon is covered by a number of small flat islets. Venice is built on pileworks driven in some of these islets as well as in the surrounding lagoon.

The "Acqua Alta"

Storm surges, locally called "acqua alta" (high water) are a centuries-old phenomenon in Venice. But never has their intensity and frequency reached the level of the last few years.

An acqua alta depends on the simultaneous action of several causes, in addition to normal tide fluctuation, that raise the sea water to an abnormally high level, thus flooding the city of Venice. Among them, atmospheric pressure, rain, wind and mass oscillations of the Adriatic sea are most important.

Tides flow in and out from the lagoon through the three inlets. The maximum tidal prism has been estimated at 260,000 acre-ft, out of which 115,000 is in the Lido basin. The corresponding water level rise averages 2 ft. At the end of the channel, especially where the "barene" have been reclaimed and substituted with embankments that suddenly stop tide flow, additional rises of two to six in. over normal tide level, basically due to kinetic energy mobilization, have been recorded. In the meandering network of narrow and shallow canals of the city, the tidal rise speed lowers to less than 20 in. per second: a minimum tide speed (although always detrimental to building foundations), is indispensable because it ensures canal cleaning and permits the town to exist without sanitary sewers.

Abnormal drops of atmospheric pressure (in Venice a minimum of less than 29 in. of mercury have been registered) can raise the sea level four

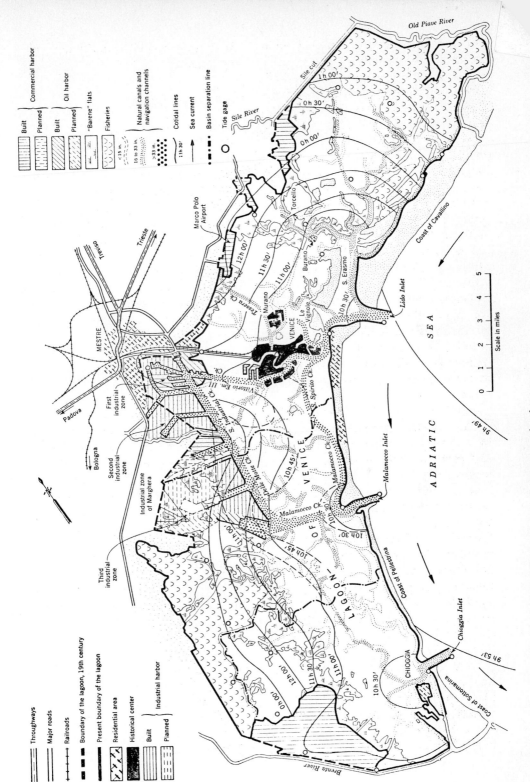

Built — Commercial harbor
Planned
Built — Oil harbor
Planned
"Barene" flats
Fisheries
Natural canals and navigation channels
<15 in.
16 to 33 in.
>33 in. — Cotidal lines
1 h 30' — Sea current
Basin separation line
Tide gage

Old Piave River

Sile cut

1 h 00'

0 h 30'

0 h 00'

Sile River

Torcello

Burano

S. Erasmo

Coast of Cavallino

Lido Inlet

Le Vignole

12 h 00'

11 h 30'

11 h 00'

Murano

Passera Ch.

Marco Polo Airport

MESTRE

Treviso

Trieste

Padova

Bologna

First industrial zone

Second industrial zone

Industrial zone of Marghera

Third industrial zone

VENICE

S. Spirito Ch.

10 h 30'

Malamocco Inlet

S. Secondo Ch.

Vittorio Em. III Ch.

S. Gerolamo Ch.

Della Monte Ch.

Malamocco Ch.

10 h 45'

10 h 30'

Malamocco Ck.

10 h 45'

11 h 00'

LAGOON OF VENICE

Coast of Pellestrina

ADRIATIC SEA

45° 49'

Scale in miles
0 1 2 3 4 5

12 h 00'

11 h 30'

11 h 00'

0 h 00'

10 h 30'

CHIOGGIA

Chioggia Inlet

Coast of Sottomarina

45° 53'

Brenta River

FIG. 1. The Venice lagoon—main hydraulic features and manmade changes.

Aerial view of St. Mark's Square, in Venice, an area subject to frequent flooding. (Photo courtesy of Italian Government Travel Office.)

to eight in. and, exceptionally, more than 12 in.

The winter concentration of precipitation on the Northern Adriatic sea and on the basins of the inflowing rivers causes seasonal rises of the sea level of about four to eight in.

The south-east wind ("scirocco"), blowing the sea waters towards the lagoon, is among the main causes of the "acqua alta." A 6 mph wind raises the water four in. and 35 mph wind, more than three ft.

And finally, in the Adriatic sea, mass oscillations—seiches—are very important and can raise the water level up to 15 to 25 in. with peaks of three ft.

The "acqua alta", as a compound effect of the above causes, reaches high levels in the measure in which the different components attain simultaneously the highest values. An estimate allows comparison of the 1966 flood with the flood that could occur in the case that the different components attain simultaneously the highest val-

Flooding of St. Mark's Square during "acqua alta" of December 1966. (United Press International.)

ues recorded (water elevation in inches above the mean sea level). (See Table 1.)

If for each component the theoretical peak value is assumed, the "acqua alta" would reach' an even higher level, 120 in.; the probability of such a disaster is every 10,000 years, but the 1966 flood probable frequency is less than 250 years, and "acque alte" of less than 60 in. are common. Since nearly 70 percent of Venice lies less than 50 in. above mean sea level, the danger is easily appreciated.

However, the greatest worry is not the "acqua alta" itself, but its greatly increasing frequency. Out of 58 "acque alte" recorded in the past hundred years, 48 occurred in the last 35 years, and 30 in the last 10 years. In other words, in the first 65 years one "acqua alta" every 5 years, in the following 25 years almost one "acqua alta" per year, and in the last 10 years three "acque alte" per year have been suffered by Venice.

Sinking of the Town

Land level is dropping relative to sea level. In reality two distinct movements, with additive effects, have been recognized: the gradual rising of sea level (eustatism), and the progres-

Table 1

	Nov. 1966	Highest "Acqua Alta" recorded
Tides	10	24
Atmospheric pressure	6	8
Rainfall	8	8
Winds	33	35
Mass oscillations	20	25
	77	100

Table 2

Period	Total subsidence in.	Mean annual subsidence in.
1908–1925	0.7	0.04
1926–1942	1.5	0.09
1943–1952	1.4	0.14
1953–1961	1.8	0.20

sive sinking of the lagoon bottom (subsidence).

Eustatism happens all over the world as a consequence of the polar caps and glaciers melting, due to the increasing of earth temperature (0.1 deg C each century). An eustatism of nearly 0.06 in. per year, i.e. about 6 in. per century, has been observed at Venice.

Subsidence (absolute, not compared with the mean sea level) is nothing new in the history of Venice lagoon. Subsidence of 140 to 240 in. has been observed, through archeological findings, from prehistoric time, and of 70 to 120 in. from Roman time. Average subsidence value for earlier centuries is estimated at 4 in. per century with a practically constant trend. In recent years, however, subsidence has acquired a progressive and definitely worrying trend. The few available data are sufficient to focus the problem; for instance, measurements conducted systematically at Palazzo Loredan (Venice Townhall) show the following subsidence values (see Table 2).

While the slow subsidence of earlier centuries has been always interpreted as the natural settlement of the loose material layers forming the lagoon's bottom for a depth of several hundred yards, no agreement has been reached on the reasons for the rapid increase of the subsidence rate observed during the last decades, and a passionate debate is still going on. Several explanations have been imagined, including movements of the deep rock formations supporting the above mentioned loose materials, natural gas exploitation in the Po river delta (carried out mainly between 1935 and 1955 and later forbidden), and man-made alterations to the lagoon's natural behavior. From a comparative analysis of the different theories, the most important cause of subsidence is likely the increased aquifer exploitation from wells located around and inside the lagoon, and in the additional load of extensive recent building, both connected with the Porto Marghera and Mestre industrial and urban development.

In effect, while detailed leveling surveys have shown that the direct influence of the gas exploitation in the Po delta stops before reaching the southern borders of the lagoon many miles from Venice, which therefore is not affected, on the contrary several observations, even if not exhaustive, show that a direct correlation exists between subsidence and aquifer exploitation (more than 200 cfs from 7,000 wells).

As a result, the aggregate of eustatism and subsidence leads to a progressive sinking of Venice, of more than 0.25 in. per year and, what is worse, this rate is steadily increasing. Seventy percent of Venice lies between 3.5 and 4.5 ft above mean sea level. Because of the town's sinking, these elevations are reached by the "acqua alta" with increasing frequency. As a consequence the surface of the flooded area goes on extending at a much higher rate than in the past, and damages become more and more heavy.

Porto Marghera Industrial Development

Marghera's port was built in the first half of this century with the distinguishing target, in comparison with other Italian ports, to allow the industrial transformation of raw materials locally. It became a clearly defined industrial port. A navigation channel between Marghera and Venice made Marghera accessible from the Lido inlet, and a first industrial zone of some 1,300 acres was built between 1919 and 1932, partly on mainland and partly on land taken from the "barene". A second industrial zone of 2,700 acres was built between 1959 and 1961.

Recently the works for a new industrial zone of 10,000 acres have begun. With the third zone the Marghera port (whose present yearly movement is more than 20 million tons) would become one of the major commercial, industrial and oil centers of South Europe; a new waterway should in fact link Venice to Milan and a motorway from Venice to Munich has been planned. The major installations of the third zone should be a new 2,000 mw

thermal power plant, a 2 million ton per year steel mill, the extension of existing aluminum and chemical industries, and an oil terminal with 16 berths and a new oil refinery. In addition to the third zone a new navigation channel is under excavation and nearly finished. This channel, 11 miles long, 600 ft wide and 50 ft deep, will link directly the industrial zones with the Malamocco inlet and so, avoiding Venice, will allow the passage of ships up to 60,000 tons.

The third industrial zone and the Malamocco channel are the object of passionate controversies. While from one side their clear economic importance is pointed out, from the other side questions are raised upon the effects that such works might have on the lagoon's internal balance. The third zone, in particular, is under discussion: the earthfilling will involve nearly 10 percent of the lagoon's surface, and almost 20 percent of the "barene" zone. There is concern about the detrimental effect the earthfilling might have on the "acqua alta"; there is fear also that new heavy installations might increase subsidence.

The Malamocco channel itself has been at first discussed, because it was said that, by increasing tide flow in the Malamocco basin, the channel would cause a shifting towards Venice of the Lido and Malamocco basins' separation line. As a consequence a possible increased tide rise in Venice together with a fall of tide speed and with pollution danger were feared. But, on the other hand, the channel has the clear advantage of detouring from Venice today's very dangerous tanker traffic passing from the Lido inlet, and this consideration has prevailed.

Perspectives and Actions

The rescue of Venice from its present progressive decay, and the contemporary development of the Porto Marghera industrial area are the factors in the complex problem of the lagoon. Everybody, and first of all the Venetians, agree on the necessity of rescuing Venice, but opinions differ on the way to do it.

On one side a numerous group of persons joined up in the "Italia Nostra" Association, are fighting a strenuous struggle, whose main objective is the preservation of the artistic patrimony of Venice, including its unique landscape. They therefore insist on the immediate cessation of the indiscriminate development of the industrial port, to avoid possible harmful impacts on the lagoon's internal equilibrium and to prevent the further spoiling of the town and its surroundings.

On the other side, the Port Authority and the Industrialists' Association of Marghera are convinced that the survival of Venice as a town without further industrial development cannot be conceived and therefore, in their opinion, the construction of the third industrial area and of the Malamocco channel should no more be delayed.

These two tendencies will have to meet to allow Venice to be saved.

Many individual initiatives have been started in the last few years for the rescue of Venice. Defenses from the sea have been studied as, for example, a system of forecast and alarm for the "acqua alta". Proposals have been made for different types of barriers such as a circular barrier inside the lagoon around Venice, translagunar barriers aiming at separating the Lido basin from that of Malamocco, belt-barriers for the defense of the entire lagoon. The possible closure of the lagoon outlets by stormgates in case of exceptional events has also been considered. Since all these works would limit the regular tide flow, the necessity would arise of providing Venice with an adequate sewerage system.

Communication between Venice and the mainland has also been the object of studies and proposals; the construction of translagunar highways has been examined but it is not likely to take place for the disfigurement which would be caused to the town. Preference is given to underground transportation and various projects provided for the construction of sublagunar highways and railways. The improvement of existing translagunar transportation as well as the linkage of the lagoon with a system of waterways on the mainland has been planned.

Various projects for the construction of new aqueducts from the mainland have been proposed with a view of substituting pumping from wells situated in the lagoon area.

The transformation of ancient Venetian palaces into office buildings without spoiling their character is also under consideration.

Unfortunately, few initiatives have so far materialized due to inadequate coordination of the different actions, and to the insufficient knowledge of the lagoon problems.

Recently, an Interministerial Study Committee for the Defense of Venice has been constituted and is already carrying out investigations in the fields of town planning and construction, sanitation and biology, sub-soil geology, geophysics and geotechnics, hydraulics (including models of the lagoon and studies on possible water table recharge), oceanography and meteorology, and administrative and legislative aspects. The studies are carried out in collaboration with various international and national bodies such as UNESCO, CNR (National Research Council), universities and private companies.

Progressive closure of existing artesian wells, prohibition of natural gas or oil drilling in proximity of Venice and, recently, submittal of any initiative in the Venice Lagoon to an appropriate Committee for control and approval are the provisional emergency measures being taken pending the findings of these investigations.

All these efforts are expected to give rise to the appropriate solutions of the various problems for Venice revival. The work is hard, and, as Venice is not only an Italian, but a worldwide patrimony, the many proposals of international scientific collaboration have been and are gratefully accepted.

Editor's note: In April 1973 the Italian government passed a long-delayed 510 million dollar law to safeguard Venice. The Bill provides funds for the restoration of paintings, frescoes, sculptures, Renaissance palaces, and unpalatial homes. There are also provisions for a sewerage system and laws against dumping noxious wastes into the surrounding water. Domestic heating units would be converted from sulphurous fuels to methane, and a 10 million dollar aqueduct from the Sile River would replace artesian wells. An 80 million dollar set of moveable dikes would be installed to hold back flood waters. Further filling of the Venetian Lagoon would be prohibited.

Author's note: In September 1975 the Italian Ministry of Public Works called for an international design competition. Participants were to design measures that assured a hydrogeologic equilibrium in the Lagoon of Venice and lower flood levels in the historical centers. Five designs, which included permanent and temporary closures of the harbor inlets, were submitted and examined by a special Commission. While acknowledging that the designs were the product of detailed studies and that they contributed substantially to the understanding and solution of the problems of protecting Venice and its lagoon, the Commission concluded that none of the designs completely satisfied all of the requirements of the program. These requirements included the gradual application, flexibility, and reversibility of the protective measures, called for by the complex and delicate nature of Venice's problems.

In January 1980, the Ministry of Public Works purchased the five designs and entrusted a board of consultants (consisting of five Italian university professors, one Dutch professor, and an expert of the National Research Council) to elaborate a unified design on the level of a generalized project (preliminary design).

By 1981, the design was complete. It includes partial closure of the lagoon, with gates at the Lido and the Malamocco inlets. The protective works are designed to protect against storm surges; at the same time, flow will not be confined to narrow, high-velocity channels that would erode easily, causing difficulties for navigation. The Commission for the Defense of Venice recently approved the plan, and final design and construction will start after the approval by the Supreme Council of Public Works and the Italian Parliament. The project will be an estimated seven years in construction and the cost of the dikes are estimated at about 515 million dollars. Funds for the work must be secured because the existing appropriations expired at the end of 1981. With pollution control and restoration of buildings and monuments, the total estimated cost of the project could exceed 735 million dollars. [June 1982]

Supplementary Readings

Adams, V., 1975, Earth Science Data in Urban and Regional Information Systems—a review, *U.S. Geological Survey Circular 712*, 29 pp.

Alfors, J., Burnett, J., and Gay, T., Jr., 1973, Urban Geology Master Plan for California, *Bull. 198, Calif. Div. Mines and Geology.*

Blair, M. and Spangle, W., 1979, Seismic Safety and Land-Use Planning—selected examples from the San Francisco Bay Region, CA, *U.S. Geological Survey Prof. Paper 941-B.*

Britton, L., Averett, R., and Ferreira, 1975, An Introduction to the Processes, Problems, and Management of Urban Lakes, *U.S. Geological Survey Circular 601-K*, 22 pp.

Coates, D., ed., 1976. Urban Geomorphology, *Geological Soc. Amer. Special Paper 174.*

Ferguson, H., ed., 1974, Geologic Mapping for Environmental Purposes, *Geological Society America Engineering Case Histories No. 10*, Geological Soc. Amer.

Grava, S., 1969, *Urban Aspects of Water Pollution Control*, Columbia Univ. Press.

Guy, H., 1970, Sediment Problems in Urban Areas, *U.S. Geological Survey Circular. 601-E.*

Hamilton, J., and Owens, W., 1972, Effects of Urbanization on Ground-Water Levels, *Bull. Assoc. Engineering Geologists*, v. 9, n. 4, pp. 327–34.

Legget, R., 1973, *Cities and Geology*, McGraw-Hill, New York, 624 pp.

Leopold, L., 1968, Hydrology for Urban Land Planning—a guidebook on the hydrologic effects of urban land-use, *U.S. Geological Survey Circular 554*, 18 p.

Leveson, D., 1980, *Geology and the Urban Environment*, Oxford Univ. Press, New York, 386 pp.

McGill, J., 1964, Growing Importance of Urban Geology, *U.S. Geological Survey Circular 487.*

Nilsen, T., et al, 1979, Relative Slope Stability and Land-Use Planning—selected examples from the San Francisco Bay Region, CA, *U.S. Geological Survey Prof. Paper 944.*

Risser, H. and Major, R., 1967, Urban Expansion—an opportunity and a challenge to industrial mineral producers, *Environmental Geology Notes Number 16.* Illinois State Geological Survey.

Savini, J., and Kammerer, J., 1961, Urban Growth and the Water Regimen, *U.S. Geological Survey Water Supply Paper 1591-A.*

Schneider, W., Rickert, D., and Spieker, A., 1973, Role of Water in Urban Planning and Management, *U.S. Geological Survey Circular 601-H.*

Schneider, W., and Goddard, J., 1974, Extent and Development of Urban Flood Plains, *U.S. Geological Survey Circular 601-J.*, 14 pp.

Thomas, H., and Schneider, W., 1970, Water as an Urban Resource and Nuisance, *U.S. Geological Survey Circular 601-D.*

Utgard, R., McKenzie, G., and Foley, D., 1978, *Geology in the Urban Environment*, Burgess Pub. Co., Minneapolis, 355 pp.

Waananen, A., et al, 1977, Flood-prone Areas and Land-Use Planning—selected examples from the San Francisco Bay Region, CA, *U.S. Geological Survey Professional Paper 942.*

Appendix A:
Classification of Rocks

Rocks are aggregates of one or more minerals. They can be characterized and classified on the basis of the number and proportions of their constituent minerals and by their texture (grain size, shape, and arrangement). Texture and mineral composition are related to rock-forming processes, and the more common schemes for classifying rocks are based on mode of origin. All rocks can be classified into three main groups: igneous, sedimentary, and metamorphic. The cyclical relationship between earth processes and the major rock types is depicted in the rock cycle (Fig. 1, p. 280).

Igneous Rocks. Most igneous rocks have solidified from molten silicate material to form an interlocking network of crystals (*crystalline texture*). The most common crystalline igneous rocks are shown in Table 1 which classifies igneous rocks on the basis of their mineral compositions (variations in the relative amounts of feldspar, quartz, and ferromagnesian minerals) and texture (finely crystalline or coarsely crystalline). Explosively ejected materials such as volcanic tuff and volcanic breccia consist of volcanic glass and rock fragments.

Sedimentary Rocks. About 75 percent of the earth's surface is covered by a thin veneer of sedimentary rocks. Most sedimentary rocks have been derived from pre-existing rocks through mechanical weathering and consist of mineral and rock fragments that have been cemented together to form a *clastic texture*. Other sedimentary rocks result from chemical precipitation in the aquatic environment. These rocks have a *crystalline texture*. The most common sedimentary rocks

Table 1 Classification of common igneous rocks

Texture	Rock name		
Intrusive coarsely crystalline rocks	Granite	Diorite	Gabbro
	Quartz and feldspar decrease ⟶ Ferromagnesian minerals increase		
Extrusive finely crystalline rocks	Rhyolite	Andesite	Basalt
Fragmental	Tuff or volcanic breccia		
Glassy	Obsidian or pumice		

Table 2 Classification of common sedimentary rocks

	Texture	Composition	Rock name
CLASTIC	Gravel-sized particles	Quartz dominant. Variable amounts of rock fragments, feldspar, and carbonate may be present.	Conglomerate and breccia
	Sand-sized particles	Quartz dominant. Variable amounts of feldspar and carbonate may be present.	Sandstone
	Sand-sized particles	Carbonates dominant. Variable amounts of quartz or clay may be present.	Clastic limestone or fossiliferous limestone
	Silt-sized particles	Quartz dominant. Variable amounts of clay may be present.	Siltstone
	Clay-sized particles	Clay minerals dominant	Shale
CRYSTALLINE		Calcite	Limestone or travertine
		Dolomite	Dolostone
		Chert	Chert
		Halite	Rocksalt
		Gypsum	Rock gypsum
		Modified plant fragments	Peat, lignite, or coal

are listed in Table 2. Sedimentary rock sequences are distinguished by surfaces (*bedding planes*) separating the layers (*strata*) of adjacent rocks. Each bedding plane marks the termination of one deposit and the beginning of another.

Metamorphic Rocks. Metamorphic rocks are the product of changes induced by heat, pressure, and chemically active fluids. The changes result in new minerals or new textures or both. The dominant texture is crystalline, but in some cases the crystals are aligned so as to resemble layering. Such alignment is called *foliation*. Any type of rock may be subject to metamorphism. The most common metamorphic rocks are listed in Table 3.

Table 3 Classification of common metamorphic rocks

Texture		Mineral composition	Rock name
Foliated	Fine ↑ ↓ Coarse	Variable	Slate Phyllite Schist Gneiss
Nonfoliated		Predominately quartz	Metaquartzite
		Calcite or dolomite	Marble
		Predominately dark silicate minerals	Hornfels

Appendix B:
Classification and Engineering Properties
of Soils and Surficial Materials

Classification. Pedologists (soil scientists) define soils as those residual products of the weathering process which can support rooted plant life. The primary factors which determine the characteristics of the soil are (1) nature of the parent material, (2) climate, (3) topography, (4) activity of organisms, and (5) time. The relative importance of each factor varies with the geologic and geographic setting. Because a large number of variables are involved in soil formation and because they operate at varying intensities, there is a seemingly infinite variety of soil types. Furthermore, soils appear to represent a continuum of materials of varying composition and size without any clearly defined natural limits within the continuum. This has posed serious problems for the recognition and classification of soils. One of the most comprehensive schemes, formulated in 1960 by the U.S. Department of Agriculture, is known as *Soil Taxonomy* or the *Seventh Approximation*. The system includes 10 orders, 40 suborders, 120 great groups, 400 subgroups, 1500 families, and 7000 series. Physical and chemical properties of the soil profile, such as soil morphology, organic content, color, and nutrient content, are emphasized. It is especially designed for agricultural purposes but it is of limited use for engineering problems.

The Department of Agriculture also classifies soils based on grain size only (Fig. 1).

Many mature soils consist of three basic layers, or horizons, which make up the *soil profile* (Fig. 2). The *A horizon,* or topsoil, contains decomposed organic matter. Soluble material is leached from this layer by downward moving water. The underlying *B horizon,* or subsoil, accumulates the leached material and is generally enriched in clay and iron oxides. The *C horizon* consists of partially altered parent material.

Engineers define soils as any solid earth material which can be removed without blasting. This includes soils as defined by pedologists *and* surficial deposits (unconsolidated transported materials) which are not capable of supporting rooted plant life (Table 1). A number of engineering soil classification schemes are based on the relative amounts of clay (< .002 mm), silt (.002 to .074 mm), sand (.074 to 2.0 mm), and gravel > 2.00 mm). *The Unified Soil Classification System* shown in Table 2 has been widely accepted.

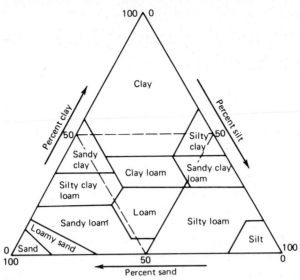

FIG. 1. U.S. Department of Agriculture classification of soil-size classes.

Table 1 Selected transported surficial deposits

Geomorphic agent	Surficial deposits
Running water	*Floodplain alluvium.* Large range in morphology and grain size. Coarser sediments are associated with the stream channel and finer sediments with the floodplain. Abandoned floodplain deposits developed during higher water levels are referred to as *terraces.*
Lakes	Includes fine- to coarse-grained clastics, organic matter, and biochemical precipitates. Silt- and clay-sized particles dominate the deposit.
Wind	*Sand dunes.* Highly variable morphology. Well-sorted and well-rounded fine sand dominate the deposit. *Loess.* Homogeneous, angular, silt-sized particles.
Waves	*Beach sands and gravel.* Tend to be well sorted. Gravel clasts rounded and disk-shaped.
Glacial ice	*Till.* Dominantly unsorted and unstratified sediments deposited directly by glacial ice with wide range of rock types and shapes. *Stratified drift.* Includes deposits reworked by glacial meltwaters. There is a wide range in rock type and in the morphology of the deposit.
Gravity	*Colluvium.* Generally poorly sorted, incoherent mixture of angular rock fragments and fine-grained materials. Chaotic internal structure. *Talus.* Coarse, angular rock fragments collecting at the base of a slope.

Appendix B: Classification and Engineering Properties of Soils and Surficial Materi

Table 2 Outline of major divisions of Unified Soil Classification System

Coarse-grained soils
  Gravels
    clean gravels (< 5 percent fines)
    dirty gravels (12 to 50 percent fines)
  Sands
    clean sands (< 5 percent fines)
    dirty sands (12 to 50 percent fines)

Fine-grained soils
  Silts (nonplastic)
  Clays (plastic)

Predominantly organics

FIG. 2. Soil profile showing A, B, and C horizons (Photo courtesy of U.S. Department of Agriculture, Soil Conservation Service.)

Although particle size is of great importance to soil performance and design, particle shape, arrangement, and composition are also important, but they are not incorporated in most classification schemes.

Engineering Properties of Soils and Surficial Deposits. The physical and chemical properties of soils and surficial deposits are of paramount importance to engineers and geologists who are concerned with excavation conditions, the stability of structural foundations, landslide potential, response to earthquake vibrations, waste disposal, water resources, and availability of sand, gravel, and aggregate. Most of the physical properties of soil are related to texture, which refers to the size, shape, and arrangement of the mineral grains. The chemical properties are related primarily to mineral composition and water content. There often is, of course, an interplay between the physical and chemical properties. It is important not only to measure these properties in their natural state but to also predict the changes that are likely to occur in response to changes in land use.

Many of these same properties apply to rocks. Other engineering properties of rocks are durability (hardness and resistance to abrasion, sudden impact, freeze and thaw, and wetting and drying), strength, and the presence of such features as bedding planes, foliation, and joints and faults.

The most significant engineering properties are included in Table 3.

Table 3 Engineering properties of soils and surficial material

Texture	Refers to size, shape, and arrangement of particles. Size distribution is usually emphasized.
Density	Refers to weight per unit volume and varies with moisture content and compactness.
Porosity	The percentage of the bulk volume occupied by interstices.
Permeability	The capacity of porous material to transmit fluids.
Moisture content	May be recorded as a percent by weight. Includes gravitational water, capillary water, and hydroscopic water.
Plasticity (Atterberg limits)	
liquid limit	The water-content boundary between liquid and plastic states.
plastic limit	The water-content boundary between the plastic and semisolid states.
plasticity index	The water-content range at which a soil is plastic.
Shrink-swell potential	Refers to the tendency to change volume through the gain or loss of water. *Expansive soils* show significant volume changes in response to changes in water content.
Compressibility	A measure of the tendency to consolidate or decrease in volume.
Cohesion	Refers to the ability of soil particles to stick together by surface forces.
Erodability	Refers to the ease with which soils can be transported by wind or water.
Ease of excavation	A measure of the procedures required to remove the soil.
Bearing capacity	The maximum load that the soil (or bedrock) can support without failing in shear.
Sensitivity	Refers to the response of soil to remolding. Sensitive soils loose shear strength upon remolding.
Corrosion potential	Generally refers to the potential for corroding various metals. Evaluated by recording soil pH and electrical resistivity.

Appendix C:
Geologic Time Scale

RELATIVE GEOLOGIC TIME			ATOMIC TIME
ERA	PERIOD	EPOCH	
Cenozoic	Quaternary	Holocene	
		Pleistocene	2-3
	Tertiary — Neogene	Pliocene	12
		Miocene	26
	Tertiary — Paleogene	Oligocene	37-38
		Eocene	53-54
		Paleocene	65
Mesozoic	Cretaceous	Late / Early	136
	Jurassic	Late / Middle / Early	190-195
	Triassic	Late / Middle / Early	225
Paleozoic	Permian	Late / Early	280
	Carboniferous — Pennsylvanian	Late / Middle / Early	
	Carboniferous — Mississippian	Late / Early	345
	Devonian	Late / Middle / Early	395
	Silurian	Late / Middle / Early	430-440
	Ordovician	Late / Middle / Early	500
	Cambrian	Late / Middle / Early	570
Precambrian			3,600 +

Appendix D:
Metric Conversions

Linear Measure

1 millimeter (mm)	= 0.039 inch
1 centimeter (cm)	= 0.39 inch
1 meter (m)	= 39.4 inches = 3.28 feet = 1.09 yards

Centimeters / Inches ruler

1 inch	= 2.54 centimeters
1 foot (ft.)	= 30.5 centimeters
1 yard (yd.)	= 0.91 meter
1 mile (mi.)	= 5,280 feet = 1.61 kilometer
1 kilometer (km)	= 1,000 meters = 0.62 mile

Area Measure

1 square yard	= 0.836 square meters
1 square mile	= 640 acres = 2.6 square kilometers
1 acre	= 43,560 square feet = 0.4 hectare
1 hectare (ha)	= 2.5 acres
1 square kilometer	= 0.4 square mile = 100 hectares

Volume

1 quart (qt.)	= 2 pints = 0.95 liters
1 liter (l)	= 1.06 quarts = 0.26 gallons
1 gallon (U. S.)	= 4 quarts = 3.8 liters
1 barrel (oil)	= 42 gallons = 159.6 liters
1 cubic yard	= 27 cubic feet = 0.76 cubic meters
1 cubic meter	= 35.3 cubic feet = 1.31 cubic yards

Weights and Masses

1 gram	= 15.43 grains = 0.035 ounce (avoirdupois)
1 kilogram (kg)	= 2.20 pounds
1 pound (lb.)	= 453.59 grams = 0.45 kilograms
1 short ton	= 2,000 pounds = 907.2 kilograms
1 long ton	= 2,240 pounds = 1,016.1 kilograms
1 metric ton	= 2,205 pounds = 1,000 kilograms

Temperature

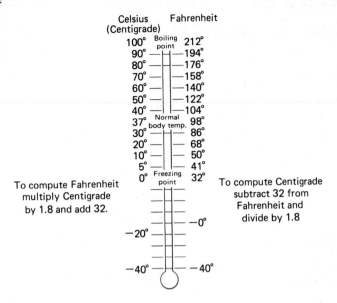

Celsius (Centigrade) / Fahrenheit

Celsius (Centigrade)		Fahrenheit
100°	Boiling point	212°
90°		194°
80°		176°
70°		158°
60°		140°
50°		122°
40°		104°
37°	Normal body temp.	98°
30°		86°
20°		68°
10°		50°
5°		41°
0°	Freezing point	32°
		−0°
−20°		
−40°		−40°

To compute Fahrenheit multiply Centigrade by 1.8 and add 32.

To compute Centigrade subtract 32 from Fahrenheit and divide by 1.8

Other Measures

1 kilowatt-hour	$= 3{,}413$ BTU $= 3.6 \times 10^{13}$ ergs $= 860{,}421$ calories
1 BTU	$= 2.93 \times 10^{-4}$ kilowatt-hours $= 1{,}0548 \times 10^{10}$ ergs $= 252$ calories
1 watt	$= 3{,}413$ BTU/hour
1 horsepower	$= 0.746$ kilowatt
1 gallon/minute	$= 8.0208$ cu. ft./min.
1 acre foot	$= 1{,}233.46$ cubic meters

Glossary

AA   A lava flow characterized by a rough, clinkery surface.

Accelerograph   An instrument for recording the acceleration in velocity of earthquake vibrations.

Adita   A tunnel or passageway by which a mine is entered.

Aerobic   Characterized by presence of free oxygen.

Aggradation   The general building up of the land by depositional processes.

Alluvium   Refers to material deposited by running water.

Anaerobic condition   Characterized by absence of air or free oxygen.

Anastomosing   Branching or interlacing with a braided appearance.

Andesitic basalt   A fine-grained extrusive igneous rock composed of plagioclase feldspars and ferromegnesian silicates.

Anthracite coal   "Hard coal," A hard, black, lustrous coal containing a high percentage of fixed carbon and a low percentage of volatile matter.

Anticline   A fold in which the rocks are bent convex upward.

Aquiclude   A relatively impermeable rock or surficial material which may absorb water slowly but cannot transmit it rapidly enough to supply a well or spring.

Aquifer   A water-bearing rock formation.

Aquitard   A rock formation which retards the flow of water. A body of impermeable material stratigraphically adjacent to an aquifer. A confining bed.

Artesian aquifer   An aquifer confined by impermeable beds.

Artesian head   The level to which water from a well will rise when confined in a standing pipe.

Artesian well   A well in which water from a confined aquifer rises above the top of the aquifer. Some wells may flow without the aid of pumping.

Asthenosphere   The layer of the earth lying between the lithosphere and mesosphere. The upper mantle.

Auger mining   A method of extracting ore by boring horizontally into a seam, much like a drill bores a hole in wood.

Batholith   A large, irregularly shaped, discordant pluton associated with the core of folded mountain ranges.

Bench mark   A mark on a fixed object indicating a particular elevation.

Benioff zone   A plane beneath deep-sea trenches along which earthquake foci cluster. The zone dips toward the continents.

Biosphere   Zone at and adjacent to the earth's surface including all living organisms.

Bituminous coal   "Soft coal." A coal which is high in carbonaceous matter and having between 15 and 50 percent volatile matter.

Blowout   An uncontrolled escape of oil or gas from an oil well.

BOD   (biochemical oxygen demand) The oxygen used in meeting the metabolic needs of aquatic aerobic microorganisms. A high BOD correlates with accelerated eutrophication.

Bomb, volcanic   Detached mass of lava or solid fragment ejected from a volcano. They range from 32 mm to several meters in length.

Bouguer anomaly   The gravity value after a correction has been made for the altitude of the station and the rock between the station and sea level.

Brackish water   Water with a salinity intermediate between that of freshwater and seawater.

Breakwater   A structure protecting a harbor, shore area, inlet, or basin from waves.

Breccia A rock made up of angular fragments. It may be produced by sedimentary, volcanic, or tectonic processes.

Breeder reactor A nuclear reactor capable of producing fissionable fuel as well as consuming it, especially one that creates more than it consumes.

Brine Refers to highly saline waters containing Ca, Na, K, Cl, and minor amounts of other elements.

BTU (British thermal unit) The quantity of heat needed to raise the temperature of one pound of water one degree Fahrenheit.

Burner reactor See *converter reactor*.

Caldera A large, semicircular volcanic depression commonly found at the summit of a volcano.

Carcinogen Any substance which tends to produce a cancer in a body.

Cesspool A pit for retaining the liquid waste from household sewage.

Cinder cone A conical structure composed of volcanic ash and cinders.

Clastic texture Refers to sedimentary rocks composed of broken fragments mechanically derived from any preexisting rocks. The individual grains are refered to as clasts.

Colluvial Loose, heterogenous material deposited at the base of a steep slope by agents of mass wasting.

Cone of depression A roughly conical depression produced in a water table by pumping.

Continental shelf The gently sloping zone bordering a continent and extending from low tide to the depth at which there is a marked increase in the slope of the ocean bottom. The greatest average depth is 183 m (600.2 ft).

Convection Refers to movement of materials as a result of heat-induced density variations.

Converter reactor (1) A nuclear reactor that produces some fissionable material, but less than it consumes. (2) A reactor that produces a fissionable material different from the fuel burned, regardless of the ratio.

Creep (1) *Gravitational creep* The slow downslope movement of soil or other surficial material. (2) *Tectonic creep* A slight, apparently continuous movement along a fault.

Crest stage The highest point of a flood.

Crystalline texture Said of rocks composed of an interlocking mosaic of mineral grains.

Culm bank Refuse coal-screenings, often piled in heaps or banks.

Curie point That temperature below which a substance ceases to be paramagnetic.

Cut Bank An erosional feature produced by lateral erosion of the outside bank of a meander.

Darcy's Law A derived formula applied to the flow of fluids.

DDT (dichloro diphenyl trichloroethane) An insecticide, one of several chlorinated hydrocarbons.

Degradation The general lowering of the land by erosional processes.

Depletion allowance A proportion of income derived from mineral production that is not subject to income tax.

Desalination Any process capable of converting saline water to potable water.

Detrital Relates to deposits formed of minerals and rock fragments transported to the place of deposition.

Diastrophism The process by which the earth's crust is deformed. Includes folding, faulting, warping, and mountain building.

Dike A tabular-shaped discordant pluton.

Dilatancy An increase in bulk volume during deformation.

Dip The angle at which a rock surface departs from a horizontal plane.

Dip slope Topographic slope conforming with the dip of the underlying bedrock.

Discharge The volume of water passing a given point within a given period of time.

Drift See *glacial drift.*

Effluent Anything that flows forth; a stream flowing out of another, a lava flow discharged from a volcanic fissure, discharge from sewage treatment facilities, and so on.

Ejecta Solid material thrown out of a volcano. It includes volcanic ash, lapilli, and bombs.

Elastic deformation A non-permanent deformation which returns to its original shape after the load is released. Elastic energy is released during return to original shape and this may produce tremors.

Electrolyte A conducting medium involving the flow of current and movement of matter.

Electrolytic hydrogen Hydrogen derived from water through the application of high current electrodes.

Endemic Refers to organisms that are restricted to a particular region or environment.

Endogenic Refers to geologic processes originating within the earth.

En echelon Offset but parallel structural features.

Enhanced recovery A general term referring to techniques for increasing the yield of oil and gas wells.

Ephemeral Said of a temporary or intermittent lake or stream.

Epicenter The point on the earth's surface directly above the point of origin of an earthquake.

Eustatic Refers to worldwide and simultaneous changes in sea level.

Eutrophication A process whereby natural bodies of water rich in plant nutrients and organisms become deficient in oxygen.

Evapotranspiration The sum of evaporation from wetted surfaces and of transpiration by vegetation.

Exogenic Refers to geological processes originating at or near the surface of the earth.

Fault scarp A cliff formed by a fault, the topographic expression of vertical displacement within the crust of the earth.

Fission See *nuclear fission.*

Flood basalt Horizontal lava flows which emanate from fissures and innundate vast areas of the earth.

Flood frequency The average interval of time between floods equal to or greater than a specified discharge or stage.

Flood peak The highest value of the stage or discharge attained by a flood.

Floodplain The area bordering a stream which becomes flooded when the stream overflows its channel.

Flood stage The stage at which overflow of the natural banks of a stream begins to cause damage in the reach in which the elevation is measured.

Floodway The channel of a water course and the adjacent land areas required to carry and discharge a flood of a given magnitude.

Fluvial Refers to rivers and to the features produced by or found in rivers.

Focus The point of origin of an earthquake.

Foliation A textural term referring to the planar arrangement of mineral grains in metamorphic rock.

Fumarole A hot spring or geyser which emits gaseous vapor.

Fusion See *nuclear fusion.*

Geodetic station A station where changes in the shape and dimensions of the earth are recorded.

Geothermal Said of heat in the earth's interior.

Glacial drift A general term applied to sedimentary material transported and deposited by glacial ice.

GNP (gross national product) The total market value of all the goods and services produced by a nation during a specified period of time.

Gouge A layer of soft material occurring along the wall of a fault.

Graben A down-faulted block. May be bounded by upthrown blocks (*horsts*).

Gradation The leveling of the land through erosion, transportation, and deposition.

Groin A shore-protection structure built to trap littoral drift. It extends perpendicularly from the shoreline out into the water. Some groins are permeable and permit the circulation of water through them.

Grout A fine mortar for finishing surfaces.

Gully erosion Erosion of topsoil by running water flowing in distinct narrow channels during or immediately after storms or snowmelt.

Horst An up-faulted block. May be bounded by downthrown blocks (*grabens*).

Hummocky Refers to uneven topography dominated by knolls and mounds.

Hydrate Refers to those compounds containing chemically combined water.

Hydraulic conductivity See *permeability coefficient*.

Hydrocarbon Organic compounds containing only carbon and hydrogen. Commonly found in petroleum, natural gas, and coal.

Hydrogenate Involves the addition of hydrogen to the molecule of an unsaturated organic compound.

Hydrologic cycle The complete cycle of phenomena through which water passes from the atmosphere to the earth and back to the atmosphere.

Hydroscopic water Soil moisture that is in equilibrium with atmospheric moisture.

Hydrosphere The aqueous portion of the earth. Includes the waters of the oceans, lakes, streams, ground water, and atmospheric water.

Hydrostatic pressure Relates to pressures exerted by liquids.

Hydrothermal Said of hot, mineral-bearing solutions.

Hypocenter (1) The region where an earthquake is initiated. (2) The point on the earth's surface directly below the center of a nuclear bomb explosion.

Impermeable Impervious to the natural movement of fluids.

Infrared sensing Detection of invisible radiation of greater wavelength than that of red light.

Isostacy A condition of balance or equilibrium in large areas of the earth's crust.

Jetty A structure extending from the mouth of a river or entrance to a bay to direct and confine stream or tidal flow to a selected channel, to help deepen or stabilize the channel.

Jökulhaup A glacial outburst flood.

Karst A type of topography characterized by closed depressions (sinkholes), caves, and subsurface streams.

Laccolith A concordant igneous pluton having a flat base and a dome-shaped upper surface.

Lahar A mudflow on the flanks of a volcano composed primarily of pyroclastic material.

Lapilli Pea-sized volcanic ejecta. Accretionary lapilli increase in size through the addition of extraneous material.

Lava Molten material derived from a volcanic eruption or a rock which solidifies from such molten material.

Lava dome A dome-shaped mountain originating from the solidification of highly fluid lavas.

Leachate The solution obtained by the leaching action of water as it percolates through soil or other materials such as wastes containing soluble substances.

Levee An embankment that confines a stream channel and limits flooding.

Leveling The process of establishing the elevations of different points on the surface of the earth by use of the surveyor's level.

Liquefaction The transformation from a solid to a liquid state.

Lithification The conversion of unconsolidated material into rock.

Lithology Refers to a description of the physical characteristics of a rock.

Lithosphere The solid or rocky portion of the earth.

Lithostatic pressure Pressure related to the weight of overlying rocks.

Littoral The environment of the sea floor between the limits of high tide and low tide.

LNG (liquefied natural gas) Natural gas which is liquefied by lowering its temperature to $-161.7°C$ ($-259°F$) The liquid occupies 1/632 the volume of the equivalent vapor.

Loess A widespread, homogeneous, windblown deposit of angular, silt-sized particles.

Longwall mining A subsurface mining technique in which parallel entries are driven into flat-lying seams of ore. From the end of the entries, workings are driven at right angles in both directions. As the workings are widened, a "longwall face" is produced and almost all the ore is recovered.

Lopolith A concordant igneous pluton having a flat upper surface and a dome-shaped base.

Magma A naturally occurring silicate melt.

Malthusian Refers to the doctrine of Malthus which states that population tends to increase at a faster rate than its means of subsistence.

Mantle (1) The layer of the earth between the crust and the core. (2) Loose, unconsolidated surficial deposits overlying bedrock (*regolith*).

Mass movement Movement of earth materials as a unit or en masse.

Meander scar An abandoned meander of a stream.

Mesosphere The lower layer of the earth's mantle.

Metamorphism The process which induces physical or compositional changes in rocks caused by heat, pressure, or chemically active fluids.

Metasomatic A replacement process by which a new mineral may grow in the body of an old mineral.

Meteoric water All forms of water particles that fall from the atmosphere.

Montmorillonite A group of clay minerals with an expanding lattice which leads to swelling on wetting.

Natural levee An embankment of sediments built by a stream on its floodplain and along both banks of its channel.

Nuclear fission The splitting of heavy nuclei and accompanying release of energy.

Nuclear fusion The combination of two light nuclei to form a heavier nucleus.

Nuclear reactor A device in which a fission chain reaction can be initiated, maintained, and controlled.

Nuée ardente A turbulent cloud of hot, expanding gas and tephra which flows as a high-speed density current down the flanks of a volcano.

OPEC (Organization of Petroleum Exporting Countries) Founded in 1960,

OPEC accounts for about 90 percent of the oil traded in the world. Its members are Saudi Arabia, Iran, Venezuela, Nigeria, Libya, Kuwait, Iraq, United Arab Emirates, Algeria, Indonesia, Qatar, and Ecuador. Gabon is an associate member. The United Arab Emirates is a federation of Abu Dhabi, Dubai, Sharjah, Ajman, Umm al-aiwain, Ras al-Khaimah, and Fujairah. Trinidad-Tobago is an observer.

Ore A "mineral" deposit which can be mined at a profit. Inlcudes metals, fossil fuels, and nonmetalliferous deposits.

Overburden (spoil) Barren bedrock or surficial material which must be removed before the underlying mineral deposit can be mined.

Oxbow A closely looping stream meander.

Oxbow lake A crescent-shaped lake formed when a meander is cut off from the main stream.

Pahoehoe Lava flow characterized by a smooth, ropy surface.

Pedology The scientific study of soil.

Pegmatite Very coarse-grained igneous rocks usually found as dikes associated with a large mass of intrusive igneous rock. Some pegmatites may contain a variety of rare minerals.

Percolation test A test used to determine the rate at which water percolates through soil.

Permafrost Permanently frozen ground.

Permeability Refers to the capacity of porous rocks or surficial material to transmit fluids. Effective porosity.

Permeability coefficient Refers to the rate of flow of water through an aquifer under standard conditions.

Pesticide Any chemical used for killing noxious organisms.

Phreatic zone See *zone of saturation.*

Physiography The descriptive study of landforms.

Piezometric surface See *potentiometric surface.*

Pillow lava Refers to lavas having a pillowlike structure which probably formed in a subaqueous environment.

Placer A surficial mineral deposit formed by the selective sorting action of waves and currents.

Planimetric map A map presenting the relative horizontal positions of natural and cultural features.

Plastic deformation Permanent deformation of the shape of a substance without rupture.

Plate tectonics A theory in which the earth's crust is divided into rigid plates which "float" on an underlying layer and move relative to each other.

Playa Desert lake basin.

Pluton A general term referring to intrusive igneous structures.

Point bar A depositional feature on the inside of a meander.

Porosity The percentage of void space in a rock.

Porphyritic A textural term for igneous rocks which contain larger crystals (phenocrysts) set in a finer matrix. A copper porphyry would contain disseminated copper minerals in a large body of porphyritic rock.

Potentiometric surface The surface to which water from a given aquifer will rise under its full head. Also referred to as the piezometric surface.

Primary wave A type of earthquake wave that moves by alternating compression and expansion of material in the direction of movement.

Proration A legal restriction of oil production to a specified fraction of potential production.

Protore Low-grade mineral deposits which can be concentrated by natural surface processes to become ore.

Pumice A light-colored volcanic froth which is cellular in texture.

Pyroclastics The solid materials which

accompany an explosive volcanic eruption.

Quick clay Deposits of clay or soil which quickly change from a solid to a liquid state when suddenly jarred.

Regolith A general term for the loose unconsolidated material overlying the bedrock.

Regulatory floodplain The part of the floodplain that would be inundated by a 100-year flood discharge. Consists of a regulatory floodway and a regulatory flood fringe.

Regulatory floodway The stream channel and unobstructed part of a floodplain that carry deep fast-moving water when inundated by the regulatory flood discharge.

Reserve The portion of an identified resource from which a usable mineral or energy commodity can be economically and legally extracted at the time of determination.

Resistivity Refers to the resistance of material to electrical current. The reciprocal of conductivity.

Resource A concentration of naturally occurring solid, liquid, or gaseous materials in or on the earth's crust in such form that economic extraction of a commodity is currently or potentially feasible.

Revetment A facing of stone or other resistant material built to protect an embankment from wave erosion.

Richter scale A scale of earthquake magnitude based on the logarithm (base 10) of the amplitudes of the deflections created by earthquake waves and recorded by a seismograph.

Rift valley A graben or elongated valley formed by down faulting.

Right-lateral movement A fault with movement parallel to strike and right-handed separation. A reference point on the side opposite an observer ap-

pears to have moved toward the right of the observer.

Riparian land Land situated along the bank of a body of water.

Riprap A layer, facing, or protective mound of stones randomly placed to prevent erosion or scour of an embankment.

Riverine Located along the banks of a river or a feature formed by a river.

Rock flour Finely ground rock fragments produced by glacial abrasion.

Room-and-pillar mining A subsurface mining technique in which the ore is mined in rooms which are separated by pillars of undisturbed ore left for roof support.

Rotary drilling The primary method for drilling deep wells in bedrock. A drill-bit, attached to drillpipe, is rotated while resting on the bottom of the hole.

Rotation slide Downslope movement along a curved slip surface, concave upward, producing a backward rotation in the displaced mass.

Rubble Loose, angular, and water-worn stones along a beach.

Runoff That part of precipitation which flows over the surface of the land as sheet wash and stream flow.

Sag pond Ponds formed by the uneven settling of the ground.

Sanitary landfill A land site where solid waste is dumped, compacted, and covered with soil in order to minimize environmental degradation.

Scarp A cliff or steep slope which may be produced by a fault in the earth's crust.

Scoria A vesicular, cindery crust on the surface of a lava flow.

Seawall A structure built along a portion of a coast primarily to prevent erosion and other damage by wave action. It retains earth against its shoreward face.

Secondary waves An earthquake wave which moves as a transverse wave and travels much more slowly than associated primary waves.

Seiche A periodic oscillation of a body of water.

Seismic activity Earth vibrations or disturbances produced by earthquakes.

Seismic sea waves See *tsunamis*.

Seismograph An instrument for recording earth vibrations (syn. seismometer).

Septic tank system An onsite disposal system consisting of an underground tank and a soil absorption field. Untreated sewage enters the tank where solids undergo decomposition. Liquid effluent moves from the tank to the absorption field via perforated pipe.

Shearing strength The internal resistance offered to tangential stress.

Shear wave A type of earthquake wave that moves by a shearing of material, so that there is movement perpendicular to the direction of propagation.

Sheet erosion Erosion of topsoil by broad continuous sheets of water.

Shield volcano Broad, gently sloping structures composed primarily of lava flows.

Shutterridge A ridge formed by displacement on a fault which cuts across ridge-and-valley topography.

Sill A tabular-shaped concordant pluton.

Sinkhole A topographic depression developed by the solution of limestone, rock salt, or gypsum bedrock.

Slump block The mass of material torn away as a coherent unit during mass movement.

Soil horizon A layer of soil distinguished from adjacent layers by such characteristics as structure, color, texture, or mineral composition.

Soil loss tolerance Refers to the maximum level of soil erosion that will permit a high level of crop productivity to be sustained economically and indefinitely.

Solifluction The slow, viscous downslope flow of saturated surficial material.

Sorb To take up and hold either by adsorption or absorption.

Sorting A dynamic gradational process which segregates sedimentary particles by size or shape. Well-sorted material has a limited size range whereas poorly sorted material has a large size range.

Spoil See *overburden*.

Stage Refers to the height of a water surface above an established datum plane.

Standard project flood The discharge that could be expected from the most severe combination of meteorologic and hydrologic conditions characteristic of an area.

Stock An irregularly shaped discordant pluton that is less than 100 km^2 in surface exposure.

Stope An underground excavation formed by the extraction of ore.

Stoping (1) Magmatic intrusion which involves detaching and engulfing country rock. (2) The process of extracting ore in an underground mine so as to remove the entire ore body.

Strain Changes in the geometry of a body which result from applied forces.

Stratabound Refers to a mineral deposit which is confined to a single rock or stratigraphic unit. If the mineral deposit is strictly coextensive with a particular layer, it may be referred to as a strataform deposit.

Stratification The structure produced by a series of sedimentary layers or beds (strata).

Stratigraphy The study of rock strata including their age relations, geographic distribution, composition, history, etc.

Stratovolcano A volcano composed of alternating layers of lava and pyroclastics.

Stress Compressional, tensional, or tortional forces that act to change the geometry of a body.

Strike The bearing or direction of the line of intersection of an inclined stratum and a horizontal plane.

Strike-slip fault A fault in which movement or slip is parallel to the strike of the fault.

Subduction The process of one crustal plate descending beneath another.

Subsidence A sinking or settling of a large part of the earth's crust.

Supergene enrichment Precipitates of ore minerals from percolating ground water.

Surficial deposit Unconsolidated transported or residual materials such as soil, alluvial, or glacial deposits.

Tailings The portions of processed ore which are too poor to be treated further.

Talus debris Unconsolidated rock fragments which form a slope at the base of a steep surface.

Tectonic Refers to deformation of the earth's crust through warping, folding, or faulting.

Tectonic creep Slight, apparently continuous movement along a fault.

Tephra A general term for volcanic pyroclastic material.

Terrace A relatively flat, horizontal, or gently inclined surface bounded by a steeper slope. May be produced by stream or wave activity.

Till Unstratified and unsorted sediments deposited by glacial ice.

Tiltmeter An instrument used to detect changes in the slope of the ground surface. Measures horizontal displacement and can be used to indicate impending volcanic or earthquake activity.

Tombolo A sandbar connecting an island with the mainland.

Transform fault A fault displaying a change in structural style; for example, strike-slip to ridgelike structures.

Translation slide Downslop movement by sliding along a bedding plane which is generally parallel to the ground surface.

Transpiration The process by which water absorbed by plants is evaporated into the atmosphere from the plant surface.

Transuranic wastes Radioactive wastes which generate little or no heat and have significantly lower radiation levels than high-level radioactive wastes.

Tsunamis Sea waves produced by large-scale disturbances of the ocean floor such as earthquakes, volcanic eruptions, or submarine slides.

Vadose zone See *zone of aeration.*

Vesicular A textural term indicating the presence of many small cavities in a rock.

Water table The surface marking the boundary between the zone of saturation and the zone of aeration. It approximates the surface topography.

Zone of aeration The zone in which the pore spaces in permeable materials are not filled (except temporarily) with water. Also referred to as unsaturated zone or vadose zone.

Zone of saturation The zone in which pore spaces are filled with water. Also referred to as phreatic zone.

Review Questions

1. What geologic data are necessary for comprehensive land-use planning? How can this data be presented for effective use by groups not trained in the earth sciences? What other types of data are necessary for effective land-use planning? Describe the status of land-use planning in your community. What role has geology played in guiding this planning?
2. What techniques are available for monitoring a volcano such as Mount St. Helens? In addition to issuing a general warning, what might be done to minimize the impact of the hazards associated with an eruption? How has development of the Cascades increased the impact of these hazards?
3. Recent advances in seismology have enabled seismologists to predict earthquakes in the U.S.S.R., New York, and California. Experience with fluid injection in Denver and Rangely, Colorado, has suggested a method of releasing strain in a fault system and thereby controlling eathquake activity. Suppose that within a decade both our ability to predict and control earthquakes improves and that there are predictions of a large earthquake in the Los Angeles area. How much publicity should be given to the prediction? What are some of the scientific, sociological, and economic considerations involved in deciding on the feasibility of initiating a strain-release program in this situation?
4. What geologic factors should one take into account before building on steep, soil-covered slopes?
5. The rainfall factor for southwestern Wisconsin is 160, and the Fayette silt loam, a common soil in the region, has a K value of 0.38. A farmer is cropping a field with a slope length of 200 ft and a steepness of 6 percent which gives an LS value of 0.95. The land is plowed in the fall and continuously row-cropped in corn, which gives a crop-management factor of 0.4. At present the farmer is not using any conservation measures (P = 1.0). Using the Universal Soil Loss Equation, calculate the expected average annual soil loss.

    If the farmer introduced contouring as a support practice (P = 0.5), what would be the average annual soil loss?

    If the corn were planted in stalk residue without tilling the soil, the C factor could be reduced to approximately 0.12. What would be the average annual erosion for no-till corn planted on the contour?

    Would any of these management systems be within the tolerable soil limits? If not, what would you advise the farmer to do?
6. Given that floodplain occupancy is a characteristic of all major industrialized and preindustrialized societies, who should bear the major cost of systems for forecasting and warning of floods? For modifications of stream channels? For watershed treatment? For floodproofing buildings? For flood insurance? For emergency assistance to flood victims? What are the implications of your

answer for further development of the assets of the floodplain? For reduction of losses when floods do occur? One popular engineering approach to flood control is the construction of flood control dams. What negative impact would this have on the area upstream from the dam? Downstream from the dam?

7. Rigorous zoning laws have been proposed by those involved in land-use planning for floodplains. Do you think that this approach should be expanded to include other geologic hazards such as volcanism, earthquakes, tectonic movements, and landslides? Are any aspects of these hazards unique so as to call for radically different approaches? Should the application of zoning restrictions be retroactive? Should resettlement be allowed after a disaster? What factors must be considered in deciding whether resettlement should be permitted?

8. The policy of some land management agencies is to attempt to control destructive natural processes. Cite some examples of the techniques which have been employed to control natural hazards along coasts.

Others involved in land management argue that natural change, even when destructive, is essential to the maintenance of ecologic systems; that attempts to interfere with the natural processes can lead to the loss of these systems and to responses which are undesirable to humans. Cite some examples which support this point of view.

Give examples to demonstrate that these two points of view can be reconciled.

9. How can nonrenewable mineral resources be replenished? What are some of the constraints to the long-term mineral supply system? What problems may accompany the development of unconventional energy resources?

10. What are some of the economic and trade problems that could arise if U.S. industry passes along the necessary costs involved in protecting the environment? Do these costs always result in a net loss, or do some safeguards or rehabilitation practices lead to marketable assets or other gains to industry?

11. In the concluding section of Reading 29, David Brew lists six major considerations involved in the decision on the trans-Alaska pipeline. List these according to your priorities and defend your decisions.

12. What are the advantages of ground water over surface water? What problems are associated with the development of ground-water reservoirs? How can these problems be solved? To what extent are these problems multidisciplinary?

13. What methods does your local community employ for disposing of its waste? What are the hydrologic implications of these methods? How safe are they? How much do they cost in comparison to other municipal services? What materials are recycled?

14. A sanitary landfill site is proposed for an abandoned sand and gravel pit located near the Rock River. The landfill is to measure 2500 × 1000 ft and the center of the landfill will be approximately 2000 ft from the river. The base of the gravel pit is at the watertable. These relationships are shown in the block

diagram in Fig. 1A. The configuration of the potentiometric surface (water table) is shown by the dashed line in the map view (Fig. 1B). Using the information included in the block diagram and map view, answer the following questions:

a. If the hydraulic conductivity of the sand and gravel is 5000 gpd/ft² and if the effective porosity is 30 percent, how fast will the ground water move? How long will it take the leachate to reach the river?

b. If the landfill is silt and fine sand (rather than gravel), with a hydraulic conductivity of 5.0 gpd/ft² and an effective porosity of 45 percent, how long would it take the groundwater to travel the same distance? (Assume the same water-table gradient.)

It is also of interest to know how much dilution will take place before the water enters the river and also *after* entering the river. To determine this, it is necessary to calculate the amount of water infiltrating the finished landfill (assume the entire pit is filled with debris).

a. Of the 30 in./year of rainfall in this area, assume that 6 in./yr will infiltrate through the landfill. How much leachate will enter the groundwater system each year (in ft³/yr)? How much will enter *each day* (in ft³/day)?

b. Before reaching the river, the leachate will be diluted by two sources: ground water passing below the landfill, and precipitation that recharges the aquifer between the landfill and the river. Calculate the daily volume of water available for diluting the leachate from each of these sources (neglect effects of lateral dispersion).

c. By what factor is the original leachate diluted before discharge into the Rock River?

d. By what factor is the original leachate diluted after entering the river if the average discharge of the Rock River is 1.12×10^9 gpd?

15. Do you think that we should continue to expand nuclear power facilities if we cannot agree on a method for handling nuclear waste materials?

16. In Reading 38, Carlo Berghinz argues that "Venice is not only an Italian but a worldwide patrimony. . . ." Do you think that this argument can be applied equally to wilderness areas? Why or why not? Would you argue for preservation of vast wilderness areas as a trust for all mankind? Cite examples of serious conflict between the needs of the present generation and its responsibility to future generations.

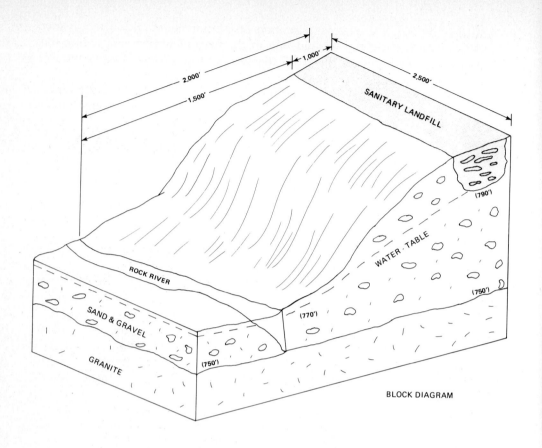

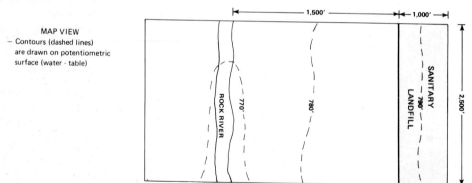

Fig. 1. Sanitary landfill site and surroundings. Above, block diagram. Below, map view. Contours drawn on potentiometric surface (water table).

Index